T0202416

Lecture Notes in Computer Science　　14492

Founding Editors

Gerhard Goos
Juris Hartmanis

Editorial Board Members

Elisa Bertino, *Purdue University, West Lafayette, IN, USA*
Wen Gao, *Peking University, Beijing, China*
Bernhard Steffen ⑩, *TU Dortmund University, Dortmund, Germany*
Moti Yung ⑩, *Columbia University, New York, NY, USA*

The series Lecture Notes in Computer Science (LNCS), including its subseries Lecture Notes in Artificial Intelligence (LNAI) and Lecture Notes in Bioinformatics (LNBI), has established itself as a medium for the publication of new developments in computer science and information technology research, teaching, and education.

LNCS enjoys close cooperation with the computer science R & D community, the series counts many renowned academics among its volume editors and paper authors, and collaborates with prestigious societies. Its mission is to serve this international community by providing an invaluable service, mainly focused on the publication of conference and workshop proceedings and postproceedings. LNCS commenced publication in 1973.

Zahir Tari · Keqiu Li · Hongyi Wu
Editors

Algorithms and Architectures for Parallel Processing

23rd International Conference, ICA3PP 2023
Tianjin, China, October 20–22, 2023
Proceedings, Part VI

 Springer

Editors
Zahir Tari
Royal Melbourne Institute of Technology
Melbourne, VIC, Australia

Keqiu Li
Tianjin University
Tianjin, China

Hongyi Wu
University of Arizona
Tucson, AZ, USA

ISSN 0302-9743 ISSN 1611-3349 (electronic)
Lecture Notes in Computer Science
ISBN 978-981-97-0810-9 ISBN 978-981-97-0811-6 (eBook)
https://doi.org/10.1007/978-981-97-0811-6

© The Editor(s) (if applicable) and The Author(s), under exclusive license
to Springer Nature Singapore Pte Ltd. 2024

This work is subject to copyright. All rights are reserved by the Publisher, whether the whole or part of the material is concerned, specifically the rights of translation, reprinting, reuse of illustrations, recitation, broadcasting, reproduction on microfilms or in any other physical way, and transmission or information storage and retrieval, electronic adaptation, computer software, or by similar or dissimilar methodology now known or hereafter developed.
The use of general descriptive names, registered names, trademarks, service marks, etc. in this publication does not imply, even in the absence of a specific statement, that such names are exempt from the relevant protective laws and regulations and therefore free for general use.
The publisher, the authors, and the editors are safe to assume that the advice and information in this book are believed to be true and accurate at the date of publication. Neither the publisher nor the authors or the editors give a warranty, expressed or implied, with respect to the material contained herein or for any errors or omissions that may have been made. The publisher remains neutral with regard to jurisdictional claims in published maps and institutional affiliations.

This Springer imprint is published by the registered company Springer Nature Singapore Pte Ltd.
The registered company address is: 152 Beach Road, #21-01/04 Gateway East, Singapore 189721, Singapore

Paper in this product is recyclable.

Preface

On behalf of the Conference Committee, we welcome you to the proceedings of the 2023 International Conference on Algorithms and Architectures for Parallel Processing (ICA3PP 2023), which was held in Tianjin, China from October 20–22, 2023. ICA3PP2023 was the 23rd in this series of conferences (started in 1995) that are devoted to algorithms and architectures for parallel processing. ICA3PP is now recognized as the main regular international event that covers the many dimensions of parallel algorithms and architectures, encompassing fundamental theoretical approaches, practical experimental projects, and commercial components and systems. This conference provides a forum for academics and practitioners from countries around the world to exchange ideas for improving the efficiency, performance, reliability, security, and interoperability of computing systems and applications.

A successful conference would not be possible without the high-quality contributions made by the authors. This year, ICA3PP received a total of 503 submissions from authors in 21 countries and regions. Based on rigorous peer reviews by the Program Committee members and reviewers, 193 high-quality papers were accepted to be included in the conference proceedings and submitted for EI indexing. In addition to the contributed papers, six distinguished scholars, Lixin Gao, Baochun Li, Laurence T. Yang, Kun Tan, Ahmed Louri, and Hai Jin, were invited to give keynote lectures, providing us with the recent developments in diversified areas in algorithms and architectures for parallel processing and applications.

We would like to take this opportunity to express our sincere gratitude to the Program Committee members and 165 reviewers for their dedicated and professional service. We highly appreciate the twelve track chairs, Dezun Dong, Patrick P. C. Lee, Meng Shen, Ruidong Li, Li Chen, Wei Bao, Jun Li, Hang Qiu, Ang Li, Wei Yang, Yu Yang, and Zhibin Yu, for their hard work in promoting this conference and organizing the reviews for the papers submitted to their tracks. We are so grateful to the publication chairs, Heng Qi, Yulei Wu, Deze Zeng, and the publication assistants for their tedious work in editing the conference proceedings. We must also say "thank you" to all the volunteers who helped us at various stages of this conference. Moreover, we were so honored to have many renowned scholars be part of this conference. Finally, we would like to thank all speakers, authors, and participants for their great contribution to and support for the success of ICA3PP 2023!

October 2023

Jean-Luc Gaudiot
Hong Shen
Gudula Rünger
Zahir Tari
Keqiu Li
Hongyi Wu
Tian Wang

Organization

General Chairs

Jean-Luc Gaudiot   University of California, Irvine, USA
Hong Shen   University of Adelaide, Australia
Gudula Rünger   Chemnitz University of Technology, Germany

Program Chairs

Zahir Tari   Royal Melbourne Institute of Technology,
     Australia
Keqiu Li   Tianjin University, China
Hongyi Wu   University of Arizona, USA

Program Vice-chair

Wenxin Li   Tianjin University, China

Publicity Chairs

Hai Wang   Northwest University, China
Milos Stojmenovic   Singidunum University, Serbia
Chaofeng Zhang   Advanced Institute of Industrial Technology,
     Japan
Hao Wang   Louisiana State University, USA

Publication Chairs

Heng Qi   Dalian University of Technology, China
Yulei Wu   University of Exeter, UK
Deze Zeng   China University of Geosciences (Wuhan), China

Workshop Chairs

Laiping Zhao	Tianjin University, China
Pengfei Wang	Dalian University of Technology, China

Local Organization Chairs

Xiulong Liu	Tianjin University, China
Yitao Hu	Tianjin University, China

Web Chair

Chen Chen	Shanghai Jiao Tong University, China

Registration Chairs

Xinyu Tong	Tianjin University, China
Chaokun Zhang	Tianjin University, China

Steering Committee Chairs

Yang Xiang (Chair)	Swinburne University of Technology, Australia
Weijia Jia	Beijing Normal University and UIC, China
Yi Pan	Georgia State University, USA
Laurence T. Yang	St. Francis Xavier University, Canada
Wanlei Zhou	City University of Macau, China

Program Committee

Track 1: Parallel and Distributed Architectures

Dezun Dong (Chair)	National University of Defense Technology, China
Chao Wang	University of Science and Technology of China, China
Chentao Wu	Shanghai Jiao Tong University, China

Chi Lin	Dalian University of Technology, China
Deze Zeng	China University of Geosciences, China
En Shao	Institute of Computing Technology, Chinese Academy of Sciences, China
Fei Lei	National University of Defense Technology, China
Haikun Liu	Huazhong University of Science and Technology, China
Hailong Yang	Beihang University, China
Junlong Zhou	Nanjing University of Science and Technology, China
Kejiang Ye	Shenzhen Institute of Advanced Technology, Chinese Academy of Sciences, China
Lei Wang	National University of Defense Technology, China
Massimo Cafaro	University of Salento, Italy
Massimo Torquati	University of Pisa, Italy
Mengying Zhao	Shandong University, China
Roman Wyrzykowski	Czestochowa University of Technology, Poland
Rui Wang	Beihang University, China
Sheng Ma	National University of Defense Technology, China
Songwen Pei	University of Shanghai for Science and Technology, China
Susumu Matsumae	Saga University, Japan
Weihua Zhang	Fudan University, China
Weixing Ji	Beijing Institute of Technology, China
Xiaoli Gong	Nankai University, China
Youyou Lu	Tsinghua University, China
Yu Zhang	Huazhong University of Science and Technology, China
Zichen Xu	Nanchang University, China

Track 2: Software Systems and Programming Models

Patrick P. C. Lee (Chair)	Chinese University of Hong Kong, China
Erci Xu	Ohio State University, USA
Xiaolu Li	Huazhong University of Science and Technology, China
Shujie Han	Peking University, China
Mi Zhang	Institute of Computing Technology, Chinese Academy of Sciences, China

Jing Gong	KTH Royal Institute of Technology, Sweden
Radu Prodan	University of Klagenfurt, Austria
Wei Wang	Beijing Jiaotong University, China
Himansu Das	KIIT Deemed to be University, India
Rong Gu	Nanjing University, China
Yongkun Li	University of Science and Technology of China, China
Ladjel Bellatreche	National Engineering School for Mechanics and Aerotechnics, France

Track 3: Distributed and Network-Based Computing

Meng Shen (Chair)	Beijing Institute of Technology, China
Ruidong Li (Chair)	Kanazawa University, Japan
Bin Wu	Institute of Information Engineering, China
Chao Li	Beijing Jiaotong University, China
Chaokun Zhang	Tianjin University, China
Chuan Zhang	Beijing Institute of Technology, China
Chunpeng Ge	National University of Defense Technology, China
Fuliang Li	Northeastern University, China
Fuyuan Song	Nanjing University of Information Science and Technology, China
Gaopeng Gou	Institute of Information Engineering, China
Guangwu Hu	Shenzhen Institute of Information Technology, China
Guo Chen	Hunan University, China
Guozhu Meng	Chinese Academy of Sciences, China
Han Zhao	Shanghai Jiao Tong University, China
Hai Xue	University of Shanghai for Science and Technology, China
Haiping Huang	Nanjing University of Posts and Telecommunications, China
Hongwei Zhang	Tianjin University of Technology, China
Ioanna Kantzavelou	University of West Attica, Greece
Jiawen Kang	Guangdong University of Technology, China
Jie Li	Northeastern University, China
Jingwei Li	University of Electronic Science and Technology of China, China
Jinwen Xi	Beijing Zhongguancun Laboratory, China
Jun Liu	Tsinghua University, China

Kaiping Xue	University of Science and Technology of China, China
Laurent Lefevre	National Institute for Research in Digital Science and Technology, France
Lanju Kong	Shandong University, China
Lei Zhang	Henan University, China
Li Duan	Beijing Jiaotong University, China
Lin He	Tsinghua University, China
Lingling Wang	Qingdao University of Science and Technology, China
Lingjun Pu	Nankai University, China
Liu Yuling	Institute of Information Engineering, China
Meng Li	Hefei University of Technology, China
Minghui Xu	Shandong University, China
Minyu Feng	Southwest University, China
Ning Hu	Guangzhou University, China
Pengfei Liu	University of Electronic Science and Technology of China, China
Qi Li	Beijing University of Posts and Telecommunications, China
Qian Wang	Beijing University of Technology, China
Raymond Yep	University of Macau, China
Shaojing Fu	National University of Defense Technology, China
Shenglin Zhang	Nankai University, China
Shu Yang	Shenzhen University, China
Shuai Gao	Beijing Jiaotong University, China
Su Yao	Tsinghua University, China
Tao Yin	Beijing Zhongguancun Laboratory, China
Tingwen Liu	Institute of Information Engineering, China
Tong Wu	Beijing Institute of Technology, China
Wei Quan	Beijing Jiaotong University, China
Weihao Cui	Shanghai Jiao Tong University, China
Xiang Zhang	Nanjing University of Information Science and Technology, China
Xiangyu Kong	Dalian University of Technology, China
Xiangyun Tang	Minzu University of China, China
Xiaobo Ma	Xi'an Jiaotong University, China
Xiaofeng Hou	Shanghai Jiao Tong University, China
Xiaoyong Tang	Changsha University of Science and Technology, China
Xuezhou Ye	Dalian University of Technology, China
Yaoling Ding	Beijing Institute of Technology, China

Yi Zhao	Tsinghua University, China
Yifei Zhu	Shanghai Jiao Tong University, China
Yilei Xiao	Dalian University of Technology, China
Yiran Zhang	Beijing University of Posts and Telecommunications, China
Yizhi Zhou	Dalian University of Technology, China
Yongqian Sun	Nankai University, China
Yuchao Zhang	Beijing University of Posts and Telecommunications, China
Zhaoteng Yan	Institute of Information Engineering, China
Zhaoyan Shen	Shandong University, China
Zhen Ling	Southeast University, China
Zhiquan Liu	Jinan University, China
Zijun Li	Shanghai Jiao Tong University, China

Track 4: Big Data and Its Applications

Li Chen (Chair)	University of Louisiana at Lafayette, USA
Alfredo Cuzzocrea	University of Calabria, Italy
Heng Qi	Dalian University of Technology, China
Marc Frincu	Nottingham Trent University, UK
Mingwu Zhang	Hubei University of Technology, China
Qianhong Wu	Beihang University, China
Qiong Huang	South China Agricultural University, China
Rongxing Lu	University of New Brunswick, Canada
Shuo Yu	Dalian University of Technology, China
Weizhi Meng	Technical University of Denmark, Denmark
Wenbin Pei	Dalian University of Technology, China
Xiaoyi Tao	Dalian Maritime University, China
Xin Xie	Tianjin University, China
Yong Yu	Shaanxi Normal University, China
Yuan Cao	Ocean University of China, China
Zhiyang Li	Dalian Maritime University, China

Track 5: Parallel and Distributed Algorithms

Wei Bao (Chair)	University of Sydney, Australia
Jun Li (Chair)	City University of New York, USA
Dong Yuan	University of Sydney, Australia
Francesco Palmieri	University of Salerno, Italy

George Bosilca	University of Tennessee, USA
Humayun Kabir	Microsoft, USA
Jaya Prakash Champati	IMDEA Networks Institute, Spain
Peter Kropf	University of Neuchâtel, Switzerland
Pedro Soto	CUNY Graduate Center, USA
Wenjuan Li	Hong Kong Polytechnic University, China
Xiaojie Zhang	Hunan University of Technology and Business, China
Chuang Hu	Wuhan University, China

Track 6: Applications of Parallel and Distributed Computing

Hang Qiu (Chair)	Waymo, USA
Ang Li (Chair)	Qualcomm, USA
Daniel Andresen	Kansas State University, USA
Di Wu	University of Central Florida, USA
Fawad Ahmad	Rochester Institute of Technology, USA
Haonan Lu	University at Buffalo, USA
Silvio Barra	University of Naples Federico II, Italy
Weitian Tong	Georgia Southern University, USA
Xu Zhang	University of Exeter, UK
Yitao Hu	Tianjin University, China
Zhixin Zhao	Tianjin University, China

Track 7: Service Dependability and Security in Distributed and Parallel Systems

Wei Yang (Chair)	University of Texas at Dallas, USA
Dezhi Ran	Peking University, China
Hanlin Chen	Purdue University, USA
Jun Shao	Zhejiang Gongshang University, China
Jinguang Han	Southeast University, China
Mirazul Haque	University of Texas at Dallas, USA
Simin Chen	University of Texas at Dallas, USA
Wenyu Wang	University of Illinois at Urbana-Champaign, USA
Yitao Hu	Tianjin University, China
Yueming Wu	Nanyang Technological University, Singapore
Zhengkai Wu	University of Illinois at Urbana-Champaign, USA
Zhiqiang Li	University of Nebraska, USA
Zhixin Zhao	Tianjin University, China

Ze Zhang   University of Michigan/Cruise, USA
Ravishka Rathnasuriya   University of Texas at Dallas, USA

Track 8: Internet of Things and Cyber-Physical-Social Computing

Yu Yang (Chair)   Lehigh University, USA
Qun Song   Delft University of Technology, The Netherlands
Chenhan Xu   University at Buffalo, USA
Mahbubur Rahman   City University of New York, USA
Guang Wang   Florida State University, USA
Houcine Hassan   Universitat Politècnica de València, Spain
Hua Huang   UC Merced, USA
Junlong Zhou   Nanjing University of Science and Technology,
                                     China
Letian Zhang   Middle Tennessee State University, USA
Pengfei Wang   Dalian University of Technology, China
Philip Brown   University of Colorado Colorado Springs, USA
Roshan Ayyalasomayajula   University of California San Diego, USA
Shigeng Zhang   Central South University, China
Shuo Yu   Dalian University of Technology, China
Shuxin Zhong   Rutgers University, USA
Xiaoyang Xie   Meta, USA
Yi Ding   Massachusetts Institute of Technology, USA
Yin Zhang   University of Electronic Science and Technology
                                     of China, China
Yukun Yuan   University of Tennessee at Chattanooga, USA
Zhengxiong Li   University of Colorado Denver, USA
Zhihan Fang   Meta, USA
Zhou Qin   Rutgers University, USA
Zonghua Gu   Umeå University, Sweden
Geng Sun   Jilin University, China

Track 9: Performance Modeling and Evaluation

Zhibin Yu (Chair)   Shenzhen Institute of Advanced Technology,
                                     Chinese Academy of Sciences, China
Chao Li   Shanghai Jiao Tong University, China
Chuntao Jiang   Foshan University, China
Haozhe Wang   University of Exeter, UK
Laurence Muller   University of Greenwich, UK

Lei Liu	Beihang University, China
Lei Liu	Institute of Computing Technology, Chinese Academy of Sciences, China
Jingwen Leng	Shanghai Jiao Tong University, China
Jordan Samhi	University of Luxembourg, Luxembourg
Sa Wang	Institute of Computing Technology, Chinese Academy of Sciences, China
Shoaib Akram	Australian National University, Australia
Shuang Chen	Huawei, China
Tianyi Liu	Huawei, China
Vladimir Voevodin	Lomonosov Moscow State University, Russia
Xueqin Liang	Xidian University, China

Reviewers

Dezun Dong	Xiaolu Li
Chao Wang	Shujie Han
Chentao Wu	Mi Zhang
Chi Lin	Jing Gong
Deze Zeng	Radu Prodan
En Shao	Wei Wang
Fei Lei	Himansu Das
Haikun Liu	Rong Gu
Hailong Yang	Yongkun Li
Junlong Zhou	Ladjel Bellatreche
Kejiang Ye	Meng Shen
Lei Wang	Ruidong Li
Massimo Cafaro	Bin Wu
Massimo Torquati	Chao Li
Mengying Zhao	Chaokun Zhang
Roman Wyrzykowski	Chuan Zhang
Rui Wang	Chunpeng Ge
Sheng Ma	Fuliang Li
Songwen Pei	Fuyuan Song
Susumu Matsumae	Gaopeng Gou
Weihua Zhang	Guangwu Hu
Weixing Ji	Guo Chen
Xiaoli Gong	Guozhu Meng
Youyou Lu	Han Zhao
Yu Zhang	Hai Xue
Zichen Xu	Haiping Huang
Patrick P. C. Lee	Hongwei Zhang
Erci Xu	Ioanna Kantzavelou

Jiawen Kang
Jie Li
Jingwei Li
Jinwen Xi
Jun Liu
Kaiping Xue
Laurent Lefevre
Lanju Kong
Lei Zhang
Li Duan
Lin He
Lingling Wang
Lingjun Pu
Liu Yuling
Meng Li
Minghui Xu
Minyu Feng
Ning Hu
Pengfei Liu
Qi Li
Qian Wang
Raymond Yep
Shaojing Fu
Shenglin Zhang
Shu Yang
Shuai Gao
Su Yao
Tao Yin
Tingwen Liu
Tong Wu
Wei Quan
Weihao Cui
Xiang Zhang
Xiangyu Kong
Xiangyun Tang
Xiaobo Ma
Xiaofeng Hou
Xiaoyong Tang
Xuezhou Ye
Yaoling Ding
Yi Zhao
Yifei Zhu
Yilei Xiao
Yiran Zhang
Yizhi Zhou

Yongqian Sun
Yuchao Zhang
Zhaoteng Yan
Zhaoyan Shen
Zhen Ling
Zhiquan Liu
Zijun Li
Li Chen
Alfredo Cuzzocrea
Heng Qi
Marc Frincu
Mingwu Zhang
Qianhong Wu
Qiong Huang
Rongxing Lu
Shuo Yu
Weizhi Meng
Wenbin Pei
Xiaoyi Tao
Xin Xie
Yong Yu
Yuan Cao
Zhiyang Li
Wei Bao
Jun Li
Dong Yuan
Francesco Palmieri
George Bosilca
Humayun Kabir
Jaya Prakash Champati
Peter Kropf
Pedro Soto
Wenjuan Li
Xiaojie Zhang
Chuang Hu
Hang Qiu
Ang Li
Daniel Andresen
Di Wu
Fawad Ahmad
Haonan Lu
Silvio Barra
Weitian Tong
Xu Zhang
Yitao Hu

Zhixin Zhao
Wei Yang
Dezhi Ran
Hanlin Chen
Jun Shao
Jinguang Han
Mirazul Haque
Simin Chen
Wenyu Wang
Yitao Hu
Yueming Wu
Zhengkai Wu
Zhiqiang Li
Zhixin Zhao
Ze Zhang
Ravishka Rathnasuriya
Yu Yang
Qun Song
Chenhan Xu
Mahbubur Rahman
Guang Wang
Houcine Hassan
Hua Huang
Junlong Zhou
Letian Zhang
Pengfei Wang
Philip Brown
Roshan Ayyalasomayajula

Shigeng Zhang
Shuo Yu
Shuxin Zhong
Xiaoyang Xie
Yi Ding
Yin Zhang
Yukun Yuan
Zhengxiong Li
Zhihan Fang
Zhou Qin
Zonghua Gu
Geng Sun
Zhibin Yu
Chao Li
Chuntao Jiang
Haozhe Wang
Laurence Muller
Lei Liu
Lei Liu
Jingwen Leng
Jordan Samhi
Sa Wang
Shoaib Akram
Shuang Chen
Tianyi Liu
Vladimir Voevodin
Xueqin Liang

Contents – Part VI

An Uncertainty-Aware Auction Mechanism for Federated Learning

Jiali Xu[1], Bin Tang[2]($\boxtimes$), Hengrui Cui[3], and Baoliu Ye[3]

[1] Software Institute, Nanjing University, Nanjing 210093, China
[2] School of Computer and Information, Hohai University, Nanjing 211100, China
cstb@hhu.edu.cn
[3] National Key Laboratory for Novel Software Technology, Nanjing University,
Nanjing 211100, China

Abstract. Federated learning enables multiple data owners to collaboratively train a shared machine learning model without the need to disclose their local training data. However, it is not practical for all clients to unconditionally contribute their resources; thus, designing an incentive mechanism in federated learning becomes an important issue. Some existing studies adopt the framework of reverse auctions for incentive design, but they do not take the uncertainty of the training time of clients into account. Consequently, a situation may arise where a client is unable to meet the training deadline, and the server cannot wait indefinitely. In this paper, we propose a reverse auction framework that takes the uncertainty of training time into account. We formulate an expected social welfare maximization problem and prove its NP-hardness. We then introduce an efficient dynamic programming-based algorithm which can find an optimal solution in pseudo-polynomial time. Building upon this, we propose a truthful auction mechanism based on the well-known Vickrey-Clarke-Groves (VCG) mechanism. Furthermore, to reduce the time complexity, we introduce an additional truthful auction mechanism based on a greedy algorithm which achieves a near-optimal performance in polynomial time. Finally, the effectiveness of the proposed two auction mechanisms is verified through simulation experiments.

Keywords: Federated learning · Incentive mechanism · Reverse auction

1 Introduction

Machine learning is one of the most rapidly advancing fields in technology today [15], finding widespread applications in various domains such as agriculture [17], physical sciences [3], biology [21] and many others. In general, the majority of machine learning processes involve aggregating large volumes of data and personal information into a central server for model training. However, this approach raises certain concerns. On one hand, as the number of mobile devices continues to increase, central servers may struggle to cope with the high computational and storage costs involved. On the other hand, valid concerns arise regarding

© The Author(s), under exclusive license to Springer Nature Singapore Pte Ltd. 2024
Z. Tari et al. (Eds.): ICA3PP 2023, LNCS 14492, pp. 1–18, 2024.
https://doi.org/10.1007/978-981-97-0811-6_1

the potential leakage of sensitive data submitted from mobile devices, thereby compromising privacy and security.

To address these challenges, the concept of federated learning (FL) was introduced, allowing data owners to collaboratively train global models in a decentralized manner [22]. A series of practical FL algorithms have been proposed [2,8,19,23]. In essence, FL is an emerging paradigm in distributed machine learning. It enables training data to remain local and avoids the necessity of exposing it. Data owners simply iteratively send model updates, trained on their local raw data, to the task publisher without needing to upload the raw data externally. This eliminates the need for the central server to acquire, store, and train on the data.

While federated learning enables collaborative learning and preserves data privacy from leakage, it encounters several significant challenges [12]. In general, clients are often hesitant to provide their data for free in the federated learning setting. [14]. This reluctance is partly due to the inherent value of their data. Additionally, participating in the training process requires clients to utilize the computational power of their own devices. Hence, designing an incentive mechanism that effectively encourages client participation has emerged as a crucial concern.

The design of the incentive mechanism in FL primarily focuses on addressing two key challenges: firstly, selecting the data owners who will participate and determining the number of data points they will contribute; and secondly, devising an appropriate payment scheme. To tackle these challenges, many existing studies adopt a reverse auction-based design [5,16,26]: initially, clients submit bids for their own data and training services, and the server selects winners based on the bidding and other relevant information. In these works, the training time is assumed to be fixed or is not even considered, which in fact is critical for the whole training efficiency. However, the training time of each client usually exhibits great randomness due to factors such as resource contention and network instability, [4,6,25], which has a significant impact on the collection of model updates of clients. Therefore, it is quite important to consider the uncertainty of training time of clients in the design of incentive mechanisms.

In this paper, we focus on designing truthful reverse auction mechanisms for FL, with the goal of simultaneously maximizing the expected social welfare. We take into account the uncertainty of clients' training time and allow flexibility in the amount of data each client contributes for training. The main contributions of this paper are summarized as follows:

- We formulate an optimization problem with the objective of maximizing the expected social welfare by selecting clients and determining the amount of training data from each selected client. Additionally, we prove that this problem is NP-hard.
- Despite its hardness, we establish a relationship between the problem and the knapsack problem. Building upon this relationship, we introduce a dynamic programming-based algorithm which can yield an optimal solution in pseudo-polynomial time. So we utilize the renowned Vickrey-Clarke-Groves (VCG)

mechanism to introduce an auction mechanism. This mechanism is both truthful and satisfies the property of individual rationality in expectation.
- To further reduce computational complexity, we introduce an auction mechanism based on a greedy algorithm, which possesses polynomial-time complexity. It also satisfies the properties of truthfulness and individual rationality in expectation, while achieving near-optimal performance in terms of social welfare.
- Finally, we substantiate the effectiveness of our proposed mechanisms through extensive simulation experiments.

The remainder of this paper is organized as follows. Section 2 discusses about related work, Sect. 3 presents the system model and problem formulation, Sect. 4 proposes the VCG-based auction mechanism, Sect. 5 discusses the greedy-algorithm-based auciton mechanism, Sect. 6 presents the performance evaluation, and Sect. 7 concludes this paper.

2   Related Work

The design of incentive mechanism for FL has been studied recently, mainly based on Stackelberg games, contract theory and auction.

By integrating the incentive mechanism with model update requirements, Zhan et al. [27] formulated a Stackelberg game for federated learning and derived its Nash equilibrium. Their algorithm, based on deep reinforcement learning, enables parameter servers and edge nodes to dynamically adjust strategies for optimizing their interests. Hu et al. proposed a two-stage stackelberg game based approach which only requires privacy budgets of the clients to ensure privacy [10].

Kang et al. [13] designed a reputation-based worker selection scheme that employs blockchain to achieve secure reputation management for employees with non-repudiation and tamper resistance characteristics in a decentralized manner. On this basis, an effective incentive mechanism combining reputation and contract theory is proposed to encourage high-reputation mobile devices with high-quality data to participate in model learning. Lim et al. [18] proposed a hierarchical incentive mechanism for FL, accounting for multiple model owners and alliances. They employed deep learning to price fresh data contributed by model training in FL, designing an optimal auction mechanism when multiple model owners vie for customer participation. Wu et al. [24] focused on information asymmetry in FL and jointly considered the task expenditure and privacy issue of federated learning.

Different from the aforementioned solutions, some studies have used the auction framework to model the FL process. Zeng et al. [26] presented FMore, an auction-based incentive framework equipped with scoring functions to evaluate clients' contributions. It selects winners based on ranked scores and provides a unique Nash equilibrium strategy for each node to maximize the expected profit. It also guides the aggregator to obtain the expected resources for some specific cases. Jiao et al. [11] applied an automatic deep reinforcement learning auction mechanism combined with graph neural network to FL in wireless communication scenarios, considering data non-IID and wireless channel sharing conflicts. Furthermore, Deng et al. [5] proposed a quality-aware FL incentive mechanism FAIR based on auction and an aggregation algorithm, considering nonideal datasets. Nguyen H. Tran et al. [16] comprehensively considered required resources, local precision, and energy costs in the wireless FL service market.

However, these existing studies do not take the uncertainty of training time of clients into account. In fact, in order to ensure the efficiency of the FL task, the central server cannot wait indefinitely for all clients to upload their model parameters. Additionally, clients might encounter difficulties in uploading their trained parameters within the specified time due to their individual training schedules. Hence, it is crucial to design an FL incentive mechanism that considers the uncertainty of clients' training time and ex-post payment.

3   System Model and Problem Definition

In this section, we first describe the system model of federated learning and specifically define the expected social welfare maximization problem. The common notations used are defined as follows in Table 1:

Table 1. Notations

Notations	Definition
$\mathcal{C}$	the set of clients
n	the number of clients
d_i	the number of data points for each $i \in \mathcal{C}$
T	the duration of a training round
c_i	the training cost per data point for each $i \in \mathcal{C}$
X_i	the startup time for each $i \in \mathcal{C}$ to train
t_i	the computation time for $i \in \mathcal{C}$ to compute with a single data point

3.1   Federated Learning

We consider a federated learning model consisting of a central server and a clients' set $C = 1, 2, ..., n$. The central server is responsible for issuing training tasks, selecting clients and completing model aggregation. It aims to train a shared global model that minimizes the global loss function $F(\theta)$. Suppose that the client $i \in \mathcal{C}$ is chosen to participate the training using x_i data points where $0 < x_i \leq d_i$. Once selected, each client downloads the shared global model θ and trains on its appointed local data. The loss function for client i can be expressed as follows:

$$F_i(\vartheta) \triangleq \frac{1}{x_i} \sum f_k(\vartheta) \tag{1}$$

Where k represents the unit of data used in training. The clients then uploads new weights to the central server, aiming to optimize $F(\theta)$. The central server sets the duration of a training round as T, and it starts timing after sending θ. Once T elapses, the central server no longer accepts parameters and initiates the model aggregation process. The set of selected clients who successfully upload their weights is denoted as $\mathcal{C}^* \subseteq \mathcal{C}$. According to the ratio of the user's data amount to the total data amount, the weight aggregation parameter is taken as the client's multi-round update method, aimed at accelerating the model convergence speed. The weight average of the central server to the client is as follows:

$$F(\vartheta) \triangleq \frac{\sum_{i \in \mathcal{C}^*} x_i F_i(\vartheta)}{\sum_{i \in \mathcal{C}^*} x_i} \tag{2}$$

The ultimate goal is to find the model parameter θ^*,

$$\vartheta^* = \arg\min F(\vartheta) \tag{3}$$

3.2   Auction Framework

We adopt reverse auction to design the incentive mechanism for FL. In our approach, the auction is conducted before the initiation of federated learning. Once selected, a client $(i \in \mathcal{C})$ remains a participant until the end. Each client submits a bid $b_i = \{d_i, c_i\}$, where d_i represents the data amount and c_i is the cost associated with participation. After receiving all bids, the central server determines which clients to select, the amount of their data to utilize, and then sends the parameters. The clients train on the selected local data points and submit the updated parameters. The central server sets the duration of a training round T and starts timing after the parameter is sent. After T, the central server ceases to accept clients' parameters, and model aggregation along with parameter updates begins. Payment is issued at the conclusion of each training round. Successful submission of updated parameters triggers payment from the central server. Our goal is to maximize the social welfare of entire auction.

In this article, we assume no client abstained or left midway. The system modeling details are as follows:

- Each client $i \in \mathcal{C}$ has d_i local data points for participating in the FL training. Assuming that the cost of each client is proportionate to the number of local data points used in the training, the unit cost of client i is expressed as c_i. If there are m data points participating in FL, the cost of the client is $c_i m$. We assume that the winning client will honestly use own data and will not make hostile attacks.
- The time for client $i \in \mathcal{C}$ to finish its local training with m data points and send the local gradient to the server, denoted by $T_i(m)$, is modeled as

$$T_i(m) = X_i + \tau_i m, \tag{4}$$

where X_i follows an exponential distribution with parameter μ_i and τ_i is a positive constant. Without loss of generality, we assume that $d_i \leq T/\tau_i$. The central server sets T before beginning. We assume that X_i and τ_i can be obtained by the central server in advance (Fig. 1).

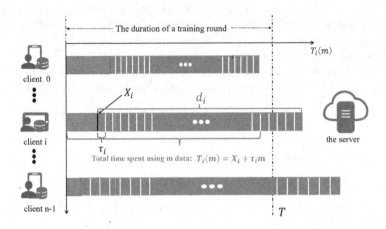

Fig. 1. Illustration of the duration of a training round.

3.3   Problem Definition

Let the client $i \in \mathcal{C}$ participate in the training using x_i data points where $x_i \leq d_i$. We define $I_i(x_i)$ as a binary random variable (0 or 1) indicating whether client i submits its results to the server in time T. The reward function of the server is an increasing, concave function of the actual data points participating the training [27], denoted by

$$g\left(\sum_{i \in \mathcal{C}} I_i(x_i) x_i\right). \tag{5}$$

The reward function $g(\cdot)$ is an increasing function of the number of data points. Hence, the social welfare S is given by

$$g\left(\sum_{i \in \mathcal{C}} I_i(x_i) x_i\right) - \sum_{i \in \mathcal{C}} c_i x_i. \tag{6}$$

When $x_i \leq T/\tau_i$,

$$
\begin{aligned}
I_i(x_i) &= \Pr(I_i(x_i) = 1) \\
&= \Pr(X_i \leq T - \tau_i x_i) \\
&= 1 - e^{-\mu_i(T - \tau_i x_i)},
\end{aligned}
$$

by the linearity of expectation, we have

$$\mathbb{E}\left[\sum_{i \in \mathcal{C}} I_i(x_i) x_i\right] = \sum_{i \in \mathcal{C}} \left(1 - e^{-\mu_i(T - \tau_i x_i)}\right) x_i \tag{7}$$

Generally speaking, the expectation of the reward of the server $\mathbb{E}\left[g\left(\sum_{i \in \mathcal{C}} I_i x_i\right)\right]$ is not equal to $g\left(\mathbb{E}\left[\sum_{i \in \mathcal{C}} I_i x_i\right]\right)$. However, under some mild assumptions, we can show by the Chernoff bound that the actual value of $\sum_{i \in \mathcal{C}} I_i x_i$ is sharply concentrated around its expectation. Hence, we use the following approximation

$$\mathbb{E}\left[g\left(\sum_{i \in \mathcal{C}} I_i x_i\right)\right] \approx g\left(\mathbb{E}\left[\sum_{i \in \mathcal{C}} I_i x_i\right]\right)$$

$$= g\left(\sum_{i \in \mathcal{C}}\left(1 - e^{-\mu_i(T - \tau_i x_i)}\right) x_i\right)$$

Then we formulate the expected social welfare maximization problem as follows:

$$(\text{P1}): \quad \max g\left(\sum_{i \in \mathcal{C}}\left(1 - e^{-\mu_i(T - \tau_i x_i)}\right) x_i\right) - \sum_{i \in \mathcal{C}} c_i x_i$$

$$s.t. \ x_i \leq d_i \quad \forall i \in \mathcal{C}$$

$$x_i \in \mathbb{N} \quad \forall i \in \mathcal{C}$$

In practice, clients might inflate their profits by misstating their bids, rendering the auction mechanism ineffective. Meanwhile, clients' utility should remain nonnegative to consistently encourage data owners' participation in FL. To guarantee market effectiveness and fairness, the incentive mechanism must adhere to the following properties:

- *Truthfulness in expectation*: For the client $i \in \mathcal{C}$, there is no way to report a fake bid c_i' for higher utility in expectation. That is, the client gets the most at the end of FL when it reports a real bid. When the bidding of other clients is fixed, it needs to satisfy

$$u_i(c_i') \leq u_i(c_i), \tag{8}$$

then it is said that the incentive mechanism has truthfulness in expectation.
- *Incentive rationality (IR) in expectation*: No client will gain negative utility due to participating in FL. The utility of one client is represented by the payment p_i received upon completing the task within the designated time, minus the cost c_i. For a client $i \in \mathcal{C}$, $u_i = (p_i - c_i)/((1 - e^{-\mu_i(T - \tau_i x_i)}) x_i) \geq 0$ is always satisfied. This ensures that the incentive mechanism has IR in expectation.

4 The VCG-Based Auction Mechanism

The VCG mechanism has been adopted in many situations under the background of game [1,7,9], owing to its effectiveness demonstrated by its established truthfulness and individual rationality properties, which align with the requirements of incentive mechanisms [20].

However, in this problem, the condition of a VCG mechanism is that the expected social welfare must be maximized rather than approximated, which has not been effectively solved in previous studies. We need to propose an algorithm capable of solving the proposed problem of maximizing the expected social welfare (P1) within polynomial time.

4.1 Problem Transformation

In order to solve the expected social welfare maximization problem, we introduce a budget B on the total training cost, i.e., $\sum_{i \in \mathcal{C}} c_i x_i \leq B$, and consider the following problem:

$$(\text{P2}): \quad \max g \left(\sum_{i \in \mathcal{C}} \left(1 - e^{-\mu_i(T - \tau_i x_i)} \right) x_i \right) - B$$

$$s.t. \quad \sum_{i \in \mathcal{C}} c_i x_i \leq B$$

$$x_i \leq d_i \quad \forall i \in \mathcal{C}$$

$$x_i \in \mathbb{N} \quad \forall i \in \mathcal{C}$$

Since the reward function $g(\cdot)$ is an increasing function, it is equivalent to the following problem:

$$(\text{P3}): \quad \max \sum_{i \in \mathcal{C}} \left(1 - e^{-\mu_i(T - \tau_i x_i)} \right) x_i$$

$$s.t. \quad \sum_{i \in \mathcal{C}} c_i x_i \leq B$$

$$x_i \leq d_i \quad \forall i \in \mathcal{C}$$

$$x_i \in \mathbb{N} \quad \forall i \in \mathcal{C}$$

Let $OPT(B)$ be the optimal value of (P3) and OPT be the optimal value of (P1). The following result establishes a relationship between $OPT(B)$ and OPT.

Lemma 1.

$$OPT = \max_{0 \leq B \leq \sum_{i \in \mathcal{C}} c_i d_i} g(OPT(B)) - B.$$

The above result shows that if we have an efficient algorithm for (P3), we can solve (P1) by applying the algorithm multiple times with different inputs B, resulting in a solution that maximizes the objective value. Regarding (P3), we have the following hardness result.

Theorem 1. *(P3) is NP-hard.*

Proof. Consider a special case of (P3) where $\tau_i = 0$ for each $i \in \mathcal{C}$ and $d_i = 1$. In this case, $e^{-\mu_i(T - \tau_i x_i)} = e^{-\mu_i T}$. Thus the objective function becomes $\sum_{i \in \mathcal{C}} (1 - e^{-\mu_i T}) x_i$. This translates into a standard 0-1 knapsack problem, where each item corresponds to a client with a weight of c_i and a value of $1 - e^{-\mu_i T}$, while the knapsack's capacity is denoted as B. As the 0-1 knapsack problem is known to be NP-hard, it follows that (P3) is also NP-hard.

4.2 Expected Social Welfare Maximization Algorithm

Despite its NP-hardness, (P3) can be solved in pseudo-polynomial time using a dynamic programming approach. Let $\mathcal{C} = \{1, 2, \ldots, n\}$ where $n = |\mathcal{C}|$. Define $f(i, b)$ be the optimal value of (P3) when $\mathcal{C}$ is replaced by $\{1, 2, \ldots, i\}$ and B is replaced by b. We have the following recursive equations.

$$f(i, b) = \begin{cases} 0 & i = 0 \\ \max_{0 \le x_i \le \lfloor b/c_i \rfloor} f(i-1, b - c_i x_i) + (1 - e^{-\mu_i(T - \tau_i x_i)}) x_i & i > 0 \end{cases}$$

Algorithm 1. Social Welfare Maximization Solution

1: **Input:** n, d_i, c_i;
2: **Output:** The clients' optimal training data points $(x_1, \ldots, x_n)$ and maximal social welfare;
3: Initialization: $x_i \leftarrow 0$, $f(0, B) \leftarrow 0$, $path(i, b) \leftarrow 0$;
4: **for** $i = 1$ to n **do**
5:     **for** $b = 0$ to B **do**
6:         **if** $k \le \lfloor b/c_i \rfloor$ **then**
7:             $f(i, b) \leftarrow \max_{0 \le k \le \lfloor T/\tau_i \rfloor} f(i-1, b - k * c_i) - k * (1 - e^{-\mu_i(T - k*\tau_i)})$;
8:             $path(i, b) \leftarrow \arg\max_{0 \le k \le \lfloor T/\tau_i \rfloor} f(i-1, b - k * c_i) - k * (1 - e^{-\mu_i(T - k*\tau_i)})$;
9:         **end if**
10:    **end for**
11: **end for**
12: $b \leftarrow \arg\max_{0 \le b \le B} g(f(n-1, b)) - b$
13: **for** $i = n$ to 1 **do**
14:    $x_i \leftarrow path(i, b); b \leftarrow b - path(i, b) * c_i$;
15: **end for**
16: **return** $W = (x_1, \ldots, x_n)$, $\max_{0 \le b \le B} g(f(n-1, b)) - b$

Algorithm 1 shows the overall process of the proposed social welfare maximization solution. Assuming that each client reports their information as $b_i = \{d_i, c_i\}$, the server determines T to ensure the effectiveness of training. Then, a dynamic programming approach is employed, involving nested iterations through n and B (line 4–11). Given that $OPT(B) \leftarrow f(n, B)$ calculates $OPT(B)_{0 \le b \le B} \leftarrow f(n, b)_{0 \le b \le B}$, we can directly use $f(i, b)_{0 \le b \le B}$ to find the optimal value for (P3) (line 12). This optimization reduces the time complexity from $O(N \cdot d[i]_{max} \cdot B \cdot B)$ to $O(N \cdot d[i]_{max} \cdot B)$. Ultimately, the server obtains the optimal list of clients' training data points denoted as W, along with the maximum social welfare value (line 16).

4.3 Payment

Since the social welfare has been maximized, the server can apply the VCG mechanism to determine the payment $p_1, p_2, \ldots, p_n$ to client $i \in \mathcal{C}$. As above, the optimal list of clients' training data points is $W = (x_1, \ldots, x_n)$. $W^*_{-i} = \left(x^*_1, \ldots, x^*_{i-1}, x^*_{i+1}, \ldots, x^*_n\right)$ is the optimal list of clients' training data points if client i doesn't participate in this federated training, which can be calculated by Algorithm 1. And $W_{-i} = (x_1, \ldots, x_{i-1}, x_{i+1}, \ldots, x_n)$ is W removed client i. Hence, the social welfare of W^*_{-i} is $S\left(W^*_{-i}\right)$ and the social welfare of W_{-i} is $S\left(W_{-i}\right)$. The payment p_i to client i is calculated as

$$p_i = \frac{\left(S\left(W^*_{-i}\right) - S\left(W_{-i}\right)\right)}{\left(\left(1 - e^{-\mu_i(T - \tau_i x_i)}\right) x_i\right)} \tag{9}$$

4.4 Properties

Given that the VCG mechanism has been proven to possess truthfulness and individual rationality, the problem addressed earlier aims to maximize the expected social welfare. It is straightforward to see that, the incentive mechanism ensures truthfulness and individual rationality in expectation.

5 Greedy Algorithm Based Auction Mechanism

Despite the VCG mechanism provides the optimal solution for maximizing social welfare, its computational complexity is proportional to the budget which could be exhibitive. In order to strike a balance between accuracy and time efficiency, in the following, we propose an auction mechanism based on a greedy algorithm.

5.1 Choice of Clients' Training Data

The algorithm of the choice of clients' training data is shown below:

1. We use v_i^k to define the value of the kth data point per training cost when the client i is selected to participate in training with k data points. Since the value decreases as the number increases for each client i, we just need to calculate the positive data point's value of each client (line 4–9) and compare them each time (line 11), which can reduce the time complexity.
2. The mechanism calculates each data point's value as follows:

$$v_i^k = \left(1 - e^{-\mu_i(T - k*\tau_i)}\right) k/c_i k = \left(1 - e^{-\mu_i(T - k*\tau_i)}\right)/c_i.$$

3. The server iteratively chooses the current biggest value and adds this data point with its corresponding client j to the current client solution for social welfare calculation (line 9–17). If current social welfare is bigger than the last result, continue the next iteration. If not, which means this selected data point cannot benefit the social welfare, the client j will quit next iterations. That is because the rest data point of client j have smaller value than this selected data point, so they cannot benefit the social welfare either. Finally, we can get the max social welfare and the corresponding client solution on path.

Algorithm 2. The choice of clients'training data

1: **Input:** n, d_i, c_i;
2: **Output:** The clients' traning data points $(x_1, ..., x_n)$ and corresponding social welfare ;
3: Initialization: $x_i \leftarrow 0$, $max\_social\_welfare \leftarrow 0$, $ignore\_clients \leftarrow \emptyset$, $pre\_client\_solution \leftarrow 0$;
4: **for** $i = 1$ to n **do**
5:     **for** $k = 1$ to $\lfloor T/\tau_i \rfloor$ **do**
6:         Caculate each $v_i^k = (1 - e^{-\mu_i(T-k*\tau_i)})/c_i$;
7:     **end for**
8: **end for**
9: **while** some i not contained in $ignore\_clients$ **do**
10:     Choose the largest value currently and record the corresponding client j;
11:     $pre\_client\_solution \leftarrow client\_soluton$ and update $client\_soluton=(x_1, ..., x_i + 1, x_n)$;
12:     Calculate $cur\_social\_welfare$ according to $client\_soluton$;
13:     **if** $cur\_social\_welfare > max\_social\_welfare$ **then**
14:         Remove the data point of the largest value from client j;
15:         $max\_social\_welfare \leftarrow cur\_social\_welfare$;
16:     **else**
17:         $client\_soluton \leftarrow pre\_client\_solution$;
18:         Add j to $pre\_client\_solution$;
19:     **end if**
20: **end while**
21: **return** $max\_social\_welfare$, $client\_soluton$

5.2 Payment

Since we choose training data one by one in Algorithm 2, the payment for one client is the sum of payments for each selected data point. Based on the algorithm, the server have selected some data points with orderly diminishing values. We assume that the client i has l ($l = d_i$) selected data points with a value list $[(v)_i^1 > (v)_i^2 > ... > (v)_i^\ell]$ (the superscript of (v) represents the order that is selected internally), the payment for each data point is

$$p_i = 1/(v)_{i'}^{\ell+1}.$$

$(v)_{i'}^{\ell+1}$ is the maximum value selected following the last selected data points of the client i in the process. It is obvious that $(v)_{i'}^{\ell+1} < (v)_i^\ell$. Finally, the payment for client i is

$$p_i x_i / \left(1 - e^{-\mu_i(T-x_i*\tau_i)}\right).$$

As we have sorted the value of each client's data points, there is no need for an additional step in the payment calculation.

5.3 Properties

Theorem 2. *The suboptimal mechanism is incentive compatible (IC) in expectation.*

To prove this theorem, the next properties need to be satisfied.

Lemma 2. *Monotonicity: If a client bid $c_i' < c_i$, its selected data points will be no less than the original.*

Proof. If a client bid $c_i' < c_i$, its data points' values will increase and have a higher ranking. The central server preferentially selects the client's data, which makes more data points likely to be selected for training.

Lemma 3. *Critical payment: For the above auction mechanism, the dominant strategy for each client i is to bid its real unit cost c_i.*

Proof. We will show that the dominant strategy for client i is to bid c_i. We just need to consider one data point of client i with value v and the lemma can be extended to its all data points. Since $v = (1 - e^{-\mu_i(T-k*\tau_i)})/c_i$, v will increase with c_i's decline. c is the bid of client if $v = (v)_{i'}^{\ell+1}$. Consider the following cases:

- Case 1: $c_i > c$. If the client bids $c_i' > c$, the data point is still unselected and the payment is still 0. If the clients bids $c_i' \leq c$ and it is not selected, ditto. We consider that if the data point is selected, its unit profit in exception is $(1/(v)_{i'}^{\ell+1} - c_i)/(1 - e^{-\mu_i(T-x_i*\tau_i)}) < (1/(v)_{i'}^{\ell+1} - 1/(v)) \leq 0$. Hence, in case 1, it is optimal for client i to bid c_i.
- Case 2: $c_i \leq c$. If the client bids $c_i' \leq c_i \leq c$, the data point is still unselected and the payment is the same. If $c_i' > c_i$, there are two cases: $\leq c$ and $> c$. If $\leq c$, the unit profit will still be the same. If $> c$ and it is selected, ditto. We consider that if the data point is not selected, the unit profit is 0. Hence, in Case 2, it is also optimal for client to bid c_i.

The proof ic accomplished.

Theorem 3. *The suboptimal mechanism is Individual Rationality (IR) in expectation.*

Proof. The unit profit of client i in expectation is

$$
\begin{aligned}
u_i &= \frac{p_i x_i - c_i x_i}{(1 - e^{-\mu_i(T-x_i*\tau_i)})} \\
&= x_i \left(\frac{p_i}{(1 - e^{-\mu_i(T-x_i*\tau_i)})} - \frac{c_i}{(1 - e^{-\mu_i(T-x_i*\tau_i)})} \right) \\
&> x_i \left(p_i - \frac{c_i}{(1 - e^{-\mu_i(T-x_i*\tau_i)})} \right) \\
&> x_i (1/v_{i'}^{\ell+1} - 1/v_i^\ell) \\
&> 0
\end{aligned}
$$

The proof is accomplished.

In Algorithm 2, generating values (line 4–8) has the time complexity $O(N \cdot d[i]_{max})$, since it needs to calculate each data point's value of each client at most. In the process of choosing clients' training data (line 9–20), its time complexity is also $O(N \cdot d[i]_{max})$, since for worst case scenario, every data point will be ergodiced. For the payment calculation in the process (line 9–20), the time complexity of the suboptimal mechanism is $O(N \cdot d[i]_{max})$, which is much smaller than the time complexity $O(N \cdot d[i]_{max} \cdot B)$ of Algorithm 1.

5.4 Further Option for Reducing Time Complexity

Although the time complexity of Algorithm 2 has already been reduced, the server will still encounter increased computation when dealing with a larger number of participating clients. In order to provide the server with additional options, we propose a further alternative based on Algorithm 2. This alternative sacrifices some accuracy in order to continue reducing the time complexity. The details are described below.

We propose that before the auction the server can set a parameter: the batch b. The definition of the batch b is similar to the one in machine learning. The server equally divide data points of each client into b batches and one batch of the client i with d_i data points has $\lceil d_i/b \rceil$ units. The value of its batch added to $queue_i$ is $\sum_{hb \leq k < (h+1)b}(1 - e^{-\mu_i(T-k*\tau_i)})k / \sum_{hb \leq k < (h+1)b} c_i k (0 \leq h < \lceil d_i/b \rceil)$. Then the server no longer selects data points one by one, but in batches. In this way, the server can freely choose b to reduce the time complexity to $O(N \cdot b)$.

6 Simulation Results

6.1 Experiment Settings

In this section, we conduct comprehensive simulation experiments to evaluate the performance of proposed auction-based incentive mechanisms. We use PyCharm11.0 as the software environment. Each clients' parameters are randomly generated within the range. Our parameter setting focuses on reflecting the differences between clients. The specific parameter settings under standard are as follows (Table 2):

Table 2. Standard Parameter Settings

T	n	d_i	c_i	μ_i	τ_i
30	80	$[40, 100]$	$[1, 10)$	$[0.04, 0.1)$	$[0.1, 1)$

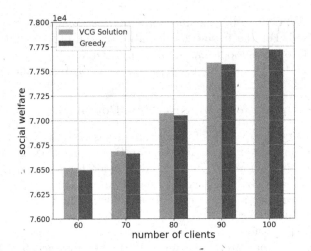

Fig. 2. VCG Solution v.s. Greedy Alg. in social welfare

6.2   Experiment Analysis

We first compare the two auction mechanisms proposed in this paper. In Fig. 2, we compare the results of the auction mechanism based on VCG and the auction mechanism based on greedy algorithm. The social welfare of the two algorithms is compared under different n. Obviously, the difference is relatively small. Therefore, it can be proved that the auction mechanism based on greedy algorithm is effective, and its results are close to the VCG mechanism. So it can be used to replace the VCG mechanism to reduce complexity in most cases (Fig. 3).

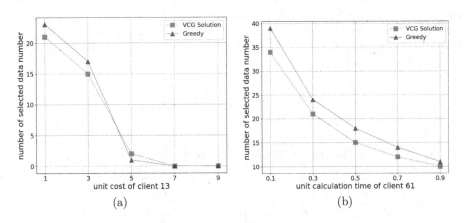

Fig. 3. Impact of unit cost c_i (a) and unit computation time τ_i (b) of one client

Then, we study the impact of the unit cost c_i and the unit computation time τ_i. Specifically, we selected Client No. 13 with a standard unit cost of 1, and varied c_i from 1 to 9 in increments of 2, while keeping the other parameters constant. We can observe that as c_i increases, the number of selected data points x_{13} decreases progressively. Similarly, we selected No. 61 Client whose unit computation time τ_i is 0.1, keep the conditions of other clients unchanged. We then modified τ_i from 0.1 to 0.9 in increments of 0.2, all with other parameters held constant. As τ_i increases, the number of selected data points x_{61} decreases. This outcome aligns with the logic of the auction mechanism. In the auction, the central server is more inclined to select clients with lower training cost and shorter training time to join in FL (Fig. 4).

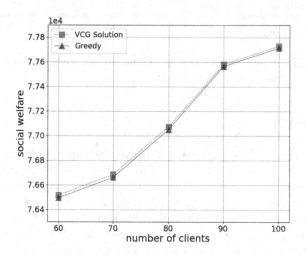

Fig. 4. Impact of the number of clients

We then compared the impact of the number of clients n on social welfare. We observe that social welfare increases with n from 60 to 100. For a central server, more clients mean more opportunities to select higher-value clients, which causes higher social welfare.

Finally, we analyze the impact of the duration of a training round T. From Fig. 5. It is obvious that when T changes from 10 to 110, both social welfare and the sum number of data selected by the central server increase correspondingly. This relationship is intuitive: lengthening T can increases the probability of receiving a client's return parameter. However, the figure also shows the impact of T won't always work; eventually, the results tend to stabilize. This is because when T is infinitely relaxed, the probability of all clients returning training parameters within the specified time approaches 1. Consequently, the result of the algorithm approaches the optimal solution of $g\left(\sum_{i \in \mathcal{C}} x_i\right) - \sum_{i \in \mathcal{C}} c_i x_i$.

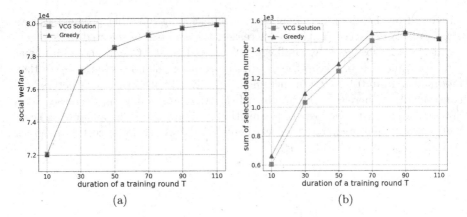

Fig. 5. Impact of the duration of a training round (a) and the social welfare (b) in the sum number of clients' selected data points

7   Conclusion

This paper focuses on the design of an incentive mechanism in federated learning to motivate data owners. A key aspect of our study is the consideration of time as an important and realistic factor in ensuring the efficiency of training. Taking into account the uncertainty of clients' training time, we formulate the problem as a reverse auction between the central server and clients, with the objective of maximizing the expected social welfare. We prove the NP-hardness of the problem and propose a VCG-based auction mechanism. To solve the NP-hard problem, we employ a dynamic programming algorithm. Additionally, we introduce an auction mechanism based on a greedy algorithm, which significantly reduces the computational complexity while still achieving close-to-optimal solutions. Both proposed auction mechanisms are demonstrated to ensure truthfulness and individual rationality while being computationally feasible in pseudo-polynomial time. Finally, we conduct experiments to validate the effectiveness of the mechanisms through simulation results.

Acknowledgements. This paper is supported by the Fundamental Research Funds for the Central Universities under Grant No. B210201053, the National Natural Science Foundation of China under Grant No. 61832005, and the Future Network Scientific Research Fund Project under Grant No. FNSRFP-2021-ZD-07.

References

1. Ahmed, K., Tasnim, S., Yoshii, K.: Simulation of auction mechanism model for energy-efficient high performance computing. In: Proceedings of the 2020 ACM SIGSIM Conference on Principles of Advanced Discrete Simulation, pp. 99–104 (2020)

2. Bonawitz, K., et al.: Towards federated learning at scale: system design. Proc. Mach. Learn. Syst. **1**, 374–388 (2019)
3. Carleo, G., et al.: Machine learning and the physical sciences. Rev. Mod. Phys. **91**(4), 045002 (2019)
4. Coleman, C., et al.: Analysis of dawnbench, a time-to-accuracy machine learning performance benchmark. ACM SIGOPS Oper. Syst. Rev. **53**(1), 14–25 (2019)
5. Deng, Y., et al.: Fair: quality-aware federated learning with precise user incentive and model aggregation. In: IEEE INFOCOM 2021-IEEE Conference on Computer Communications, pp. 1–10. IEEE (2021)
6. Duan, Y., Wang, N., Wu, J.: Minimizing training time of distributed machine learning by reducing data communication. IEEE Trans. Netw. Sci. Eng. **8**(2), 1802–1814 (2021)
7. Dütting, P., Henzinger, M., Starnberger, M.: Valuation compressions in VCG-based combinatorial auctions. ACM Trans. Econ. Comput. (TEAC) **6**(2), 1–18 (2018)
8. Ghosh, A., Chung, J., Yin, D., Ramchandran, K.: An efficient framework for clustered federated learning. Adv. Neural. Inf. Process. Syst. **33**, 19586–19597 (2020)
9. Gu, Y., Hou, D., Wu, X., Tao, J., Zhang, Y.: Decentralized transaction mechanism based on smart contract in distributed data storage. Information **9**(11), 286 (2018)
10. Hu, R., Gong, Y.: Trading data for learning: incentive mechanism for on-device federated learning. In: GLOBECOM 2020-2020 IEEE Global Communications Conference, pp. 1–6. IEEE (2020)
11. Jiao, Y., Wang, P., Niyato, D., Lin, B., Kim, D.I.: Toward an automated auction framework for wireless federated learning services market. IEEE Trans. Mob. Comput. **20**(10), 3034–3048 (2020)
12. Kairouz, P., et al.: Advances and open problems in federated learning. Found. Trends® Mach. Learn. **14**(1–2), 1–210 (2021)
13. Kang, J., Xiong, Z., Niyato, D., Xie, S., Zhang, J.: Incentive mechanism for reliable federated learning: a joint optimization approach to combining reputation and contract theory. IEEE Internet Things J. **6**(6), 10700–10714 (2019)
14. Kim, H., Park, J., Bennis, M., Kim, S.L.: Blockchained on-device federated learning. IEEE Commun. Lett. **24**(6), 1279–1283 (2019)
15. Kravets, P., et al.: Markovian learning methods in decision-making systems. In: Babichev, S., Lytvynenko, V. (eds.) ISDMCI 2021. LNDECT, vol. 77, pp. 423–437. Springer, Cham (2022). https://doi.org/10.1007/978-3-030-82014-5_28
16. Le, T.H.T., et al.: An incentive mechanism for federated learning in wireless cellular networks: an auction approach. IEEE Trans. Wireless Commun. **20**(8), 4874–4887 (2021)
17. Liakos, K.G., Busato, P., Moshou, D., Pearson, S., Bochtis, D.: Machine learning in agriculture: a review. Sensors **18**(8), 2674 (2018)
18. Lim, W.Y.B., et al.: Hierarchical incentive mechanism design for federated machine learning in mobile networks. IEEE Internet Things J. **7**(10), 9575–9588 (2020)
19. Mohri, M., Sivek, G., Suresh, A.T.: Agnostic federated learning. In: International Conference on Machine Learning, pp. 4615–4625. PMLR (2019)
20. Roughgarden, T.: Algorithmic game theory. Commun. ACM **53**(7), 78–86 (2010)
21. Tarca, A.L., Carey, V.J., Chen, X.W., Romero, R., Drăghici, S.: Machine learning and its applications to biology. PLoS Computat. Biol. **3**(6), e116 (2007)
22. Tran, N.H., Bao, W., Zomaya, A., Nguyen, M.N., Hong, C.S.: Federated learning over wireless networks: optimization model design and analysis. In: IEEE INFOCOM 2019-IEEE Conference on Computer Communications, pp. 1387–1395. IEEE (2019)

23. Wang, H., Yurochkin, M., Sun, Y., Papailiopoulos, D., Khazaeni, Y.: Federated learning with matched averaging. arXiv preprint arXiv:2002.06440 (2020)
24. Wu, M., Ye, D., Ding, J., Guo, Y., Yu, R., Pan, M.: Incentivizing differentially private federated learning: a multidimensional contract approach. IEEE Internet Things J. 8(13), 10639–10651 (2021)
25. Yan, F., Ruwase, O., He, Y., Chilimbi, T.: Performance modeling and scalability optimization of distributed deep learning systems. In: Proceedings of the 21th ACM SIGKDD International Conference on Knowledge Discovery and Data Mining, pp. 1355–1364 (2015)
26. Zeng, R., Zhang, S., Wang, J., Chu, X.: FMore: an incentive scheme of multidimensional auction for federated learning in MEC. In: 2020 IEEE 40th International Conference on Distributed Computing Systems (ICDCS), pp. 278–288. IEEE (2020)
27. Zhan, Y., Li, P., Qu, Z., Zeng, D., Guo, S.: A learning-based incentive mechanism for federated learning. IEEE Internet Things J. 7(7), 6360–6368 (2020)

Persistent Sketch: A Memory-Efficient and Robust Algorithm for Finding Top-k Persistent Flows

Ziqi Sun[1], Yu-E Sun[2,3(✉)], Yang Du[1(✉)], Jia Liu[4], and He Huang[1]

[1] School of Computer Science and Technology, Soochow University, Suzhou 215006, China
duyang@suda.edu.cn
[2] School of Rail Transportation, Soochow University, Suzhou 215137, China
sunye12@suda.edu.cn
[3] Key Laboratory of Embedded System and Service Computing (Tongji University), Ministry of Education, Shanghai 201804, China
[4] State Key Laboratory for Novel Software Technology, Nanjing University, Nanjing 210023, China

Abstract. Finding top-k persistent flows in high-speed network traffic is crucial for applications like click-fraud detection and covert attacker detection. The prior studies either do not separate persistent and non-persistent flows during online traffic processing and waste significant space to record numerous non-persistent flows, or only realize unstable separation that is not robust to flow frequency. We proposes Persistent Sketch (PE-Sketch), the first memory-efficient and robust algorithm for finding top-k persistent flows. The basic idea is accurately separating persistent flows and then tracking them. Because it is difficult to perform separation by persistence directly, PE-Sketch introduces the concept of event sampling to sample the persistence increment events (each flow's first arrival in every time window) with a pre-defined probability, where the number of sampled events is proportional to flow persistence. Then we design a memory-efficient candidate matrix to accurately separate and track the flows with the most sampled events, *i.e.*, persistent flows. With the two key techniques, we find persistent flows regardless of their frequencies, attaining robust and accurate estimation results. Experimental results demonstrate that, compared to the state-of-the-art (On-Off Sketch), PE-Sketch is robust, and it can improve the precision by up to 15.6 times and reduce the error by up to 2 orders of magnitude when using the same space.

Keywords: Network traffic measurement · Sketch · Sampling · Top-k persistent flows

1 Introduction

Finding top-k persistent flows is a fundamental traffic measurement task over big streaming data in high-speed networks, where we aim to find the flows which persist to occur in the stream over a long timespan. Different from simply counting

© The Author(s), under exclusive license to Springer Nature Singapore Pte Ltd. 2024
Z. Tari et al. (Eds.): ICA3PP 2023, LNCS 14492, pp. 19–38, 2024.
https://doi.org/10.1007/978-981-97-0811-6_2

the appearances of flows (*i.e.*, frequency problem), this task additionally considers the temporal characteristic of flows. We divide time into T non-overlapping and continuous time slots (called *periods*) and use flow persistence to represent the number of *distinct* periods in which a flow occurs in the traffic. The top-k persistent flows refer to the k flows with the largest persistence among all flows, which exhibit a repeated and periodic pattern of arrival [1], and are significant targets for applications like click-fraud detection [2,3], stealthy DDoS detection [4–7], and APT attacker detection [8–11]. For instance, in malicious behavior detection, finding the top-k persistent flows enables the firewall to discriminate covert attackers as they demonstrate uncustomary high persistence, *e.g.*, an APT attacker prefers using a small probing rate for a long period of time to discover system vulnerabilities.

It is challenging to find top-k persistent flows in real time because we need to catch up with the increasing high speed of big streaming data (*e.g.*, 1.5M packets per second [12]) and maintain accuracy simultaneously. To achieve high speed, existing approaches [1,13–15] choose to store the data structure entirely in high-speed cache memory (such as SRAM) on chip such that only caches are accessed during processing. To address the challenge of counting *distinct* periods for flows, majority prior methods add a flag for each counter in memory to ensure that the persistence values of the occurred and recorded flows only increase 1 in single period. However, due to the limitation of memory size, the number of flows we can record is far less than the flow number in the traffic, and we should decide whether to record the new arrival unrecorded flow. Hence, the frequent but not persistent flows would constantly try to replace the recorded flows and the persistent but not frequent flow may be replaced before it occurs in the latest period, resulting in estimation error. Some prior studies have tried to eliminate the interference of flow frequency, but they did not do it well. We will introduce it in Sect. 2.

We aim to address the limitations of the prior studies by accurately and robustly separating persistent and non-persistent flows during online processing, using a small space to track and finally report persistent flows. However, it is tricky to directly separate flows by persistence because we cannot predict flow persistence when streaming flows arrive. Besides, it is difficult to count flow persistence (*i.e.*, count distinct periods on the flow level) with small space. To address these challenges, we follow *sampling methods* [16–20] and introduce the concept of *event sampling*: We regard all packets in traffic as a sequence of flow arrival events (arrival event for short) and particularly regard every flow's first arrival in every period as a persistence increment event (increment event for short). For each flow, an increment event represents increasing its persistence by 1. Event sampling takes in arrival events while only samples increment events with a pre-defined probability. Therefore, a sampled increment event refers to a fixed persistence increment of the flow, and then we can distinguish persistent flows by tracking the number of sampled increment events. By this means, we reduce persistence-based flow separation to sampled-event-based flow separation, which separates flows by the number of sampled events, enabling us to apply frequency-based separation methods [21–25]. With these two designs, the number of sampled events is directly proportional to flow's persistence, and the flows we track are persistent ones. Therefore, our solution is robust and is applicable to various traffic scenarios (*i.e.*, not influenced by flow frequency).

In this paper, we propose the Persistent Sketch (PE-Sketch), the first memory-efficient and robust algorithm for accurately finding top-k persistent flows in network traffic. PE-Sketch consists of two stages of processing: *event sampling* and *candidate tracking*. In the first stage, we propose a novel Flip Filter to perform event sampling with a pre-defined probability p. In stage two, we place a small candidate matrix to perform sampled-event-based flow separation and track the persistent flows, where we follow the idea of Unbiased Space-Saving [25] and develop the *probabilistic replacement strategy*, ensuring that flows with higher persistence can be tracked in candidate matrix with higher probabilities. With these designs, PE-Sketch can work with small memory usage and accurately find the top-k persistent flows from network. Meanwhile, PE-Sketch is robust to the influence of flow frequency. Our main contributions are concluded as below:

- We propose a novel two-stage sketch named Persistent Sketch to find the top-k persistent flows from high-speed network traffic. To the best of our knowledge, PE-Sketch is the first memory-efficient and robust algorithm that can achieve high accuracy with small memory usage for traffic with different skewness.
- We introduce the concept of event sampling to reduce the studied persistence problem to a problem of finding the k flows with the most sampled events (a frequency problem). We also design a memory-efficient candidate matrix to accurately separate and track the persistent flows, *i.e.*, those with larger numbers of sampled events.
- We conduct extensive experiments on real-world and synthetic datasets. The experimental results demonstrate that compared to the state-of-the-art, PE-Sketch is more robust, and can improve the precision by up to 15.6 times and reduce the error by up to 2 orders of magnitude within the same space.

2   Related Work

Many prior studies can be applied to find top-k persistent flows. In this section, we will briefly introduce four recent arts and show their shortcomings.

PIE [14] creates a separate Space-Time Bloom Filter (STBF) for every period to record the appeared flows and determines a flow's persistence based on its existences in the STBFs of T periods. Small-Space(SS) [1] samples a fraction of packets and records the flow labels from sampled packets in a hash table to exactly count their persistence. Hence, the memory requirements of these two algorithms are closely positively correlated with the number of distinct flows in .the traffic, where they both require a large space to record flows they have seen if there exist abundant flows.

To address this problem, Adaptive Sampling (AS) [13] gradually decreases the sampling rate as the period number increases, and decreases the value of each recorded flow in the hash table by a geometrical random number when the sampling rate changes. Therefore, AS can timely weed out the non-persistent flows so as to improve space efficiency. However, the dynamic sampling rate and the random decreasing value make it impossible to accurately estimate flow's persistence. AS can only answer which flows are persistent, not how many periods they have appeared. On-Off Sketch [15] (the state-of-the-art) tries to identify the persistent flows on the fly and track their persistence within a small space. It maintains an array of l counters to record persistence, randomly maps each flow label to one of the counters, and assigns each counter with w cells to store the labels and persistence for the w most persistent flows this counter has seen. Because multiple flows can be mapped to the same counter and cause overestimation of persistence, On-Off Sketch only allows the first flow arrives at each counter in every period to increase counter value (i.e., persistence) by 1 and replace the smallest tracked flow among the corresponding w cells if the counter value is larger. However, this strategy is not robust to the influence of flow frequency: When there are numerous non-frequent flows in the traffic, a persistent but not frequent flow may never arrive first at a counter in any period, i.e., it will not be identified by On-Off Sketch.

3 Problem Statement

We consider a network traffic processing model as follows: Time is divided into T non-overlapping and continuous periods. The streaming data S consists of a sequence of M packets, where each packet has a flow label e (e.g., IP address or defined arbitrarily based on one or a combination of fields in packet headers) and a period index $t \in [1, T]$ denoting the index of the period in which it appears. Suppose there are N distinct flows in S, where $N \leq M$, we denote the set of flow labels as $\mathbb{E} = \{e_1, e_2, ..., e_N\}$. Therefore, the network traffic can be abstracted as a sequence of label-index tuples, formally $S = \{\langle e^{(1)}, t^{(1)} \rangle, \langle e^{(2)}, t^{(2)} \rangle, \cdots, \langle e^{(M)}, t^{(M)} \rangle\}$, where $e^{(i)} \in \mathbb{E}$ and $t^{(i)} \in [1, T]$ denote the flow label and period index of the i-th flow in the sequence. We use n_e to represent the persistence of flow e, which is the number of distinct periods in which e has appeared. We reorder all flows as $\{e_{i_1}, e_{i_2}, \cdots, e_{i_N}\}$ such that we have $n_{e_{i_1}} \geq \cdots \geq n_{e_{i_N}}$. The studied top-$k$ persistent flows detection problem is defined as follows:

Given a network traffic stream $S = \{\langle e^{(1)}, t^{(1)} \rangle, \langle e^{(2)}, t^{(2)} \rangle, \cdots, \langle e^{(M)}, t^{(M)} \rangle\}$ and a parameter k, the task of finding top-k persistent flows is to determine the k flows with the largest persistence among all flows, i.e., $\{e_{i_1}, e_{i_2}, \cdots, e_{i_k}\}$.

4 The PE-Sketch Design

4.1 Approach Overview

The key idea of PE-Sketch is to separate flows by persistence and use a small space to track the flow labels and persistence of persistent flows. The descriptive architecture is shown in Fig. 1. Due to the difficulty in directly separating flows by persistence, PE-Sketch divides the task into two stages: *event sampling* and *candidate tracking*. In the first stage, we design a Flip Filter to sample increment events with a fixed probability p. The sampled events will be passed to the second stage, where we develop a memory-efficient candidate matrix to separate flows based on the number of sampled events and then only track the persistent flows. Finally, when answering the query, PE-Sketch reports the k flows with the largest persistence in the candidate matrix. The benefit of this design is two-fold. First, it reduces the studied problem to the problem of finding the flows with the most sampled events, enabling us to follow mature frequency-based separation methods and guarantee memory efficiency. Second, the sampling and separation process only depends on flow persistence and is robust to the influence of flow frequency.

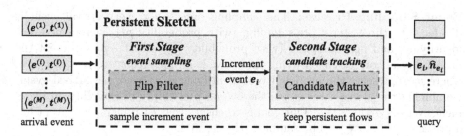

Fig. 1. The workflow of the PE-Sketch.

4.2   Event Sampling Based on Flip Filter

We design Flip Filter, a memory-efficient method to perform event sampling, which samples each flow's increment event in every period with the same preset probability of p. To achieve this goal, Flip Filter deploys a bit array to identify flows' increment events (first arrival in every period). Unlike traditional duplicate filtering algorithms (such as Bloom filter [26]) statically setting bits from '0' to '1' and re-initializing the bit array when each period ends, Flip Filter develops the *flag flip operation* to toggle dynamically between setting bits to '1' and setting bits to '0' in different periods according to the distribution of '0' and '1' in the bit array, so as to eliminate the reset overhead, ensuring that the measurement task can be carried out continuously without interruption.

Data Structure: As shown in Fig. 2, Flip Filter includes a bit array B of m bits, a *flag* bit f, and a counter c recording the number of '1's in B. It also maintains z hash functions $H_1(e, t), H_2(e, t), \cdots H_z(e, t)$, whose output ranges are $\{0, 1, \cdots, m-1\}$, to pseudo-randomly map flow e in period t on z bits. We use z bits of $flag$ to indicate the flow existence, where we also use *flag flip operation* to determine when we should flip the $flag$ bit from '1' to '0' (or from '0' to '1') to make sure Flip Filter can work across periods. The bits in B are initialized to '0' or '1' with the same probability (*i.e.*, 50%). The $flag$ bit is initialized to '1', and $c = m/2$.

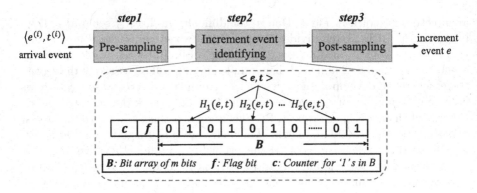

Fig. 2. The workflow of Flip Filter.

Event Sampling Process: The sampling process for every arrival event is divided into three steps: pre-sampling (with probability p_1), increment event identifying, and post-sampling (with probability p_2). Let $1 - p_f$ represent the probability of correctly identifying the event as an increment event in step two, where p_f represents the probability of identifying it as a normal arrival event. Combining the above three steps, the overall probability of sampling an increment event is $p_1(1 - p_f)p_2$. Our goal is to ensure that each increment event has a pre-set probability p to be sampled. Hence, we have

$$p_1(1 - p_f)p_2 = p \tag{1}$$

Note that p_1, p_f, p_2 and p are positive numbers and no more than 1. The above equation should always hold during sampling. In this way, Flip Filter uniformly and fairly samples increment events for each flow with probability p regardless of flow's arrival time in every period.

The specific process of sampling is as follows: Given an arrival event $\langle e, t \rangle$, in the first step, Flip Filter samples $\langle e, t \rangle$ with a stable probability p_1 to decelerate the change speed of c in bit array B. The sampling is performed as follows: Flip Filter performs a hash $h' = H'(e, t)$ where H' is a hash function whose range is $[0, D)$. If and only if $h' \leq p_1 D$, the arrival event $\langle e, t \rangle$ passes through sampling and enters into the next step. The second stage is used to identify whether $\langle e, t \rangle$ is an increment event. Flip Filter maps e to arbitrary z bits in B with z hash functions, sets the z bits to $flag$, and updates the value of c. If e sets bits successfully (i.e., the value of c changes), Flip Filter identifies the incoming event as an increment event and sends it to step three. The third step is increment event sampling with a variable probability p_2, where the Flip Filter calculates p_2 from Eq. 1 and performs sampling with another hash function H'', following the same approach in the first step.

Shortly, we will discuss how to calculate p_2. For each increment event, it will be mistaken as a normal arrival flow if the values of z bits it mapped to are $flag$. Thus we have $p_f = (c/m)^z$ if $flag = 1$. Otherwise, $p_f = (1 - c/m)^z$. Substitute p_f to Eq. 1, and we can get p_2. Note that the input of all hash functions we use in

Flip Filter is flow label e and period number t. The hash values of the same flow label are different in distinct periods.

Flag Flip Operation: When each period ends, we use the flag flip operation to decide $flag$ for the next period, eliminating bit array reset overhead. We flip $flag$ in the same way regardless of $flag = 0$ or $flag = 1$. For convenience, just set $flag = 1$ in period t. Substitute $p_f = (c/m)^z$ to Eq. 1, we have

$$p_1(1 - (c/m)^z)p_2 = p \tag{2}$$

Therefore, in pace with inserting incoming flows to B in one period, c is monotonically increasing, so does p_2. As p_2 can be at most 1, we further have deformation as inequality:

$$p_1(1 - (\frac{c}{m})^z) \geq p \tag{3}$$

Thus we have $c \leq m\sqrt[z]{1 - p/p_1}$. It means that up to $m\sqrt[z]{1 - p/p_1}$ bits in B can be set to '1'. We denote this maximum value of c as c_{thres}. Recall that we initialize $c = m/2$. Hence c_{thres} must exceed $m/2$, and we have

$$\frac{1}{2} < \sqrt[z]{1 - \frac{p}{p_1}} \tag{4}$$

Besides, the maximum increment of c in one period should not exceed $c_{thres} - m/2$ (denoted as c_{max}). To avoid termination in next period, *i.e.*, c exceeds c_{thres}, Flip Filter flips $flag$ from '1' to '0' if $c + c_{max} \geq c_{thres}$ when each period ends. Similarly, when $flag = 0$, Flip Filter turns $flag$ to '1' if $c - c_{max} \leq m - c_{thres}$. In Sect. 5.1, we will show how to get the optimal values of p_1, p and z.

4.3 Candidate Tracking

After event sampling, the problem we study is transformed into finding the k flows that are sampled the most increment events. We develop a compact data structure called candidate matrix to accomplish this task. Since we cannot record all flows within a small space, we propose the *probabilistic replacement strategy* to perform sampled-event-based flow separation and track the persistent candidates, where flows with higher persistence will be recorded in the matrix with higher probability.

Data Structure: As shown in Fig. 3, the candidate matrix is comprised of d buckets, and each bucket is comprised of w cells. Each cell stores a key-value (KV) pair: the flow label e_i and its value f_i, *i.e.*, the number of increment events e_i is sampled. For convenience, we use S_i to represent the ith bucket and use $S_i[e_j]$ to represent the value of e_j. The matrix is associated with one hash function $H'''(e)$, whose range is $[0, d)$, mapping each flow into one bucket. Initially, all fields in each bucket are zero.

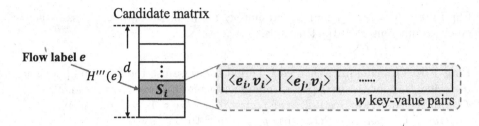

Fig. 3. Data structure of candidate matrix.

Probabilistic Replacement Strategy: A bucket S_i in the matrix can only record w KV pairs, which is far less than the number of flows mapped to S_i. Therefore, the matrix has to determine whether to replace a recorded flow with the new arrival flow e or to still maintain the recorded flows. Let e' be the smallest flow in S_i with value $f_{e'}$ ($f_{e'} = 0$ if S_i has empty cells). If e is not recorded, we try to replace e' by e with probability P_r:

$$P_r = \frac{1}{f_{e'} + 1} \tag{5}$$

and then update $S_i[e]$ to $f_{e'} + 1$ if successfully replacing. Specifically, we use a hash function $G(e)$ to perform the replacement, where the output range of $G(e)$ is $[0, D)$. If and only if $G(e) \leq P_r D$, e' will be replaced by e. The successful replacement indicates that e has sampled $f_{e'} + 1$ events and is likely to be more persistent than e'. We set $S_i[e] = f_{e'} + 1$. This method has three advantages. First, it has low computation overheads. Second, the abortive replacement will not affect the value of recorded flows *i.e.*, a failure replacement introduces no estimation error to recorded flows. Third, a new flow expects to succeed through $f_{e'} + 1$ attempts, which equals its estimated number of sampled events after the successful replacement. It ensures that the persistent flows arriving later are still expected to have an accurate estimation.

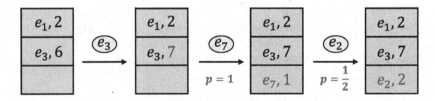

Fig. 4. Examples of candidate matrix insertion.

Event Insertion: For each sampled event (belongs to e), our matrix first inserts e into one bucket S_i, where $i = H'''(e)$. There are two cases for insertion. If e has been recorded in S_i, increase $S_i[e]$ by 1. Otherwise, the matrix tries to replace the smallest flow e' in S_i with e through the probability replacement strategy.

An insertion example is depicted in Fig. 4. Let $w = 3$. When e_3 arrives, the candidate matrix first maps e_3 to S_i and finds that e_3 has been stored, it directly updates $S_i[e_3]$ to 7. To insert e_7, because that e_7 is not recorded in S_i and the smallest value in S_i is 0 (the value of the empty cell), the probability of replacement is 1. Insert e_7 to S_i and set $S_i[e_7] = 1$. After that, to insert e_2, the matrix finds e_2 is not stored either. Now the smallest value in S_i is 1, and the replacement probability P_r is 1/2. Suppose that e_7 is replaced with e_2, increase $S_i[e_2]$ to 2. Otherwise, the matrix does nothing.

Query Top-k Persistent Flows. When answering queries, PE-Sketch traverses all KV pairs in the candidate matrix, sorts and selects the top-k largest values, and then reports the k corresponding flow as top-k persistent flows. Give an estimation persistence $\widehat{n_e}$ for a recorded flow e ($S_i[e] = f_e$) as $\widehat{n_e} = f_e/p$ (ignoring integrality).

5   Performance Analysis

In this section, we discuss how to determine the appropriate parameters of Flip Filter (Sect. 5.1) and provide a performance analysis for the candidate matrix on finding persistent flows (Sect. 5.2).

5.1   Parameters of Flip Filter

For arbitrary flow e, suppose that e's actual persistence is n_e, and e is sampled f_e times. We estimate e's persistence as: $\widehat{n_e} = f_e/p$, where p is the pre-set constant sampling probability. The sampling process can be regarded as a standard Bernoulli process, where we have $f_e \sim B(n_e, p)$. Thus the variance of f_e is $Var(f_e) = n_e p(1 - p)$ and the mathematical expectation is $E(f_e) = n_e p$. The relative standard deviation of $\widehat{n_e}$ is $\sigma_r = \sqrt{Var(\widehat{n_e})}/E(\widehat{n_e}) = \sqrt{(1 - p)/n_e p}$. We find that for any persistence value n_e, the relative standard deviation σ_r gets smaller as p increases. Therefore, the goal of parameter selection is to maximize p to reduce error.

Regardless of the value of $flag$, Flip Filter performs event sampling in the same way and thus will follow the same analysis process. We set $flag = 1$ and omit the analysis for $flag = 0$ to save space. Let c_{start}^t/c_{end}^t represent the value of c at the beginning/end of period t. Consider an arbitrary bit $B[i]$ in Flip Filter, when period t begins, the probability of $B[i] = 0$ is $(1 - c_{start}^t)/m$. Each increment event has a probability of $(1 - 1/m)^z$ not being hashed to it and then set $B[i]$ to 1. Therefore, the probability of $B[i] = 0$ after recording $N p_1$ distinct flows (N denotes the number of distinct flows in period t) is $(1 - c_{start}^t/m)(1 - 1/m)^{zN p_1}$, which means that the probability of hashing none flow to $B[i]$ is $(1 - c_{start}^t/m)(1 - 1/m)^{zN p_1}$. The p_f of the Flip Filter can be written as $(c/m)^z$ and is continuously increasing in each period, which reaches its maximum when period t ends. Then, we have p_f in B when period t ends:

$$p_f = (\frac{c_{end}^t}{m})^z \le (1 - (1 - \frac{c_{start}^t}{m})(1 - \frac{1}{m})^{zN p_1})^z \tag{6}$$

Applying $(1 - 1/m)^{zNp_1} \approx e^{-zNp_1/m}$ to Eq. 6, we have

$$(\frac{c_{end}^t}{m})^z \le (1 - (1 - \frac{c_{start}^t}{m})e^{\frac{-zNp_1}{m}})^z \tag{7}$$

From Eq. 3, we have $1 - (c_{end}^t/m)^z \ge p/p_1$. Combining it and Eq. 7, the following inequality should hold

$$1 - \frac{p}{p_1} \ge (1 - (1 - \frac{c_{start}^t}{m})e^{\frac{-zNp_1}{m}})^z \tag{8}$$

To further narrow Eq. 8, we set c_{start}^t to its maximum $m/2$.

$$p \le p_1(1 - (1 - \frac{1}{2}e^{\frac{-zNp_1}{m}})^z) \tag{9}$$

Given a fixed memory space m and an estimated N from historical information, we aim to get the maximum p to improve accuracy. The optimization problem of p is formally defined as follows by combining Eq. 4 and Eq. 9.

$$maximize \quad p$$

$$s.t. \begin{cases} p < p_1 - \frac{p_1}{2^z}, \\ p \le p_1(1 - (1 - \frac{1}{2}e^{\frac{-zNp_1}{m}})^z), \\ 0 < p < p_1 \le 1, \\ z = \{1, 2, ..., m\}. \end{cases} \tag{10}$$

Hence, the optional value of p can be solved from Eq. 10 by using grid search.

5.2   Estimation Error Bound of Candidate Matrix

In this part, we will analyze the estimation error of tracking persistent flows in the candidate matrix. Note that, after event sampling, a persistent flow will have a large number of sampled events. For convenience, we use 'frequency' to represent the number of sampled events in the following.

Theorem 1. *If flow e replaces another flow, the estimation error caused by replacement is unbiased.*

Proof. Let $\widehat{f_e}, f_e$ denote the estimated and actual frequency of e, respectively. Suppose that e' is the flow with the smallest recorded frequency f_{min} in the bucket. Once e tries to replace e', the estimated frequency of e will be incremented by $f_{min} + 1$ with probability $1/(f_{min} + 1)$ for an expected increment by 1, which proves the unbiasedness.

Theorem 2. *If e is replaced by other flows, it will cause a downward estimation error X on e's frequency. The expectation of X is*

$$E(X) \le l \tag{11}$$

where l is the number of times other flows try to replace e.

Proof. Suppose that e is the smallest flow with frequency $f(t)$ in the bucket at time t and $l(t)$ flows try to replace e before e occurs again. Let $X(t)$ be the downward estimation error on $f(t)$. Obviously, if all $l(t)$ flows fail, e is still recorded, and $X(t) = 0$. If e is replaced, $\widehat{f_e}$ is updated to 0 and $X(t) = f(t)$. The probability that all replacements fail is $(1 - 1/(f(t) + 1))^{l(t)}$, thus we have

$$E(X(t)) = f(t)(1 - (1 - \frac{1}{f(t) + 1})^{l(t)}) \le l(t) \tag{12}$$

Assuming that $l(t)$ is independent of $f(t)$, we have

$$E(X) = \sum_t E(X(t)) \le \sum_t l(t) = l \tag{13}$$

Let Z be the sum of the sampled increment events of all flows in the network traffic, and e_i has the ith largest frequency among all sampled N distinct flows, whose frequency is f_i. We have the frequency estimation $\widehat{f_i}$ of e_i:

$$\widehat{f_i} = f_i - X_i + Y_i' - Y_i'' \tag{14}$$

X_i is the decrement from being replaced. Y_i' and Y_i'' are upward and downward estimation errors from replacing other flows, respectively.

Theorem 3. *For any $\epsilon > 0$, the upper bound of estimation error for e_i is*

$$\Pr\{\widehat{f_i} - f_i \ge \epsilon Z\} \le \frac{1}{\epsilon}(\frac{1}{2d} + \frac{1}{2Z}) \tag{15}$$

Proof. Let F_{min} be the random variable representing the smallest frequency in the bucket when e_i swaps in successfully. The maximum of Y_i' is f_{min}, where the replacement happens in the first occurrence of e_i. Thus We have

$$E(Y_i'|F_{min} = f_{min}) \le f_{min} \tag{16}$$

Let F_{min}' be the smallest value in the bucket when querying. F_{min} is uniformly distributed within $[1, F_{min}']$. We have

$$E(Y_i'|F_{min}' = f_{min}')$$
$$= \sum_{f_{min}=1}^{f_{min}'} E(Y_i'|F_{min} = f_{min})\frac{1}{f_{min}'} \le \frac{f_{min}' + 1}{2} \tag{17}$$

We further obtain the law of total expectation

$$E(Y_i') = \sum_{f_{min}'} E(Y_i'|F_{min}' = f_{min}')\Pr\{F_{min}' = f_{min}'\}$$
$$\le \frac{Z}{2d} + \frac{1}{2} \tag{18}$$

Then, we use Markov inequality to transform the bound of expectation into the bound of possibility:

$$\Pr\{\widehat{f_i} - f_i \geq \epsilon Z\} = \Pr\{-X_i + Y_i' - Y_i'' \geq \epsilon Z\}$$
$$\leq \Pr\{Y_i' \geq \epsilon Z\}$$
$$\leq \frac{1}{\epsilon}(\frac{1}{2d} + \frac{1}{2Z}) \quad (19)$$

Theorem 4. *Given a flow e_i with i^{th} largest frequency (number of sampled events), for any $\epsilon > 0$, the lower estimation error bound is*

$$\Pr\{f_i - \widehat{f_i} \geq \epsilon Z\}$$
$$\leq \frac{1}{\epsilon}(\frac{1}{d}\binom{i-1}{\lambda-1}(\frac{1}{d})^{\lambda-1}(1 - \frac{1}{d})^{i-\lambda} + \frac{1}{2d} + \frac{1}{2Z}) \quad (20)$$

Proof. If and only if e_i is the smallest flow in the bucket, it may be replaced by other flows. Let U be the random variable representing the number of flows whose frequencies are larger than e_i and mapped to the same bucket with e_i. U follows a binomial distribution $B(i-1, 1/d)$. Suppose that $\lambda - 1$ flows among the $i - 1$ flows are mapped to the same bucket with e_i. We have

$$\Pr\{U = \lambda - 1\} = \binom{i-1}{\lambda-1}(\frac{1}{d})^{\lambda-1}(1 - \frac{1}{d})^{i-\lambda} \quad (21)$$

Let L be the random variable representing the number of flows that try to replace e_i. As proved in Theorem 2, the expectation of decrement caused by being replaced is $E(X_i | L = l) \leq l$. If e_i is the smallest flow in the bucket, the number of replacement attempts is equal to the number of increment events of flows whose frequencies are smaller than e_i. Thus, we have

$$E(L | U = \lambda - 1) = \frac{1}{d}(\sum_{j=i+1}^{N} f_j) \leq \frac{Z}{d} \quad (22)$$

The expectation of L is

$$E(L) = E(L | U = \lambda - 1)\Pr\{U = \lambda - 1\}$$
$$\leq \frac{Z}{d}\binom{i-1}{\lambda-1}(\frac{1}{d})^{\lambda-1}(1 - \frac{1}{d})^{i-\lambda} \quad (23)$$

We finally get expectation of X_i as

$$E(X_i) = \sum_l E(X_i | L = l)\Pr\{L = l\}$$
$$\leq \sum_l l\Pr\{L = l\} \leq E(L) \quad (24)$$

From Theorem 1, the estimation error is unbiased when e_i replaces others. Therefore, the expectation of downward error equals that of upward error.

$$E(Y_i'') = E(Y_i') \leq \frac{Z}{2d} + \frac{1}{2} \quad (25)$$

We use Markov inequality to get the lower estimation error bound.

$$
\begin{aligned}
\Pr\{f_i - \widehat{f_i} \geq \epsilon Z\} &= \Pr\{X_i - Y_i' + Y_i'' \geq \epsilon Z\} \\
&\leq \Pr\{X_i + Y_i'' \geq \epsilon Z\} \\
&\leq \frac{1}{\epsilon}\Big(\frac{1}{d}\binom{i-1}{\lambda-1}\Big(\frac{1}{d}\Big)^{\lambda-1}\Big(1 - \frac{1}{d}\Big)^{i-\lambda} \\
&\quad + \frac{1}{2d} + \frac{1}{2Z}\Big)
\end{aligned}
\tag{26}
$$

We can find the lower estimation error bound is tightly relevant to the number of cells in each bucket and the rank of flow's number of sampled events, indicating that candidate matrix can provide more accurate estimations for flows with higher persistence.

6 Experimental Evaluation

6.1 Experiment Setup

We use a real-world dataset (IP Trace [27]) and several synthetic datasets [15] to evaluate the algorithms. The IP Trace Dataset is 1-min anonymized Internet traces downloaded from CAIDA. There are around 1M distinct flows and 36M packets in total. Each flow label (8 bytes) contains a source IP address (4 bytes) and a destination IP address (4 bytes). Besides, we generate 4 synthetic datasets following the Zipf [28] distribution with various skewness (from 0.3 to 1.2 with a step of 0.3). Each dataset has 32M packets in total, while the number of distinct flows depends on the skewness. The length of each flow label is 8 bytes. We follow the experimental setup of On-Off Sketch, dividing datasets into T time windows (periods) and setting $T = 1600$.

We implement all solutions in C++, using the same hash function (mmh3 [29]) with distinct initial seeds. All algorithms we implemented are single-thread. All experiments are conducted on a machine with one 8-core processor (16 threads, Intel(R) Core(TM) i7-10700 CPU @ 2.9 GHz) and 16 GB DRAM memory. We compare the PE-Sketch to the prior arts for their performance on top-k persistent items detection in terms of **Average Absolute Error** (AAE), **Average Relative Error** (ARE), **precision** and **throughput**. We use **Million of insertions per second** (Mips) to measure the throughput.

6.2 Evaluation on Event Sampling

We first evaluate the performance of our Flip Filter to conduct event sampling, where we fix memory to 10 KB, set $z = 3$, and directly offload the sampled flows to a hash table. Then we use p to estimate their persistence in the same way as in the candidate matrix. The experimental results are shown in Fig. 5(a)–(c), where each point represents a flow. The x-axis (y-axis) denotes the actual (estimated) persistence. Thus, the closer the point is to $y = x$, the more accurate the estimation result is. From the plots, we find that our algorithm is capable of generating persistence increment events uniformly for all flows. Next, we vary p from 0.1 to 0.7, and fix memory to 10 KB and 20 KB. The experimental results are shown in Fig. 5(d). Our results show that $p = 0.7$ reduces the AAE of $p = 0.1$ by around 78.31%, demonstrating that the estimation accuracy increases as p increases.

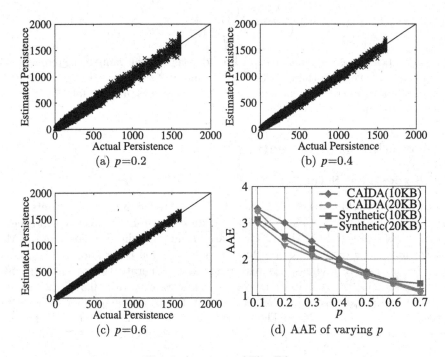

(a) $p=0.2$

(b) $p=0.4$

(c) $p=0.6$

(d) AAE of varying p

Fig. 5. Accuracy of Flip Filter.

6.3 Finding Top-k Persistent Flows

For finding top-k persistent flows, we compare our solution with On-Off sketch (OO) [15], Adaptive Sampling (AS) [13], and Small-Space (SS) [1]. We do not compare our PE-Sketch to PIE because PIE requires recovering flow labels and does not support real-time querying. We introduce a parameter μ ($0 < \mu < 1$) to control the memory budget of the Flip Filter in the PE-Sketch, where μ is the ratio of the memory size of the Flip Filter to the total memory. We set $\mu = 0.2$, $z = 3$ for Flip Filter. Set $w = 8$ both for the candidate matrix and On-Off sketch and set $k = 1000$ for all algorithms. Both our solution and On-Off sketch perform well on more than 200 KB of memory space, *i.e.*, the precision exceeds 0.96 and AAE is less than 10.0. To save space, we omitted these experimental results from the comparative experiments.

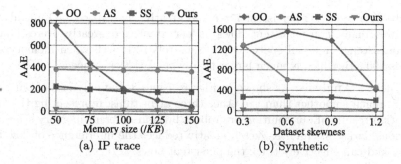

Fig. 6. AAE on finding top-k persistent flows.

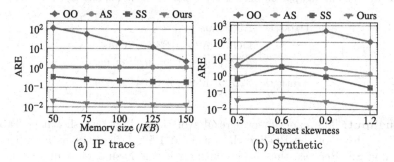

Fig. 7. ARE on finding top-k persistent flows.

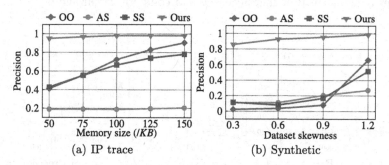

Fig. 8. Precision on finding top-k persistent flows.

Compared with Related Work: For experiments based on synthetic datasets, we fix the memory size 50 KB. As shown in Fig. 6, our solution reduces the AAE of OO, AS and SS by on average 95.9%, 91.2%, 86.6%, respectively. As shown in Fig. 7, PE-Sketch reduces the ARE of prior methods by around 2 orders of magnitudes. As shown in Fig. 8, the precision of PE-Sketch is up to 15.61, 9.33, 10.59 times higher than that of OO, AS and SS. Our experimental results show that our PE-Sketch achieves high accuracy and robustness in small memory (as low as 50 KB). The major reason is that our sketch effectively distinguishes flows by persistence and only records the persistent flows, which both reduces the interference of non-persistent flows and greatly saves space. OO mistakenly recognizes many non-persistent flows as persistent and greatly overestimates their persistence in small space. It performs relatively well when memory size exceeds 200 KB, which is much larger than what PE-Sketch needs.

Robustness Analysis: As shown in Figs. 6, 7 and 8, the polylines of our solution are much smoother than the other three algorithms, representing that the estimation results of our solution are robust both in different memory sizes and in various datasets. Our PE-Sketch greatly reduces the interference of flow frequency and can effectively track the persistent flows.

Table 1. Throughput on finding top-k persistent flows, memory = 50 KB.

Datasets	Algorithm			
	OO	AS	SS	Ours
IP trace	35.461	24.814	25.403	32.462
Synthetic	32.195	13.982	14.204	29.429

Throughput: The experimental results of throughput are shown in Table 1. The throughput of the PE-Sketch is around 1.59, 1.56 times as large as that of AS and SS. Because that hash collisions in the hash table will increase the number of memory access. Compared to the fastest algorithm OO, our sketch achieves a significantly higher accuracy and competitive throughput (91.5% of OO).

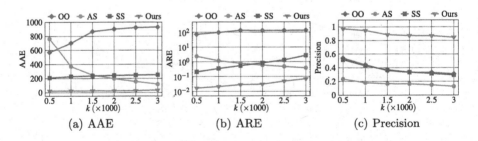

(a) AAE   (b) ARE   (c) Precision

Fig. 9. Influence of k setting (IP Trace).

Effect of k: To evaluate the influence of k on performance, we fix the memory 50 KB, vary k from 500 to 3000, and keep all the other parameters unchanged. The experimental results are shown in Fig. 9. Compared with OO, AS and SS, PE-Sketch reduces the AAE by 96.1%, 97.0%, 88.9% when $k = 500$, and by 96.4%, 71.5%, 86.6% when $k = 3000$. The precision of our sketch gradually decreases as k increases, yet it still is 1.74, 5.67, 1.88 times higher than OO, AS and SS when $k = 3000$. Due to dynamically adjusting the sampling probability, AS can only track the labels of flows whose persistence exceed the pre-set threshold, but not estimate persistence accurately. For the flows with higher persistence, the estimation error increases.

6.4   Parameter Setting

We mainly focus on two parameters, μ and w, to evaluate the influence of parameter setting. We fix the total memory usage of PE-Sketch and vary μ and w, respectively.

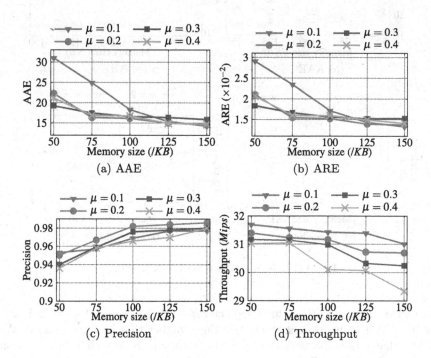

Fig. 10. Influence of varying μ, set $w = 8$.

Effect of μ: The experimental results are shown in Fig. 10. Our results show that, in the IP Trace dataset, the AAE(ARE) of $\mu = 0.1$ is the highest, which is around 1.23(0.22), 1.21(0.2), 1.22(0.19) times higher than that of $\mu = 0.2$, $\mu = 0.3$ and $\mu = 0.4$. The gap among precision under different parameters is small, which does not exceed 0.02. The throughput of $\mu = 0.1$ is around 0.12, 0.37, 1.04 higher than that of $\mu = 0.2$, 0.3 and 0.4. We find that PE-Sketch achieves the best balance between accuracy and throughput when $\mu = 0.2$. If μ is too small, we have to set p lower in Flip Filter, reducing the sampling accuracy. However, with μ increases, the candidate matrix will be smaller, and the randomness of mistakenly recognizing a non-persistent flow as a persistent one increases. Besides, the throughput decreases as μ increases because Flip Filter will process more events.

Effect of w: The experimental results are shown in Fig. 11. Our results show that, the AAE(ARE) of $w = 2$ is 1.83(29.38), 2.04(32.49), 2.09(32.91) times higher than that of $w = 4$, $w = 8$ and $w = 32$, and its throughput is 0.95, 1.52,

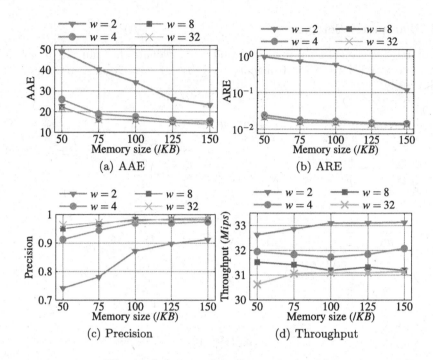

Fig. 11. Influence of varying w, set $\mu = 0.2$.

1.98 higher than the other three respectively. The precision of $w = 32$ is the highest, which is up to 0.29, 0.05, 0.01 times higher than $w = 2$, 4 and 8. We can find that estimation accuracy increases with w increases while the insertion throughput decreases. Because the number of memory accesses is proportional to w, where PE-Sketch searches the w cells of the bucket a flow mapped to.

Summarizing, we suggest setting $\mu = 0.2$, $w = 8$, considering the trade-off between accuracy and throughput.

7 Conclusion

This paper proposes Persistent Sketch, the first memory-efficient and robust algorithm for finding top-k persistent flows in network traffic. Based on the new concept of event sampling, the proposed sampling method reduces the studied problem to the problem of finding the flows with the most sampled increment events. It avoids repeatedly counting persistence for the same flow in a single period, guaranteeing the robustness of persistence estimation in various traffic scenarios. Moreover, the proposed novel candidate matrix is capable of tracking the persistent flows with high accuracy within a small space. Experimental results show that, compared to the state-of-the-art, PE-Sketch is more robust and can improve the precision by up to 15.6 times and reduce the error by up to 2 orders of magnitude when using the same space.

Acknowledgements. The work of Yu-E Sun and He Huang is supported in part by the National Natural Science Foundation of China (NSFC) under Grant 62332013, Grant 62072322, and Grant U20A20182, and in part by the Open Project of Tongji University Embedded System and Service Computing of Ministry of Education of China under Grant ESSCKF 2022-05. The work of Yang Du is supported in part by NSFC under Grant 62202322 and in part by the Natural Science Foundation of Jiangsu Province under Grant BK20210706. The work of Jia Liu is supported by NSFC under Grant 62072231.

References

1. Lahiri, B., Chandrashekar, J., Tirthapura, S.: Space-efficient tracking of persistent items in a massive data stream. In: Proceedings of the 5th ACM International Conference on Distributed Event-Based System, pp. 255–266 (2011)
2. Oentaryo, R., et al.: Detecting click fraud in online advertising: a data mining approach. J. Mach. Learn. Res. **15**(1), 99–140 (2014)
3. Zhu, T., Meng, Y., Hu, H., Zhang, X., Xue, M., Zhu, H.: Dissecting click fraud autonomy in the wild. In: Proceedings of the 2021 ACM SIGSAC Conference on Computer and Communications Security, New York, NY, USA, pp. 271–286 (2021)
4. Xiao, Q., Qiao, Y., Zhen, M., Chen, S.: Estimating the persistent spreads in high-speed networks. In: 2014 IEEE 22nd International Conference on Network Protocols, pp. 131–142 (2014)
5. Xu, Z., Wang, X., Zhang, Y.: Towards persistent detection of DDoS attacks in NDN: a sketch-based approach. IEEE Trans. Dependable Secure Comput. **20**, 1–17 (2022)
6. Zhou, Y., Zhou, Y., Chen, M., Chen, S.: Persistent spread measurement for big network data based on register intersection. SIGMETRICS Perform. Eval. Rev. **45**(1), 67 (2017)
7. Javed, M., Paxson, V.: Detecting stealthy, distributed SSH brute-forcing. In: Proceedings of the 2013 ACM SIGSAC Conference on Computer & Communications Security, New York, NY, USA, pp. 85–96 (2013)
8. Ghafir, I., Přenosil, V.: Advanced persistent threat attack detection: an overview. Int. J. Adv. Comput. Netw. Secur. **4**(4), 50–54 (2014)
9. Wang, P., Jia, P., Tao, J., Guan, X.: Detecting a variety of long-term stealthy user behaviors on high speed links. IEEE Trans. Knowl. Data Eng. **31**(10), 1912–1925 (2019)
10. Albanese, M., Jajodia, S., Venkatesan, S.: Defending from stealthy botnets using moving target defenses. IEEE Secur. Priv. **16**(1), 92–97 (2018)
11. Alshamrani, A., Myneni, S., Chowdhary, A., Huang, D.: A survey on advanced persistent threats: techniques, solutions, challenges, and research opportunities. IEEE Commun. Surv. Tutor. **21**(2), 1851–1877 (2019)
12. Zhong, Z., Yan, S., Li, Z., Tan, D., Yang, T., Cui, B.: BurstSketch: finding bursts in data streams. In: Proceedings of the 2021 International Conference on Management of Data, pp. 2375–2383 (2021)
13. Chen, L., Phan, R.C.W., Chen, Z., Huang, D.: Persistent items tracking in large data streams based on adaptive sampling. In: IEEE Conference on Computer Communications, pp. 1948–1957 (2022)
14. Dai, H., Shahzad, M., Liu, A.X., Zhong, Y.: Finding persistent items in data streams. Proc. VLDB Endow. **10**(4), 289–300 (2016)

15. Zhang, Y., et al.: On-off sketch: a fast and accurate sketch on persistence. In: Proceedings of the VLDB Endowment, vol. 14, pp. 128–140 (2020)
16. Du, Y., et al.: Short-term memory sampling for spread measurement in high-speed networks. In: IEEE Conference on Computer Communications, pp. 470–479 (2022)
17. Du, Y., Huang, H., Sun, Y.E., Chen, S., Gao, G., Wu, X.: Self-adaptive sampling based per-flow traffic measurement. IEEE/ACM Trans. Netw. **31**(3), 1010–1025 (2023)
18. Sun, Y.E., Huang, H., Ma, C., Chen, S., Du, Y., Xiao, Q.: Online spread estimation with non-duplicate sampling. In: IEEE Conference on Computer Communications, pp. 2440–2448 (2020)
19. Huang, H., et al.: Spread estimation with non-duplicate sampling in high-speed networks. IEEE/ACM Trans. Networking **29**(5), 2073–2086 (2021)
20. Huang, H., et al.: An efficient k-persistent spread estimator for traffic measurement in high-speed networks. IEEE/ACM Trans. Networking **28**(4), 1463–1476 (2020)
21. Zhou, Y., et al.: Cold filter: a meta-framework for faster and more accurate stream processing. In: Proceedings of the 2018 International Conference on Management of Data, pp. 741–756 (2018)
22. Zhao, B., Li, X., Tian, B., Mei, Z., Wu, W.: DHS: adaptive memory layout organization of sketch slots for fast and accurate data stream processing. In: Proceedings of the 27th ACM SIGKDD Conference on Knowledge Discovery & Data Mining, pp. 2285–2293 (2021)
23. Yang, T., et al.: HeavyKeeper: an accurate algorithm for finding top-k elephant flows. IEEE/ACM Trans. Networking **27**(5), 1845–1858 (2019)
24. Yang, T., Gong, J., Zhang, H., Zou, L., Shi, L., Li, X.: HeavyGuardian: separate and guard hot items in data streams. In: Proceedings of the 24th ACM SIGKDD International Conference on Knowledge Discovery & Data Mining, pp. 2584–2593 (2018)
25. Ting, D.: Data sketches for disaggregated subset sum and frequent item estimation. In: Proceedings of the 2018 International Conference on Management of Data, pp. 1129–1140 (2018)
26. Bloom, B.H.: Space/time trade-offs in hash coding with allowable errors. Commun. ACM **13**(7), 422–426 (1970)
27. CAIDA: Anonymized Internet Traces 2016. https://catalog.caida.org/dataset/passive_2016_pcap. Accessed 20 June 2022
28. Powers, D.M.W.: Applications and Explanations of Zipf's Law. In: Proceedings of the Joint Conferences on New Methods in Language Processing and Computational Natural Language Learning, pp. 151–160 (1998)
29. Appleby, A.: Murmurhash. https://sites.google.com/site/murmurhash/. Accessed 9 June 2022

FaCa: Fast Aware and Competition-Avoided Balancing for Data Center Network

Haiyang Jiang[1], Yuchao Zhang[1(✉)], Haoqiang Huang[1], Lei Wang[1], Xirong Que[1], Zhuo Jiang[2(✉)], and Wendong Wang[1]

[1] Beijing University of Posts and Telecommunications, Beijing, China
{jianghaiyang,yczhang,hhq_erii,lwang,rongqx,wdwang}@bupt.edu.cn
[2] ByteDance, Beijing, China
zjiang@bytedance.com

Abstract. Nowadays, the scale of business data is expanding at an unprecedented rate. To cater to the needs of large businesses, data center networks (DCNs) have been widely deployed and are continuing to expand. However, the influx of large-scale concurrent flows into DCNs often results in network congestion due to the concurrent competition for resources. While existing load balancing mechanisms can handle concurrent competition, they often do so at the cost of time. As a result, there is currently no ideal solution that effectively addresses both time consumption and concurrent competition issues. In this paper, we present a novel load balancing solution called FaCa, which runs on the host-end in a completely software-based manner. FaCa incorporates Inband Network Telemetry (INT), leveraging traffic transmission within the network to swiftly obtain a partial global view of network load. Additionally, we propose the Flowing&Jumping algorithm to mitigate concurrent path competition by introducing an element of randomness to load balancing process. FaCa is easy to deploy and has demonstrated superior performance compared to other mechanisms. Our evaluation on production DCN reveals that FaCa incurs minimal additional time overhead while achieving better load balancing results compared to existing approaches. Specifically, it resulted in a 14.28% reduction in congestion and a 22.5% increase in host throughput.

Keywords: Multi-Node HPC Cluster · DCN · Load Balancing

1 Introduction

Given the rapid expansion of the Internet, the volume of generated data has reached unprecedented levels, placing significant strain on data processing capabilities. According to a report by the International Data Corporation (IDC) [34],

This work is supported in part by the National Natural Science Foundation of China (NSFC) under Grant 62172054 and 62072047, the National Key R&D Program of China under Grant 2019YFB1802603, and the BUPT-ByteDance Research Project.

© The Author(s), under exclusive license to Springer Nature Singapore Pte Ltd. 2024
Z. Tari et al. (Eds.): ICA3PP 2023, LNCS 14492, pp. 39–58, 2024.
https://doi.org/10.1007/978-981-97-0811-6_3

the total volume of data is projected to reach 163ZB until 2025. To effectively handle this data pressure, enterprises have been establishing high-performance computing (HPC) clusters within their data center networks (DCNs). To optimize the performance of HPC clusters, researchers have developed high-speed communication technologies, such as the All-Reduce [27] communication mode, to enhance the performance of HPC clusters. With the consideration of reliability, DCN also adopts multi-path design to connect hosts, along with interconnection architectures such as the Clos architecture [5] in networking. Based on existing resources, the requirement for an efficient load balancing method in DCNs is urgent to enhance overall performance. Nevertheless, the majority of existing load balancing methods are primarily designed for conventional cloud-scale networks and might not be well suited for DCNs with HPC clusters. This is due to the unique characteristics of DCNs with HPC clusters, which differ from traditional networks and necessitate specialized load balancing approaches [16]. One of the challenges in load balancing for DCNs is the presence of concurrent flows. In a DCN, when multiple hosts initiate communications simultaneously, a significant volume of data is concurrently transmitted within the network. These concurrent flows exhibit characteristics such as simultaneous generation, partial inclusion of large flows, and bursts. Existing load balancing algorithms often face challenges in effectively managing such flows.

The load balancing methods can be categorized into two types: **centralized** and **distributed**. The centralized mechanism [3,10,16,30,38,39] collects network information periodically to calculate globally optimal load balancing strategies. However, it experiences delays due to the central controller waiting for and retrieving information from network devices. This delay significantly slows down the load balancing process, especially in DCNs with concurrent flows. The other approach is based on a distributed method. Some of them focus on hardware upgrades such as Ananta [28], which upgrades multiplexers (Mux), and SilkRoad [24], which utilizes load balancing ASICs. However, these designs require costly development of dedicated load balancing hardware, limiting their adoption in DCNs.

The most common distributed solutions [9,11–14,18,19,26,32,33,36] focus on strategies for achieving load balancing by leveraging existing devices, such as programmable switches [12] and back-end servers [26]. This type of approach selects the path with the lowest load for distributing elephant flows, but it overlooks the issue of "concurrency competition". When two nodes simultaneously detect the same path with the lowest load, conflicts arise on shared links when both nodes transmit traffic concurrently. This leads to a significantly higher load on the common link compared to other links and creates the potential for network congestion.

Designing load balancing for DCN is a challenging task that involves balancing **fast reaction** and **competition avoidance** while preserving network performance. Collecting network load information typically requires communication between central controllers and remote network devices, which often incurs ≥ 1 RTTs [3,30,38,39]. Load balancing scheduling introduces a time delay that impacts the timeliness of the mechanism. In a distributed algorithm, nodes receive identical network load information within the same network environment. This can lead to nodes making identical scheduling decisions, resulting in concurrent flows competing for bandwidth on shared links and causing network congestion.

In this paper, we introduce *FaCa*, a load balancing solution that addresses the aforementioned problems. FaCa utilizes In-band Network Telemetry (INT) to quickly respond to network imbalance by leveraging network-transmitted packets to probe real-time network load without additional time consumption. In addition, we propose Flowing&Jumping, a distributed load balancing algorithm that mitigates concurrent competition. Flowing&Jumping schedules flows in a manner resembling water flow and introduces random "jump" to prevent path competition. This simple integration of randomness effectively avoids conflicts. Furthermore, FaCa is designed as software-based and built upon the foundation of Equal-Cost Multi-Path (ECMP) [14], which ensures easy deployment and will be explained in Section IV. The **main contributions** of this paper are as follows:

- We identify and highlight the existing problems in load balancing that can potentially diminish network performance.
- We propose FaCa, a load balancing method that utilizes INT to enable quick congestion reaction, and introduces Flowing&Jumping to mitigate concurrent competition.
- We develop a prototype of FaCa within a production DCN and carry out performance evaluations. The findings showcase that FaCa achieved a substantial 22.5% increase in throughput and a noteworthy 14.28% reduction in congestion.

The paper is structured as follows. In Sect. 2, we introduce prior research on network load balancing. In Sect. 3, we identified existing issues that degrade network performance. In Sect. 4, we present insights on resilience and concurrent competition. In Sect. 5, we explain the load balancing system FaCa, including its modular introduction. In Sect. 6, we discuss the concrete challenges and implementation environment of FaCa. Finally, we prototype and evaluate FaCa's performance in Sect. 7 and conclude in Sect. 8.

2   Related Work

This section briefly introduces existing load balancing methods and their limitations. Load balancing approaches can be categorized into two groups: centralized and distributed methods.

2.1   Centralized Load Balancing

Centralized load balancing mechanisms typically employ logically centralized controllers to gather information on links (such as bandwidth, utilization, etc.)

and switches (including buffer size, queue length, etc.). Using this comprehensive network information, the centralized controllers distribute flows to less utilized paths.

Mahout [10] improves on Hedera by using SDN technology to reduce the overhead of identifying elephant flows. It also brings about time consumption of collecting network information. Freeway [39] divides links into low-latency oriented links (LOL) and high-throughput orient links (HOL) by estimating the link utilization. Different from the methods that distribute flows based on the amount of bytes, FDALB [38] divides flows into long-lived flows and short-lived flows. For long-lived flows, FDALB marks them on end-hosts and globally schedules such flows with a greedy and polling algorithm. For short-lived flows, it simply continues to use ECMP. MicroTE [3] utilizes OpenFlow DCN to forecast traffic patterns. It classifies flows into predictable and unpredictable categories. Predictable flows are assigned load balancing tasks by centralized controllers, while unpredictable flows make use of the default balancing mechanisms of the network, such as ECMP and WCMP [45]. Fastpass [30] achieves load balancing at the packet level by determining transmission time slots and paths for each packet using centralized controllers.

While centralized methods are theoretically capable of achieving optimal performance, the practical implementation is hindered by the time-consuming process of information collection and the overhead of redundant communication. Additionally, the fixed interval at which information is collected limits the ability to quickly respond to changes in network topology due to failures or hardware restarts. In real-world scenarios, centralized methods often struggle with coarse-grained scheduling, making it challenging to meet the desired level of accuracy in load balancing mechanisms.

2.2 Distributed Load Balancing

Distributed mechanisms are locally deployed on hosts and switches, which often make decisions based on local information, resulting in much faster reaction compared to centralized mechanisms. Nevertheless, they are not immune to competition issues.

As the first method to introduce Mux, Ananta [28] designs a multi-layer structure. It provides ECMP in the third layer and schedules connections in the forth layer. Duet [12] manages to develop load balancing with the API on programmable switches and combines it with Mux to provide a resilience to network failures. Silkroad [24] addresses the shortcomings of stateful load balancing by presenting a stateless solution based on switching ASIC development. This approach offloads load balancing from switches and leverages dedicated hardware for efficient and high-performance load balancing. However, the deployment overhead and scalability remain as challenges. FLARE [18] introduces the load balancing method called FLowlet and highlights a characteristic of TCP. It states that if the interval between sending two packets is greater than the delay of parallel paths, packet transportation on these paths will not cause out-of-order problems. Building upon the concept of Flowlet, numerous works have been developed. CONGA [2] addresses the limitations of local congestion-aware load balancing in asymmetric network topologies and simplifies the complexity by introducing a leaf-to-leaf mechanism. LetFlow [36] explores the relationship between sub-flows and path load, and introduces a Markov model to demonstrate the high performance of Flowlet in multi-path load balancing. MLAB [11] proposes a modularized solution for multi-path DCN. In essence, MLAB practically implements Flowlet within a Fat-Tree topology. OLTEANU [26] makes use of back-end servers to offload connection states from switches, which has inspired our work to maintain the state on hosts instead of switches. However, in OLTEANU, severs receive a significant number of packets that belong to other connections, and this redundancy hinders performance. Google proposes PLB [32], which is built on the basis of ECMP/WCMP. It detects whether it is experiencing congestion through the TCP protocol with a threshold judgment and selects an available path to redirect the connection by assigning a new flow label for subsequent outbound packets. Although it can mitigate switch link load imbalance and reduce switch packet loss, it still lacks in concurrent competition.

Distributed load balancing mechanisms have the advantage of being able to react quickly to congestion in DCNs. However, one of their challenges is the potential for concurrent competition when blindly selecting the path with the lowest load for distribution. This issue arises because distributed mechanisms often have access to only partial information about the network, which may results in that multiple nodes simultaneously choose the same low-load path, leading to congestion and performance degradation. The above are the problems with existing load balancing work, some existing work [6,17,44] on network routing and link failure recovery in DCN, has provided us with assistance in solving the above problems.

3   Background

With substantial improvement of equipment performance, data center has the ability to complete complex tasks, such as data storage [8,42], distributed machine learning (DML) [7,20–22,37] and content distribution [1,4,29]. Meanwhile, with the development of host performance, network performance increases as well. However, the development of load balancing, as one of the factors important to network performance, is still marking time. Despite its shortcomings, ECMP has been a convenient and widely used balancing mechanism for a long time. The state-of-the-art work comprehensively performs poorly.

Slow Reaction. With the increasing scale of data and device access in DCNs, network congestion has become a significant challenge. Simply increasing the number of devices is not enough to alleviate congestion in large-scale networks. The key to effective load balancing lies in the ability to quickly detect congestion and react accordingly. However, centralized mechanisms used in traditional load balancing approaches rely on periodic information collection from the entire network. This results in delayed response times to congestion events and longer information collection times, especially in rapidly expanding network environments. Both factors contribute to wasted time and ultimately degrade network performance.

High Concurrent Network Environment. DCN is inundated with concurrent flows. Numerous applications operate simultaneously on the network, resulting in a substantial volume of data transmission at all times. With the prevalence of massive businesses running on DCN, concurrency has become a common occurrence. Surveys about traffic patterns in data center [35,40,43] indicate that burst flows and concurrent competition often lead to congestion and serious packet loss. However, concurrent traffic has received relatively limited research attention. The focus on achieving optimal balancing often poses a challenge in effectively addressing concurrent competition. Assume that load balancing mechanism always chooses a minimum load path for each flow and what would happen? Most of the flows would be scheduled into a minority of paths! We call it *concurrent path competition*. The competition can cause even more serious unbalance and congestion. Introducing a centralized scheduler could potentially be helpful, but the slow reaction time of the scheduler can degrade the performance of DCN.

4   Motivation

In this section, we introduce the principle of ECMP and discuss its shortcomings. We also highlight a key insight using an example.

4.1   Why Not ECMP?

ECMP is considered the most adaptable load balancing solution in DCNs, which operates on a hop-by-hop basis and uses flow-based load balancing. When there are multiple links available to reach the same destination address, ECMP employs a specific strategy (such as Hash, Round robin, etc.) to distribute the flow across these paths [15]. It avoids the limitation of using a single link and improves available bandwidth by distributing traffic across multiple links. ECMP switches typically use the five-tuple of each data packet as a hash key and hash it to a random path. Additionally, ECMP does not take into account the size of flows, resulting in a rough flow scheduling based solely on hashing. This can lead to "hash collisions", where long-lived large flows are mapped to the same path, resulting in imbalances and congestion. Moreover, ECMP operates independently on each switch without coordination, meaning that forwarding decisions made in one switch are not influenced by those in other switches. Consequently, ECMP's lack of coordination between switches results in upstream switches not considering the capacity of downstream switches. This can lead to congestion when upstream switches send large flows that exceed the capacity of downstream switches.

Indeed, ECMP has gained popularity in DCN due to its cost-effectiveness and robustness, but it falls short in terms of network performance. To illustrate

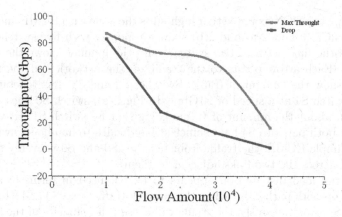

Fig. 1. Throughput "Drop" in DCN.

this, we deploy a 2-layer Leaf-Spine [23] DCN architecture, where the hosts are divided into two sides of the Clos network. Each side of the network sends flows to the other side, allowing us to observe the limitations of ECMP in practice. We evaluate the performance of ECMP in our DCN by measuring the max throughput and the metric named "Drop" of switch ports and is recorded on one of the layer-1 switches. The "Drop" represents the difference between the maximum and minimum throughput, it helps evaluate the distribution of throughput. As shown in Fig. 1, as the number of flows increases, the maximum throughput of the network tends to decrease. When the number of flows increases to 40,000, the max throughput decreases down to almost a quarter. The drop in throughput also indicates that the overall network performance is at a low level. The fundamental factor is that ECMP is prone to collisions in a large concurrent environment.

4.2 A Motivation Example

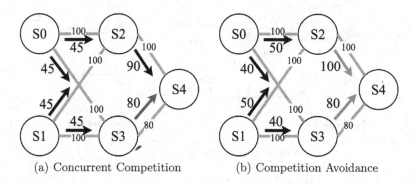

(a) Concurrent Competition   (b) Competition Avoidance

Fig. 2. The Path Competition in ECMP.

Figure 2 depicts a DCN example that highlights the issue with ECMP and the key insight of FaCa. Figure 2(a) and 2(b) show a topology with five switches. Each blue line is the link between two switches and the number shows the capacity of the link. Each arrow points to the traffic on the network and the numbers on arrows show the amount of traffic. Switches S0 and S3 simultaneously send traffic to switch S4 at a speed of 90 Gbps. In Fig. 2(a), switch S0 does not have information about the amount of traffic being sent by switch S1. However, S0 knows that both paths to S4 have sufficient bandwidth to handle its own traffic. With the simple ECMP, the traffic from switches S0 and S3 is evenly split and distributed across the two links adjacent to them.

The initial load distribution appears to be balanced in terms of available bandwidth on both paths. But when the network traffic from S0 and S1 arrive at link $S3 \rightarrow S4$ simultaneously, the traffic will exceed the capacity of the link, and a competition also occurs. To address the potential congestion issue, a solution is proposed where a portion of the traffic originating from S0 and S1 is redirected to the other available path, as depicted in Fig. 2(b). Certainly, implementing such a solution requires access to network information in order to make informed decisions about traffic redirection. The details of obtaining and utilizing network information will be discussed in subsequent sections.

5   Design

The section presents the system design of FaCa, which is a modularized system and consists of three parts: network probing, load estimates and balancing decisions.

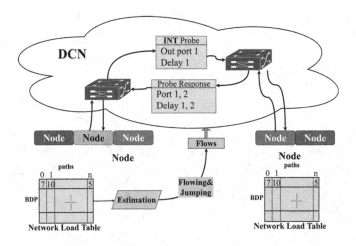

Fig. 3. System Design of FaCa.

5.1   Overview

FaCa implements all the parts on hosts and needs no changes on DCN. To acquire the load information and mapping relationships between paths and five-tuples, hosts send INT probe packets to each other. The information forms a network load table and on this basis, FaCa estimates short-term network load with Exponentially Weighted Moving-Average (EWMA). Derived by estimation, hosts make a decision on flow schedule with the help of Flowing&Jumping.

5.2   Network Model

We consider the network topology as $G = (V, E)$, where V indicates the hosts and Layer3 switches, and E indicates edges between them. The link from host/switch i to host/switch j is simply denoted as ij. P_{ij} is the set of paths between host i and host j and P is the union of all different P_{ij}s. A path p_{ij}^k that connects host i and host j is an orderly permutation of links. For instance, host i and host j is connected by switch l, and then we can refer to it as $p_{ij}^k = (i, l, j)$. The bandwidth of the path is denoted as B_{ij}^k. Note that, B_{ij}^k is the minimum bandwidth of links that the path p_{ij}^k traverses. Similarly, we use D_{ij}^k to denote the delay of p_{ij}^k. F_{ij} is denoted as the set of flow demands from host i to host j and f_{ij}^c is the c-th flow demand in F_{ij}, F is the set of all different F_{ij}.

5.3   Partly Global Vision with INT

Probe packets help hosts acquire a global vision of network status information and this probe technique is INT. In the beginning, INT is used only under an in-network way. We draw inspiration from the design concept of INT and successfully implement it at the host end. Firstly, hosts encapsulate the business packets into probe packets and send to other hosts which are the destinations of these packets. When probe packets go through a switch, the switch adds information at the end of the probe packets, including the network status information (such as delay, bandwidth, etc.) and the path that the packet traverses. When a host receives probe packets, it analyzes the information and updates it in the network load table. Meanwhile, the host answers a copy of the probe packets to the sender and parses them. Likewise, when the sender receives a response from a probe packet, it initiates an analysis of the information and updates the network status message table. Besides, INT only focuses on end-to-end path information, and every host builds up a "partly" global vision in this way. That is, each host only has the bandwidth and delay of the path to the destination host, rather than stores the information of the whole network, which reduces the time to wait for the global convergence of the network. As shown in Sect. 4.2, FaCa manages to be elastic to asymmetry with the global vision.

By obtaining the network status information in near real-time, as described above, we can achieve faster response when congestion occurs in the network. Compared to collecting information at regular intervals, this approach significantly reduces the time required and allows for more efficient decision-making processes.

5.4   Load Estimate

What FaCa gains is the delay and bandwidth of paths but it is far from path load. FaCa introduces bandwidth-delay product (BDP) to describe path load. BDP is formulated as follows:

$$\beta_{ij}^k = B_{ij}^k \cdot D_{ij}^k \tag{1}$$

where β_{ij}^k is the BDP of path p_{ij}^k. BDP measures the in-flight traffic and the remaining maximum bandwidth of a path. Nevertheless, there is a little time interval between updating a probe result and making a schedule decision. The interval makes it inaccurate to describe the realistic path load with BDP in load table. In the field of communication, EWMA is mainly used to estimate and smooth the state parameters of the network. Consequently, FaCa introduces EWMA to fill the gap and the estimation of load is formulated as follows:

$$E_{ij}^k(t) = \alpha E_{ij}^k(t - \Delta t) + (1 - \alpha)\beta_{ij}^k \tag{2}$$

where $E_{ij}^k(t)$ is the estimation of BDP β_{ij}^k at time t and Δt is the time interval between latest record update time and t. The $E(t)$ in the algorithm is the set of all $E_{ij}^k(t)$ where i and j are different. The α is a fitting coefficient to control the proportion of estimation and probe information, which is calculated by the attenuation function model in Newton's cooling law [25]. The specific formula is as follows:

$$\alpha = 1/e^{k*\Delta t} \tag{3}$$

The e and k in the above formula are constants. With BDP and EWMA, FaCa estimates load of paths makes a schedule decision.

5.5   Flowing&Jumping

Load balancing treats flows as running water and flows pour into the light load path until there is no imbalance. In this process, two details are worth thinking. **Path Selection.** FaCa is totally software-based and it seems hard to select which path to transport the flow. In traditional network, routers decide routing and it is transparent to hosts. The turning point appears in the layer 3 switches that widely deployed in DCN. Switches schedule flows with a stable hashing algorithm ECMP, and actually, there is a hidden relationship between paths and five-tuple: *every five-tuple maps to a path*. The five-tuple of a probe packet maps to the path it traverses and it is the basis of our load balancing path selection. Just a modification to the source port in five-tuples is able to choose a selected path between sources and destinations, which accomplishes routing decision on host end.

Competition Avoidance. Numerous applications are running large and concurrent communications in DCN, such as database synchronization and DML. Numerous hosts start large flows in the same time and select paths with light load, and competition occurs. The main reason for concurrent competition is

---

Algorithm 1: Flowing&Jumping

Input: Paths P and Loads $E(t)$, Flow Demands F
Output: Flow Demand Schedule Result

foreach $F_{ij} \in F$ *in parallel* **do**
   while $\|F_{ij}\| > 0$ **do**
      randomly a select path p_{ij}^k;
      while $E_{ij}^{k-1}(t) > E_{ij}^k$ and $E_{ij}^{k+1}(t) > E_{ij}^k$ **do**
         select a flow from F_{ij}, schedule to p_{ij}^k;
         update E_{ij}^k and F_{ij}^k;
      while $E_{ij}^{k-1}(t) < E_{ij}^k(t)$ **do**
         select a flow from F_{ij}, schedule to p_{ij}^{k-1};
         update $E_{ij}^k(t)$ and F_{ij}^k;
      while $E_{ij}^{k+1}(t) > E_{ij}^k(t)$ **do**
         select flows from F_{ij}, schedule to p_{ij}^{k+1};
         update $E_{ij}^k(t)$ and F_{ij}^k;

---

that hosts are lack of the load balancing decisions from other hosts, and the decision would never be known before it is taken into practice. And that is the reason why concurrent competition is never considered in DCN load balancing. A centralized controller decides the load balancing schedule uniformly. While in DCN, it is not a perfect solution due to the hysteresis of centralization.

In this case, our solution is to sacrifice the "optimal" balancing in exchange for competition avoidance through adding a little randomness. Algorithm 1 explains the main process of Flowing&Jumping. For flow demands from host i to j, it randomly selects a path p_{ij}^k, and compares it with the two adjacent paths (the "adjacent" indicates continuous numbering). When p_{ij}^k is the lightest load path, flows are scheduled to it. While not, flows would be scheduled to the adjacent paths with lighter load. Either way, the load of paths is not allowed to exceed the adjacent ones. It seems like the water, which schedules flow demands to descend to the sunken place. We add randomness by randomly select the preferable path at every iteration, in case that flows are always scheduled to the lightest load path and the algorithm turns back to an optimal load balancing.

To show the process of Flowing&Jumping more clearly, we illustrate an example in Fig. 4. It shows the balancing process from the perspective of a source host, which it has flow demands to send to a destination host. The source host links to eight paths and we number the paths from 0 to 7 with black number. For convenience, we assume that size of every flow demand is one unit and the size of bandwidth of each path is 10 units. At the beginning six flow demands need to be scheduled and we randomly select path 1. Compared to the adjacent paths, load of path 1 is less than path 0 but greater than path 2. Hence, we schedule flows to path 2 until load of path 2 reaches the level of path 1. Again, with remaining flow demands, we randomly select another path 6 and compare with

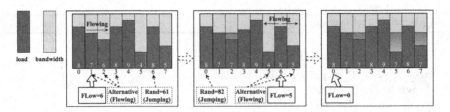

Fig. 4. An example of Flowing&Jumping. Each bar indicates a path and the black number below the bar denotes the rank of path. Above path numbers, red bars are the load of the paths and light yellow bars are the bandwidths. On the bottom of bars, the number with white color denotes the specific value of load. The solid box with flow demands points to the path randomly selected and the two dashed boxes separately points to flowing paths and jumping paths. (Color figure online)

the adjacent paths 5 and 7, while load of path 6 is higher than the that of other two paths. Thus, we schedule flows to path 5 and 7 until their loads reach the level of path 6. At this time flow demands are completely scheduled out, and the algorithm process stops.

Throughout this process, there is a little trick on random selection: *treat it like a ring*. In Fig. 4, when selecting the next random path after path 1, there is a random number 61. While the total amount of paths are 8, we select path $(61 + 1) \bmod 8 = 6$ as the next one. The modular calculation makes the path tail to head. Similarly, the adjacent paths of path 7 is $(7 + 1) \bmod 8 = 0$ and $(7 - 1) \bmod 8 = 6$. It balances the situation on head paths and tail paths.

6   Implementation

This section describes the deployment environment of FaCa and the challenges encountered in implementing it.

6.1   Productive Experiment Setting

We utilize a production cluster in DCN operating under Clos topology, as shown in Fig. 5. The Clos network comprises three layers, namely two access layers (S0, S3) and a core layer (S1). There are 16 switches in the core layer, and in each access layer, there are 8 switches, with one of them connecting five hosts, and the links between the hosts and S0 are unique. Specifically, each host is equipped with multiple RDMA [41] NICs (RNIC), and each port of the S0 switches is uniquely connected to one RNIC. The connection between the access layer and the core layer is fully connected. We develop a prototype based on perftest, which is a RDMA performance test tool and used to evaluate on this DCN.

Ideally, the Clos network can tolerate the whole traffic from hosts under the same S0 and forward it to the hosts under S3. However, once unbalance happens on the out ports of S0, it is possible to cause a congestion. We design a simple

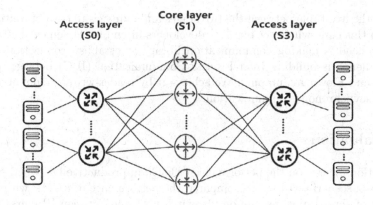

Fig. 5. Production Sub-Network (Clos) Topology.

perftest evaluation, which ten RNICs under one S0 switch send elephant flows to hosts under S3. And details are listed as follows:

Flow Setting. We choose RNICs under one S0 to send and RNICs under S3 to receive. Each RNIC sends two flows and initial sending rate of each flow is 100 Gbps. For each flow we distribute enough buffer on both senders and receivers to ensure no congestion are caused by out of buffer.

IP Pair. We limit that one RNIC uniquely connects to another RNIC. For the reason that every RNIC has the unique IP, we call the connection "IP Pair". Such design is able to prevent incast [46] from disturbing experiment results.

Data Collection. Before flows run on the DCN, connection establishment consumes a little bandwidth. Flows last until the end of experiments and therefore, measurement indexes would be stable after establishment. Consequently, we collect experiment data when the load of paths is stable.

6.2 Implementation Challenges

The first challenge is exploration probe. INT only detects the load of switches and paths of probe packets, while information of five-tuples is still unknown. FaCa acquires it through hook function, which is able to probe flow information from system kernel. Note that, although RDMA is kernel-bypass, it still needs TCP or UDP to exchange connection information before establishing it. Therefore, it is easy to hook connection information in establishment with a daemon.

It requires extra work to provide a software-based load balancing that supports for lightweight implementation. Our solution is to develop a communication library that provides Flowing&Jumping on hosts. Based on InfiniBand (IB) Verbs [31], the establishment of RDMA connection covers several steps and the five-tuple information is hidden in Queue Pair (QP) modification. In establishment stage, QP status goes through the following process: RESET → INIT → Ready To Receive (RTR) → Ready to Send (RTS). Between INIT and RTR, senders and receivers exchange the information of five-tuple. Consequently, FaCa updates the five-tuple between the two stages and manages to control routing on hosts. In this case, whenever business developers intend to implement FaCa, the only overhead is making communication library to establish connections. The final challenge is building Inter-Process Communication (IPC) between probe daemon and Flowing&Jumping. We achieve it by reader-writer model and support an asynchronous communication.

7   Evaluation

This section focuses on the performance of FaCa in productive DCN and micro-benchmark scenarios. First, we compared the performance of ECMP with FaCa in terms of throughput, mainly on S0 switch port and between IP pairs. Then, we compared their congestion in different experiments to verify FaCa's ability to handle concurrent competition. Finally, to better demonstrate our performance, we add comparison with PLB and WCMP. Due to the limitations of implementing PLB on a productive environment, we conducted a micro-benchmark simulation for experiments.

7.1   Throughput Performance

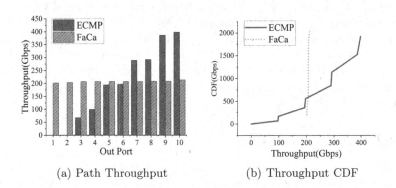

(a) Path Throughput   (b) Throughput CDF

Fig. 6. Throughput of one S0 switch.

As shown in Fig. 6, we measure the throughput of ten out ports in one of the S0 switches which receives totally 10×100 Gbps flows and forwards to S1 switches. The red bars in Fig. 6(a) show that ECMP forwards at least four 100 Gbps flows to out port 10, the same to port 9. It leads to an unbalanced on port 0–4, while port 0 sends nearly no traffic. This asymmetric network resources causes unbalance and performance down. The situation in FaCa achieves a better balance. In the coarse-grained elephant flows situation, Flowing&Jumping schedules two flows to each path. More precisely, Flowing&Jumping schedules the flows to go through different output ports of S0 on average. If we use the standard deviation of throughput to evaluate the balance performance, the value in ECMP is 148.06, while in FaCa, it is 2.96, which represents an improvement of almost 49 times.

As shown in Fig. 6(b), to make flow distribution more distinct, we collected the throughput data and plotted it as a Cumulative Distribution Function (CDF). The red curve illustrates that the flow distribution in ECMP is quite dispersed, whereas the green curve depicts a denser flow distribution in FaCa, indicating better load balancing performance. However, due to flow collisions, the total throughput of ECMP is slightly lower than that of FaCa. Note that the experiment is based on elephant flows, and thus in business scenarios, the flow collision problem is likely to be less noticeable when the number of mouse flows is high. Additionally, because FaCa is based on ECMP, it may also achieve slightly higher performance.

During the experiment, we also measured the throughput of IP pairs. To the best of our knowledge, the imbalance problem in switches does not always result in a performance degradation on hosts. This is the reason why improvements in load balancing often do not significantly enhance network performance. As shown in Fig. 7(a), performance difference on host RNIC is not as distinct as that on S0 out ports. The flow collision and unbalance on ECMP finally lead to a performance downgrade in host, as the red bars show. Compared to the throughput of ECMP, FaCa maintains a load-balanced performance and helps RNICs achieve full bandwidth. Precisely, the throughput "gap" of RNICs in FaCa load balancing is 6 Gbps (191 Gbps vs 197 Gbps), while that in ECMP is 38 Gbps (160 Gbps vs 198 Gbps), which is almost 6.3 times improvement. Meanwhile, the standard deviation of throughput in ECMP is 13.61, while that in FaCa is 1.78, which is a 7.08 times improvement. It is much less compared to the improvement in switches.

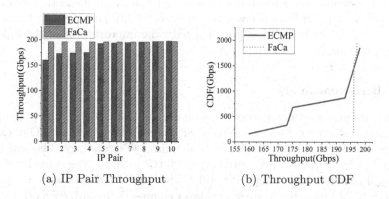

(a) IP Pair Throughput   (b) Throughput CDF

Fig. 7. Throughput of IP Pairs.

Similar to the performance in switches, FaCa achieves a denser CDF curve than that in ECMP, as shown in Fig. 7(b). And the total throughput in FaCa is also slightly higher than ECMP. Generally, the average of flow throughput in FaCa outperforms that in ECMP with 22.5%.

7.2 Concurrent Competition

Simultaneously, we test the resilience to concurrent competition and illustrate in Fig. 8. Precisely, competition in ECMP is caused by hash collision rather than concurrency. Nevertheless, ECMP is still an ideal baseline because the comparison to ECMP directly reflects the improvement in production DCN for the wide deployment of ECMP.

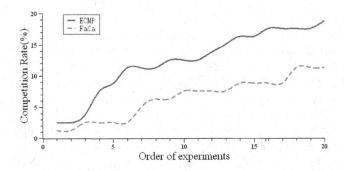

Fig. 8. Path competition in concurrent flows over twenty experiments.

In every experiment, we send persistent flows with random size in hosts and then concurrently send temporary flows. We evaluate path competition resilience by the proportion of the amount of competitive paths. In all congestion scenarios, the congestion rate of Faca is lower than that of ECMP. On average, FaCa outperforms ECMP with 5.88%. Associated with the improvement of throughput, slightly path competition brings serious unbalance.

7.3 Micro-benchmark

In order to better evaluate the performance of FaCa, we conduct a micro-benchmark environment to compare it with some other methods. Our experiment is simulated using a two-layer Leaf-spine DCN topology. The leaf layer consisted of 16 switches with a capacity of 200 Gbps, while the spine layer consisted of 4 switches with a capacity of 380 Gbps and 4 switches with a capacity of 400 Gbps. The switches in the access layer communicate with each other using a point-to-point communication mode. In each iteration of the experiment, the access switches simultaneously transmit fixed flows to their connected pairs.

In the scenario of continuously sending concurrent streams, we record all the congested traffic on the core switch over a duration of 20 s. Figure 9(b) shows the corresponding change trend. FaCa demonstrates significant improvements

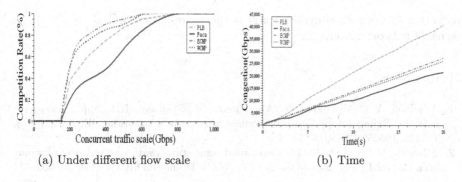

(a) Under different flow scale   (b) Time

Fig. 9. Congestion on core switches.

compared to other methods. Due to the emphasis on balance in ECMP/WCMP, while we focus on randomness, the ECMP-based strategy can exhibit slightly better performance than the latter method under the traffic scale that the link can carry. However, this phenomenon gradually disappears as the traffic increases, and FaCa become more evident, congestion in the network under the FaCa increases at a slower rate compared to other methods. In the scenario where all nodes send traffic evenly, it is possible for ECMP to appear more efficient than PLB. However, although this scenario is not typically encountered in practical network environments, it still proves the superiority of FaCa.

Additionally, we also compare the congestion rates of various methods under different levels of sending traffic. Figure 9(a) displays the congestion rates of each algorithm by increasing the size of concurrent streams. As the scale of concurrent streams continued to increase, ECMP and WCMP quickly experience congestion, followed by PLB. FaCa can fully utilize lower-load links while avoiding conflicts, thereby reducing the occurrence of congestion. Compared to evenly balancing the utilization of all paths, the probability of congestion occurring is much lower. Specifically, FaCa achieves a nearly 50% improvement in congestion rate compared to ECMP, and a 31.47% improvement compared to PLB. These results highlight the effectiveness of FaCa in mitigating network congestion.

8   Conclusion

In this paper, we address the primary challenges associated with implementing load balancing methods on multi-node HPC cluster networks and introduce FaCa as a solution. FaCa is designed for ease of implementation and is built upon a software design approach, leveraging the widely deployed ECMP mechanism as its foundation. By introducing INT probe technology, FaCa establishes a partially global view of the network, reducing the time required for information collection and enabling fast response to network congestion. To address concurrent path competition, FaCa introduces a load balancing algorithm called Flowing&Jumping, which incorporates a small amount of randomness into the flow scheduling process. Our evaluation demonstrates that FaCa achieves 14.28%

reduction in network congestion and outperforms native ECMP by 22.5% in terms of network throughput.

References

1. Abudaqa, A.A., Mahmoud, A., Abu-Amara, M., Sheltami, T.R.: Super generation network coding for peer-to-peer content distribution networks. IEEE Access **8**, 195240–195252 (2020)
2. Alizadeh, M., et al.: CONGA: distributed congestion-aware load balancing for datacenters. In: Proceedings of the 2014 ACM Conference on SIGCOMM, pp. 503–514 (2014)
3. Benson, T., Anand, A., Akella, A., Zhang, M.: MicroTE: fine grained traffic engineering for data centers. In: Proceedings of the Seventh Conference on Emerging Networking Experiments and Technologies, pp. 1–12 (2011)
4. Bilal, K., Khalid, O., Erbad, A., Khan, S.U.: Potentials, trends, and prospects in edge technologies: fog, cloudlet, mobile edge, and micro data centers. Comput. Netw. **130**, 94–120 (2018)
5. Clos, C.: A study of non-blocking switching networks. Bell Syst. Tech. J. **32**(2), 406–424 (1953)
6. Cong, P., et al.: DIT and beyond: inter-domain routing with intra-domain awareness for IIoT. IEEE Internet Things J., 1 (2023). https://doi.org/10.1109/JIOT.2023.3293500
7. Cong, P., Zhang, Y., Wang, W., Xu, K.: SOHO-FL: a fast reconvergent intradomain routing scheme using federated learning. IEEE Netw., 1–8 (2023). https://doi.org/10.1109/MNET.132.2200505
8. Corbett, J.C., et al.: Spanner: Google's globally distributed database. ACM Trans. Comput. Syst. (TOCS) **31**(3), 1–22 (2013)
9. Cui, W., Qian, C.: DiFS: distributed flow scheduling for adaptive routing in hierarchical data center networks. In: 2014 ACM/IEEE Symposium on Architectures for Networking and Communications Systems (ANCS), pp. 53–64. IEEE (2014)
10. Curtis, A.R., Kim, W., Yalagandula, P.: Mahout: low-overhead datacenter traffic management using end-host-based elephant detection. In: 2011 Proceedings IEEE INFOCOM, pp. 1629–1637. IEEE (2011)
11. Fan, F., Hu, B., Yeung, K.L.: Routing in black box: modularized load balancing for multipath data center networks. In: IEEE INFOCOM 2019-IEEE Conference on Computer Communications, pp. 1639–1647. IEEE (2019)
12. Gandhi, R., et al.: Duet: cloud scale load balancing with hardware and software. ACM SIGCOMM Comput. Commun. Rev. **44**(4), 27–38 (2014)
13. Ghorbani, S., Godfrey, B., Ganjali, Y., Firoozshahian, A.: Micro load balancing in data centers with drill. In: Proceedings of the 14th ACM Workshop on Hot Topics in Networks, pp. 1–7 (2015)
14. Hopps, C.: Analysis of an equal-cost multi-path algorithm. RFC (2000)
15. Hsu, K.F., Tammana, P., Beckett, R., Chen, A., Rexford, J., Walker, D.: Adaptive weighted traffic splitting in programmable data planes. In: Proceedings of the Symposium on SDN Research, SOSR 2020, pp. 103–109. Association for Computing Machinery (2020)
16. Hu, J., et al.: Adjusting switching granularity of load balancing for heterogeneous datacenter traffic. IEEE/ACM Trans. Netw., 2367–2384 (2021)

17. Huang, H., et al.: FRAVaR: a fast failure recovery framework for inter-DC network. In: 2023 IEEE Wireless Communications and Networking Conference (WCNC), pp. 1–6 (2023). https://doi.org/10.1109/WCNC55385.2023.10119088
18. Kandula, S., Katabi, D., Sinha, S., Berger, A.: Dynamic load balancing without packet reordering. ACM SIGCOMM Comput. Commun. Rev. **37**(2), 51–62 (2007)
19. Katta, N., Hira, M., Kim, C., Sivaraman, A., Rexford, J.: HULA: scalable load balancing using programmable data planes. In: Proceedings of the Symposium on SDN Research, pp. 1–12 (2016)
20. Kraska, T., Talwalkar, A., Duchi, J.C., Griffith, R., Franklin, M.J., Jordan, M.I.: MLbase: a distributed machine-learning system. In: CIDR, vol. 1, pp. 1–2 (2013)
21. Li, M., Andersen, D.G., Smola, A.J., Yu, K.: Communication efficient distributed machine learning with the parameter server. In: Advances in Neural Information Processing Systems, vol. 27 (2014)
22. Li, X., Gomena, S., Ballard, L., Li, J., Aryafar, E., Joe-Wong, C.: A community platform for research on pricing and distributed machine learning. In: 2020 IEEE 40th International Conference on Distributed Computing Systems (ICDCS), pp. 1223–1226 (2020)
23. Luo, S., Xing, H., Li, K.: Near-optimal multicast tree construction in leaf-spine data center networks. IEEE Syst. J. **14**, 2581–2584 (2020)
24. Miao, R., Zeng, H., Kim, C., Lee, J., Yu, M.: SilkRoad: making stateful layer-4 load balancing fast and cheap using switching ASICs. In: Proceedings of the Conference of the ACM Special Interest Group on Data Communication, pp. 15–28 (2017)
25. Newton, I.: Mathematical Principles of Natural Philosophy. In: London: Printed for Benjamin Motte (1729)
26. Olteanu, V., Agache, A., Voinescu, A., Raiciu, C.: Stateless datacenter load-balancing with beamer. In: 15th USENIX Symposium on Networked Systems Design and Implementation (NSDI 2018), pp. 125–139 (2018)
27. Patarasuk, P., Yuan, X.: Bandwidth optimal all-reduce algorithms for clusters of workstations. J. Parallel Distrib. Comput. **69**(2), 117–124 (2009)
28. Patel, P., et al.: Ananta: cloud scale load balancing. ACM SIGCOMM Comput. Commun. Rev. **43**(4), 207–218 (2013)
29. Peng, G.: CDN: content distribution network. arXiv preprint cs/0411069 (2004)
30. Perry, J., Ousterhout, A., Balakrishnan, H., Shah, D., Fugal, H.: Fastpass: a centralized "zero-queue" datacenter network. In: Proceedings of the 2014 ACM Conference on SIGCOMM, pp. 307–318 (2014)
31. Pfister, G.F.: An introduction to the infiniband architecture. High Perform. Mass Storage Parallel I/O **42**(617–632), 102 (2001)
32. Qureshi, M.A., et al.: PLB: congestion signals are simple and effective for network load balancing. In: Proceedings of the ACM SIGCOMM 2022 Conference, pp. 207–218 (2022)
33. Raiciu, C., Barre, S., Pluntke, C., Greenhalgh, A., Wischik, D., Handley, M.: Improving datacenter performance and robustness with multipath TCP. ACM SIGCOMM Comput. Commun. Rev. **41**(4), 266–277 (2011)
34. Reinsel, D., Gantz, J., Rydning, J.: IDC white paper: Data age 2025: "the evolution of data to life-critical" (2017)
35. Suryavanshi, M., Yadav, J.: Mitigating TCP incast in data center networks using enhanced application layer technique. Int. J. Inf. Technol. **14**, 1–9 (2022)
36. Vanini, E., Pan, R., Alizadeh, M., Taheri, P., Edsall, T.: Let it flow: resilient asymmetric load balancing with flowlet switching. In: 14th USENIX Symposium on Networked Systems Design and Implementation (NSDI 2017), pp. 407–420 (2017)

37. Verbraeken, J., Wolting, M., Katzy, J., Kloppenburg, J., Verbelen, T., Rellermeyer, J.S.: A survey on distributed machine learning. ACM Comput. Surv. (CSUR) **53**(2), 1–33 (2020)

38. Wang, S., Zhang, J., Huang, T., Pan, T., Liu, J., Liu, Y.: FDALB: flow distribution aware load balancing for datacenter networks. In: 2016 IEEE/ACM 24th International Symposium on Quality of Service (IWQoS), pp. 1–2. IEEE (2016)

39. Wang, W., Sun, Y., Zheng, K., Kaafar, M.A., Li, D., Li, Z.: Freeway: adaptively isolating the elephant and mice flows on different transmission paths. In: 2014 IEEE 22nd International Conference on Network Protocols, pp. 362–367. IEEE (2014)

40. Wang, Z., Li, Z., Liu, G., Chen, Y., Wu, Q., Cheng, G.: Examination of WAN traffic characteristics in a large-scale data center network. In: Proceedings of the 21st ACM Internet Measurement Conference, IMC 2021, pp. 1–14. Association for Computing Machinery (2021)

41. Yan, S., Wang, X., Zheng, X., Xia, Y., Liu, D., Deng, W.: ACC: automatic ECN tuning for high-speed datacenter networks. In: Proceedings of the 2021 ACM SIGCOMM 2021 Conference, SIGCOMM 2021, pp. 384–397. Association for Computing Machinery (2021)

42. Yang, C.T., Shih, W.C., Huang, C.L., Jiang, F.C., Chu, W.C.C.: On construction of a distributed data storage system in cloud. Computing **98**(1), 93–118 (2016)

43. Zhang, J., Ren, F., Lin, C.: Modeling and understanding TCP incast in data center networks. In: 2011 Proceedings IEEE INFOCOM, pp. 1377–1385. IEEE (2011)

44. Zhang, Y., Zhang, H., Cong, P., Wang, W., Xu, K.: Grandet: cost-aware traffic scheduling without prior knowledge in SD-WAN. In: 2023 IEEE/ACM 31st International Symposium on Quality of Service (IWQoS), pp. 1–10 (2023). https://doi.org/10.1109/IWQoS57198.2023.10188706

45. Zhou, J., et al.: WCMP: weighted cost multipathing for improved fairness in data centers. In: Proceedings of the Ninth European Conference on Computer Systems, pp. 1–14 (2014)

46. Zhu, Y., et al.: Congestion control for large-scale RDMA deployments. ACM SIGCOMM Comput. Commun. Rev. **45**(4), 523–536 (2015)

Optimizing GNN Inference Processing on Very Long Vector Processor

Kangkang Chen, Huayou Su$^{(\boxtimes)}$, Chaorun Liu, and Yalin Li

National University of Defense and Technology, Changsha, Hunan, China
shyou@nudt.edu.cn, liuchaorun@nudt.edu.cn

Abstract. Graph Neural Network (GNN) has shown great success in graph learning. However, within the complexity of the real-world tasks and the big graph datasets, current GNN models become increasingly bigger and more complicated to enhance learning ability and prediction accuracy, which poses a huge challenge to the computation of GNN. The accelerated optimization work for GNN mainly focuses on typical architecture, such as GPU and CPU, with relatively little research on other architectures. In this paper, we focus on accelerating the GNN inference on a very long vector processor. There are several problems in deploying GNN models effectively on long vector architectur, including the lack of an efficient library for GNN on long vector architecture, the very low memory access bandwidth of DDR, and the heterogeneous scheduling issues. To address these challenges, we proposed several strategies to accelerate GNN inference on long vector architecture. Specifically, we build an efficient GNN operator library targeting on the long vector architecture. Secondly, we designed an operator fusion strategy to improve on-chip memory utilization and alleviate the pressure of off-chip memory access. Finally, we implemented a heterogeneous multi-threaded scheduling strategy for long vector architecture to eliminate the overhead of kernel launch. The experimental results show that compared to DGL on Phytium 2000+, the proposed GNN inference on a very long vector processor achieve up to 6.47x speedups.

Keywords: GNN · Long Vector Architecture · Accelerated Inference · Deep Learning Optimization

1 Introduction

In recent years, many fields have flourished with the introduction of Deep Neural Networks (DNN). GNNs have become the main method for learning from graphs, providing a better representation of nodes while revealing the relationships between nodes in the graph. They are efficient in fields such as physical systems [41]. However, compared to traditional graph analysis algorithms, GNNs

---

K. Chen and H. Su—These authors contributed equally to this work.

© The Author(s), under exclusive license to Springer Nature Singapore Pte Ltd. 2024
Z. Tari et al. (Eds.): ICA3PP 2023, LNCS 14492, pp. 59–77, 2024.
https://doi.org/10.1007/978-981-97-0811-6_4

have a high computational cost because a node must collect and aggregate all its neighbors' feature vectors in its receptive field to compute a forward channel. To speed up GNNs training, for example, the number of neighbors can be reduced by using a random node sampling technique [6,7,13,39] to train a graph of a certain size in just a few seconds. However, many GNN applications need help to perform inference when deployed because performing a full forward pass on all neighbors during inference leads to a high memory footprint and latency. The number of nodes and edges of the graph is getting larger and larger as time goes by, resulting in a larger and larger memory footprint and higher latency. Notably, there are many fine-grained, homogeneous and independent data operations in GNNs. The parallelism of these operations increases with the number of nodes in the graph and the number of neighboring nodes. These characteristics are more applicable to long vector architecture, so it is a worthwhile direction to study GNN inference on long vector architecture.

However, there are still many challenges in inferring GNNs on long vector architecture. Firstly, the efficiency of the operator library directly affects the performance of GNN inference. On different architectures, the operator library cannot be reused or is inefficient to be reused, so it is necessary to reconstruct the operator library for long vector architecture. At the same time, inference devices are memory-constrained. As the number of nodes and edges in the graph increases, more and more devices are unable to meet the memory requirements of the dataset, requiring memory optimization in inference. In addition, some of the GNN operations are more logical. However, many architectures, such as GPUs and long vector architecture, need to improve at handling logical operations. Hence, they need to be computed with the powerful logic processing power on the CPU. This approach also presents a serious problem in that there is a large kernel launch overhead, which results in a loss of performance, and as the number of nodes increases, the kernel launch overhead increases, as well as the memory footprint.

To address the above challenges, firstly, we implement all operators in the Graph Convolution Network (GCN) [23] and Graph Attention Networks (GAT) [31] using manual assembly, and we design an operator fusion strategy that reduces the memory footprint while reducing the data transfer, and implement a heterogeneous multi-threaded cooperative strategy with CPU-long vector architecture to hide the kernel launch overhead as much as possible. The main contributions of this paper are:

(1) We proposed the first GNN operator library based on long vector architecture, which can support the construction and efficient execution of typical CNN models.
(2) We designed an operator fusion strategy that places the temporary results in on-chip memory as much as possible, which reduces the memory footprint while increasing the overall computational efficiency and improves the ability to handle larger graph data.

(3) We implemented a heterogeneous multi-threaded scheduling strategy for long vector architecture that is based on the programming characteristics of very long vector processor and significantly reduces kernel launch overhead.

The remainder of the paper is structured as follows: Sect. 2 introduces the background of GNNs and very long vector processor, Sect. 3 presents the related work on GNN optimization, Sect. 4 presents the details of accelerated GNN inference, Sect. 5 shows and analyzes the experimental results, and Sect. 6 shares the conclusions.

2 Background

2.1 GNN

In recent years, many deep learning tasks have required graph data processing. GNNs have also become a widely used neural network model for capturing dependencies and modeling complex relational networks through message passing between nodes and have become the main method for learning knowledge from graphs.

In the GNN model, the graph G consists of nodes and edges. Each node n in G has an m_1-dimensional matrix that represents the node's characteristics. Edges describe the relationships between nodes (e.g., in a social graph, edges represent friendships) and are usually represented by an m_2-dimensional feature matrix. Figure 1 shows a typical graph structure where the adjacency matrix A stores the relationships between nodes, h represents the feature vector of the node, h' represents the updated feature vector of the node, and e represents the feature vector of the edge. Like a CNN model, a GNN model can have many layers. The processing of each layer is divided into two main types of operations: graph operations and neural operations. In graph operations, two phases are involved, aggregation and update. In the aggregation phase, each node receives the characteristics of its neighbors and the corresponding edge information and performs some operations similar to sparse matrix multiplication. In the update phase, the node is notified of the updated status based on the aggregation result. Neural operations are performed on the features of the updated node. Equations 1 and 2 describe GNN's graph operations and neural operations, respectively, where *reduce* can be a sum, average, minimum, or maximum operation. $\odot$ can be addition, subtraction, multiplication, division, left/right reservation acquisition, or point multiplication. W^l represents the learnable parameters of layer l, N_v represents the neighboring nodes of node v. e_{uv} represents the edge features on edge uv. h_u^l represents the node features at layer l on node u.

$$h_v^l = reduce_{u \in N_v}(e_{uv} \odot h_u^l) \tag{1}$$

$$h_v^{l+1} = \sigma(W^l h_v^l) \tag{2}$$

We focus on the GCN [23] and GAT [31] of GNNs, which show breakthrough performance on many deep learning tasks. where GCN's neural operators are

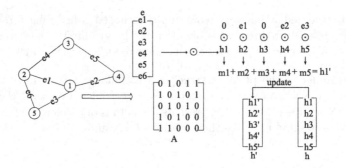

Fig. 1. GNN graph operations.

mainly General Matrix Multiplication (GEMM), while GAT's neural operators are mainly GEMM, General Matrix-vector Multiplication (GEMV) LeakyReLU, softmax.

2.2 Brief Introduction to the Long Vector Architecture Processor

Phytium Matrix is a processor based on a long vector architecture similar to the FT-m7032 [38]. However, each cluster includes 24 accelerator cores that share 6 MB of on-chip Global Shared Memory (GSM), and each core contains 768 KB of Array Memory (AM) and 64 KB of Scalar Memory (SM). GSM and Double Data Rate (DDR) data access via Direct Memory Access (DMA) mode. However, the access bandwidth of the DDR is much smaller than that of the GSM. Data should be transferred as much as possible between on-chip memory to maximize the utilization of access bandwidth, such as GSM and AM, rather than between DDR and on-chip memory. The Scalar Processing Unit (SPU) is responsible for scalar computation and flow control. It provides broadcast instructions to broadcast data from the scalar registers to the vector registers of the Vector Processing Unit (VPU), which provides the main vector computation capability with 16 integrated vector processing units (VPEs), each containing a local register file of 64-bit registers. Each VPE contains a local register stack of 64-bit registers, while all local registers of the same VPE logically form a long vector register of 1024 bits. The processor core supports the simultaneous transmission of 11 variable-length VLIW instructions, and the instruction execution package supports up to five scalar instructions and five vector instructions. Each VPE contains three FMAC operation units, one BP unit, and two L/S data access units. The vector instructions assigned to each VPE are executed independently simultaneously, and each clock cycle supports the concurrent execution of three FMAC instructions and two vector L/S instructions.

3   Related Works

GNNs have a number of features that distinguish them from traditional CNNs [37,40], making existing machine learning libraries for CNNs (e.g. PyTorch [27], and Tensorflow [1], etc.) and hardware platforms [37,40] inefficient. These features include: (1) The diversity of aggregation and update functions of Eq. 1 and Eq. 2 allows for various GNN models. (2) The computation depends on the properties of the input graph in terms of size, sparsity, clustering, or length of the associated feature vector. (3) The unique combination of computational properties of deep learning and graph processing leads to alternate execution modes. The former is computationally constrained and involves dense matrix multiplication [30], while the latter is memory constrained and involves sparse algebra [12]. (4) Large graphs with up to billions of edges must be extended in some applications.

Given these challenges, GNN requires new solutions on both the software and hardware sides. On the software side, several libraries have been proposed to improve support for GNN and handle its multiple variants, extensions of popular libraries such as PyG [10] and DGL [33]. On the hardware side, new accelerator architectures have recently emerged [11,24] to address the flexibility and scalability challenges of GNNs.

Graphs and sparse matrices are intrinsically related, and their duality [22] has been used in many graph algorithms [3,4,28], and libraries [5,8,21,26,29] were exploited. PYG [10] and DGL [33] are two GNN frameworks that researchers widely use; both provide message passing programming interfaces to optimize graphics processing and have been integrated into mainstream deep learning frameworks such as PyTorch. Both libraries abstract GNN processing into Sparse-dense Matrix Multiplication (SpMM) and Sampled Dense-Dense Matrix Multiplication (SDDMM) and rely on optimized cuSPARSE on GPUs, MKL and LIBXSMM [16] on CPUs. DGL combines off-the-shelf SpMM (e.g., csrmm2 in cuSparse) for simple and reduced aggregation [14,23] and uses its own CUDA kernel for more complex aggregation schemes with edge properties [32,36]. Feat-Graph [17] efficiently implements SDDMM and SPMM for CPUs and GPUs. They [18] have experimentally investigated the performance of GNNs on GPUs and summarized the five factors that limit GNNs' performance. Based on these insights, they developed several optimizations to bridge the gap. Yuke Wang [34] et al. proposed GNNAdvision, which is an optimized acceleration of GNN model training using a hybrid parallelism strategy, and Yidi Wu [35] et al. proposed Seastar, which is a point-centered programming model that proposes a complete set of solutions from front-end expression, intermediate expression, and back-end execution; however, they all focus on GNN training rather than inference. Xiaotian Han [15] et al. proposed MLPInit, which uses the weights of the MLP to initialize the GNN, but there is no operator-level speedup. GraphTensor [19] is an open source framework that can run on GPUs and run different graph neural network (GNN) models in a destination-centric, feature-wise manner. Tim Kaler [20] et al. scrutinize the performance bottlenecks specific to GNNs in batch preparation and data transfer, and propose corresponding solution strategies. In addition to the GPU-oriented research mentioned above, there are also some FPGA-oriented research, such as HP-GNN [25], which automatically performs hardware mapping to the target CPU-FPGA platform and enables high-throughput GNN training on a given CPU-FPGA platform, and GenGNN [2], which is a general-purpose GNN acceleration framework for fast inference and scaling up of models.

4   Efficient GNNs Inference on Long Vector Architecture

The GNN inference process consists of two main types of operations: graph operations and neural operations. According to the calculation mode of the GNN model, we first construct an efficient GNN operator library based on a long vector processor to ensure its high efficiency at the operator level. In addition, to improve the reusability of on-chip memory and data transfer efficiency, we implement an operator fusion strategy for graph and neural operations and between neural operators. Also, considering that some logical operations in GNN computation require the CPU and long vector processor to cooperate, we design a heterogeneous multi-threaded cooperative strategy to reduce kernel launch overhead fully. In summary, through the optimization measures proposed in this paper, the inference efficiency of the GNN model has been greatly improved. At the same time, more large graph data can be inferred, which has a broader application prospect.

4.1   Efficient GNN Operator Library

To optimize the performance of GNN for Phytium Matrix's heterogeneous architecture, we construct an efficient GNN operator library based on long vector architecture. We introduce the implementation of the graph operations based on long vector architecture, take the graph operations of the GCN model as an example, and illustrate the design ideas of the operator. In the GCN model, the node features of all neighbors need to be aggregated, and the values of these node features need to be accumulated. Since the computation of the nodes does not interfere with each other, we adopt the idea of node-level parallelism to implement the proposed operator. We use a matrix $nodeFeature$ of the size of the number of nodes multiplied by the features, where $nodeFeature_i$ represents the feature of the ith node. Also, the graph node and edge information is transformed into a Compressed Sparse Row (CSR) format, where $numRows$ represents the number of nodes in the graph, the ith element in $indptr$ represents the sum of the number of neighbors of the previous $i - 1$ nodes and $indices$ represents the node's neighbor node labels.

Algorithm 1 gives the pseudo-code for one of the graph operations ($reduce$ is the summation, $\odot$ is taking left values) implemented in this section. The symbol & indicates that the instruction is executed in the same cycle as the previous instruction. Firstly, transferring the required node feature data into AM, initiated by all cores in the cluster, requires transferring a larger amount of data from DDR to AM; therefore, its performance is limited by DDR bandwidth (Lines 2–7). Next is the calculation of the graph operator (Lines 8–25). It involves the *vload*, *vstore*, and *vmuladd* instructions and a vector data register, VR_i, where *vload* and *vstore* are used to transfer data from AM/SM and vector registers VR_i to the specified memory, and *vmuladd* is used to compute multiply-and-add instructions. Due to the limitation of the number of on-chip registers, we need to block the $m \times n$ node features (Line 9), grouping them by 512. We read the node features into the vector registers VR_0 to VR_{15} (Lines 11–13), and we read the node features of the neighbor nodes from AM into VR_{32}, VR_{33} and VR_{34} in turn, and compute and save them into the corresponding VR_0 to VR_{15} (Lines 15–21). Finally, we save the computed data to the memory of the new node features (Lines 22–24). By loop unrolling, we can make full use of the computing units of the accelerator core to achieve maximum efficiency (Lines 15–21). We implement *reduce* as the summation operation, $\odot$ as the left/right value taking, addition operations for graph operations, the LeakyReLU operator, the softmax, the GEMV operator, and the GEMM operator for neural operations.

We focus on the specific design of a compilation of GEMM operators for neural operations. The parameters of the typical GNN model can be found in Table 2 of Sect. 5.1. For example, for the first layer, the input matrix A of size 3×1024 can be viewed as a GEMM operation with a weight matrix B of size 1024×512 to get the output matrix C. From the introduction of GNN in Sect. 2.1, it can be seen that the output of the graph operations can be used as the input of the neural operations, and the location of the output data of the graph operations can be stored on DDR or GSM. We will introduce the GEMM operator implementation algorithm based on the scheme after the operator fusion strategy (Fig. 3(b)) in Sect. 4.2, that is, A and B are stored on GSM.

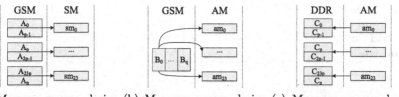

(a) Memory access design (b) Memory access design (c) Memory access design
for matrix A for matrix B for matrix C

Fig. 2. Memory access design for GEMM operators.

Assuming an average arrangement of p nodes per group, Fig. 2(a) shows the data transfer process in A, where each A_i represents a matrix of size 3×1024, gradually passing the matrix from the GSM to SM of the corresponding core via DMA. Figure 2(b) shows the data transfer process in B. Because the size of GSM and AM is limited, matrix B is divided and transferred to AM of all cores in turn via DMA, and then, the data in SM and AM are multiplied by the corresponding matrices A and B, and the results are stored in AM of the corresponding cores.

Algorithm 1. The graph operations *reduce* is summation, $\odot$ is taking left values

---

Require: Graph CSR format information: *numRows*, *indptr* and *indices* and node feature matrix *nodeFeature*, current node processing needs node number set *nodeSet*, node feature size $m \times n$

Ensure: New node feature matrix *newNodeFeature*

1: **while** *nodeSet* is not empty **do**
2:     Take a node *node* from *nodeSet* that has yet to be processed, and remove it from the set.
3:     Transfer the data $nodeFeature_{node}$ from DDR to AM via DMA.
4:     **for** $j = indptr[node]; j < indptr[node + 1]; j + +$ **do**
5:         Transfer the data $nodeFeature_j$ from DDR to AM via DMA.
6:     **end for**
7:     Record the number of neighbors of node *node* as d.
8:     Set the value of VR_{50} to 1.
9:     **for** $cnt = 0; cnt < m \times n/512; cnt + +$ **do**
10:       $am = cnt \times 512$
11:       **for** $i = 0; i < 16; i + +$ **do**
12:         vload am++[1] VR_i
13:       **end for**
14:       $am+ = m \times n$
15:       **for** $i = 0; i < d; i + +$ **do**
16:         **for** $j = 0; j < m \times n/32; j+ = 3$ **do**
17:           vload am++[3],VR_{32},VR_{33},VR_{34}
18:           vmuladd MAC1 VR_{32},VR_{50},VR_j,VR_j
               & vmuladd MAC2 VR_{33},VR_{50},VR_{j+1},VR_{j+1}
               & vmuladd MAC3 VR_{34},VR_{50},VR_{j+2},VR_{j+2}
19:         **end for**
20:         $am+ = m \times n$
21:       **end for**
22:       **for** $i = 0; i < 16; i + +$ **do**
23:         vstore $newNodeFeature + +[1]$ VR_i
24:       **end for**
25:     **end for**
26: **end while**

Figure 2(c) shows the data transfer process in C. The sub-matrix C_i is obtained from AM to DDR via DMA as input to the next layer of graph operators.

We observe that the rows of matrix A are small for a typical GCN model. For a typical GAT model, a GEMV operator exists, which will not fully use the computational resources if computed using the GEMM operator. Therefore, we redesigned the assembly microkernel for the typical GNN models, an assembly microkernel with $3 \times 4k \times v\_size$ for GEMM, and an assembly microkernel with $1 \times 8k \times v\_size$ for GEMV, assuming the maximum vector length v_size that the processor can handle simultaneously.

Algorithm 2 shows the pseudo-code for the serial implementation of GEMM. First, we load the first value of the first row of matrix A with the scalar load instruction, then we vectorize it with the low-bit extension instruction and the broadcast instruction, multiply and add it to the first row of matrix B, and store it in VR_6–VR_8.

The first value of the second and third rows of matrix A is then multiplied and added to the first row of matrix B in the same pattern and stored in VR_9–VR_{14}. Then the second values of the first, second, and third rows of matrix A are multiplied and added to the second row of matrix B, and so on, so that when the loop of line 13 ends, the 36 VRs that hold the results will be stored, which will be called the initial result, and then the loop will be performed on the columns of matrix A to multiply and add the 36 VRs of the initial result. Finally, the 36 VRs are added together at intervals of 4 to obtain the final 9 VRs. That is the data of $3 \times v\_size$ length is stored in matrix C. The $1 \times 8k \times v\_size$ assembly microkernel is similar, except that more vector registers are needed to store temporary intermediate results.

Algorithm 2. Assembly for $3 \times 4k \times v\_size$ GEMM algorithm

---

Require: The maximum vector length v_size that can handle simultaneously, SM of a matrix A of size $3 \times k_A$, and the AM_B of a matrix B of size $k_A \times v\_size$.

Ensure: AM_C of a matrix C of size $3 \times v\_size$

1: **for** $k = 0; k < k_A; k{+} = 4$ **do**
2:   **for** $i = 0; i < 4; i{+}{+}$ **do**
3:     **for** $j = 0; j < 3; j{+}{+}$ **do**
4:       sload $SM[j \times k_n + (i+k)]$ R_0
5:       The data in R_0 is extended by the low-bit and stored in R_1.
6:       Broadcast the data in R_1 to the vector VR_0
7:       vload $AM_B[(i+k) \times 3 + 0)]$ VR_1
8:       vload $AM_B[(i+k) \times 3 + 1)]$ VR_2
9:       vload $AM_B[(i+k) \times 3 + 2)]$ VR_3
10:       vmuladd $VR_0, VR_1, VR_{6+i \times 9+j \times 3}, VR_{6+i \times 9+j \times 3}$
11:       vmuladd $VR_0, VR_2, VR_{7+i \times 9+j \times 3}, VR_{7+i \times 9+j \times 3}$
12:       vmuladd $VR_0, VR_3, VR_{8+i \times 9+j \times 3}, VR_{8+i \times 9+j \times 3}$
13:     **end for**
14:   **end for**
15: **end for**
16: **for** $i = 0; i < 3; i{+}{+}$ **do**
17:   **for** $j = 0; j < 3; j{+}{+}$ **do**
18:     vadd $VR_{6+i \times 3}, VR_{6+i \times 3+(j+1) \times 9}$
19:     vadd $VR_{7+i \times 3}, VR_{7+i \times 3+(j+1) \times 9}$
20:     vadd $VR_{8+i \times 3}, VR_{8+i \times 3+(j+1) \times 9}$
21:   **end for**
22: **end for**
23: **for** $i = 0; i < 3; i{+}{+}$ **do**
24:   vstore $VR_{6+i \times 3}, AM_C[i \times 3 + 0]$
25:   vstore $VR_{7+i \times 3}, AM_C[i \times 3 + 1]$
26:   vstore $VR_{8+i \times 3}, AM_C[i \times 3 + 2]$
27: **end for**

After instruction pipeline rearrangement, the computation utilization of $3 \times 4k \times v\_size$ assembly microkernel is 98.6% when $k = 256$, 97.3% when $k = 128$, and 94.7% when $k = 64$. At the same time, we can also compute the neural operations in batches by first computing the graph operations of two nodes and then merging the output data using an assembly microkernel of $6 \times 2k \times v\_size$, which computes the graph operations several times and then computes the neural operations in batches, it can also achieve a high utilization rate.

4.2   GNN Operator Fusion Strategy

Due to the sparsity of the graph, GNN inference is a data-intensive application, and its memory bandwidth becomes the performance bottleneck of GNN. The bandwidth constraint of Phytium Matrix's memory means that, in most cases, the processor cores are not utilizing their computational power due to waiting for data transfer. Maximizing the number of data reuses is necessary to minimize the time processor cores cannot perform computation due to waiting for data operations with limited bandwidth resources.

Figure 3(a) shows the data flow of the original case GCN operator on Phytium Matrix. The data are the node feature values required to calculate graph operations. They are generally stored in DDR as input data for the calculation of a layer of the model, noted as Da, and transferred directly from DMA to AM of the corresponding computing core in the original case. After the data has been transferred and computed, it is transferred back to DDR via the DMA from AM, noted as Db. For neural operations, the input data includes the node feature data transferred from Db to the SM via DMA and the weight matrix transferred from the DDR to the GSM and the AM via DMA. When the data transfer is completed, it is computed by the corresponding computational core. The output data will be stored on AM and transferred to DDR via DMA to get the new node feature value data.

Due to the limited size of GSM, the data needs to be grouped, where $decnt$ represents the number of degrees of the node, requiring a total of $48decnt + 1 + 48decnt = 96decnt + 1$ data interactions between DDR and GSM/AM/SM, and $24decnt$ data interactions between the GSM and AM. The above analysis shows many data interactions between DDR and on-chip memory. However, The access bandwidth of the DDR is much smaller than that of the GSM, so reducing the number of data interactions between DDR and GSM/SM/AM can improve the inference performance of GNN on long vector architecture. It is clear from Sect. 2.1 that the output data of graph operations is the input data of neural operations. We have designed an operator fusion strategy to reduce the number of data interactions between DDR and GSM/SM/AM by fusing the graph and neural operations and putting the feature data calculated by the graph operations into GSM. Figure 3(b) shows the direction of data flow after using the operator fusion. DDR requires $48decnt + 1$ DDR and GSM/AM/SM data interactions, while GSM requires $72decnt$ of GSM and AM/SM data interactions. Compared to the original data flow process, this reduces the number of DDR and GSM/AM/SM data interactions by $48decnt$ and increases the number of GSM

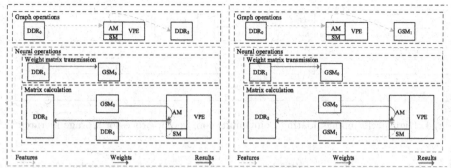

(a) The direction of data flow before optimization of the GNN operator

(b) The direction of data flow after optimization of the GNN operator

Fig. 3. The direction of data flow of the GNN operator.

and AM/SM data interactions by $48decnt$. Due to the bandwidth difference, the operator fusion strategy will improve overall efficiency significantly. When inference the GAT model, more fine-grained fusion can be achieved, such as LeakyReLU and the e^x in softmax, which are elementwise operators that can be computed in AM directly without being transferred to DDR or GSM.

4.3 Heterogeneous Multi-threaded Cooperative Strategy for Long Vector Architecture

As described in Sect. 2.2, Phytium Matrix uses a multi-core architecture with heterogeneous CPUs and accelerators. Some operations are more logical for cooperative computing, and accelerators, including GPUs, are not good at handling logical operations. CPUs are better at logical operations than accelerators.

We design and implement a heterogeneous multi-threaded cooperative strategy to better cooperate the computational power of the CPU and the accelerator, which can better cooperate the task scheduling between the CPU and accelerator to improve the overall performance. Based on this, a heterogeneous cooperative strategy is used to perform cooperative computation using the CPU for the computational characteristics of the GCN. Due to the difference in the storage hierarchy between CPUs and accelerator cores, memory data consistency is not guaranteed for CPUs and accelerator cores.

Figure 4 shows that the CPU is connected to DDR via the L1 and L2 caches, while the accelerator cores are connected to the GSM via AM/SM and GSM is connected to DDR via DMA. The CPU and accelerator cores can perform different tasks independently of each other. It is critical to improve the overall performance by building a heterogeneous multi-threaded cooperative strategy using the operator library and architecture to enhance the cooperative computing capability between the CPU and the accelerator. We can build a heterogeneous multi-threaded cooperative strategy by taking advantage of the fact that both CPUs and accelerators can read data from DDR. Specifically, the CPU and

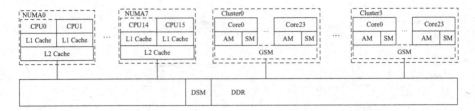

Fig. 4. High performance floating point multi-core vector processor storage hierarchy.

the accelerator read and write data from the shared memory, called *DSM*, to communicate with each other at the program's start. However, for the CPU architecture, when the CPU executes the store instruction and hits the cache, we only update the data in the cache. Each cache line will have a bit to record whether the data has been modified, called a dirty bit. We set the dirty bit to 1. The data in DDR will only be updated when the cache line is replaced or explicitly cleared. Therefore, the modified data may be in the cache rather than the DDR, resulting in inconsistent data. To ensure that the CPU's modifications to the *DSM* data are written to memory on time, the data in the cache needs to be flushed directly into the *DSM*. Specifically, Algorithm 3 and Algorithm 4 shows the pseudo-code, and Algorithm 3 runs on the CPU side, and Algorithm 4 runs on the accelerator side. The state of each accelerator core is stored using *state* of *DSM*.

Algorithm 3. CPU cooperative code

Require: *id* is the core of the current thread, *core* is the mapping between the CPU cores and the managed accelerator cores, *state* is the status code of the shared memory

1: **while** true **do**
2:     **while** Retrieve a managed accelerator core number from core[id], as *sid* **do**
3:         **switch** (state[sid])
4:         **case 0:**
5:             Retrieve tasks from the queue, prepare completion data, and set state[sid]=1
6:         **case 1:**
7:             No processing
8:         **case 2:**
9:             The accelerator completes the calculation, takes out the data, and set state[sid]=0
10:     **end switch**
11:     **end while**
12:     Flush CPU cache
13:     **if** All tasks have been completed **then**
14:         Set all states=3 and break
15:     **end if**
16: **end while**

Algorithm 4. Accelerator cooperative code

Require: *sid* is the core number of the accelerator,*state* is the status code of the shared memory
```
 1: while true do
 2:    switch (state[sid])
 3:    case 0:
 4:       Waiting for CPU to assign tasks without any processing
 5:    case 1:
 6:       Get the data and perform the corresponding calculation operations, and set
          state[sid]=2
 7:    case 2:
 8:       Maintain current status
 9:    case 3:
10:       Break
11:    end switch
12: end while
```

Table 1 shows the meaning of each state. The CPU side is responsible for distributing tasks and preparing the data for a task before the accelerator is busy. The accelerator side is responsible for computing the tasks the CPU distributes and putting the data into the specified memory.

Table 1. Accelerator core status code

Status code	CPU cores	Accelerator cores
0	Prepare data required for accelerator tasks	Wait
1	Wait or perform other tasks	CPU distribution tasks
2	Fetch accelerator completion data	Wait
3	Task all done, accelerator exit	Exit

5   Experiment

5.1   Experimental Setup

We use a model consisting of three layers as a test case. Table 2 shows the model's parameters. The experiments focus on simulating this three-layer GCN and GAT model for inference, where random functions generate node feature values and weight values.

Table 3 shows the main parameters of the six graph datasets from Open Graph Benchmark (OGB), including the number of nodes (Nodes), the number of edges (Edges), and the average degree of each node (Average).

Table 2. GNN model parameters

Layers	Node feature dimension	Weights dimension
1	(3, 1024)	(1024, 512)
2	(3, 512)	(512, 256)
3	(3, 256)	(256, 128)

Table 3. Dataset characteristics

Dataset	Nodes	Edges	Average
arxiv	169K	1.2M	7
collab	236K	2.4M	10
proteins	133K	79M	597
ddi	4K	2.1M	501
ppa	576K	30M	73.7
biokg	93K	5M	47.5

5.2 Overall Performance

Table 4 shows the experimental results based on the performance of a single acceleration cluster on Phytium Matrix and Phytium 2000+ [9], where DGL and OpenBLAS are used on the Phytium 2000+ and our proposed strategies are used on the Phytium Matrix.

Table 4. Time(s) comparison between Phytium 2000+ and Phytium Matrix

Datasets	GCN			GAT		
	Phytium 2000+	Phytium Matrix	Speedup	Phytium 2000+	Phytium Matrix	Speedup
arxiv	8.67	1.34	6.47	8.08	1.89	4.28
collab	11.22	2.03	5.53	11.69	2.94	3.98
proteins	31.59	43.92	0.72	142.49	34.81	4.09
ddi	1.23	1.21	1.01	1.16	0.96	1.20
ppa	33.48	24.35	1.37	588.74	OOM	/
biokg	5.94	3.16	1.88	10.70	3.02	3.54

We can observe that Phytium Matrix outperforms Phytium 2000+ on five datasets for GCN and GAT inference, where speedups of up to 6.47x and 4.28x are achieved on GCN and GAT, respectively. Because the operator fusion strategy and the cooperative strategy work together. However, for GCN inference on the proteins dataset, Phytium Matrix does not perform as well as Phytium 2000+ because the proteins dataset has a higher average degree and has more access pressure. In contrast, the access bandwidth of the Phytium Matrix is much smaller than that of the Phytium 2000+. For GAT inference on the ppa dataset, an Out Of Memory (OOM) error occurs on the Phytium Matrix, while it normally runs on the Phytium 2000+ because the memory size of Phytium Matrix is much smaller than that of Phytium 2000+.

5.3 Operator Fusion Strategy

The efficiency of the operator library directly determines the performance of GNN inference. We designed an operator fusion strategy for GNN based on a reconfigured operator library, which allows for a significant reduction in the number of data interactions from DDR to GSM/AM/SM. Figure 5 shows the performance improvement of the operator fusion strategy on several datasets.

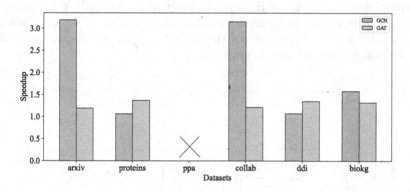

Fig. 5. GCN and GAT operator fusion performance improvement in six datasets.

We can observe that the average degree of the nodes in arxiv and collab is small for GCN inference, the operator fusion strategy can achieve 3.19x and 3.15x speedups, respectively, and the average degree of the nodes in the ddi and proteins datasets is large, the operator fusion strategy can achieve around 1.06x speedups. Because the operator fusion strategy reduces the transfer of intermediate results from on-chip memory to DDR, when the average degree of the nodes is large, the most important factor affecting the overall performance is the in-core data handling and the computation of the neural operations. When the average degree of the nodes is small, the most important factor affecting the overall performance is the data transmission of the graph operation operators. For GAT inference, the neural operator is more time-consuming than the graph operator, and the performance improvement basically comes from the operator fusion of the LeakyReLU and softmax operators. For the ppa dataset for GCN inference, the number of nodes and edges is too large for the initial version, which leads to OOM and cannot be executed. In contrast, the operator fusion strategy can be successfully executed on the ppa dataset. In conclusion, the operator fusion strategy can reduce the number of data interactions between on-chip memory and DDR, which improves the overall performance but is more suitable for data sets with a relatively small average degree of the nodes, and it can reduce the memory footprint, thus allowing larger graph datasets to be run in the limited memory.

5.4 Heterogeneous Multi-threaded Cooperative Strategy

For heterogeneous architectures, in addition to kernel runtime, the overall performance is also affected by kernel launch overhead. The heterogeneous multi-threaded cooperative strategy can effectively reduce kernel launch overhead. We adopt the three-layers GCN model from Sect. 5.1, which treats the calculation of the feature values of each node as a task as the baseline version and sends a task to each accelerator core for each heterogeneous call to mimic the scenario of heterogeneous multi-core performing different tasks. Similarly, the number of multi-threads is set to 1 to ensure fairness, which means one CPU core manages 24 accelerator cores. Figure 6 shows the speedups of the heterogeneous multi-threaded cooperative strategy in several datasets.

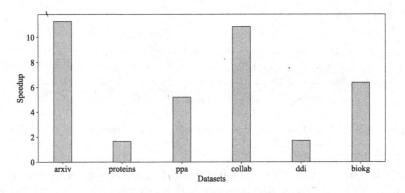

Fig. 6. Performance improvement of GCN multi-threaded cooperative strategy in six datasets

We can observe that the heterogeneous multi-threaded cooperative strategy achieves 1.65x–11.28x speedups. For the arxiv and collab datasets, the cooperative strategy achieves around 11x speedups. For the ppa and biokg datasets, the cooperative strategy achieves around 5x speedups. For the proteins and ddi datasets, the cooperative strategy only achieves around 1.6x speedups. Because the cooperative strategy mainly reduces the kernel launch overhead for calls to graph datasets with short kernel execution time. The arxiv and collab datasets have a smaller average degree of the nodes, resulting in a short kernel execution time. The ddi and proteins datasets have an average degree of the nodes of 500, which results in a longer kernel execution time, and the kernel launch overhead is relatively small. Hence, the cooperative strategy could be more effective.

6 Conclusion

In this paper, we study the optimization of inference graph neural networks on very long vector processor, and we design a GNN operator library that can efficiently process GEMM, GEMV, LeakyReLU, softmax and other operators on very long vector processor. We designed an operator fusion method to reduce memory usage and improve computational efficiency to inference larger graphs. We design a heterogeneous multi-threaded cooperative strategy for multi-level storage of very long vector processor, which can reasonably reduce the kernel launch overhead. The experimental results show that our proposed GNN inference method on very long vector processor achieves up to 6.47x speedup compared to using DGL on Phytium 2000+.

Acknowledgements. We thank the Foundation of Xiang Jiang Laboratory No. 22XJ01012 for supporting this research.

References

1. Abadi, M., et al.: TensorFlow: a system for large-scale machine learning. In: Proceedings of the 12th USENIX conference on Operating Systems Design and Implementation, pp. 265–283 (2016)
2. Abi-Karam, S., He, Y., Sarkar, R., Sathidevi, L., Qiao, Z., Hao, C.: GenGNN: a generic FPGA framework for graph neural network acceleration. CoRR abs/2201.08475 (2022). https://arxiv.org/abs/2201.08475
3. Azad, A., Buluç, A., Gilbert, J.R.: Parallel triangle counting and enumeration using matrix algebra. In: 2015 IEEE International Parallel and Distributed Processing Symposium Workshop, IPDPS 2015, Hyderabad, India, 25–29 May 2015, pp. 804–811. IEEE Computer Society (2015)
4. Azad, A., Pavlopoulos, G.A., Ouzounis, C.A., Kyrpides, N.C., Buluç, A.: HipMCL: a high-performance parallel implementation of the markov clustering algorithm for large-scale networks. Nucleic Acids Res. **46**(6), e33–e33 (2018)
5. Buluç, A., Gilbert, J.R.: The combinatorial BLAS: design, implementation, and applications. Int. J. High Perform. Comput. Appl. **25**(4), 496–509 (2011)
6. Chen, J., Ma, T., Xiao, C.: FastGCN: fast learning with graph convolutional networks via importance sampling. arXiv e-prints (2018)
7. Chen, J., Zhu, J., Song, L.: Stochastic training of graph convolutional networks with variance reduction (2017)
8. Davis, T.A.: Algorithm 1000: Suitesparse: Graphblas: graph algorithms in the language of sparse linear algebra. ACM Trans. Math. Softw. **45**(4), 44:1–44:25 (2019)
9. Fang, J., Liao, X., Huang, C., Dong, D.: Performance evaluation of memory-centric armv8 many-core architectures: a case study with phytium 2000+. J. Comput. Sci. Technol. **36**(1), 33–43 (2021). https://doi.org/10.1007/s11390-020-0741-6
10. Fey, M., Lenssen, J.E.: Fast graph representation learning with PyTorch Geometric (2019)
11. Geng, T., et al.: AWB-GCN: a graph convolutional network accelerator with runtime workload rebalancing. In: 53rd Annual IEEE/ACM International Symposium on Microarchitecture, MICRO 2020, Athens, Greece, pp. 922–936. IEEE (2020)

12. Gui, C., et al.: A survey on graph processing accelerators: challenges and opportunities. J. Comput. Sci. Technol. **34**(2), 339–371 (2019)
13. Hamilton, W.L., Ying, R., Leskovec, J.: Inductive representation learning on large graphs (2017)
14. Hamilton, W.L., Ying, Z., Leskovec, J.: Inductive representation learning on large graphs. In: Advances in Neural Information Processing Systems 30: Annual Conference on Neural Information Processing Systems 2017, Long Beach, CA, USA, pp. 1024–1034 (2017)
15. Han, X., Zhao, T., Liu, Y., Hu, X., Shah, N.: MLPInit: embarrassingly simple GNN training acceleration with MLP initialization. In: The Eleventh International Conference on Learning Representations, ICLR 2023, Kigali, Rwanda, 1–5 May 2023. OpenReview.net (2023). https://openreview.net/pdf?id=P8YIphWNEGO
16. Heinecke, A., Henry, G., Hutchinson, M., Pabst, H.: LIBXSMM: accelerating small matrix multiplications by runtime code generation. In: Proceedings of the International Conference for High Performance Computing, Networking, Storage and Analysis, SC 2016, Salt Lake City, UT, USA, pp. 981–991. IEEE Computer Society (2016)
17. Hu, Y., et al.: FeatGraph: a flexible and efficient backend for graph neural network systems. In: Proceedings of the International Conference for High Performance Computing, Networking, Storage and Analysis, SC 2020, Virtual Event/Atlanta, Georgia, USA, p. 71. IEEE/ACM (2020)
18. Huang, K., Zhai, J., Zheng, Z., Yi, Y., Shen, X.: Understanding and bridging the gaps in current GNN performance optimizations. In: PPoPP 2021: 26th ACM SIGPLAN Symposium on Principles and Practice of Parallel Programming, Virtual Event, Republic of Korea, pp. 119–132. ACM (2021)
19. Jang, J., Kwon, M., Gouk, D., Bae, H., Jung, M.: GraphTensor: comprehensive GNN-acceleration framework for efficient parallel processing of massive datasets. In: IEEE International Parallel and Distributed Processing Symposium, IPDPS 2023, St. Petersburg, FL, USA, 15–19 May 2023, pp. 2–12. IEEE (2023). https://doi.org/10.1109/IPDPS54959.2023.00011
20. Kaler, T., et al.: Accelerating training and inference of graph neural networks with fast sampling and pipelining. In: Marculescu, D., Chi, Y., Wu, C. (eds.) Proceedings of Machine Learning and Systems 2022, MLSys 2022, Santa Clara, CA, USA, August 29 - September 1 2022. mlsys.org (2022). https://proceedings.mlsys.org/paper/2022/hash/35f4a8d465e6e1edc05f3d8ab658c551-Abstract.html
21. Kepner, J., et al.: Mathematical foundations of the graphBLAS. In: 2016 IEEE High Performance Extreme Computing Conference, HPEC 2016, Waltham, MA, USA, pp. 1–9. IEEE (2016)
22. Kepner, J., Gilbert, J.R.: Graph Algorithms in the Language of Linear Algebra, Software, Environments, Tools, vol. 22. SIAM (2011)
23. Kipf, T.N., Welling, M.: Semi-supervised classification with graph convolutional networks. In: 5th International Conference on Learning Representations, ICLR 2017, Toulon, France (2017)
24. Liang, S., et al.: EnGN: a high-throughput and energy-efficient accelerator for large graph neural networks. IEEE Trans. Comput. **70**(9), 1511–1525 (2021)
25. Lin, Y., Zhang, B., Prasanna, V.K.: HP-GNN: generating high throughput GNN training implementation on CPU-FPGA heterogeneous platform. In: Adler, M., Ienne, P. (eds.) FPGA 2022: The 2022 ACM/SIGDA International Symposium on Field-Programmable Gate Arrays, Virtual Event, USA, 27 February 2022–1 March 2022, pp. 123–133. ACM (2022). https://doi.org/10.1145/3490422.3502359

26. Malewicz, G., et al.: Pregel: a system for large-scale graph processing. In: Proceedings of the ACM SIGMOD International Conference on Management of Data, SIGMOD 2010, Indianapolis, Indiana, USA, pp. 135–146. ACM (2010)
27. Paszke, A., et al.: PyTorch: an imperative style, high-performance deep learning library, vol. 32 (2019)
28. Shun, J., Blelloch, G.E.: Ligra: a lightweight graph processing framework for shared memory. In: ACM SIGPLAN Symposium on Principles and Practice of Parallel Programming, PPoPP 2013, Shenzhen, China, 23–27 February 2013, pp. 135–146. ACM (2013)
29. Sundaram, N., et al.: GraphMat: high performance graph analytics made productive. Proc. VLDB Endow. **8**(11), 1214–1225 (2015)
30. Sze, V., Chen, Y., Yang, T., Emer, J.S.: Efficient processing of deep neural networks: a tutorial and survey. Proc. IEEE **105**(12), 2295–2329 (2017)
31. Velickovic, P., Cucurull, G., Casanova, A., Romero, A., Liò, P., Bengio, Y.: Graph attention networks. In: 6th International Conference on Learning Representations, ICLR 2018, Vancouver, BC, Canada, April 30 - May 3, 2018, Conference Track Proceedings. OpenReview.net (2018). https://openreview.net/forum?id=rJXMpikCZ
32. Velickovic, P., et al.: Graph attention networks. Stat **1050**(20), 10–48550 (2017)
33. Wang, M., et al.: Deep graph library: towards efficient and scalable deep learning on graphs. CoRR abs/1909.01315 (2019)
34. Wang, Y., et al.: GNNAdvisor: an adaptive and efficient runtime system for GNN acceleration on GPUs. In: Brown, A.D., Lorch, J.R. (eds.) 15th USENIX Symposium on Operating Systems Design and Implementation, OSDI 2021, 14–16 July 2021, pp. 515–531. USENIX Association (2021). https://www.usenix.org/conference/osdi21/presentation/wang-yuke
35. Wu, Y., et al.: Seastar: vertex-centric programming for graph neural networks. In: Barbalace, A., Bhatotia, P., Alvisi, L., Cadar, C. (eds.) EuroSys 2021: Sixteenth European Conference on Computer Systems, Online Event, United Kingdom, 26–28 April 2021, pp. 359–375. ACM (2021). https://doi.org/10.1145/3447786.3456247
36. Xu, K., Hu, W., Leskovec, J., Jegelka, S.: How powerful are graph neural networks? In: 7th International Conference on Learning Representations, ICLR 2019, New Orleans, LA, USA (2019)
37. Yan, M., et al.: Characterizing and understanding GCNs on GPU. IEEE Comput. Archit. Lett. **19**(1), 22–25 (2020)
38. Yin, S., Wang, Q., Hao, R., Zhou, T., Mei, S., Liu, J.: Optimizing irregular-shaped matrix-matrix multiplication on multi-core DSPs. In: IEEE International Conference on Cluster Computing, CLUSTER 2022, Heidelberg, Germany, 5–8 September 2022, pp. 451–461. IEEE (2022). https://doi.org/10.1109/CLUSTER51413.2022.00055
39. Zeng, H., Zhou, H., Srivastava, A., Kannan, R., Prasanna, V.: GraphSAINT: graph sampling based inductive learning method (2020)
40. Zhang, Z., Leng, J., Ma, L., Miao, Y., Li, C., Guo, M.: Architectural implications of graph neural networks. IEEE Comput. Archit. Lett. **19**(1), 59–62 (2020)
41. Zhou, J., et al.: Graph neural networks: a review of methods and applications. AI Open **1**, 57–81 (2020)

GDTM: Gaussian Differential Trust Mechanism for Optimal Recommender System

Lixiao Gong[1], Guangquan Xu[1], Jingyi Cui[1(✉)], Xiao Wang[1,2,3], Shihui Fu[4], Xi Zheng[5], and Shaoying Liu[6]

[1] Tianjin University, Tianjin 300350, China
cuijingyi@tju.edu.cn
[2] Tianjin University of Finance and Economics, Tianjin 300221, China
[3] Guangdong Provincial Key Laboratory of Novel Security Intelligence Technologies, Guangdong 518055, China
[4] Delft University of Technology, Delft, The Netherlands
[5] Macquarie University, Sydney, NSW 2109, Australia
[6] Hiroshima University, Higashihiroshima 739-8511, Japan

Abstract. As recommender systems have become increasingly popular in providing users with personalized recommendations, researchers have implemented protective measures to safeguard users' privacy. However, the implementation of such mechanisms is extremely difficult to ensure both recommendation accuracy and privacy protection. In this paper, we propose a novel protective mechanism that addresses this challenge. Our approach introduces the concept of differential trust, which integrates matrix factorization and the combination theorem of differential privacy. We then propose the Gaussian Differential Trust Mechanism, which protects users' historical ratings while maintaining recommendation accuracy to a certain extent. The rationality of our proposed mechanism is verified by theoretical explanation and experimental evaluation. The experiment results demonstrate that our method effectively balances the competing goals of recommendation accuracy and privacy preservation, providing a solution to the challenges faced by recommender systems.

Keywords: Differential trust · Collaborative optimization system · Privacy protection

1 Introduction

With the development of big data, recommender systems have become an important tool for online businesses to enhance their income and tackle the issue of information overload [1]. In addition to giving attackers access to information, the huge quantity of data also makes it possible for them to infer users' personal information from relevant data using background knowledge [2]. Therefore, one of the key research directions in the field of recommendations is the development of reliable recommendation methods.

© The Author(s), under exclusive license to Springer Nature Singapore Pte Ltd. 2024
Z. Tari et al. (Eds.): ICA3PP 2023, LNCS 14492, pp. 78–92, 2024.
https://doi.org/10.1007/978-981-97-0811-6_5

Traditional recommendation algorithms typically rely on historical user behaviors to make predictions, and most algorithms assume users are independent and identically distributed with a strong reliance on user behavior. To overcome the limitation, some research incorporated the trust factor into the recommender systems, which increases the effectiveness of recommendations in some way. However, due to concerns about personal information exposure, most users are reluctant to disclose explicit trust, leading to the issue of sparse trust [3]. Distrust factors have been identified in social networks, which are often not taken into account in recommendation systems [4]. Besides, to address the issue of user information leakage, social networks commonly adopt privacy protection technologies such as access control and anonymization. Nevertheless, access control and anonymization techniques have limitations. De-anonymization techniques can compromise user privacy, and access control may not strike the right balance between recommendation accuracy and user privacy [5].

To address the above problems, we propose the Gaussian Differential Trust Mechanism, which is aimed at collaborative optimization to ensure the accuracy and security of recommender systems. Our main contributions are threefold.

Firstly, we propose the new concept of differential trust based on trust recommendation and the Moments Accountant theorem of differential privacy.

Secondly, we extend the differential trust and propose the Gaussian Differential Trust Mechanism to enhance data security by protecting against attacks that attempt to identify specific user ratings and their existence.

Thirdly, we use theoretical explanations to justify the rationality of differential trust. Additionally, we conduct experiments to evaluate the performance of GDTM. The experiment results show that GDTM can achieve a good trade-off between the accuracy and security in recommender systems that take into account the influence of both explicit and implicit trust in rating prediction.

The remainder of this paper is organized as follows. Section 2 describes the related work on trust-based recommendation and privacy-protection methods in recommender systems. Section 3 explores our trade-off mechanism. Section 4 and Sect. 5 presents our experiments and performance analysis. Section 6 concludes this paper and discusses our future work.

2   Related Work

2.1   Trust-Based Recommendation

The sustainable development of recommender systems has benefited from the rise of social networks [6]. According to research in psychology and sociology, people tend to make choices relying on those of their friends in social circles, so a trust-based recommender system generates personalized recommendations by aggregating the opinions of other users in the trust network [7]. Ma et al. [8] proposed a factor analysis approach based on probabilistic matrix factorization called SoRec. Following that, they proposed a novel probabilistic factor analysis framework called RSTE [9], which combined the preferences of the users and their trusted friends. They also proposed a matrix factorization framework with social regularization called SoReg [10]. Jamali et al. [11] proposed a recommendation model called SocialMF in social networks, which introduced a trust propagation mechanism into it. Zhang et al. [12] proposed the CUNE for extracting implicit and reliable social information from user feedback.

However, the trust-based recommendation algorithms mentioned above primarily focus on explicit trust in social networks, while ignoring the impact of implicit trust on recommender systems. Guo et al. [13] proposed a trust-based recommendation model called TrustSVD, which incorporated both the explicit and implicit influence of user trust and item ratings. Xu et al. [14] proposed TBSVD, a social trust and behavior method based on the Singular Value Decomposition algorithm. Hu et al. [15] proposed the SSL-SVD to mine sparse trust between users and improve the performance of recommender systems. It integrated social trust and sparse trust into SVD++, which effectively exploited the explicit and implicit influence of trust for rating prediction in recommender systems. One of the primary goals of this paper is to leverage users' trust data to enhance the performance of the recommender system in social networks.

2.2   Recommended Protection Mechanism

Improper data collection, storage, and transmission data have increased the probability of users' sensitive data leakage [16]. The victim will suffer irreparable harm if sensitive data is made available to malevolent parties. Therefore, many researchers have recently sought privacy protection methods for recommender systems to maintain their security and accuracy.

Most recent research on differential privacy recommendation has concentrated on recommender systems based on matrix factorization. Friedman et al. [17] proposed a general differential privacy framework based on matrix factorization recommendations, and suggested system aspects that should be addressed when applying differential privacy to privacy-preserving solutions. Shin et al. [18] proposed a matrix factorization method under local differential privacy. Aiming at ensuring the privacy of their personal items and ratings, users need to randomize data to satisfy their own differential privacy requirements before sending the perturbed data to aggregate, ensuring the privacy of their personal items and ratings. Zhang et al. [19] proposed a personalized differential privacy recommendation scheme based on the probabilistic matrix factorization model. It employed a modified sample mechanism and could satisfy users' privacy requirements specified at the item level. Zhou et al. [20] proposed a lightweight matrix factorization recommendation method trained on local IoT devices, making it possible for users to train big data models locally.

As demonstrated by the methods described above, despite the fact that differential privacy is widely used in recommender systems, it cannot strike an appropriate balance between accuracy and security. The influence of trust factor is not considered when predicting ratings either. Therefore, how to compromise the accuracy and security while protecting user data in recommender systems is the main objective of our research work.

3 Trade-Off Mechanism

In this section, we define the definitions of related trust and extend the trust recommendation. On this basis, we propose the concept of differential trust and construct the Gaussian Differential Trust Mechanism. The overall framework is shown in Fig. 1.

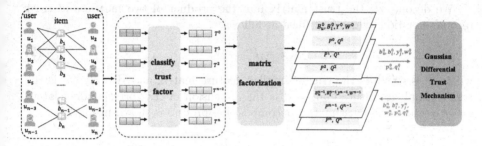

Fig. 1. The framework of the Gaussian Differential Trust Mechanism for recommender system.

3.1 Trust-Related Definition

It is impossible to define trust in a unified way due to its highly subjective characteristics in the field of social recommendation [21]. Therefore, combined with previous work, we consummate and put forward the definitions of related trust.

Trust is reflected in the preference similarity and behavior similarity between users in recommender systems, indicating the degree of their reliance on evaluation provided by others [22]. On the other hand, distrust indicates a connection between users with minimal or no likelihood of dependence [23]. And trust cold start refers to the provision of too little trust data in trust networks, resulting in extreme trust sparseness [24]. Besides, most of the data have no obvious relationship with each other and are in a fuzzy state. We define fuzzy trust as the absence of direct information to prove whether users have a trust or distrust relationship.

Assuming that trust matrix $T = [t_{u,v}]_{m \times m}$, matrix element $t_{u,v}$ represents the trust value of $user_u$ to $user_v$. The following is an example of the trust matrix.

$$T = \begin{bmatrix} 0 & 1 & 0 \\ 0 & 0 & -1 \\ -1 & 0 & 0 \end{bmatrix} \tag{1}$$

The value of trust is 1, distrust is -1 and fuzzy trust is 0. m refers to the total user number in trust networks. The rows represent trustors and the columns represent trustees. The elements represent the one-way trust between users. Classifying trust factor increases the utilization rate of trust information so as to improve the effectiveness of social recommendation.

3.2 Trust Recommendation

Trust recommendation is a method that leverages trust relationships to construct a social network among users [25]. It offers pertinent recommendations based on the ratings of users with direct or indirect social connections to the target user. In this paper, we define that trust recommendation as a method for applying explicit trust to recommender systems through matrix factorization.

We decompose the trust matrix into the product of two matrices in trust recommendation, which is defined as follows,

$$T \approx P^T \times W \tag{2}$$

where $P \in \mathbb{R}^{d \times m}$ is the trustor implicit feature matrix and $W \in \mathbb{R}^{d \times m}$ is the trustee implicit feature matrix. d is the dimension of the corresponding latent factor vector. The trust matrix decomposition process is shown in Fig. 2.

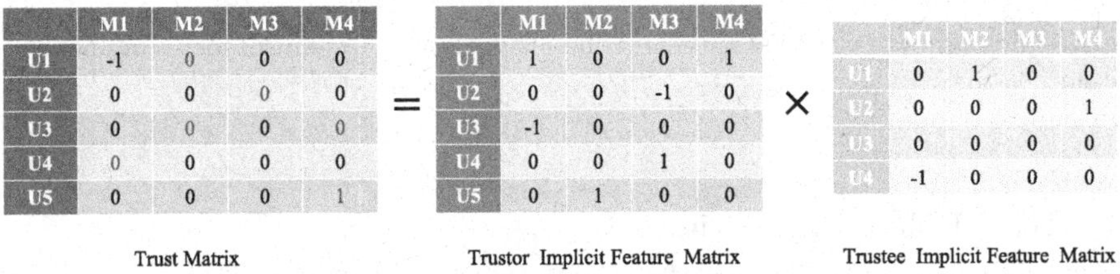

	M1	M2	M3	M4
U1	-1	0	0	0
U2	0	0	0	0
U3	0	0	0	0
U4	0	0	0	0
U5	0	0	0	1

Trust Matrix

$=$

	M1	M2	M3	M4
U1	1	0	0	1
U2	0	0	-1	0
U3	-1	0	0	0
U4	0	0	1	0
U5	0	1	0	0

Trustor Implicit Feature Matrix

$\times$

	M1	M2	M3	M4
U1	0	1	0	0
U2	0	0	0	1
U3	0	0	0	0
U4	-1	0	0	0

Trustee Implicit Feature Matrix

Fig. 2. The example of Trust matrix factorization.

According to Eq. 2, the prediction matrix is defined as follows.

$$\hat{T} \approx P^T \times W \tag{3}$$

The loss function of trust matrix decomposition is as follows,

$$L = \frac{1}{2} \sum_u \sum_{v \in T_u} \left(w_v^T p_u - t_{u,v} \right)^2 + \frac{\lambda}{2} \left(\sum_u \|p_u\|_F^2 + \sum_v \|w_v\|_F^2 \right) \tag{4}$$

where w_v is the latent factor vector of $trustee_v$. p_u is the latent factor vector of $trustor_u$. λ is the regularization factor.

Considering the influence of rating biases, SVD++ [26] introduced bias terms into matrix factorization to create a more reasonable and balanced model. Moreover, researchers have introduced the trust factor to enhance the performance of SVD++. Its rating prediction is shown in Eq. 5,

$$\hat{r}_{u,i} = b_i + b_u + \mu + q_i^T \left(p_u + |I_u|^{-\frac{1}{2}} \sum_{j \in I_u} y_j + |T_u|^{-\frac{1}{2}} \sum_{v \in T_u} w_v \right) \tag{5}$$

where μ is the average of item ratings in the system. b_u represents the deviation of average rating by $trustor_u$ from average rating of all items and b_i represents the deviation of average rating by $item_i$ from average rating of all items. q_i and p_u represent the item implicit feature vector and user implicit feature vector respectively. $q_i^T \times p_u$ is the matrix factorization without bias. I_u is the set of items rated by $trustor_u$. y_{ij} is the implicit feedback of $item_j$ on $item_i$. $|I_u|^{-\frac{1}{2}} \sum_{j \in I_u} y_j$ represents the implicit influence of historical rating of $trustor_u$ on $item_i$. T_u is the set of users in explicit trust marked by $trustor_u$. w_v is the latent feature vector of users trusted by $trustor_u$. $|T_u|^{-\frac{1}{2}} \sum_{v \in T_u} w_v$ represents the influence of the trustee on the rating prediction. The relative objective function is as follows.

$$
\begin{aligned}
L = &\frac{1}{2} \sum_u \sum_{i \in I_u} (\hat{r}_{u,i} - r_{u,i})^2 + \frac{\lambda}{2} (\sum_u b_u^2 + \sum_i b_i^2 \\
&+ \sum_u \|p_u\|_F^2 + \sum_i \|q_i\|_F^2 + \sum_j \|y_j\|_F^2 + \sum_v \|w_v\|_F^2)
\end{aligned}
\tag{6}
$$

To extend trust, we extract the implicit influence of trusted users on the item separately. Since user scopes of trust networks and rating systems are consistent, the user feature space is shared by the rating matrix and the trust matrix under matrix factorization. Therefore, traditional rating recommendations and trust recommendations can be effectively integrated through the user feature space. The rating prediction is shown in Eq. 7.

$$
\hat{r}_{u,i}^T = q_i^T |T_u|^{-\frac{1}{2}} \sum_{v \in T_u} w_v
\tag{7}
$$

The corresponding objective function is as follows,

$$
L^T = \frac{1}{2} \sum_u \sum_{v \in T_u} \left(\hat{t}_{u,v} - t_{u,v} \right)^2
\tag{8}
$$

where $\hat{t}_{u,v} = w^T p_u$ represents the predicted trust value of $trustor_u$ to $trustee_v$ in the explicit trust network.

According to Eq. 6, the weighted regularization method is adopted after extracting trust. The objective function obtained is shown as follows,

$$
\begin{aligned}
L = &\frac{1}{2} \sum_u \sum_{i \in I_u} (\hat{r}_{u,i} - r_{u,i})^2 + \frac{\lambda_t}{2} \sum_u \sum_{v \in T_u} \left(\hat{t}_{u,v} - t_{u,v} \right)^2 \\
&+ \frac{\lambda}{2} \sum_u |I_u|^{-\frac{1}{2}} b_u^2 + \frac{\lambda}{2} \sum_i |U_i|^{-\frac{1}{2}} b_i^2 \\
&+ \sum_u (\frac{\lambda}{2} |I_u|^{-\frac{1}{2}} + \frac{\lambda_t \alpha}{2} |T_u|^{-\frac{1}{2}}) \|p_u\|_F^2 \\
&+ \frac{\lambda}{2} \sum_i |U_i|^{-\frac{1}{2}} \|q_i\|_F^2 + \frac{\lambda}{2} \sum_j |U_j|^{-\frac{1}{2}} \|y_{ij}\|_F^2 \\
&+ \frac{\lambda}{2} |T_v^+|^{-\frac{1}{2}} \|w_v\|_F^2
\end{aligned}
\tag{9}
$$

where U_i, U_j represent the sets of users rated $item_i$, $itemj$. T_v^+ is the sets of users trusted to $user_v$ in explicit trust network.

3.3 Gaussian Differential Trust Mechanism

The privacy budget spent by Moments Accountant on differential privacy is smaller than both the Naive Composition Theorem and Strong Composition Theorem when setting the same noise size [27]. Therefore, we introduce the Moments Accountant theorem into differential trust to control the privacy budget. We define differential trust as follows.

Definition 1. *Differential trust is the method of selecting a protection mechanism based on differential privacy to protect data information in social recommender systems. The process of it is shown in Fig. 3.*

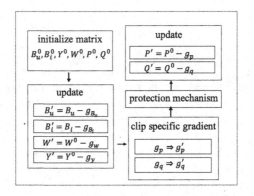

Fig. 3. The framework of differential trust.

According to differential trust, Gaussian Mechanism is applied to construct the Gaussian Differential Trust Mechanism to prevent attackers from analyzing the predicted ratings to infer whether users have rated items. Firstly, GDTM initializes user rating bias matrix B_u, item rating bias matrix B_i, user latent factor matrix P, item latent factor matrix Q, other item implicit feedback matrix Y and other user implicit feedback matrix W. It calculates matrix gradient g_{BU}, g_{BI}, g_P, g_Q, g_Y, g_W respectively by stochastic gradient descent, then updates parameters except g_P and g_Q. Then GDTM crops the gradient g_P and g_Q to satisfy differential privacy, which is formulated as follows,

$$\bar{g}_P = \frac{g_P}{\max(1, \sqrt{\|g_P\|_2^2 + \|g_Q\|_2^2}/C_{PQ})} \tag{10}$$

$$\bar{g}_Q = \frac{g_Q}{\max(1, \sqrt{\|g_P\|_2^2 + \|g_Q\|_2^2}/C_{PQ})} \tag{11}$$

where C_{PQ} is the threshold of the sum of P and Q. Finally, after using the Gaussian mechanism, GDTM calculates the mean of cropped gradients as follows,

$$\tilde{g}_P = \frac{1}{N_{now}} \left(\sum_{N_{now}} \bar{g}_P + \mathcal{N}\left(0, \sigma^2 C_{PQ}^2\right) \right) \tag{12}$$

$$\tilde{g}_Q = \frac{1}{N_{now}} \left(\sum_{N_{now}} \bar{g}_Q + \mathcal{N}\left(0, \sigma^2 C_{PQ}^2\right) \right) \tag{13}$$

where $\mathcal{N}\left(0, \sigma^2 C_{PQ}^2\right)$ is Gaussian distribution with probability density. Finally, update the user latent factor matrix P and the item latent factor matrix Q.

The algorithm uses the historical rating information and explicit trust relationship to train implicit feature matrices of users and items and then generates the predicted ratings based on their product. The sharing of user feature space closely links rating information and trust information in trust recommendation. Therefore, the differential trust mechanism is used in the user characteristic matrix solving function to protect not only the rating matrix but also the user feature in trust networks.

Algorithm 1. Gaussian Difference Trust Algorithm

Input: User rating data, Explicit trust network, Learning rate α, Total number of iterations N_{iter}, User-Item gradient, Gradient Threshold C_{PQ}

Output: Rating prediction results

1: set $r(t) = x(t)$
2: Initialize matrix: B_u, B_i, P, Q, Y, W;
3: **for** i from 1 to N_{iter} **do**
4:    Random sample N_{now} pieces of data, then calculate g_{BU}, g_{BI}, g_P, g_Q, g_Y, g_W;
5:    Update part of parameters: B_u, B_i, Y, W;
6:    Clip gradients;
7:    Add Gaussian mechanism;
8:    Update parameters:
      $P = P - \tilde{g}_P$, $Q = Q - \tilde{g}_Q$;
9: **end for**
10: Substitute the parameters into Eq. 5 to rate prediction.

---

4   Theoretical Analysis

In this section, we demonstrate the viability of our method in terms of protecting privacy. GDTM adopts the Gaussian Mechanism based on the Moments Account theorem, corresponding to approximate differential privacy $(\varepsilon, \delta) - DP$. Since the discussion on the combination theorem of differential privacy is based on privacy loss, we do not discuss the privacy budget. Therefore, we aim to demonstrate that the proposed method in this paper satisfies $(\varepsilon, \delta) - DP$.

We first explain the fundamental concept of the Moments Account theorem and then demonstrate that our approach satisfies $(\varepsilon, \delta) - DP$ [27].

Theorem 1. *Differential privacy is equivalent to constraining the difference between two distributions and measuring the difference by Max-Divergence,*

$$c(o) = \ln \frac{\Pr[\mathcal{M}(aux, X) = o]}{\Pr[\mathcal{M}(aux, X') = o]} \tag{14}$$

where aux is the auxiliary input.

Theorem 2. *The maximum constraint on the moment generating function $\gamma_{\mathcal{M}}(\lambda)$ is obtained by traversing all viable auxiliary inputs and neighboring datasets.*

$$\gamma_{\mathcal{M}}(\lambda) \triangleq \max_{aux, X, X'} \gamma_{\mathcal{M}}(\lambda; aux, X, X') \tag{15}$$

Theorem 3. *Assume that the random mechanism $\mathcal{M}$ consists of a series of adaptive mechanisms $\mathcal{M}_1, ..., \mathcal{M}_k$, where $\mathcal{M}_i : \prod_{j=1}^{i-1} \mathcal{R}_j \times \mathcal{X} \to \mathcal{R}_i$, for any λ, exists:*

$$\gamma_{\mathcal{M}}(\lambda) \leq \sum_{i=1}^{k} \gamma_{\mathcal{M}_i}(\lambda) \tag{16}$$

If differential privacy is not satisfied, use the Markov inequality to derive the moment-generating function, which is formulated as follows.

$$\Pr_{o \sim \mathcal{M}(X)}[c(o) \geq \varepsilon] = \Pr_{o \sim \mathcal{M}(X)}[e^{(\lambda c(o))} \geq e^{(\lambda \varepsilon)}]$$

$$\leq \frac{\mathbb{E}_{o \sim \mathcal{M}(X)}[e^{(\lambda c(o))}]}{e^{(\lambda \varepsilon)}} \tag{17}$$

$$\leq e^{(\gamma - \lambda \varepsilon)}$$

The privacy loss theorem of $(\varepsilon, \delta) - DP$ states that the minimum value of the moment generating function is subject to the constraint δ. The value of σ depends on the obtained in the previous step, leading to the following two theorems.

Theorem 4. *For the randomized algorithmic mechanism $\mathcal{M}$, $\forall \varepsilon > 0$, the sufficient condition for $\mathcal{M}$ to satisfy $(\varepsilon, \delta) - DP$ is*

$$\delta = \min_{\lambda} e^{(\gamma_{\mathcal{M}}(\lambda) - \lambda \varepsilon)} \tag{18}$$

Theorem 5. *For a stochastic algorithm mechanism $\mathcal{M}$, if there exist two constants c_1 and c_2, known sampling probability q, algorithm iteration number T, for any $\varepsilon < c_1 q^2 T$ and $\delta > 0$, the sufficient condition for $\mathcal{M}$ to satisfy $(\varepsilon, \delta) - DP$ is*

$$\sigma \geq c_2 \frac{q\sqrt{T \log(1/\delta)}}{\varepsilon}, \tag{19}$$

the formula for solving privacy budget ε can be obtained from the transformation of the above theorem when δ is known:

$$\varepsilon(\delta) = \min_{\lambda} \frac{\gamma_{\mathcal{M}}(\lambda) - \ln \delta}{\lambda} \tag{20}$$

Theorem 6. *Provided that the random mechanism $\mathcal{M}_1$ satisfies $(\varepsilon, \delta) - DP$, for any mechanism $\mathcal{M}_2$, $\mathcal{M}_2(\mathcal{M}_1(\cdot))$ also satisfies $(\varepsilon, \delta) - DP$.*

This is our proof process.

Proof: In each iteration, the sum of cropped gradients $\bar{g}$ satisfy:

$$\bar{g} = \sqrt{\|\bar{g}_P\|_2^2 + \|\bar{g}_Q\|_2^2} \leqslant C_{PQ} \tag{21}$$

Let $\mathcal{M}$ is the sum of cropped gradients and Gaussian Mechanism

$$\mathcal{M}(X) = \sum_{N_{now}} \bar{g} + \mathcal{N}\left(0, \sigma^2 C_{PQ}^2\right) \tag{22}$$

where $\mathcal{N}(0, \sigma^2 C_{PQ}^2)$ is the Gaussian probability density function with mean 0 and variance $\sigma^2 C_{PQ}^2$.

Assuming that

$$\mu_0 = \mathcal{N}(0, \sigma^2 C_{PQ}^2) \tag{23}$$

$$\mu_1 = \mathcal{N}(1, \sigma^2 C_{PQ}^2) \tag{24}$$

where $\mu = (1 - q)\mu_0 + q\mu_1$ and $p = N_{now}/N$. N is the total number of training data.

Let $\sum_{N_{now}} \bar{g}$ of X as the baseline. $\mathcal{M}(X)$ and $\mathcal{M}(X')$ are formulated as follows.

$$\mathcal{M}(X) \sim \mu_0 = \mathcal{N}(0, \sigma^2 C_{PQ}^2) \tag{25}$$

$$\mathcal{M}(X') \sim \mu_1 = (1 - q)\mathcal{N}(0, \sigma^2 C_{PQ}^2) + q\mathcal{N}(1, \sigma^2 C_{PQ}^2) \tag{26}$$

Let moment generating function $\gamma(\lambda) = \ln max(E_1, E_2)$,

$$E_1 = \mathbb{E}_{z \sim \mu_0}\left[(\mu_0(z)/\mu(z))^\lambda\right] \tag{27}$$

$$E_2 = \mathbb{E}_{z \sim \mu}\left[(\mu(z)/\mu_0(z))^\lambda\right] \tag{28}$$

According to Eq. 15,

$$\gamma_{\mathcal{M}_i}(\lambda) \leq \gamma(\lambda) \tag{29}$$

According to Theorem 3 and 4, both user latent factor matrix P and item latent factor matrix Q satisfy $((\varepsilon, \delta) - DP$. The rating prediction matrix is obtained by multiplying P and Q. Our method satisfies $(\varepsilon, \delta) - DP$ by Theorem 6.

5   Experiment Summary

5.1   Research Questions

To evaluate the effectiveness of GDTM, we conducted experiments to answer the following research questions:

- RQ1: To what extent can the use of the GDTM improve the accuracy of recommendation algorithms compared to the original algorithms?
- RQ2: Compared to the original algorithms, how effective is the GDTM in improving the cold start problem compared to the original recommendation algorithms?

Our experimental design consists of the following procedures. Firstly, we conduct our experiments using TensorFlow and PyTorch frameworks with two publicly available datasets: FilmTrust[1] and Douban[2]. In this paper, we treat all trust relationships as directed, regardless of the network type in the original data.

5.2   Benchmark Model

We select a set of benchmark models to evaluate the performance of GDTM, including FunkSVD [28], PMF [29], SVD++ [26], SoRec [8], RSTE [9], SocialMF [11], SoReg [10], TrustSVD [13], CUNE [12] and SSL-SVD [15]. These models are representative examples of recommendation algorithms based on matrix factorization or social recommendation. To ensure results are as reliable as possible, we consult the original papers or previous research work to select optimal parameters for each model:

- Learning Rate: 0.01
- Model Training Threshold: 0.0001
- Maximum iterations: 100
- Latent Factor Dimension: 5/10

5.3   Evaluation Metrics

The efficacy of the proposed method is reflected using the Mean Absolute Error (MAE) and the Root Mean Squared Error (RMSE).
The metric MAE is defined as:

$$\text{MAE} = \frac{1}{n} \sum_{i=1}^{n} |\hat{y}_i - y_i| \tag{30}$$

The metric RMSE is defined as:

$$\text{RMSE} = \sqrt{\frac{1}{n} \sum_{i=1}^{n} (\hat{y}_i - y_i)^2} \tag{31}$$

where $\hat{y}_i$ is the true target value for test instance y_i and n is the number of test instances.

---

[1] From a website where we can download the data set directly. (http://trust.mindswap.org/FilmTrust).

[2] From an anonymized Douban dataset. Two files are included in this Douban dataset, the user-item rating file and the user social friend network file. (https://www.cse.cuhk.edu.hk/irwin.king.new/pub/data/douban).

Table 1. Comparison of results before and after differential trust algorithm for all user mode in the FilmTrust dataset. w/o is the algorithm without GDTM. w/ is the algorithm with GDTM.

Algorithm (d = 5)	Evaluation metrics			
	w/o MAE	w MAE	w/o RMSE	w RMSE
SVD++	0.6471	1.4289	0.8638	1.8219
SoRec	0.6655	1.5904	0.8761	1.9547
SocialMF	0.6586	1.4968	0.8606	1.8680
TrustSVD	0.6524	1.9627	0.8458	2.2660
CUNE	0.6731	0.8663	0.8542	1.1664
SSL-SVD	**0.5646**	**0.6019**	**0.7308**	**0.7505**

5.4 Performance Analysis

RQ1: Recommendation Accuracy Comparison

To evaluate the efficacy of our proposed method, we conduct a comparative analysis between the original algorithms and those with GDTM in both all user mode and cold start mode, using the FilmTrust dataset, as presented in Table 1 and Table 2. The results demonstrate that SSL-SVD with GDTM outperforms other algorithms, particularly in all user mode. Our speculation is that SSL-SVD rightfully mines sparse trust between users to efficiently leverage the effects of implicit and explicit trust, resulting in an enhanced recommendation accuracy and increased trust density degree of datasets, thereby decreasing the impact of noise. We find it intriguing that TrustSVD with GDTM performed worse than other algorithms, whereas the original TrustSVD did not register such a decrease in performance when $d = 5$. TrustSVD with GDTM improves when $d = 10$, as seen from Fig. 4, which suggests that the latent factor vector's dimensions significantly impact some algorithms' performance and its optimal size may vary from one algorithm to another.

RQ2: Cold Start Improvement Capability

In the analysis of research question 2 (RQ2), we refer to Fig. 4. It can be concluded that Sorec, SocialMF, and TrustSVD with GDTM perform better in cold start mode than in all user mode when $d = 5$. These improved results suggest that the three algorithms can reduce the cold start problem by incorporating differential trust. This improvement might be due to social factors being unaffected by trust factors, leading to a more effective recommendation in cold start mode. Conversely, other algorithms with GDTM perform poorly in cold start mode. Although the original SVD++ mitigates the cold start problem effectively, SVD++ with GDTM is unable to do the same. This could be attributed to the fact that SVD++ does not belong to the trust recommendation model, the impact of differential trust cannot be compensated for by the trust factor. Combined with the rest of the subgraphs, due to original SSL-SVD's strong cold-start response ability, the performance difference between it and SSL-SVD with

Table 2. Comparison of results before and after differential trust algorithm for cold start mode in the FilmTrust dataset. w/o is the algorithm without GDTM. w/ is the algorithm with GDTM.

Algorithm (d = 5)	Evaluation metrics			
	w/o MAE	w MAE	w/o RMSE	w RMSE
SVD++	0.6352	1.6916	0.8765	2.1050
SoRec	0.7294	1.4797	0.9184	1.9088
SocialMF	0.7193	1.4635	0.9129	1.8958
TrustSVD	0.6611	1.8956	0.8542	2.2105
CUNE	0.6840	0.8957	0.8615	1.2596
SSL-SVD	**0.5091**	**0.6059**	**0.6450**	**0.7513**

GDTM is minimal in cold start mode. In summary, our proposed method has the ability to cope with the cold start problem.

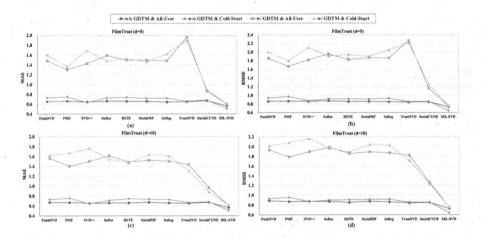

Fig. 4. Comparison of mode before and after GTDM algorithm in FilmTrust dataset. (a) is MAE and (b) is RMSE when d = 5. (c) is MAE and (d) is RMSE when d = 10.

6 Conclusion

The recommender systems provide users with the services in a convenient manner, but it is difficult to break through the bottleneck of data sparsity and cold start. Although the social recommendations have addressed the aforementioned issues to some extent, their protective measures inevitably greatly weaken the accuracy of recommendation systems. To address the mentioned issue, we propose the differential trust and extend it to propose the Gaussian Differential

Trust Mechanism to protect users' own private information. Through theoretical explanations, we have demonstrated the feasibility of our method in protecting privacy. Our experiment results also indicate that the proposed protective mechanism makes an efficient trade-off between security and accuracy.

For future work, we intend to enhance the current trade-off between privacy protection and recommendation accuracy.

Acknowledgements. This work is supported in part by the National Key RD Program of China under No. 2022YFB3102100, the National Science Foundation of China under Grants U22B2027, 62172297 and 61902276, the Key Research and Development Project of Sichuan Province under Grant 2021YFSY0012, Tianjin Intelligent Manufacturing Special Fund Project under Grants 20211097, and Guangdong Provincial Key Laboratory of Novel Security Intelligence Technologies under No. 2022B1212010005.

References

1. Gao, R., Shah, C.: Counteracting bias and increasing fairness in search and recommender systems. In: Proceedings of the 14th ACM Conference on Recommender Systems, pp. 745–747 (2020)
2. Gong, N.Z., Liu, B.: Attribute inference attacks in online social networks. ACM Trans. Priv. Secur. (TOPS) **21**(1), 1–30 (2018)
3. Xu, G., et al.: Soprotector: safeguard privacy for native so files in evolving mobile IoT applications. IEEE Internet Things J. **7**(4), 2539–2552 (2020). https://doi.org/10.1109/JIOT.2019.2944006
4. Pei, F., He, Y.W., Yan, A., Zhou, M., Chen, Y.W., Wu, J.: A consensus model for intuitionistic fuzzy group decision-making problems based on the construction and propagation of trust/distrust relationships in social networks. Int. J. Fuzzy Syst. **22**, 2664–2679 (2020)
5. Ouaddah, A., Mousannif, H., Abou Elkalam, A., Ouahman, A.A.: Access control in the internet of things: big challenges and new opportunities. Comput. Netw. **112**, 237–262 (2017)
6. Zhang, Q., Lu, J., Jin, Y.: Artificial intelligence in recommender systems. Complex Intell. Syst. **7**, 439–457 (2021)
7. Ozsoy, M.G., Polat, F.: Trust based recommendation systems. In: Proceedings of the 2013 IEEE/ACM International Conference on Advances in Social Networks Analysis and Mining, pp. 1267–1274 (2013)
8. Ma, H., Yang, H., Lyu, M.R., King, I.: Sorec: social recommendation using probabilistic matrix factorization. In: Proceedings of the 17th ACM Conference on Information and Knowledge Management, pp. 931–940 (2008)
9. Ma, H., King, I., Lyu, M.R.: Learning to recommend with social trust ensemble. In: Proceedings of the 32nd International ACM SIGIR Conference on Research and Development in Information Retrieval, pp. 203–210 (2009)
10. Ma, H., Zhou, D., Liu, C., Lyu, M.R., King, I.: Recommender systems with social regularization. In: Proceedings of the Fourth ACM International Conference on Web Search and Data Mining, pp. 287–296 (2011)
11. Jamali, M., Ester, M.: A matrix factorization technique with trust propagation for recommendation in social networks. In: Proceedings of the Fourth ACM Conference on Recommender Systems, pp. 135–142 (2010)

12. Zhang, C., Yu, L., Wang, Y., Shah, C., Zhang, X.: Collaborative user network embedding for social recommender systems. In: Proceedings of the 2017 SIAM International Conference on Data Mining, pp. 381–389. SIAM (2017)
13. Guo, G., Zhang, J., Yorke-Smith, N.: TrustSVD: collaborative filtering with both the explicit and implicit influence of user trust and of item ratings. In: Proceedings of the AAAI Conference on Artificial Intelligence, vol. 29 (2015)
14. Xu, X., Yuan, D.: A novel matrix factorization recommendation algorithm fusing social trust and behaviors in micro-blogs. In: 2017 IEEE 2nd International Conference on Cloud Computing and Big Data Analysis (ICCCBDA), pp. 283–287. IEEE (2017)
15. Hu, Z., et al.: SSL-SVD: semi-supervised learning-based sparse trust recommendation. ACM Trans. Internet Technol. (TOIT) 20(1), 1–20 (2020)
16. Huang, W., Liu, B., Tang, H.: Privacy protection for recommendation system: a survey. In: Journal of Physics: Conference Series, vol. 1325, p. 012087. IOP Publishing (2019)
17. Friedman, A., Berkovsky, S., Kaafar, M.A.: A differential privacy framework for matrix factorization recommender systems. User Model. User-Adap. Inter. 26, 425–458 (2016)
18. Shin, H., Kim, S., Shin, J., Xiao, X.: Privacy enhanced matrix factorization for recommendation with local differential privacy. IEEE Trans. Knowl. Data Eng. 30(9), 1770–1782 (2018). https://doi.org/10.1109/TKDE.2018.2805356
19. Zhang, S., Liu, L., Chen, Z., Zhong, H.: Probabilistic matrix factorization with personalized differential privacy. Knowl.-Based Syst. 183, 104864 (2019)
20. Zhou, H., Yang, G., Xiang, Y., Bai, Y., Wang, W.: A lightweight matrix factorization for recommendation with local differential privacy in big data. IEEE Trans. Big Data 9(1), 160–173 (2023). https://doi.org/10.1109/TBDATA.2021.3139125
21. Xu, G., et al.: TT-SVD: an efficient sparse decision-making model with two-way trust recommendation in the AI-enabled IoT systems. IEEE Internet Things J. 8(12), 9559–9567 (2020)
22. Wang, F., Zhu, H., Srivastava, G., Li, S., Khosravi, M.R., Qi, L.: Robust collaborative filtering recommendation with user-item-trust records. IEEE Trans. Comput. Soc. Syst. 9(4), 986–996 (2021)
23. Fang, H., Guo, G., Zhang, J.: Multi-faceted trust and distrust prediction for recommender systems. Decis. Support Syst. 71, 37–47 (2015)
24. Nie, P., Xu, G., Jiao, L., Liu, S., Liu, J., Meng, W., Wu, H., Feng, M., Wang, W., Jing, Z., et al.: Sparse trust data mining. IEEE Trans. Inf. Forensics Secur. 16, 4559–4573 (2021)
25. Cutillo, L.A., Molva, R., Strufe, T.: Safebook: a privacy-preserving online social network leveraging on real-life trust. IEEE Commun. Mag. 47(12), 94–101 (2009)
26. Koren, Y.: Factor in the neighbors: scalable and accurate collaborative filtering. ACM Trans. Knowl. Discov. Data (TKDD) 4(1), 1–24 (2010)
27. Abadi, M., et al.: Deep learning with differential privacy. In: Proceedings of the 2016 ACM SIGSAC Conference on Computer and Communications Security, pp. 308–318 (2016)
28. Funk, S.: Netflix update: try this at home (2006)
29. Mnih, A., Salakhutdinov, R.R.: Probabilistic matrix factorization. In: Advances in Neural Information Processing Systems, vol. 20 (2007)

Privacy-Enhanced Dynamic Symmetric Searchable Encryption with Efficient Searches Under Sparse Keywords

Lingyun Cao[✉] and Xiang Li

School of Software Engineering, Tongji University, Shanghai 201804, China
68cptn@gmail.com

Abstract. Dynamic Searchable Symmetric Encryption (DSSE) enables users to search and update an encrypted database stored on a semi-honest server. It preserves the confidentiality of the data. Currently, many works focus on achieving forward and backward privacy and efficiency performance. However, in sparse keywords scenarios, a large document collection consists of a small number of distinct keywords, it is still challenging to design DSSE schemes that achieve both high security and practical performance. To address these issues, we propose a scheme called SGsse-F, which satisfies forward update privacy (FuP) and weak forward search privacy (FsP), and a scheme named SGsse-FB that satisfies FuP, weak FsP and TYPE-II backward privacy (BP). Both schemes employ partitioning technique, hash and pseudo-random functions to construct encryption entities and effectively support parallel keyword search. Experimental results demonstrate that the SGsse-F and SGsse-FB schemes exhibit better search performance compared to Dual, Mitra, FSSE and FSSE*. Additionally, the client storage for both schemes is suitable.

Keywords: Dynamic Searchable Symmetric Encryption · Forward Update Privacy · Forward Search Privacy · Backward Privacy

1 Introduction

With the continuous generation of a large amount of data be stored in databases, data security is becoming increasingly important. While encryption helps protect the security of user data [1], secure and efficient searches on encrypted data remains a significant challenge. To protect user privacy in the data and enable efficient querying, researchers have proposed DSSE [6], which enables efficient keyword searching in encrypted indexes. It allows the server to store encrypted indexes and documents, but the server has minimal leaked information about the indexes and encrypted documents.

The DSSE system framework is illustrated in Fig. 1. It involves client and server. The client is assumed to be honest and does not leak any information, while the server is semi-honest (honest but curious). The client uploads the encrypted files and update tokens to the server. During the query process, the

© The Author(s), under exclusive license to Springer Nature Singapore Pte Ltd. 2024
Z. Tari et al. (Eds.): ICA3PP 2023, LNCS 14492, pp. 93–113, 2024.
https://doi.org/10.1007/978-981-97-0811-6_6

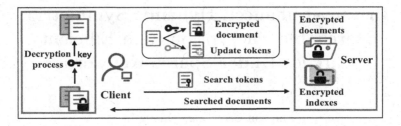

Fig. 1. System model

client generates search tokens using the key and the query keywords, then sends tokens to the server. The server utilizes the encrypted indexes and search tokens to compute and returns the corresponding file identifiers and encrypted files to the client. In our system model, inverted indexes [5] are employed to enable efficient search queries in the format of (*wkey*, *wvalue*) pairs. *wkey* represents a keyword, and *wvalue* comprises a list of document identifiers that contain specific keyword. Utilizing keyword tokens, we can efficiently retrieve all the documents that match the given keyword on the server-side. The system model of DSSE about document encryption and decryption process are not the primary focus of this study, as they are inherent parts in DSSE schemes.

Based on the system model, this study researches sparse keyword queries (In Sect. 2.1, sparse keywords can be described as $|DB|$ is large, while $|DB(W)|$ is relatively smaller) on personal privacy data. For example, as preference queries, they are defined as queries that prioritize specific parts of a certain attribute field (a attribute field represents one column of data) in a database. These type of queries are similar to "LIKE" queries commonly used in databases such as MySQL. In scenario, the Chinese ID number has unique characteristics, and different parts of the ID number have specific meanings. For instance, a randomly generated ID number "310109197502259352" from the Faker library [2], "31" represents Shanghai, "3101" represents the urban district of Shanghai, and "310109" represents the Hongkou District of Shanghai. When a database stores Chinese ID numbers, data owners may prefer to query who belongs to a specific province or who shares the same birthdate. Data owners require efficient queries with minimal information leakage. The challenges in sparse keywords scenarios are as follows:

- Sparse keywords means a smaller plaintext keywords space. This situation makes the system vulnerable to file injection attacks [11,12] and keywords statistical attacks based on the DSSE scheme.
- As the number of data records in the database increases, the list of document identifiers corresponding to search keywords become longer when using inverted indexes. Efficient and secure searches are required.

Our contributions are as follows:

- We present the SGsse-F scheme that satisfies FuP and weak FsP and the scheme SGsse-FB, which satisfies FuP, weak FsP, and Type-II BP. The

Table 1. Comparison Several SSE Schemes.

Scheme	FuP	FsP	BP	Computation		Communication		Storage	
				Search	Update	Search	Update	Client	Server
Dual [29]	✓	✗	✗	$O(a_w - d_w)$	$O(1)$	$O(n_w)$	$O(1)$	$O(KlogD)$	$O(N)$
Khon [22]	✓	✓	III	$O(n_w)$	$O(1)$	$O(n_w)$	$O(1)$	$O(mlogD + DlogK)$	-
FSSE [14]	✓	✗	✗	$O(a_w)$	$O(1)$	$O(a_w)$	$O(1)$	$O(KlogD)$	$O(N^+)$
Mitra [18]	✓	✗	II	$O(a_w)$	$O(1)$	$O(a_w)$	$O(1)$	$O(KlogD)$	$O(N^+)$
Bestie [24]	✓	✗	III	$O(a_w)$	$O(1)$	$O(a_w - d_w)$	$O(1)$	$O(KlogD)$	$O(N)$
Fides [16]	✓	✗	II	$O(a_w)$	$O(1)$	$O(a_w)$	$O(1)$	$O(KlogD)$	$O(N^+)$
Diana$_{del}$ [16]	✓	✗	III	$O(a_w)$	$O(loga_w)$	$O(n_w + d_w loga_w)$	$O(1)$	$O(KlogD)$	-
Janus [16]	✓	✗	III	$O(n_w a_w)$	$O(1)$	$O(n_w)$	$O(1)$	$O(KlogD)$	-
TWORAM [19]	✓	✗	✗	$\tilde{O}(a_w logN + log^3 N)$	$\tilde{O}(log^2 N)$	$\tilde{O}(a_w logN + log^3 N)$	$\tilde{O}(log^3 N)$	$O(1)$	$O(p + N)$
SGsse-F	✓	✓	✗	$O(a_w - d_w)$	$O(1)$	$O(n_w)$	$O(1)$	$O(KlogD)$	$O(N)$
SGsse-BF	✓	✓	II	$O(a_w)$	$O(1)$	$O(a_w)$	$O(1)$	$O(KlogD)$	$O(N^+)$

* K is the total number of distinct keywords, m is the total number of sub-keywords, D is the total number of documents. N is the total number of keyword/document pairs and p is the number of keywords in the database. For keyword w, n_w is the number of matched query, a_w is total number of updates, d_w is the number of deleted entries for w. N^+ is total keyword/document pairs historically stored in the database. FuP is forward update privacy. FsP is Forward search Privacy. BP is backward privacy. The partial comparative results are quoted from the references [22, 29].

SGsse-F scheme supports direct physical deletion, while the SGsse-FB scheme achieves logical deletion.

- To reduce information leakage, the SGsse-F and SGsse-FB schemes divide the inverted index into separate partitions and assign a sub-keyword to each partition. Both schemes utilize these sub-keywords to construct partitions that satisfy privacy security. partitions are independent of each other, enabling efficient parallel keyword search on longer inverted index. Furthermore, the client-side storage requirements of both schemes are suitable.
- Our proposed schemes, SGsse-F and SGsse-FB, outperform existing solutions, particularly in terms of query efficiency for longer inverted index. As indicated in Table 1, our schemes offer nearly optimal complexity and security from every perspective, in comparison to other approaches.

2 Preliminaries

2.1 Notations

We denote the security parameters $\lambda, \iota, \varsigma \in \mathbb{N}$. $k \xleftarrow{\$} \{0,1\}^\lambda$ represents uniformly random sample generated from λ bits of 0 or 1. PPT is probabilistic polynomial-time. For a set M, $|M|$ means the cardinality of the set M. Let m be one entry of set M, M/m means the set M drop out m. Operator $\parallel$ denotes the concatenation of strings.

A function $negl : \mathbb{N} \to \mathbb{N}$ is negligible in λ, for all positive polynomial function $p(\lambda)$ where λ is large, then $negl(\lambda) < 1/p(\lambda)$.

The document f is defined as (ind, W_{ind}), where ind represents the document identifier and W_{ind} represents the set of distinct keywords in the document. The database is denoted as $DB = (ind_i, W_{ind_i})_{i=1}^n$. The set of all keywords in the database is $DB(W) = \cup_{i=1}^n W_{ind_i}$, and the set of documents containing the same keyword w is represented as $DB(w) = \{ind_i | w \in W_{ind_i}\}$.

Pseudo-random functions (PRF). Let G be a pseudo-random function family: $\{0,1\}^\lambda \times \{0,1\}^\iota \to \{0,1\}^\varsigma$. $K \leftarrow \{0,1\}^\lambda$ and $G_K(x)$ is a secure pseudo-random function under probabilistic polynomial-time (PPT) and advantage Adv, an adversary cannot break the pseudo-random function $G_K(x)$: $|Pr[Adv_K^G(\cdot)(1^\lambda) = 1]$-$Pr[Adv^{F(\cdot)}(1^\lambda) = 1]| \le negl(\lambda)$, where F is a random function:$\{0,1\}^\iota \to \{0,1\}^\varsigma$.

2.2 Dynamic Searchable Symmetric Encryption

A DSSE scheme $\Pi =$ (Setup, Search, Update) contains three algorithms [22]. The descriptions are as follows:

- **Setup(λ):** The client generates the encryption key K_Π using the security parameter λ and initializes the storage EDB of encrypted data, the encryption key K_Π stores in local state σ.
- **Search(σ, w; EDB):** In client, input local state σ and search keyword w to generate search tokens. The server utilizes the search tokens and the encrypted data database EDB to compute and return the corresponding file identifiers or encrypted files.

For forward privacy scheme, the algorithm Update consists of Add and Del processes [29]:

- **Add(σ, W_{ind}, ind; EDB):** When the client inputs the document f consists of its identifier ind and the distinct keywords W_{ind}, it uses the distinct keywords W_{ind} in f to generate addition tokens, then updates the local state and transmits the addition tokens to the server's EDB.
- **Del(σ, ind; EDB):** The client generates deletion tokens using the document identifier ind to remove all keywords associated with the document. The server uses the deletion tokens to delete the keywords in EDB.

For both forward and backward privacy scheme, the Update algorithm [18] is as follows:

- **Update(σ, w, op, ind; EDB):** The client generates update tokens with the encryption key K_Π, client-stored state information σ, keyword w, operation type op and the corresponding file identifier ind. The server adds update tokens to the encrypted data database EDB.

Adaptive Security of DSSE. A commonly approach for defining the adaptive security of DSSE is by establishing the indistinguishability for adversaries between a real DSSE game and an ideal DSSE game. In these two games, adversaries can adaptively perform update and queries operations. In the real game, actual data and protocols are used, whereas in the ideal game, a simulator responds to adversary's queries using only leakage functions [24].

Definition 1:A DSSE scheme Π involves the leakage functions $\mathcal{L} = (\mathcal{L}_{Setup}, \mathcal{L}_{Search}, \mathcal{L}_{Update})$. It is $\mathcal{L}$-adaptively secure with parameter λ if for *any PPT adversary* $\mathcal{A}$, *there exists a PPT simulator* $\mathcal{S}$, *satisfies that* $|Pr[\boldsymbol{Real}_{\mathcal{A}}^{\Pi}(\lambda) = 1] - Pr[\boldsymbol{Ideal}_{\mathcal{A},\mathcal{S},\mathcal{L}}^{\Pi}(\lambda) = 1]| \leq negl(\lambda)$, *where* $\boldsymbol{Real}_{\mathcal{A}}^{\Pi}$ *and* $\boldsymbol{Ideal}_{\mathcal{A},\mathcal{S},\mathcal{L}}^{\Pi}$ *are as below:*

- $\boldsymbol{Real}_{\mathcal{A}}^{\Pi}$: *The DSSE scheme run on real data and protocols. The adversary* $\mathcal{A}$ *obtains the EDB created using the Setup algorithm.* $\mathcal{A}$ *can choose a keyword* w *and the corresponding query tokens generated using the keyword* w *and Search algorithm, then the tokens return to* $\mathcal{A}$. *In the Update process,* $\mathcal{A}$ *choose a document, then the Update algorithm is executed. Then encrypted date will be return to* $\mathcal{A}$.
- $\boldsymbol{Ideal}_{\mathcal{A},\mathcal{S},\mathcal{L}}^{\Pi}$: *The adversary* $\mathcal{A}$ *interacts with a simulated secure DSSE protocol. The ideal game constructs a PPT simulator* $\mathcal{S}$ *who is aware of the leakage function* $\mathcal{L}$. $\mathcal{S}$ *initializes the corresponding local state and encrypted date using the leakage function* $\mathcal{L}_{Setup}$, *then send encrypted date to* $\mathcal{A}$. $\mathcal{S}$ *constructs update transcripts through* $\mathcal{L}_{Update}$ *and returns them to* $\mathcal{A}$. *When* $\mathcal{A}$ *performs search operations, the transcripts generated from* $\mathcal{L}_{Search}$ *by the simulator will be received.*

The adversary $\mathcal{A}$ *interacts with the two games, analyzes the corresponding return results, and Eventually outputs a bit 0 or 1.*

Forward Privacy. In terms of forward update privacy, A query state $\mathcal{Q}$ keeps track of the query keyword w and its timestamp j [13]. For a query keyword w, the search pattern based on the query state $\mathcal{Q}$ is defined as: $sp(w) = \{t|(t,w) \in \mathcal{Q}\}$. Let $TimeDB(w)$ contains all documents (excluding deleted documents) that match w, along with their insertion timestamps in the database [18]. And $w_1, ..., w_i$ are the derived sub-keywords from w. We follow the formal definition in [22].

Definition 2: A $\mathcal{L}$*-adaptively secure DSSE scheme is forward update privacy and weak forward search privacy, if the update leakage function* $\mathcal{L}_{update}$ *and the search leakage function* $\mathcal{L}_{search}$ *can be defined as:*

$$\mathcal{L}_{Update}(ind, W_{ind}) = \mathcal{L}'(ind, |W_{ind}|)$$
$$\mathcal{L}_{Search}(w_i) = \mathcal{L}''(TimeDB(w_i))$$

where $\mathcal{L}'$ *and* $\mathcal{L}''$ *are stateless.*

Backward Privacy. The updates on keyword w are defined as $Updates(w)$, which is a list of timestamps for updates on w [18].

Definition 3: An $\mathcal{L}$*-adaptively secure DSSE scheme satisfies TYPE-II backward privacy, such that:*

$$\mathcal{L}_{Update}(op, w, ind) = \mathcal{L}'(op, w)$$
$$\mathcal{L}_{Search}(w) = \mathcal{L}''(TimeDB(w), Updates(w))$$

where $\mathcal{L}'$ *and* $\mathcal{L}''$ *are stateless.*

3   Forward Privacy DSSE: SGsse-F

We utilize Partitioning technique [22] to partition the keyword-document identifiers inverted index into multiple partitions. Each partition is associated with a partition identifier. We construct a sub-keyword using a keyword w_1 and the block identifier $part_{cnt}$ under PRF. For each partition, there is a maximum insertion limit, and when the number of insertions exceeds this limit, we insert the data into the next block. Additionally, we keep track of the insertion status add_{cnt} for the current partition. This state add_{cnt} is associated with the sub-keywords and ensures that the inserted data in current partition satisfies FuP. This construction is well-suited for utilizing multiple threads to perform query operations on different partitions concurrently. To achieve FsP, we define a search state $srch_{cnt}$. After each completion of a keyword query, we set $srch_{cnt}$ to $part_{cnt}$ to indicate that the $srch_{cnt}$ partitions have been queried. If a new entity is added, it needs to be inserted into the next available partition, such that the queried blocks and the newly added partition will no longer be associated with each other, thereby satisfying the FsP security. The above usage idea is shown in Fig. 2.

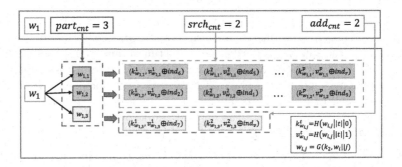

Fig. 2. Basic idea

3.1   Construction of SGsse-F

The SGsse-F scheme consists of four algorithms. The algorithms are described as follows:

Setup: In Algorithm 1, it takes the security parameter λ as input and generates the keys. The client initializes three empty maps (Cnt, EDB, GDB) and the maximum size of the partition. Specifically, EDB actually consists of two storage entities: $edb\_file\_key, edb\_key\_value$, more like Dual [29].

Add: As shown in Algorithm 1, key_{ind} is computed for the document identifier ind using the PRF function F and k_1, and the file deletion state cnt_{ind} is initialized (line 3–4). For each keyword in the set W_{ind}, the client updates the state of ($part_{cnt}, search_{cnt}, add_{cnt}$) based on whether w is included in Cnt (line 7–10). Then the partition identifier $part_{cnt}$ and the insertion count add_{cnt} are updated, if the entity inserted in the partition reaches the limit length or if the current partition has been queried (line 11–13). Otherwise, only the insertion count add_{cnt}

is updated (line 15). Using k_2, w, and $part_{cnt}$, the sub-keyword w_i is generated. The client calculates two hashes, $H(w_i\|add_{cnt}\|0)$ and $H(w_i\|add_{cnt}\|1)$, where the first hash is used as k. The second hash is XORed with ind, and the XOR result v becomes the value for k. For the current addition keyword w, fst for key_{ind} and cnt_{ind} is computed. When the server receives $AddTk$, it parses one of entry in $AddTk$ as (fst, k, v), then edb_file_key stores (fst, k), edb_key_value stores (k, v) (line 2). Line 7 of algorithm Add, "Find" function means using keyword w to check if this value exists in Cnt.

Algorithm 1. SGsse-F.Setup and Add

Setup(λ)

Client:

1: $k_1, k_2, k_3 \xleftarrow{\$} \{0,1\}^{\lambda}$
2: $Cnt \leftarrow \varnothing$
3: $EDB \leftarrow \varnothing$
4: $GDB \leftarrow \varnothing$
5: Initialize p as partition size
6: $\sigma \leftarrow (Cnt, k_1, k_2, k_3, p)$
7: send EDB, GDB to Server

Add(σ, ind, W_{ind}; EDB)

Client:

1: $AddTK \leftarrow \varnothing$
2: $(part_{cnt}, srch_{cnt}, add_{cnt}) \leftarrow (0, 0, 0)$
3: $key_{ind} \leftarrow F(k_1, ind)$
4: $cnt_{ind} \leftarrow 1$
5: **while** $|W_{ind}| \neq \varnothing$ **do**
6: $\quad w \leftarrow W_{ind}$; $W_{ind} \leftarrow W_{ind}/w$
7: $\quad$ **if** $Cnt.\text{Find}(w) = \bot$ **then**
8: $\quad\quad (part_{cnt}, srch_{cnt}, add_{cnt}) \leftarrow (1, 0, 1)$
9: $\quad$ **else**

10: $\quad\quad (part_{cnt}, srch_{cnt}, add_{cnt}) \leftarrow Cnt[w]$
11: $\quad\quad$ **if** $add_{cnt} \bmod p = 0$ or $part_{cnt} = srch_{cnt}$ **then**
12: $\quad\quad\quad part_{cnt} \leftarrow part_{cnt} + 1$
13: $\quad\quad\quad add_{cnt} \leftarrow 1$
14: $\quad\quad$ **else**
15: $\quad\quad\quad add_{cnt} \leftarrow add_{cnt} + 1$
16: $\quad\quad$ **end if**
17: $\quad$ **end if**
18: $\quad w_i \leftarrow G(k_2, w\|part_{cnt})$
19: $\quad k \leftarrow H(w_i\|add_{cnt}\|0)$
20: $\quad v \leftarrow H(w_i\|add_{cnt}\|1) \oplus ind$
21: $\quad fst \leftarrow H(key_{ind}\|cnt_{ind})$
22: $\quad cnt_{ind} \leftarrow cnt_{ind} + 1$
23: $\quad Cnt[w] \leftarrow (part_{cnt}, srch_{cnt}, add_{cnt})$
24: $\quad AddTK \leftarrow AddTK \cup (fst, k, v)$
25: **end while**
26: $\sigma \leftarrow Cnt[w], k_1, k_2, p$
27: send $AddTK$ to server

Server:

1: receive $AddTK$
2: $EDB \leftarrow EDB \cup AddTK$

Search: In Algorithm 2, we add GDB to store searched keywords information in partitions. Therefore we use Search $(\sigma, w; EDB, GDB)$ instead of Search(σ, w; EDB). The client utilizes the search keyword w to get the state from the local storage (line 3). The client initializes cnt_{srch} as p for the current query partition (line 8). then the client computes the sub-keyword w_i and I_{w_i} (line 9–10). If the current partition has not been queried before, the state cnt_{srch} is updated as add_{cnt}(line 11–12). Since all blocks of this keyword have been queried, the $srch_{cnt}$ is updated to $part_{cnt}$(line 16). In server, for each query token tuple $(w_i, I_{w_i}, cnt_{srch})$, if I_{w_i} not exists in GDB, (I_{w_i}, cnt_{srch}) is inserted into GDB. Then, if it is successfully got, cnt_{srch} is updated to $GDB[I_{w_i}]$ (line 5–8). The hash values $H(w_i\|j\|0)$ for j from 1 to cnt_{srch} are computed, and using these values, st_v can be obtained from EDB(line 10–13). st_v is XORed with $H(w_i||j||1)$ to obtain the document identifier data ind, which will be added to res. Line 2 of algorithm in the client about "Find" is to check w if exist in Cnt and the same with line 4 in the server-side. Line 12 of Search in the server,"Find" function is showed as using st_k to check it if exist in edb_key_value, and in line 13, the server uses st_k to get st_v in edb_key_value. In Search Algorithm 2, line 7–15 of client and line 3–18 of server can be decomposed into concurrent queries, and each thread can handle a sub-keyword query.

Del: The Del process in Algorithm 2 is identical to Dual [29]. The client computes ind to obtain fst. In server, Using fst, all the keywords in document ind can be deleted. In algorithm Del, the server uses fst to search in edb_file_key (line 5) and the process of line 8 is two steps: first step, the server uses fst to get it's value in edb_file_key and then remove the pair; step two, uses the value to delete the record in edb_key_value.

Algorithm 2. SGsse-F.Search and Del

Search(σ,w; EDB,GDB)

Client:

1: $srchTK \leftarrow \varnothing$
2: **if** $Cnt.\text{Find}(w) \neq \bot$ **then**
3: $\quad (part_{cnt}, srch_{cnt}, add_{cnt}) \leftarrow Cnt[w]$
4: **else**
5: $\quad$ return
6: **end if**
7: **for** $i \leftarrow 1$ to $part_{cnt}$ **do**
8: $\quad cnt_{srch} \leftarrow p$
9: $\quad w_i \leftarrow G(k_2, w||i)$
10: $\quad I_{w_i} \leftarrow G(k_3, w||i)$
11: $\quad$ **if** $i = part_{cnt}$ and $part_{cnt} \neq srch_{cnt}$ **then**
12: $\quad\quad cnt_{srch} \leftarrow add_{cnt}$
13: $\quad$ **end if**
14: $\quad srchTK \leftarrow srchTK \cup (w_i, I_{w_i}, cnt_{srch})$
15: **end for**
16: $srch_{cnt} \leftarrow part_{cnt}$
17: $Cnt[w] \leftarrow (part_{cnt}, srch_{cnt}, add_{cnt})$
18: $\sigma \leftarrow Cnt[w], k_2, k_3, p$
19: send $srchTK$ to server

Server:

1: receive $srchTK$
2: $res \leftarrow \varnothing$
3: **for** each in $srchTK$ **do**
4: $\quad (w_i, I_{w_i}, cnt_{srch}) \leftarrow each$
5: $\quad$ **if** $GDB.\text{Find}(I_{w_i}) \neq \bot$ **then**
6: $\quad\quad cnt_{srch} \leftarrow GDB[I_{w_i}]$

7: $\quad$ **else**
8: $\quad\quad GDB[I_{w_i}] \leftarrow cnt_{srch}$
9: $\quad$ **end if**
10: $\quad$ **for** $j \leftarrow 1$ to cnt_{srch} **do**
11: $\quad\quad st_k \leftarrow H(w_i||j||0)$
12: $\quad\quad$ **if** $EDB.\text{Find}(st_k) \neq \bot$ **then**
13: $\quad\quad\quad st_v \leftarrow EDB.\text{Get}(st_k)$
14: $\quad\quad\quad ind \leftarrow st_v \oplus H(w_i||j||1)$
15: $\quad\quad\quad res \leftarrow res \cup ind$
16: $\quad\quad$ **end if**
17: $\quad$ **end for**
18: **end for**
19: send res to Client

Del(σ, ind; EDB)

Client:

1: $key_{ind} \leftarrow F(k_1, ind)$
2: send key_{ind} to Server

Server:

1: receive key_{ind}
2: $cnt_{ind} \leftarrow 1$
3: **while** 1 **do**
4: $\quad fst \leftarrow H(key_{ind}||cnt_{ind})$
5: $\quad$ **if** $EDB.\text{Find}(fst) = \bot$ **then**
6: $\quad\quad$ break
7: $\quad$ **else**
8: $\quad\quad$ remove fst in EDB
9: $\quad\quad cnt_{ind} \leftarrow cnt_{ind} + 1$
10: $\quad$ **end if**
11: **end while**

3.2 An Example of SGsse-F

We introduce an example to clearly explain the algorithm of SGsse-F with the following two parts.

In Fig. 3 The keyword w_1 includes three sub-keywords, and $w_{1,1}$ and $w_{1,2}$ have been queried, while $w_{1,3}$ remains unqueried. When file ind_v is added, the client sets add_{cnt} to 3 and adds the $(k^3_{w_{1,3}}, v^{3,v}_{w_{1,3}})$ pair to the server's EDB. When the keyword w_1 is queried again, the client generates query tokens based on $Cnt[w_1]$. Then the server retrieves the current entity numbers p for $w_{1,1}$ in its corresponding partition from GDB with $I_{1,1}$, and similarly, the entity numbers for $w_{1,2}$ in its partition as q (despite some entities have been deleted), then inserts $(I_{w_{1,2}}, 3)$ into GDB. By utilizing $(w_{1,1}, p)$, $(w_{1,2}, q)$, and $(w_{1,3}, 3)$, document identifiers can be obtained. Finally, the state of $Cnt[w_1]$ will be change to $(srch_{cnt} = 3, add_{cnt} = 1)$. Due to the physical deletion involved in SGsse-F, GDB reduces the server-side data query time complexity from $(3*p)$ to $p + q + 3$, where p is the partition's maximum insertion limit.

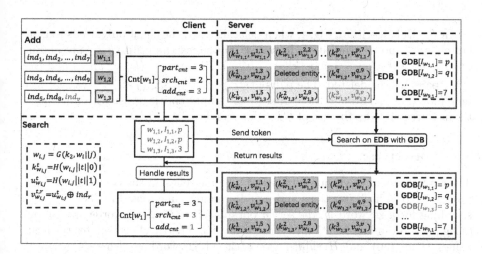

Fig. 3. Example of SGsse-F about Add and Search algorithms. In order to clearly demonstrate the example, this figure does not focus on the process of generating tokens for file deletion.

In the Add algorithm, when file ind_v is added, the client also generates tokens for file deletion. In Fig. 4, where $cnt_{ind} = 6$ indicates that ind_v contains 6 keywords, the client generates 6 pairs, $(fst^1_{key_{ind_v}}, k^3_{w_{1,3}}),...,(fst^6_{key_{ind_v}}, k^1_{w_{8,1}})$, based on the document keywords. $fst^{cnt_{ind_m}}_{key_{ind_m}}$ represents the document deletion label and $k^t_{w_{i,j}}$ corresponds to the added pair $(k^t_{w_{i,j}}, v^{t,r}_{w_{i,j}})$. (Note that cnt_{ind} will not be stored by the client.)

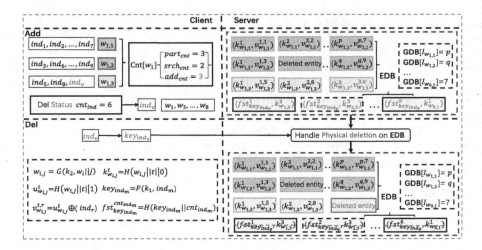

Fig. 4. Example of SGsse-F about Add and Del algorithms

In the Del algorithm, Fig. 4, when deleting ind_v, the client only needs to generate key_{ind} and transmit it to the server. The server can utilize key_{ind} to obtain $(fst^1_{key_{ind_v}}, ..., fst^6_{key_{ind_v}})$ and $(k^1_{w_{1,3}}, ..., k^1_{w_{8,1}})$. As a result, $(k^3_{w_{1,3}}, v^{3,v}_{w_{1,3}})$ and the other pairs will be deleted. The pairs $(fst^1_{key_{ind_v}}, k^3_{w_{1,3}}), ..., (fst^6_{key_{ind_v}}, k^1_{w_{8,1}})$ will be removed last. Through these operations, empty entities exist within the partitions, necessitating the use of GDB for searches.

3.3 Correctness

The correctness of the proposed SGsse-F scheme relies on the collision-resistant property of the hash functions and pseudo-random functions. In SGsse-F, when searching for a keyword w, the server obtains a set of sub-keywords and tokens using pseudo-random function G. The client then repetitively calculates the hash values $H(w_i||i||0)$ and $H(w_i||i||1)$ for each sub-keyword w_i, where i increases from 1 to cnt_{srch} and obtains cnt_{srch} distinct values. These values ensure the retrieval of all matching ciphertexts from the EDB, and obtain the correct document identifiers. In this process, if the H and G functions produce the same output for different inputs, it will impact the correctness of the scheme. The correctness of deletion process can be described in the same way.

3.4 Security Analysis

Let w be the search keyword, and $w_1, ..., w_i$ represent the sub-keywords from keyword w. We have $sp(w) = \{sp(w_1), ..., sp(w_i)\}$, $TimeDB(w) = \{TimeDB(w_1), ..., TimeDB(w_i)\}$ and $Update(w) = \{Update(w_1), ..., Update(w_i)\}$.

Theorem 1. *SGsse-F is $\mathcal{L}$-adaptively secure under the security parameter λ with FuP and FsP. The leakage function $\mathcal{L}$ is defined as $\mathcal{L}^{SGsse-F} = (\mathcal{L}_{Setup}^{SGsse-F}, \mathcal{L}_{Search}^{SGsse-F}, \mathcal{L}_{Add}^{SGsse-F}, \mathcal{L}_{Del}^{SGsse-F})$, where:*

$$\mathcal{L}_{Setup}^{SGsse-F}(\lambda) = \bot$$
$$\mathcal{L}_{Search}^{SGsse-F}(w) = (sp(w), TimeDB(w))$$
$$\mathcal{L}_{Add}^{SGsse-F}(W_{ind}, ind) = |W_{ind}|$$
$$\mathcal{L}_{Del}^{SGsse-F}(ind) = ind$$

Proof. We demonstrate the security of our proposal through a series of games. In this series of games, we reduce the security of SGsse-F to the security of pseudo-random functions and hash functions.

Game$_0$. It represents the real experiment game $\mathbf{Real}_{\mathcal{A}}^{SGsse-F}(\lambda)$ of SGsse-F. We associate the real *Setup*, *Add* and *Del* algorithms with the adversary $\mathcal{A}$. For each query operation, we interact with $\mathcal{A}$ using the *Search* algorithm. We have:

$$Pr[\mathbf{Real}_{\mathcal{A}}^{SGsse-F}(\lambda) = 1] = Pr[\mathbf{Game_0} = 1]$$

Game$_1$. We use a table T_F to represent the pseudo-random function F. For each input document identifier *ind*, T_F stores a uniformly random string of secure length generated instead of directly computing it using the function F. If $\mathcal{A}$ can distinguish the difference between **Game$_1$** and **Game$_0$**, It means that $\mathcal{A}$ can distinguish the replacement of the pseudo-random function F. We have:

$$|Pr[\mathbf{Game_1} = 1] - Pr[\mathbf{Game_0} = 1]| \leq Adv_{F,\mathcal{A}}^{prf}(\lambda)$$

Game$_2$. The system maintains a table T_G to perform the pseudo-random function G. For each input keyword w, the system picks a uniformly random string of the corresponding length. T_G stores the random string for reuse. Then, we have:

$$|Pr[\mathbf{Game_2} = 1] - Pr[\mathbf{Game_1} = 1]| \leq Adv_{G,\mathcal{A}}^{prf}(\lambda)$$

Game$_3$. During the game, we maintain a table R_H to handle queries from a random oracle to replace the hash function H. R_H keeps track of responses for $H(w_i||add_{cnt}||0)$, $H(w_i||add_{cnt}||1)$ and $H(key_{ind}||cnt_{ind})$. The Add algorithm updates tokens k and v by generating random strings instead of using the hash function H. Additionally, the token fst, used for document deletion, is replaced with a random string instead of computing the hash function $H(key_{ind}||cnt_{ind})$. The random oracle with R_H is then programmed to ensure that $H(w_i||add_{cnt}||0) = k$, $H(w_i||add_{cnt}||1) = v$ and $H(key_{ind}||cnt_{ind}) = fst$. Each time the hash function H is called, we maintain the returned transcripts through table R_H. If $\mathcal{A}$ can distinguish **Game$_3$** from **Game$_2$**, it implies that $\mathcal{A}$ can differentiate between H and a random oracle. We have:

$$|Pr[\textbf{\textit{Game}}_3 = 1] - Pr[\textbf{\textit{Game}}_2 = 1]| \leq Adv_{H,\mathcal{A}}^{hash}(\lambda)$$

Simulator. $\textbf{\textit{Game}}_3$ and $\textbf{\textit{Ideal}}_{\mathcal{A},\mathcal{S},\mathcal{L}}^{SGsse-F}$ are identical. In $\textbf{\textit{Ideal}}_{\mathcal{A},\mathcal{S},\mathcal{L}}^{SGsse-F}$, we restrict the privilege of $\mathcal{A}$, let $\mathcal{A}$ only interact with the simulator $\textbf{\textit{S}}$. $\textbf{\textit{S}}$ can access the leakage information $\mathcal{L}$ from $\mathcal{A}$ in the model Setup, Add, Search, and Del processes. We have:

$$Pr[\textbf{\textit{Ideal}}_{\mathcal{A},\mathcal{S},\mathcal{L}}^{SGsse-F}(\lambda) = 1] = Pr[\textbf{\textit{Game}}_3 = 1]$$

In conclusion, during the above games, SGsse-F achieves FuP and FsP under $\mathcal{A}$ with $\mathcal{L}$ such that:

$$|Pr[\textbf{\textit{Real}}_{\mathcal{A}}^{SGsse-F}(\lambda) = 1] - Pr[\textbf{\textit{Ideal}}_{\mathcal{A},\mathcal{S},\mathcal{L}}^{SGsse-F}(\lambda) = 1]|$$
$$\leq Adv_{F,\mathcal{A}}^{prf} + Adv_{G,\mathcal{A}}^{prf} + Adv_{H,\mathcal{A}}^{hash}$$

4   Forward and Backward Privacy DSSE: SGsse-FB

In this section, we propose another DSSE scheme called SGsse-FB. The construction idea of SGsse-FB is consistent with SGsse-F. Compared to SGsse-F, SGsse-FB also achieves Type-II BP, but SGsse-FB doesn't need GBD. Additionally, SGsse-FB only supports logical deletion, which leads to lower storage requirements on the server-side. And SGsse-FB has three algorithms in Algorithm 3.

Setup: In contrast to the SGsse-F scheme, SGsse-BF utilizes security parameters to generate two keys, k_1 and k_2. The client initializes two empty maps, Cnt and EDB. EDB stores the encrypted entities. The storage of the EDB differs from SGsse-F in that EDB only contains a map.

Update: To support deletion operations and achieve backward privacy, we encrypt the entries $(ind||op)$ using the key k_1 and conceal their encrypted forms (line 16). The rest of the algorithm is similar to Algorithm 1 Add, but it does not include the physical deletion operations of documents based on ind.

Search: The difference between this algorithm and Algorithm 2 Search is that the server in this algorithm uses search tokens to compute and then gets encrypted entries $(ind||op)$. When the client receive entries, it decrypts each encrypted entry with k_1. Then, it adds document identifiers with op equal to add to $inds$ and removes document identifiers with op equal to del. The algorithm also does not involve GDB to search. As described in Sect. 3.1, the Search process in Algorithm 3, line 7–10 of client and line 3–16 of server can conduct concurrent queries using sub-keywords.

4.1   Correctness

As demonstrated in Sect. 3.3, the correctness of SGsse-FB is similarly reliant on the collision-resistant property of the hash functions and pseudo-random functions, as well as the correctness of the encryption function Enc. In SGsse-FB, the

Algorithm 3. SGsse-FB

Setup(λ)

Cient:

1: $k_1, k_2 \xleftarrow{\$} \{0,1\}^\lambda$
2: $Cnt \leftarrow \varnothing$
3: $EDB \leftarrow \varnothing$
4: Initialize p as partition size
5: $\sigma \leftarrow (Cnt, k_1, k_2, p)$
6: send EDB to Server

Update(σ,w,op,ind;EDB)

Client:

1: $(part_{cnt}, srch_{cnt}, add_{cnt}) \leftarrow (0,0,0)$
2: $ei \leftarrow \text{Enc}(k_1, ind||op)$
3: if $Cnt.\text{Find}(w) = \bot$ then
4:    $(part_{cnt}, srch_{cnt}, add_{cnt}) \leftarrow (1,0,1)$
5: else
6:    $(part_{cnt}, srch_{cnt}, add_{cnt}) \leftarrow Cnt[w]$
7:    if $add_{cnt} \bmod p = 0$ or $part_{cnt} = srch_{cnt}$ then
8:      $part_{cnt} \leftarrow part_{cnt} + 1$
9:      $add_{cnt} \leftarrow 1$
10:    else
11:      $add_{cnt} \leftarrow add_{cnt} + 1$
12:    end if
13: end if
14: $w_i \leftarrow G(k_2, w||part_{cnt})$
15: $k \leftarrow H(w_i||add_{cnt}||0)$
16: $v \leftarrow H(w_i||add_{cnt}||1) \oplus ei$
17: $Cnt[w] \leftarrow (part_{cnt}, srch_{cnt}, add_{cnt})$
18: $\sigma \leftarrow Cnt[w], k_1, k_2, p$
19: send (k,v) to server

Server:

1: receive (k,v)
2: $EDB \leftarrow EDB \cup (k,v)$

Search(σ,w; EDB)

Client:

1: $srchTK \leftarrow \varnothing$
2: if $Cnt.\text{Find}(w) \neq \bot$ then

3:    $(part_{cnt}, srch_{cnt}, add_{cnt}) \leftarrow Cnt[w]$
4: else
5:    return
6: end if
7: for $i \leftarrow 1$ to $part_{cnt}$ do
8:    $w_i \leftarrow G(k_2, w||i)$
9:    $srchTK \leftarrow srchTK \cup w_i$
10: end for
11: $srch_{cnt} \leftarrow part_{cnt}$
12: $Cnt[w] \leftarrow (part_{cnt}, srch_{cnt}, add_{cnt})$
13: $\sigma \leftarrow Cnt[w], k_2, p$
14: send $srchTK$ to server

Server:

1: receive $srchTK$
2: $res \leftarrow \varnothing$
3: for w_i in $srchTK$ do
4:    $i \leftarrow 1$
5:    while (1) do
6:      $st_k \leftarrow H(w_i||i||0)$
7:      if $EDB.\text{Find}(st_k) = \bot$ then
8:        break
9:      else
10:        $st_v \leftarrow EDB.\text{Get}(st_k)$
11:        $ei \leftarrow st_v \oplus H(w_i||i||1)$
12:        $res \leftarrow res \cup ei$
13:      end if
14:      $i \leftarrow i + 1$
15:    end while
16: end for
17: send res to server

Client:

1: receive res
2: $inds \leftarrow \varnothing$
3: for $i \leftarrow 1$ to $|res|$ do
4:    $(ind||op) \leftarrow \text{Dec}(k_1, res[i])$
5:    $inds \leftarrow inds \cup (ind||op)$
6: end for
7: remove $inds$ that have been deleted

Enc function encrypts both the document identifier ind and the operation-type op. When the server returns the corresponding encrypted ciphertexts, the correctness of SGsse-FB depends on the correctness of the encryption and decryption procedures.

4.2   Security Analysis

Theorem 2. *SGsse-FB is $\mathcal{L}$-adaptively secure with FuP, FsP and TYPE-II BP under the security parameter λ, if the leakage function is defined as $\mathcal{L}^{SGsse-FB} = (\mathcal{L}_{Setup}^{SGsse-FB}, \mathcal{L}_{Search}^{SGsse-FB}, \mathcal{L}_{Update}^{SGsse-FB})$, where:*

$$\mathcal{L}_{Setup}^{SGsse-FB}(\lambda) = \perp$$

$$\mathcal{L}_{Search}^{SGsse-FB}(w) = (sp(w), TimeDB(w), Updates(w))$$

$$\mathcal{L}_{Update}^{SGsse-FB}(op, w, ind) = \perp$$

Updates(w) is showed in Sect. 3.4

Proof. The proof method for SGsse-FB is similar to SGsse-F in Sect. 3.4. The difference between SGsse-F and SGsse-FB is that in SGsse-F, the server can obtain *ind* when using search tokens, while in SGsse-FB, the server can only obtain the encrypted entity of *ind*||*op*. Therefore, the simulator $\mathcal{S}$ needs to replace *ind*||*op* with random strings in the games. Additionally, the encrypted entity of *ind*||*op* is concealed by using a random strings in games, making it indistinguishable from random entry observed by the server during the update process. Moreover, the server is unaware of the type of *op*. As for the backward security mentioned in Sect. 2.2, the server only knows the number of entries related to keyword *w* and the corresponding update times.

5   Related Works

Song et al. [3] proposed the first static searchable symmetric encryption scheme in 2000. Goh [4] used Bloom filters to construct a document secure searchable encrypted index. Curtmola et al. [5] proposed static search encryption schemes SSE1 and SSE2 using the technique of keyword-document identifiers inverted index. They also proposed the concept of adaptively secure. In 2012, Kamara et al. [6] designed a DSSE scheme with a sub-linear search time complexity, which also satisfies dynamic addition and physical deletion. Cash et al. [7] showed a multi-keyword searchable encryption scheme based on cyclic groups, extending single-keyword to multi-keyword queries. Cash et al. [8] also proposed a DSSE scheme for large-scale data, which has higher efficiency in large-scale data encryption and queries, but only supports dynamic addition.

As a result of the vulnerability of DSSE schemes [6,9,10] to file injection attacks [11,12], Bost et al. [13] proposed the security definition for forward privacy and introduced a forward privacy scheme. It is based on public key cryptography, and the update efficiency is low and server-side storage is relatively large. Therefore, Kim et al. [29] proposed an effective searchable encryption scheme Dual based on symmetric encryption is designed, which has greatly improved encryption efficiency and storage, and also supports physical deletion. Wei et al. [14] proposed the forward privacy searchable encryption scheme FSSE based on chains, which constructs an entity chain to link the collection of different document identifiers for the same keyword. Etemad et al. [15] implemented forward privacy by replacing the key that is publicly revealed to the server each time a search is performed. In the research focusing on forward privacy, Bost et al. [16] proposed the theoretical framework for backward privacy security and constructed the schemes Fides, Diana, Janus. Janus satisfies both forward and backward privacy security using the asymmetric puncturable technology.

Sun et al. [17] suggested an scheme called Janus++ based on symmetric puncturable encryption, which improves the efficiency of deletion. Ghareh et al. [18] presented three different schemes that satisfy forward privacy and three types of security for backward privacy. In that paper, a scheme with TYPE-I backward privacy security was implemented using ORAM technology named Orion. However, ORAM-based schemes [19–21] require more computation and storage. Li et al. [22] gave the Khons scheme, where they introduced the concepts of FuP and FsP for the first time. To reduce client-side storage, He et al. [23] constructed the schemes, CLOSE-F and CLOSE-FB, which maintain a global update state to ensure constant client-side storage. Chen et al. [24] presented a more efficient DSSE scheme, Bestie, that satisfies forward privacy and achieves physical deletion under TYPE-III backward privacy security. Sun et al. [25] introduced the Symmetric Revocable Encryption (SRE) primitive and implemented an Non-interactive forward and backward privacy scheme based on this primitive. Alongside the focus on scheme security, some research [26–28] has also considered the impact of I/O on searchable encryption schemes.

6   Benefits and Drawbacks of DSSE Schemes Analysis

In this section, we compare and analyze the schemes Dual [29], Khon [22], Mitra [18], and FSSE [14] with our proposed schemes. We discuss their similarities, differences, and the implications of these variations on the results. And Table 1 also illustrates a comparison with the above solutions in terms of efficiency and security.

The similarities between the Dual scheme and the proposed SGsse-F scheme are that SGsse-F draws inspiration from the deletion algorithm of Dual, enabling physical deletion and making the way of storage on the server side similar to Dual. The difference lies in the fact that SGsse-F satisfies both FuP and FsP, whereas Dual only satisfies FuP. Additionally, the Dual scheme re-encrypts the obtained document identifiers during the query process, which is not done in SGsse-F. This results in SGsse-F having a storage size that is approximately 0.8 times that of Dual.

The Khon scheme and our proposed SGsse-F and SGsse-FB schemes all utilize partitioning techniques [22] to divide the inverted indexes and support concurrent queries on single keyword. The difference lies in the fact that SGsse-FB satisfies FuP, FsP, and TYPE-II BP, while Khon satisfies FuP, FsP, and TYPE-III BP. The Khon scheme requires larger client-side storage, while the SGsse-F and SGsse-FB schemes only store the keywords along with the corresponding states of three numbers. Khon encrypts and pads the partitions that have been queried but not yet filled with data. In contrast, our schemes do not require padding. Due to the need for partition padding, the server-side storage of the Khon scheme is larger than that of the SGsse-FB scheme. Additionally, the Khon scheme incurs lower query efficiency compared to the SGsse-FB scheme, as the padded partition blocks also need to be searched during the query process.

Both the Mitra scheme and the SGsse-FB scheme satisfy FuP and TYPE-II BP. But SGsse-FB also satisfies FsP and supports concurrent queries on single keywords. SGsse-FB provides higher security and better query efficiency. However, SGsse-FB increases client-side storage in order to achieve higher security.

The FSSE scheme utilizes chain technology to construct indexes and only achieves FuP. The chain technology ensures that the update status of client storage is limited by the exhaustion of states, in contrast to the $\sum o\varphi o\varsigma$ [13] scheme. Compared to the SGsse-F and SGsse-FB schemes proposed in this paper, our scheme offers strong privacy security and keyword query efficiency.

All the aforementioned schemes are employed hash and pseudo-random functions, which ensure the efficiency of encrypted queries.

7 Evaluation

7.1 Experimental Setup

To demonstrate the effectiveness of schemes SGsse-F and SGsse-FB, we compared them with other schemes, Dual, FSSE, Mitra and FSSE*. We repeated the time-related experiments three times and took the average values as the final results.

Implementation. We have rewritten the code for Dual and developed a modified version of the FSSE scheme without remote communication based on the FSSE scheme [14] and its code [30]. We made modifications for FSSE to achieve forward privacy and TYPE-II backward privacy, namely FSSE* and the specific implementation details are that we encrypted $ind||op$ using AES-128 in the FSSE Update algorithm(line 6) where b is showed as $(id*, "ind||op"||key||ind) \oplus mask)$ and we replaced the tree-based storage mode $\mathcal{T}$ in FSSE with a $(key, value)$ storage format. By considering the TYPE-II BP discussed in Sect. 2.2, we can understand that due to hiding the operation type op, the server does not know which deletion corresponds to which addition. Thus, FSSE* satisfies TYPE-II BP. For Mitra, we used the code by the author [31].

Programming Environment. For the experiments, we chose an Aliyun ecs.u1 c1m8.x-large Instance as the physical machine, which has a 4-core Intel(R) Xeon(R) Platinum 2.5 GHz CPU, 32 GB RAM and a 196 GB SSD. The operating system is Ubuntu 18.04, and the code compiler used is GCC 4.8.5. The encryption library is OpenSSL 1.0.2c. All schemes were organized using the C++ standard library STL's *unordered_map* for client-side storage, while RocksDB 7.2.5 was used for server-side storage. None of our schemes or comparison schemes implemented remote communication.

Cryptographic Primitives. We applied the SHA256 from the OpenSSL library for hash functions in our proposed schemes and comparison schemes. HMAC-SHA256 was used for the pseudo-random functions, and AES-128 was used for encryption. In the case of the Mitra scheme, we replaced the pseudo-random functions implemented with AES-128, as stated in their scheme, with a pseudo-random function implemented with HMAC-SHA256 to make an effective comparison with our proposed schemes and its implementation.

Dataset. To generate the sparse keywords data, we have written Python code that utilizes the Faker library [2] to randomly generate ten million Chinese ID card numbers. For each ID number, we extracted the first 2 digits, first 4 digits, first 6 digits, and digits 7–13 as experimental data for privacy preference queries and assigned a document identifier to each record. This means that each data record contains four keyword-document identifier pairs.

Table 2. Keywords occurrences in different datasets

records/occurrences	≤ 64	≤ 256	≤ 1024	≤ 4096	≤ 16384	≤ 655536	≤ 262144	≤ 524288
2×10^6	1875	26281	29833	29922	30178	30188	30206	30206
4×10^6	0	26281	26282	29837	30133	30182	30206	30206
6×10^6	0	25329	26281	29836	30005	30180	30199	30206
8×10^6	0	59	26281	29833	29923	30178	30188	30206
1×10^7	0	0	26281	29833	29863	30178	30184	30203

For the different numbers of data records: the first two million, four million, six million, eight million and ten million data records, we extracted the keywords of data records and conducted experiments. We used Python code to count the occurrences of the keywords in the five datasets mentioned above. From the Table 2, we can observe that the keywords counts remains constant at 30,206 across five datasets. Additionally, In the first two million, four million and six million data records, most of the keywords occurrences are less than 256 but greater than 64. For the ten million data records, most of the keywords occurrences are less than 1024 but greater than 256. Moreover, in ten million date records, there exist 3 keywords have file identifiers beyond 524288.

7.2 Experimental Analysis

Client Storage. Client storage is compared among six schemes in five dataset. (As the number of data records increases, the total number of keywords remains constant.) In our schemes SGsse-F and SGsse-FB, the client storage includes keywords and their corresponding three count values. Dual scheme's client storage includes keywords, and their corresponding two count values and two keys. Mitra scheme only includes keywords and only one count value of earch keyword in client. FSSE and FSSE* schemes store keywords, and IDs, and keys of earch keyword as shown in the Table 3. Mitra has the smallest client storage, followed by our schemes SGsse-F and SGsse-FB, while the Dual scheme has the largest client storage.

Table 3. Client Storage

Schemes	SGsse-F	SGsse-FB	Dual	Mitra	FSSE	FSSE*
Size(KB)	820	820	2384	464	2148	2148

Server Storage. Table 4 shows the server storage for six schemes. SGsse-F and Dual require more server storage compared to other schemes. And SGsse-FB requires nearly minimum storage in server-side. This is mainly due to SGsse-F and Dual supporting physical deletion, which necessitates the storage of an EDB containing the (*document key, keyword label*) pairs. Additionally, Dual involves re-encryption of the keyword-document identifier, resulting in storage of (*keyword label, (document key, encrypted value)*) pairs [29]. Therefore, Dual exhibits larger server-side storage requirements.

The server-side storage for SGsse-FB is similar to Mitra, with slight variations in specific storage values due to direct measurement of RocksDB storage files. However, FSSE* requires more server-side storage compared to SGsse-FB and Mitra. This is because the storage format of FSSE* includes (*keyword label, encrypted value*) where the *encrypted value* contains $ind||op$, as well as the ID and key.

Table 4. Server Storage (MB)

Records/Schemes	SGsse-F	SGsse-FB	Dual	Mitra	FSSE	FSSE*
2×10^6	1100	549	1334	550	845	905
4×10^6	2234	1131	2698	1146	1717	1836
6×10^6	3366	1686	4088	1704	2586	2792
8×10^6	4479	2231	5451	2230	3458	3723
1×10^7	5581	2777	6785	2778	4321	4646

Addition and Deletion Time. We employed multi-threading to insert the encrypted data of all schemes into RocksDB. We utilized RocksDB's WriteBatch to perform batched disk writes and reduce the number of I/O operations. As shown in Fig. 5a, SGsse-F and Dual demonstrate similar insertion times, while SGsse-FB, FSSE, Mitra and FSSE* exhibit faster insertion time. Regarding data deletion, we only compared SGsse-F and Dual since other schemes employ logical deletion. From the experimental results of the deletion process in Table 5, SGsse-F and Dual exhibit comparable performance.

Table 5. Deletion Time (seconds)

Scheme/Records	2×10^6	4×10^6	6×10^6	8×10^6	1×10^7
SGsse-F	153.5	374.3	576.6	928.6	1231.6
Dual	160.2	357.6	574.3	903.0	1192.7

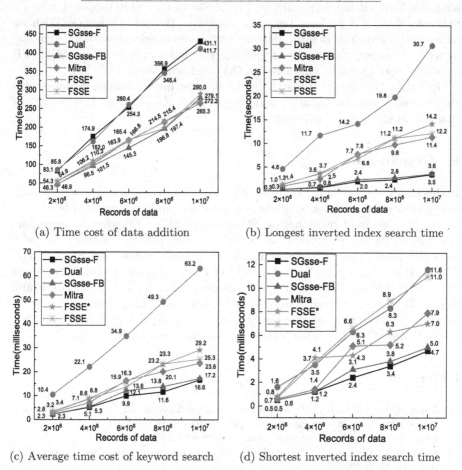

(a) Time cost of data addition

(b) Longest inverted index search time

(c) Average time cost of keyword search

(d) Shortest inverted index search time

Fig. 5. Time-related experimental results

Search Time. In this study, we utilized the C++ Pthread library to perform parallel keyword queries on the sub-keywords generated from the search keyword as described in Sect. 3.1 about Search algorithm. We performed queries for each keyword and recorded the corresponding search time. The average search time was then calculated to evaluate the time required for different datasets. As shown in Fig. 5c, it can be observed that our proposed schemes SGsse-F and SGsse-FB exhibit the optimal average query time, while the scheme Dual incurs the longest query time and the query times of other five schemes are relatively close to each other. Dual involves re-encryption, which requires *get, deletion, insertion* operations in RocksDB for each keyword-identifier search, while other schemes only involve *get* operations.

We conducted query operation experiments on the longest keyword-document identifier inverted index. Results in Fig. 5b shows that Our proposed schemes SGsse-F, SGsse-FB perform better compared to the other schemes. With an increase in the records of data, the query time also experiences a corresponding increase. The query time for the longest inverted index in SGsse-F and SGsse-FB increases more gradually compared to the other schemes.

In Fig. 5d, the keyword search on the shortest inverted index demonstrated that our solutions are equally applicable to general scenarios. And the search time for all schemes are relatively fast.

8   Conclusion

In this paper, we conducted analyses of keywords searches security and efficiency in sparse keyword set scenarios where a large document set contains fewer keywords. We proposed two schemes, SGsse-F and SGsse-FB, that leverage partitioning technique and effectively support parallel keyword search. They also have enhanced privacy. Experimental results demonstrated that SGsse-F and SGsse-FB exhibited near-optimal search efficiency compared to Dual [29], Mitra [18], FSSE [14] and FSSE* (Sect. 7.1). In terms of client storage, SGsse-F and SGsse-FB have storage sizes only slightly larger than Mitra, surpassing other schemes. The results from the shortest inverted index analysis indicated that SGsse-F and SGsse-FB are applicable to general scenarios.

References

1. Popa, R.A., Redfield, C.M., Zeldovich, N., Balakrishnan, H.: CryptDB: protecting confidentiality with encrypted query processing. In: SOSP (2011)
2. Faker. https://github.com/joke2k/faker. Accessed 27 May 2023
3. Song, D.X., Wagner, D., Perrig, A.: Practical techniques for searches on encrypted data. In: S&P (2000)
4. Goh, E.J.: Secure indexes. Cryptology ePrint Archive, Report 2003/216 (2003). https://eprint.iacr.org/2003/216
5. Curtmola, R., Garay, J., Kamara, S., Ostrovsky, R.: Searchable symmetric encryption: improved definitions and efficient constructions. In: CCS (2006)
6. Kamara, S., Papamanthou, C., Roeder, T.: Dynamic searchable symmetric encryption. In: CCS (2012)
7. Cash, D., Jarecki, S., Jutla, C., Krawczyk, H., Rosu, M.C., Steiner, M.: Highly-scalable searchable symmetric encryption with support for boolean queries. In: CRYPTO (2013)
8. Cash, D., et al.: Dynamic searchable encryption in very-large databases: data structures and implementation. In: NDSS (2014)

9. Kamara, S., Papamanthou, C.: Parallel and dynamic searchable symmetric encryption. In: Eyal, I., Garay, J. (eds.) Financial Cryptography and Data Security 2013. LNCS, vol. 13411, pp. 258–274 (2013). https://doi.org/10.1007/978-3-642-39884-1_22

10. Stefanov, E., Papamanthou, C., Shi, E.: Practical dynamic searchable encryption with small leakage. In: NDSS (2014)

11. Zhang, Y., Katz, J., Papamanthou, C.: All your queries are belong to us: the power of file-injection attacks on searchable encryption. In: USENIX (2016)

12. Wang, G., Cao, Z., Dong, X.: Improved file-injection attacks on searchable encryption using finite set theory. Comput. J. $64(8)$, 1264–1276 (2020)

13. Bost, R.: $\sum o\varphi o\varsigma$: forward secure searchable encryption. In: CCS (2016)

14. Wei, Y., Lv, S., Guo, X., Liu, Z., Huang, Y., Li, B.: FSSE: forward secure searchable encryption with keyed-block chains. Inf. Sci. **500**, 113–126 (2019)

15. Etemad, M., Küpçü, A., Papamanthou, C., Evans, D.: Efficient dynamic searchable encryption with forward privacy. arXiv preprint arXiv:1710.00208 (2017)

16. Bost, R., Minaud, B., Ohrimenko, O.: Forward and backward private searchable encryption from constrained cryptographic primitives. In: CCS (2017)

17. Sun, S.F., et al.: Practical backward-secure searchable encryption from symmetric puncturable encryption. In: CCS (2018)

18. Ghareh Chamani, J., Papadopoulos, D., Papamanthou, C., Jalili, R.: New constructions for forward and backward private symmetric searchable encryption. In: CCS (2018)

19. Garg, S., Mohassel, P., Papamanthou, C.: TWORAM: efficient oblivious RAM in two rounds with applications to searchable encryption. In: CRYPTO (2016)

20. Hoang, T., Ozmen, M.O., Jang, Y., Yavuz, A.A.: Hardware-supported ORAM in effect: practical oblivious search and update on very large dataset. Proc. Priv. Enh. Technol. **2019**(1), 172–191 (2019)

21. Naveed, M.: The fallacy of composition of oblivious ram and searchable encryption. Cryptology ePrint Archive, 2015/668 (2015). https://eprint.iacr.org/2015/668

22. Li, J., et al.: Searchable symmetric encryption with forward search privacy. IEEE Trans. Dependable Secure Comput. **18**(1), 460–474 (2019)

23. He, K., Chen, J., Zhou, Q., Du, R., Xiang, Y.: Secure dynamic searchable symmetric encryption with constant client storage cost. IEEE Trans. Inf. Forensics Secur. **16**, 1538–1549 (2020)

24. Chen, T., Xu, P., Wang, W., Zheng, Y., Susilo, W., Jin, H.: Bestie: very practical searchable encryption with forward and backward security. In: ESORICS (2021)

25. Sun, S.F., et al.: Practical non-interactive searchable encryption with forward and backward privacy. In: NDSS (2021)

26. Demertzis, I., Papadopoulos, D., Papamanthou, C.: Searchable encryption with optimal locality: achieving sublogarithmic read efficiency. In: CRYPTO (2016)

27. Miers, I., Mohassel, P.: IO-DSSE: scaling dynamic searchable encryption to millions of indexes by improving locality. In: NDSS (2017)

28. Minaud, B., Reichle, M.: Dynamic local searchable symmetric encryption. In: CRYPTO (2022)

29. Kim, K.S., Kim, M., Lee, D., Park, J.H., Kim, W.H.: Forward secure dynamic searchable symmetric encryption with efficient updates. In: CCS (2017)

30. Implementation of FSSE. https://gitlab.com/suifengrudao/ffse.git. Accessed 2019

31. Implementation of Mitra, Orion, Horus, Fides, and DianaDel. https://github.com/jgharehchamani/SSE. Accessed 2018

CBA Sketch: A Sketching Algorithm Mining Persistent Batches in Data Streams

Qian Zhou[1], Yu-E Sun[2,3(✉)], He Huang[1], and Yifan Han[1]

[1] School of Computer Science and Technology, Soochow University, Suzhou, Jiangsu, China

[2] School of Rail Transportation, Soochow University, Suzhou, Jiangsu, China
sunye12@suda.edu.cn

[3] Laboratory of Embedded System and Service Computing (Tongji University), Ministry of Education, Shanghai, China

Abstract. Batch is a vital data pattern commonly observed in data streams, representing a group of identical items that occur closely together. However, existing works primarily focus on the periodicity mining of batches, neglecting other numerous essential patterns. In this paper, we introduce the concept of *persistent batch*, a particular pattern in data streams where multiple occurrences of the same batch happen in at least k out of t measurement periods. Mining persistent batches holds significance in applications such as APT detection, DDoS detection, and Click Fraud detection, *etc.* To fill up the gap of the prior art, we propose *CBA Sketch*, a memory-efficient sketching algorithm that effectively mines persistent batches from data streams. The CBA Sketch utilizes a Circular-Time Sketch (CT-Sketch) to accurately calculate item intervals and capture batches with limited memory resources. We incorporate the carefully designed Bloom Filter-based Existence Recorder (BE Recorder) and Approximate Size Recorder (AS Recorder) to preserve batch information. Additionally, we introduce a novel metric called *dual-mean size* to provide measurements for persistent batch sizes. Extensive experiments demonstrate that our CBA Sketch outperforms the strawman solution about 62× in terms of average relative error and 2× in terms of throughput.

Keywords: Data Streams · Persistent Batch · Sketch

1 Introduction

1.1 Background and Motivation

Batch mining is a crucial task in data streams, as a batch represents a sequence of identical items with the same ID that arrive consecutively with intervals less than a predefined threshold T [1,2]. Batches find applications in various domains, including cache management [3], burst detection [4], and machine learning [5].

© The Author(s), under exclusive license to Springer Nature Singapore Pte Ltd. 2024
Z. Tari et al. (Eds.): ICA3PP 2023, LNCS 14492, pp. 114–132, 2024.
https://doi.org/10.1007/978-981-97-0811-6_7

Extracting batches with specific patterns holds significant practical importance, such as identifying periodic batches in financial transaction streams for detecting potential money laundering activities [2]. However, mining periodic batch patterns in data streams may fail to detect sophisticated attackers dynamically adjusting their periodicity. Hence, there is a need for more robust batch patterns to help uncover suspicious actions in dynamic data streams. Here we list several practical scenarios where periodic patterns exhibit certain limitations while a new pattern (i.e., persistence of batches) can make a real difference.

Case 1 - APT Detection: Advanced Persistent Threat (APT) is a sophisticated network attack utilized by malicious actors to acquire sensitive information from individuals and organizations illicitly [6,7]. APT attackers persistently infiltrate target systems and continuously extract and disclose private data. Traditional detection approaches relying on periodic patterns become ineffective in discerning suspicious APT activities. However, by treating packets sent to the same destination as identical items in APT attacks, the influx of leaked packets will form a data stream comprising numerous large-sized persistent batches. By leveraging the detection of persistent batch patterns, it becomes feasible to detect and respond to potential APT attacks promptly.

Case 2 - DDoS Detection: Distributed Denial of Service (DDoS) attacks involve coordinated efforts from multiple sources to disrupt service availability [8–10]. Traditional DDoS flooding attacks continuously inundate target hosts or servers with high traffic volumes, depleting network resources and causing network congestion. Stealthy DDoS attackers may employ a clever strategy by deliberately skipping some measurement periods at random to evade detection. In this case, the attack flows will not show periodic patterns. Consequently, DDoS detections built upon recognizing periodic batches may prove ineffective in identifying these stealthy attacks. However, by considering packets destined for the same destination as identical items, akin to the approach utilized for APT detection, the attack traffic of DDoS can be viewed as a data stream comprising numerous persistent batches of substantial sizes. The identification of these persistent batches can contribute to the recognition of potential stealthy DDoS attack streams.

Case 3 - Click Fraud Detection: Click Fraud refers to the deliberate clicks on pay-per-click advertisements with malicious intent [11,12]. Fraudsters engage in this activity to increase their earnings by creating artificial spikes in click volumes, often sustaining this behavior over an extended period. More specifically, click fraudsters deliberately sustain lower activity after generating concentrated clicks to avoid their malicious click spikes being detected. At the same time, the intervals remain irregular and unpredictable. Conventional detection methods relying on identifying periodic click patterns are ineffective in uncovering the actual fraudsters due to irregular intervals between consecutive concentrated clicks. However, by treating clicks on the same advertisement link as identical items, Click Fraud generates numerous persistent batches with considerable sizes. Detection of such persistent batches can assist in identifying potential click fraudsters and mitigating financial losses for advertisers.

To overcome the limitations of periodic batch detection in various scenarios, we introduce a novel pattern called **persistent batch**. By monitoring a data stream across t measurement periods, we identify a group of batches sharing the same ID as a persistent batch if they occur in at least k out of the t periods, where $1 \le k \le t$. Here, the batch ID is defined by the ID of included items. The detection of persistent batches, particularly those with large sizes (indicating numerous items), holds significant importance as they may represent APT, DDoS attacks, Click Fraud, or other abnormal behaviors. We should note that persistent batch detection cannot directly identify the specific type of abnormal behaviors in the data stream. However, it can assist us in promptly monitoring anomalies and notifying administrators to take timely action. If a separate detection scheme is deployed for each type of abnormal behavior, it may result in the failure to detect other abnormal behaviors that do not have a specific scheme deployed.

Remarkably, there is currently a lack of research focusing explicitly on the persistence of batches. This paper aims to fill the gap in the field of mining persistent batches in data streams, encompassing their detection and accurate size estimation. However, this task poses several challenges. Firstly, detecting batches itself is already a challenging task [2]. Batch is a time-related data pattern requiring the arrival time of each item to compute intervals of adjacent items and detect batches. However, storing complete timestamps for every arrived item demands unacceptable memory capacity. For instance, preserving the complete 32-bit timestamp for each batch in a one-minute network traffic trace CAIDA would consume about 3MB memory [13], while memory resources are limited [14–16]. Therefore, limited memory resources necessitate the development of a novel timestamp-encoded solution to reduce memory usage while ensuring detection accuracy. Secondly, determining the criteria for defining the persistence of batch remains a significant challenge. Furthermore, regardless of how we define persistent batches, detecting persistent batches requires historical batch information from previous measurement periods. However, all existing approaches overlook the utilization of such historical information [1,2]. Thirdly, the evaluation of sizes becomes crucial for upper-layer applications once persistent batches are successfully detected. Storing the sizes of each batch is impractical, considering the potential presence of multiple batches within a persistent batch and the high volume of batches in the data stream. Furthermore, even with sufficient memory, summarizing the overall size to assess batch detection and size estimation accuracy poses a challenge. Therefore, devising an effective method of recording batch size information to provide an evaluation of persistent batches is essential. To address these challenges, we aim to design an algorithm with high accuracy and low memory overhead to detect persistent batches and evaluate their sizes.

1.2   Our Solution

For efficient mining of persistent batches from data streams, we propose **CBA Sketch**, an algorithm consisting of three main components: Circular-Time Sketch (**CT-Sketch**), Bloom Filter-based Existence Recorder (**BE Recorder**) and

Approximate Size Recorder (*AS Recorder*). Our CBA Sketch takes three phases to detect persistent batches and provide measurements. In phase 1, we divide the timeline into circular-time slices and encode the timestamp of each item. This approach minimizes space overhead while ensuring accuracy. We also design the CT-Sketch to preserve encoded temporal information by utilizing space-sharing techniques, reducing memory consumption. In phase 2, we introduce the BE Recorder to track the occurrence of each batch within the current period. Additionally, we develop the AS Recorder based on the Count-Min Sketch to store relevant size information about captured batches. In phase 3, we aggregate historical measurement·results to answer queries related to persistent batches. We introduce a new metric called *dual-mean size* to effectively evaluate sizes of persistent batches, providing reliable statistical support for upper-layer applications.

1.3 Key Contributions

- We are the first to establish the problem of mining persistent batches in data streams, which is crucial for various practical applications.
- We propose an efficient and accurate solution, **CBA Sketch**, to tackle this problem, requiring low memory footprints while maintaining high precision.
- Within the CBA Sketch, we employ the CT-Sketch, a carefully designed mechanism that detects batches to overcome memory limitations. Additionally, leveraging historical information ignored by existing approaches, CBA Sketch achieves precise detection and evaluation of persistent batches. We also introduce a novel metric named *dual-mean size* to evaluate the size of persistent batches.
- Through extensive simulations using real Internet traces, we demonstrate the effectiveness of our approach. With a memory usage of just 600 KB, the Average Relative Error (ARE) of CBA Sketch is reduced by 98.39% compared to the strawman method, and the throughput is increased by 79.15%.

2   Problem Formulation and Related Work

2.1   Problem Formulation

Before formally defining our problem, we introduce some relevant notations. We denote the data stream as $\Gamma = \{e_1, d_2, e_3, e_4, d_5, ..., e_i, ...\}$, where the subscript i represents the i-th item in the data stream while e and d denotes the ID of items. It is important to note that an item in Γ can appear multiple times. The arrival time of the i-th item is denoted as t_i. The interval between two items is represented by δ. Specifically, $\delta_{i,j}$ signifies the interval between the i-th and j-th item. Our problem can be applied to various scenarios, such as APT/DDoS detection, where each item represents a network packet, or click fraud, where each item represents a click on the advertisement link. For brevity, we may omit the subscript i in the subsequent sections of this paper.

Definition 1. *Batch:* *Given a subsequence of data stream* $\{e_i, e_{i+1}, ..., e_j\} \in \Gamma$, *which consists of items with the same ID e. It forms a batch* B_e *if the following conditions are met: (1)* $\delta_{i-1,i} > T$ *and* $\delta_{j,j+1} > T$, *indicating that there is a time gap larger than* T *before* e_i *and after* e_j, *and (2)* $\delta_{k,k+1} < T$ *holds for any k ranging from i to j − 1, ensuring that all consecutive items within the sequence have time gaps smaller than* T. *We use the item ID e as the subscript to denote the formed batch B. The size of batch* B_e *is defined as the number of items it contains, represented by* $j − i + 1$.

We use the data flow depicted in Fig. 1 as an illustrative example. In this example, we assign the ID e to the yellow items and the ID d to the green items. The items in the data stream are numbered from left to right. Notably, the intervals between e_6 and e_7, as well as between e_{14} and e_{15}, are both smaller than the predefined threshold T. However, the interval between e_7 and e_{14}, denoted as $\delta_{7,14}$, exceeds the threshold T. Consequently, e_6 and e_7, as well as e_{14} and e_{15}, form separate batches denoted as B_e. Therefore, the total number of batches marked as B_e is 2. Similarly, we identify three batches marked as B_d, which correspond to the pairs d_3 and d_5, d_{10} and d_{11}, and d_{16}, respectively.

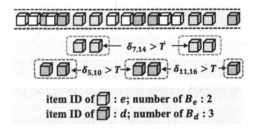

item ID of ▱ : *e*; number of B_e : 2
item ID of ▱ : *d*; number of B_d : 3

Fig. 1. Examples of batches in the data stream.

Definition 2. *Persistent Batch:* *Given t measurement periods, those batches with the same ID occur in at least k out of t measurement periods are regarded as a persistent batch, where* $1 \leq k \leq t$.

Inspired by [17], we define a persistent batch as a series of batches that appear in at least k out of t measurement periods instead of all. This definition is more general as it can decrease false negatives due to deliberate absence during some periods and improve the detecting precision.

2.2   Related Work

To the best of our knowledge, there is no prior work specifically focused on persistent batches. Therefore, this section reviews related works on batch mining and persistent patterns in data streams.

Item Batches Detection. The concept of item batches is initially defined in [1]. The authors propose a framework called Clock-Sketch to handle different measurement tasks related to batches (*e.g.*, batch size, batch activeness). Clock-Sketch incorporates a clock cell for each unit in the basic sketch to determine the expiration of items. A decaying pointer scans and decrements the clock cells, indicating outdated batch information. However, Clock-Sketch only preserves batch information within a fixed time window, failing to maintain historical information and detect persistent batches. Additionally, maintaining a decaying pointer requires an additional scanning thread, resulting in low efficiency. Moreover, deploying multiple Clock-Sketches for different tasks lead to redundant measurement information and diminished memory efficiency.

Recent research, periodic batches [2], has focused on detecting specific batch patterns. A sketching algorithm called HyperCalm is proposed to measure periodic batches in real-time. HyperCalm consists of two components: the HyperBF algorithm for batch detection and the CalmSS algorithm for identifying top-k recording periodic batches (*i.e.*, reporting k groups of periodic batches with the k largest periodicities). In HyperBF, the timeline is divided into multiple time slices of length T, and the start of a batch is determined based on whether two adjacent items span a complete time slice. The distinction between Clock-Sketch and HyperCalm lies in their approach to determining the expiration of items. Clock-Sketch utilizes an additional thread to sweep expired items, whereas HyperCalm eliminates expired items by comparing approximate time slices. However, this time division method in HyperCalm is coarse-grained with T as the precision and is unsuitable for detecting the persistent batches we are interested in.

Mining Persistent Items. There is a growing interest in the research on mining persistent patterns of single items in data streams [17–23]. Persistence refers to long-term patterns in data streams. Measuring item persistence involves striking a balance between accuracy and memory efficiency, as it is challenging to accurately preserve the historical information of numerous items within limited space.

Xiao *et al.* [23] proposed a method to measure the persistent spread of network flows, which can be used to detect long-term stealthy malicious activities. For example, the persistent spread of a destination host is defined as the number of distinct sources that have contacted it persistently in all predefined t measurement periods. However, this definition assumption lacks universality [17]. In specific scenarios like stealthy DDoS attacks, attackers may intentionally drop malicious packets during specific periods to evade detection. To address this, Huang *et al.* [17] proposed a more general definition of persistence: an element is considered persistent if it occurs in at least k out of t measurement periods, where k is an adjustable parameter. We argue that this latter definition of persistence is more suitable for our problem. For instance, in the case of click fraud, the actions of fraudsters are unpredictable. To avoid detection, they may pause for a certain period after a series of concentrated clicks, making themselves appear as legitimate users. If we only consider clickers that occur in all measurement periods as persistent clickers, these fraudulent actions would be overlooked. Therefore, inspired by [17], we define persistent batches as batches that occur in at least k out of t measurement periods instead of requiring occurrence in all t periods.

3 The CBA Sketch

Overview (Fig. 2): Our CBA Sketch contains three main components, namely Circular-Time Sketch (**CT-Sketch**), Bloom Filter-based Existence Recorder (**BE Recorder**), and Approximate Size Recorder (**AS Recorder**). The workflow of CBA Sketch consists of three phases: 1) The CT-Sketch measures intervals between adjacent items to mine batches, 2) The BE Recorder and AS Recorder preserve batch information, and 3) A reporter answers the queries of persistent batches based on historical measurements. Additionally, we propose a novel metric, *dual-mean size*, to measure the size of persistent batches accurately.

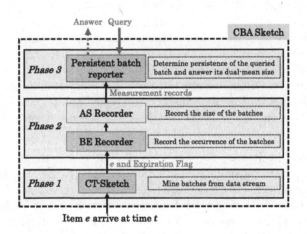

Fig. 2. CBA Sketch workflow.

When an item e arrives at time t, it initially enters the CT-Sketch to calculate intervals and generates an expiration flag, indicating whether e marks the start of a new batch. Subsequently, e and the flag are transmitted to phase 2, where crucial information of batch B_e is preserved. In order to mitigate the impact of noise resulting from small batches, we insert e into the BE Recorder with a probability p. Additionally, the AS Recorder performs specific update operations based on the flag. Once t measurement periods finish, phase 3 becomes available for querying persistent batches. In response to a query regarding batch B_e, we first consult the BE Recorders to confirm its persistence. If confirmed, the batch is reported as a persistent batch, and we proceed to query the AS Recorders to provide size evaluation for B_e. Otherwise, we return 0, indicating that B_e is not a persistent batch.

3.1   Circular-Time Sketch

Rationale: We propose the Circular-Time Sketch (CT-Sketch) algorithm in phase 1 to mine batches, which provides a more precise calculation of intervals between consecutive arriving items while consuming less memory.

Previous approaches like HyperCalm [2] divide the timeline into several time slices of length T and record the arrival time of items compactly using time slices. However, this fuzzy perception of time leads to inaccurate calculation of intervals, resulting in frequent misjudgments and omissions of batches. In contrast, CT-Sketch employs a more fine-grained solution by dividing the timeline into consecutive time slices of length L and using a more precise time point to calculate the interval within each time slice. We first transform the original continuous timestamp t into a recurring approximate time point s. Then, as shown in Fig. 3, there are two cases for computing intervals in CT-Sketch. In case 1, when two adjacent items arrive within the same time slice, the interval is computed as $\delta_{0,1} = s_1 - s_0$. In case 2, when two adjacent items appear in different time slices, resulting in $s_1 < s_0$, the interval is calculated as $\delta_{0,1} = L - s_0 + s_1$. The choice of parameter L affects the space required to store item arrival times. Hence, to reduce space overhead without compromising accuracy, we set L to approximately twice the value of T (*i.e.*, $L = \lceil 2T \rceil$). This configuration ensures the accurate calculation of batch intervals within a time slice.

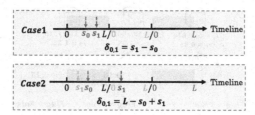

Fig. 3. Examples of circular-time slices of CT-Sketch.

Furthermore, CT-Sketch employs the concept of space sharing to store time points, allowing different items to utilize the same space for recording temporal information. This approach yields two significant benefits. On the one hand, temporal information of expired items will be automatically overwritten by new ones due to space sharing, obviating the need for an additional thread to clean expired items as required by Clock-Sketch [1]. On the other hand, items that arrive consecutively within a short period are mapped to the same location. According to the short-term memory of the data stream, we are free from the storage of item IDs [24], further reducing the space overhead.

Data Structure: CT-Sketch utilizes an array $\mathcal{C}$ consisting of m x-bit cells, denoted as $\mathcal{C}[1], \mathcal{C}[2], ..., \mathcal{C}[m]$, for storing the approximate arrival time point s of each item. The time point s is stored with the same precision as T, discarding any extra fractional parts to minimize storage overhead. Additionally, each item e is mapped to a specific cell $\mathcal{C}[h(e)]$ using a hash function $h(\cdot)$. All cells in $\mathcal{C}$ are initialized to 0 at the start of each measurement period.

Insert: For each arriving item e at time t, we first calculate its approximate time point $s_{now} = (t \bmod L)$, in which L is the length of a time slice. We then apply the hash function $h(e)$ to obtain the hashed cell $\mathcal{C}[h(e)]$. Based on the relationship between s_{now} and the time point of previous item s_{pre} stored in $\mathcal{C}[h(e)]$, there will be two cases:

Case 1: ($s_{now} - s_{pre} > T$) or ($s_{now} < s_{pre}$ and $L - s_{pre} + s_{now} > T$). It all means the arriving item e is a start of a new batch, and the batch stored in the $\mathcal{C}[h(e)]$ is outdated. So we set the expiration flag as *True*.

Case 2: Otherwise, it means the arriving item is still a member of the batch stored in the $\mathcal{C}[h(e)]$. So we set the expiration flag as *False*.

After identifying the start and end of batches, we should update the time point in $\mathcal{C}[h(e)]$.

Example 1: For an incoming item e_1 arriving at time $t_1 = 6.36044$, we first calculate the current time point $s_{now} = 0.36$. Then we get the hashed cell $\mathcal{C}[2]$ by $\mathcal{C}[h(e_1)]$ and obtain the time point of previous item $s_{pre} = 1.64$. We find that s_{now} is less than s_{pre}, so e appears in a new time slice. Therefore, the interval is computed as $2 - s_{pre} + s_{now}$, which is less than the threshold T and means a start of a new batch. According to **Case 1**, we set the expiration flag as *False*. Then, we write s_{now} to $\mathcal{C}[2]$ to update the records (Fig. 4).

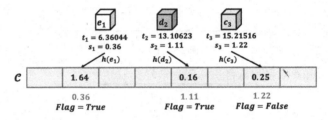

Fig. 4. Examples of insertions in CT-Sketch ($m = 8, T = 0.72, L = \lceil 2T \rceil = 2$).

Example 2: For an incoming item d_2 arriving at time $t_2 = 13.10623$, we first calculate the current time point $s_{now} = 1.11$. Then we get the hashed cell $\mathcal{C}[5]$ by $\mathcal{C}[h(d_2)]$ and obtain the time point of previous item $s_{pre} = 0.16$. Since $s_{now} - s_{pre} > T$ (*i.e.*, **Case 2**), we set the expiration flag as *True*. Then, we update $\mathcal{C}[5]$ to s_{now}.

Example 3: For an incoming item c_3 arriving at time $t_3 = 15.21516$, we first calculate the current time point $s_{now} = 1.22$. Then we get the hashed cell $\mathcal{C}[7]$ by $\mathcal{C}[h(c_3)]$ and obtain the time point of previous item $s_{pre} = 0.25$. Here, $s_{now} > s_{pre}$ and $2 - s_{pre} + s_{now} > T$. So, according to **Case 1**, a new batch was generated. We set the flag as *True* and write s_{now} to $\mathcal{C}[7]$.

3.2   Bloom Filter-Based Existence Recorder

The existence recorder in our CBA Sketch, called BE Recorder, utilizes a Bloom Filter [25] to track the presence of each batch within the current measurement period. This recording is done with a sampling probability denoted as p. By sampling, the BE Recorder effectively filters out noises, specifically batches with small sizes. Consequently, the accuracy of detecting and measuring batches with larger sizes is significantly improved.

Data Structure: The BE Recorder comprises a Bloom Filter $\mathcal{B}$, which is an array of u bits associated with q hash functions $g_i(\cdot)$, where $1 \leq i \leq q$. Additionally, we employ a sampling strategy to select each arriving item with a probability p, enabling effective noise filtering. All bits in $\mathcal{B}$ are initialized to 0 at the start of each measurement period.

Insert: When an incoming item e is encountered in the BE Recorder, a random decimal r is generated. If r less than the sampling probability p, indicating that the item is to be sampled. Hence, q hash functions $g_i(e)$ are computed to obtain the q corresponding hashed bits $\mathcal{B}[g_i(e)]$ in the Bloom Filter, which are then set to 1. Conversely, if r is greater than p, indicating that the item will be discarded, the algorithm proceeds to process the next item in the stream.

3.3 Approximate Size Recorder

Rationale: To address the challenge of handling a data stream with a potentially large and unknown number of batches without pre-allocating excessive memory, we introduce a novel metric called the *dual-mean size* (to be discussed in Sect. 3.4) to evaluate the size of persistent batches accurately. Instead of storing the exact size information of each batch, we use AS Recorder to estimate the dual-mean size by recording the total size and number of mined batches. By utilizing this approach, we can effectively evaluate the size of persistent batches while efficiently managing memory usage.

Data Structure: The AS Recorder adopts a structure based on Count-Min Sketch [26], which consists of d bucket arrays, $\mathcal{A}_1$, $\mathcal{A}_2$, ..., $\mathcal{A}_d$. Each array $\mathcal{A}_i$ is associated with one hash function $h_i(\cdot)$ and comprises w buckets. Each bucket in the arrays has two fields: $\mathcal{A}_i[h_i(\cdot)].size$, which records the number of items hashed to the bucket, and $\mathcal{A}_i[h_i(\cdot)].num$, which records the number of batches recorded in the bucket. This structure allows for efficient estimation of batch size while conserving memory resources. Both of the two fields in each bucket are initialized to 0 at the start of each measurement period.

Insert: When processing an incoming item e with its associated expiration flag, we compute d hash functions $h_i(e)$ to determine the corresponding buckets in the arrays. We then update each hashed bucket $\mathcal{A}_i[h_i(e)]$, where $1 \leq i \leq d$, based on the following three cases:

Case 1: If $\mathcal{A}_i[h_i(e)]$ is empty, regardless of the expiration flag, we just set $\mathcal{A}_i[h_i(e)].size = 1$ and $\mathcal{A}_i[h_i(e)].num = 1$. It means e first occurs in the current measurement period, generating a new batch B_e.

Case 2: If the expiration flag is *False* and $\mathcal{A}_i[h_i(e)]$ is not empty, we increment $\mathcal{A}_i[h_i(e)].size$ by 1. It means e is still a member of the current batch stored in the bucket, so we just increment the total size to record this item.

Case 3: If the expiration flag is *True* and $\mathcal{A}_i[h_i(e)]$ is not empty, we increment $\mathcal{A}_i[h_i(e)].size$ and $\mathcal{A}_i[h_i(e)].num$ by 1 separately. This indicates that the latest batch stored in the bucket has expired, and the arrival of item e marks the start of a new batch. We update the number of batches in the bucket to reflect the generation of a new batch, and we also increase the total size to account for the inclusion of this item (Fig. 5).

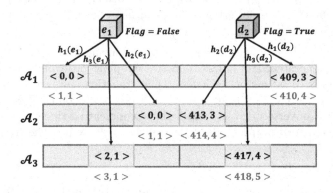

Fig. 5. Examples of insertions in AS Recorder ($d = 3, w = 6$).

Example 1: For an incoming item e_1, we get 3 hashed buckets $\mathcal{A}_1[h_1(e_1)]$, $\mathcal{A}_2[h_2(e_1)]$, $\mathcal{A}_3[h_3(e_1)]$. Since both $\mathcal{A}_1[h_1(e_1)]$ and $\mathcal{A}_2[h_2(e_1)]$ are empty, we directly modify them to $< 1, 1 >$. For $\mathcal{A}_3[h_3(e_1)]$, we set it to $< 3, 1 >$ because the expiration flag is *False*.

Example 2: When an incoming item d_2 arrives, we calculate the hashed buckets $\mathcal{A}_1[h_1(d_2)]$, $\mathcal{A}_2[h_2(d_2)]$, and $\mathcal{A}_3[h_3(d_2)]$. Since all three buckets are not empty and the flag is *True*, we need to record the generation of a new batch and add the new item to the count value. Therefore, for $\mathcal{A}_1[h_1(d_2)]$, we set it to $< 410, 4 >$. For $\mathcal{A}_2[h_2(d_2)]$, we set it to $< 414, 4 >$. Lastly, for $\mathcal{A}_3[h_3(d_2)]$, we set it to $< 418, 5 >$.

3.4 Query

After t measurement periods, CBA Sketch can receive and respond to queries regarding persistent batches. When given a query for batch B_e, the first step is to ascertain its persistence by examining the BE Recorders from the t periods. The BE Recorder corresponding to the i-th period employs q hash functions to locate the respective q bits of batch B_e. If all of these bits have a value of '1', it indicates that B_e occurred during the i-th measurement period. By performing this verification across all t periods, we can determine the number of periods in which B_e existed, denoted as M.

If $M \geq k$, B_e is reported as a persistent batch. To accurately measure the size of persistent batches, we introduce a novel metric called *dual-mean size*. The motivation behind employing this metric is as follows: Since the number of batches with the identical ID within a period is uncertain, allocating sufficient space in advance to store each batch individually is not feasible. Instead, we adopt a simple yet effective approach to record the average size of multiple batches sharing the same ID. This approach enables us to store only the total size and the number of batches associated with a specific ID, resulting in significant space savings. Moreover, calculating the average size helps mitigate the impact of small-sized batches on the performance of application tasks caused by data fluctuations.

To compute the dual-mean size of each persistent batch, we examine M AS Recorders corresponding to the periods in which batch B_e occurred. For each AS Recorder of the i-th period, we utilize hash functions to determine the d hashed buckets and report $\frac{A_j[h_j(e)].size}{A_j[h_j(e)].num}$, where $1 \leq j \leq d$, as the average batch size of B_e in the i-th period, where $A_j[h_j(e)]$ is the bucket with the smallest num cell among the d hashed buckets. Finally, we calculate the average value of the M average batch sizes across the M periods, which serves as the dual-mean size of persistent batch B_e.

If $M < k$, we return 0, indicating that the queried batch B_e is a non-persistent batch.

4   Evaluation

In this section, we comprehensively evaluate the performance of our proposed CBA Sketch by conducting extensive experiments using real Internet traffic traces. As no existing work focuses on mining persistent batches, we first introduce a strawman approach as a baseline and then outline the experimental setup. Subsequently, we conduct a sensitivity analysis to assess the impact of various parameters on the performance of the CBA Sketch. Finally, we evaluate the effectiveness and efficiency of CBA Sketch in the task of mining persistent batches.

4.1   Strawman Approach

The initial algorithm for measuring item batches was introduced in [1], encompassing two key algorithms: Clock-CM for measuring item batch sizes and Clock-BF for detecting item batch activeness. In order to provide a comparative analysis with our proposed CBA Sketch in mining persistent batches, we adopt these two algorithms as the foundation for our strawman method. Additionally, since Clock-Sketch cannot save historical records, the strawman method incorporates a CM Sketch [26] to aggregate batch size information and a BF [25] to record the presence of batches in each measurement period.

Specifically, the strawman algorithm utilizes Clock-CM to capture the size information of active batches. Once the recorded batches expired, their sizes stored in Clock-CM are transferred to the CM Sketch. However, Clock-CM does not store batch IDs. So the batch ID corresponding to each size must be available to record it in the CM Sketch upon detecting batch expiration. Hence, we append an additional cell after each bucket in Clock-CM to store the complete batch ID.

Furthermore, the strawman approach utilizes Clock-BF to track the existence of active batches. When a batch expires, it is inserted into the BF. After t measurement periods, we can query these t BFs to determine if a batch is persistent. Similar to Clock-CM, Clock-BF does not store batch IDs. Therefore we also add a cell after each bucket in Clock-BF to store the batch ID.

4.2   Experimental Setup

Platform and Implementations: For the sake of comparison with the strawman method, all methods are written in C++ and evaluated on a server equipped with two Intel Xeon E5-2643 v4 @3.40GHz CPU and 256GB RAM. Besides, the hash functions are implemented by a well-known fast hash function, Murmur Hash[1].

Datasets: We utilize two real Internet traces, CAIDA-2016 and CAIDA-2019, each lasting for 5 min, as our datasets. These traces were downloaded from CAIDA and collected in 2016 and 2019, respectively [13,27]. In our experiments, we assign the destination IP (4 bytes) as the ID for each item in the datasets. For both CAIDA-2016 and CAIDA-2019, we set the batch threshold T to 0.72 s, following the recommendation in [2]. With this configuration, the CAIDA-2016 dataset has approximately 160,453 batches per minute, while the CAIDA-2019 dataset has around 770,523 batches per minute in the data stream.

Default Parameters: For our CBA Sketch, we set $x = 8$ in CT-Sketch, $p = 0.0005$ in BE Recorder, and $d = 3$ in AS Recorder, by default. For the strawman approach, we set the size of each clock cell in Clock-CM as 8 bits and the size of each clock cell in Clock-BF as 2 bits. We perform a total of five periods of measurements (*i.e.*, $t = 5$), each lasting for one minute. According to the definition, batches the same ID are referred to as a persistent batch if they are observed to occur in at least four measurement periods (*i.e.*, $k = 4$). The following experiments will be based on default parameters unless otherwise specified.

Evaluation Metrics: (1) **Average Relative Error (ARE):** Let n_e denotes the real dual-mean size of batch B_e, and $\hat{n}_e$ denotes the estimate dual-mean size of batch B_e. Ψ is the query set, each e in it is the ID of a batch. So the average relative error is denoted as $\frac{1}{|\Psi|} \sum_{e \in \Psi} \frac{|n_e - \hat{n}_e|}{n_e}$. (2) **Throughput:** We record the total time to insert all items and then calculate the throughput. The throughput is defined as $\frac{N}{\Delta}$, where N is the total number of items, and Δ is the total time used to insert them. We use Million of insertions per second (Mps) to measure the throughput.

---

[1] The source code of Murmur Hash is at https://github.com/aappleby/smhasher/tree/master/src/MurmurHash3.cpp.

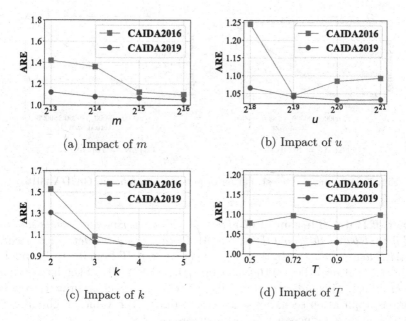

(a) Impact of m (b) Impact of u

(c) Impact of k (d) Impact of T

Fig. 6. Sensitivity analysis of CBA Sketch.

4.3   Sensitivity Analysis

Impact of the Length of CT-Sketch (m) (Fig. 6a): *The experimental results show that with the increase of m, the performance exhibits a continuous improvement.* In this experiment, we vary m from 2^{13} to 2^{16} with a step of 1024. We find that the ARE decreases continuously as m increases both in CAIDA-2016 and CAIDA-2019. The reason is evident because as m increases, the probability of different items mapping to the same cell decreases. Fewer hash collisions result in more accurate detection of the starts and ends of batches, thereby reducing subsequent estimation errors.

Impact of the Length of BE Recorder (u) (Fig. 6b): *The experimental results show that the optimal length of BE Recorder is 2^{19} and 2^{20} in CAIDA-2016 and CAIDA-2019 respectively.* In this experiment, we systematically vary the parameter u from 2^{18} to 2^{21} with an increment of 1024. We observe that as u increases, the ARE initially decreases significantly but then gradually increases.

Impact of the Persistence Threshold (k) (Fig. 6c): *The experimental results demonstrate that regardless of how the parameter k changes, the ARE of CBA Sketch consistently remains below 1.55.* In this experiment, we systematically vary k from 2 to 5 with a step of 1. We observe that as k increases, the ARE decreases, indicating that CBA Sketch becomes more accurate in identifying persistent batches. Therefore, the longer a batch persists in the data stream, the more reliably CBA Sketch can detect it as a persistent batch.

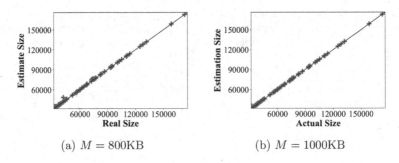

(a) $M = 800$KB   (b) $M = 1000$KB

Fig. 7. Dual-mean size estimation of persistent batches in CAIDA-2016.

Impact of Batch Threshold (T) (Fig. 6d): *The experimental results show that for different batch threshold T, CBA Sketch can always achieve high accuracy with ARE lower than 1.1.* In this experiment, we vary T from 0.5 to 1 randomly. We observe that regardless of the value of T, the ARE of CBA Sketch consistently stays around 1. This indicates that CBA Sketch can deliver accurate estimations in various application scenarios that require different T values, highlighting its versatility and robustness across different cases.

4.4   Experiments on ARE

To assess the accuracy of persistent batch detection under different memory constraints, we conducted experiments by varying the memory size from 600KB to 1200KB, with intervals of 200KB, for both two datasets. Figure 7 depicts the estimation of the dual-mean size of persistent batches by CBA Sketch with a memory size of 800KB and 1000KB for the CAIDA-2016 dataset. The x-axis represents the true values of dual-mean sizes of persistent batches, while the y-axis represents the corresponding estimated values. The solid line in the figure represents the function $y = x$, and the closer the data points are to this line, the higher the estimation accuracy. Notably, CBA Sketch exhibits exceptional accuracy, with minimal deviations from the ideal line, indicating its robustness in achieving highly accurate estimations even under constrained memory resources.

ARE in CAIDA-2016 vs. Memory Size (Fig. 8): In the CAIDA-2016 dataset, we observe significant differences in the ARE between the strawman and CBA Sketch for persistent batches with different dual-mean size ranges. When considering memory sizes of 600KB, 800KB, 1000KB, and 1200KB, the strawman exhibits ARE values that are about 62 times, 37 times, 19 times, and 15 times larger than those of CBA Sketch, respectively, for persistent batches with dual-mean sizes smaller than 5000. Similarly, for persistent batches with dual-mean sizes between 5000 and 20000, the strawman has ARE values about 4 times, 7 times, 10 times, and 7 times larger than CBA Sketch. Finally, for persistent batches with dual-mean sizes larger than 60000, the strawman has ARE values about 227 times, 179 times, 637 times, and 363 times larger than those of CBA Sketch. These results demonstrate that CBA Sketch outperforms the

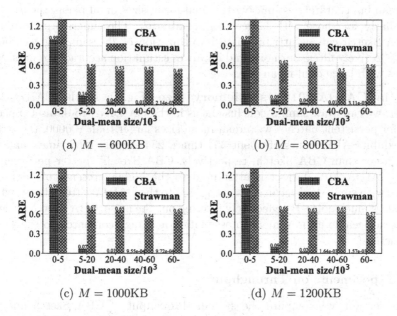

Fig. 8. The ARE of different dual-mean size range under different memory size in CAIDA-2016.

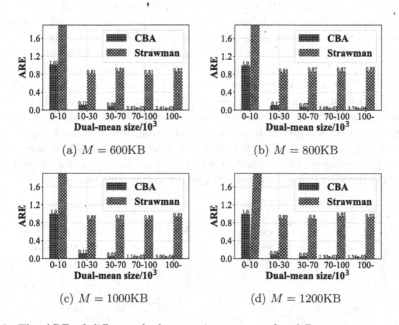

Fig. 9. The ARE of different dual-mean size range under different memory size in CAIDA-2019.

strawman in accurately estimating the dual-mean sizes of all persistent batches, particularly for those with larger dual-mean sizes. The superior performance of CBA Sketch can be attributed to its utilization of the sampling technique, which effectively filters out noise and improves estimation accuracy for persistent batches with large dual-mean sizes.

ARE in CAIDA-2019 vs. Memory Size (Fig. 9): In the CAIDA-2019 dataset, we can draw similar conclusions as in the CAIDA-2016 dataset. Specifically, for persistent batches with dual-mean sizes larger than 100000, the strawman exhibits ARE values of about 354 times, 2346 times, 1866 times, and 690 times larger than CBA Sketch, respectively. CBA Sketch's better performance in the CAIDA-2019 dataset is attributed to the higher presence of persistent batches with small dual-mean sizes, compared to the CAIDA-2016 dataset, which introduces much noise. By effectively filtering out the noise, CBA Sketch achieves a more noticeable improvement in estimation accuracy compared to the strawman method.

4.5   Experiments on Throughput

In this section, we compare the system throughput of CBA Sketch and the strawman method by varying the memory size from 600KB to 1200KB, focusing on the insertion speed. Our analysis indicates that CBA Sketch outperforms the strawman method regarding the processing speed.

As depicted in Fig. 10, the throughput comparison between CBA Sketch and the strawman method reveals that CBA Sketch consistently outperforms the strawman method in terms of speed. When using 600KB of memory, the average throughput of CBA Sketch is about 4.99 Mps for CAIDA-2016, which is approximately 1.8 times higher than that of the strawman method. Similarly, for CAIDA-2019, the average throughput of CBA Sketch is about 4.72 Mps with 600KB of memory, which is approximately 2 times higher than that of the strawman method. The superior performance of CBA Sketch in processing speed can be attributed to its lower computational overhead during each insertion compared to the strawman method. In contrast, the strawman method requires processing on two separate Clock-based data structures and relying on a separate thread to clean expired cells, significantly impeding its speed.

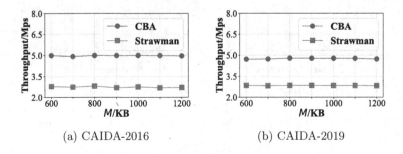

(a) CAIDA-2016   (b) CAIDA-2019

Fig. 10. Throughput comparison between CBA Sketch and the strawman method.

5   Conclusion

This paper introduces a novel pattern in data streams named persistent batches, which holds great significance for numerous applications, yet has not been explored. We propose the CBA Sketch, an efficient method for mining persistent batches in data streams, to bridge this research gap. The CBA Sketch comprises three phases, each designed to tackle specific challenges. In phase 1, we use a carefully designed CT-Sketch to identifies batches within limited memory space accurately. Phase 2 aims to record the information of mined batches in the proposed AS Recorder and BF Recorder. Finally, in phase 3, we synthesize past measurements to respond to queries and provide precise measurements effectively. Through extensive experiments, we demonstrate the effectiveness and efficiency of our approach, thereby highlighting its immense potential for practical applications in real-world scenarios.

Acknowledgements. This work is supported in part by the National Natural Science Foundation of China (NSFC) under Grant 62332013, Grant 62072322, Grant U20A20182, and Grant 62202322, in part by the Open Project of Tongji University Embedded System and Service Computing of Ministry of Education of China under Grant ESSCKF 2022-05, and in part by the Natural Science Foundation of Jiangsu Province under Grant BK20210706.

References

1. Chen, P., Chen, D., Zheng, L., Li, J., Yang, T.: Out of many we are one: measuring item batch with clock-sketch. In: Proceedings of the ACM SIGMOD International Conference on Management of Data, pp. 261–273 (2021)
2. Liu, Z., et al.: Hypercalm sketch: one-pass mining periodic batches in data streams. In: Proceedings of the International Conference on Data Engineering, pp. 14–26. IEEE (2023)
3. Zhao, F., Li, S., Zhou, B.B., Jin, H., Yang, L.T.: Hcache: a hash-based hybrid caching model for real-time streaming data analytics. IEEE Trans. Serv. Comput. **14**(5), 1384–1396 (2018)
4. Kotozaki, S., Tamura, K., Kitakami, H.: Identifying local burstiness in a sequence of batched georeferenced documents. Int. J. Electron. Commer. Stud. **6**(2), 269–288 (2015)
5. Choi, Y., Kim, Y., Rhu, M.: Lazy batching: an SLA-aware batching system for cloud machine learning inference. In: Proceedings of the IEEE International Symposium on High-Performance Computer Architecture, pp. 493–506. IEEE (2021)
6. Alshamrani, A., Myneni, S., Chowdhary, A., Huang, D.: A survey on advanced persistent threats: techniques, solutions, challenges, and research opportunities. IEEE Commun. Surv. Tutor. **21**(2), 1851–1877 (2019)
7. Virvilis, N., Gritzalis, D.: The big four-what we did wrong in advanced persistent threat detection? In: Proceedings of the International Conference on Availability, Reliability and Security, pp. 248–254. IEEE (2013)
8. Zargar, S.T., Joshi, J., Tipper, D.: A survey of defense mechanisms against distributed denial of service (DDoS) flooding attacks. IEEE Commun. Surv. Tutor. **15**(4), 2046–2069 (2013)

9. Yan, Q., Yu, F.R., Gong, Q., Li, J.: Software-defined networking (SDN) and distributed denial of service (DDoS) attacks in cloud computing environments: a survey, some research issues, and challenges. IEEE Commun. Surv. Tutor. **18**(1), 602–622 (2015)

10. Jing, X., Yan, Z., Pedrycz, W.: Security data collection and data analytics in the internet: a survey. IEEE Commun. Surv. Tutor. **21**(1), 586–618 (2018)

11. Pooranian, Z., Conti, M., Haddadi, H., Tafazolli, R.: Online advertising security: issues, taxonomy, and future directions. IEEE Commun. Surv. Tutor. **23**(4), 2494–2524 (2021)

12. Zhu, F., Zhang, C., Zheng, Z., Al Otaibi, S.: Click fraud detection of online advertising-LSH based tensor recovery mechanism. IEEE Trans. Intell. Transp. Syst. **23**(7), 9747–9754 (2021)

13. CAIDA: Anonymized internet traces 2019. https://catalog.caida.org/details/dataset/passive_2019_pcap. Accessed 27 Dec 2021

14. Sun, Y.E., Huang, H., Ma, C., Chen, S., Du, Y., Xiao, Q.: Online spread estimation with non-duplicate sampling. In: Proceedings of IEEE Conference on Computer Communications, pp. 2440–2448. IEEE (2020)

15. Huang, H., et al.: Spread estimation with non-duplicate sampling in high-speed networks. IEEE/ACM Trans. Netw. **29**(5), 2073–2086 (2021)

16. Du, Y., Huang, H., Sun, Y.E., Chen, S., Gao, G.: Self-adaptive sampling for network traffic measurement. In: Proceedings of IEEE Conference on Computer Communications, pp. 1–10. IEEE (2021)

17. Huang, H., et al.: You can drop but you can't hide: k-persistent spread estimation in high-speed networks. In: Proceedings of the IEEE Conference on Computer Communications, pp. 1889–1897. IEEE (2018)

18. Chen, L., Phan, R.C.W., Chen, Z., Huang, D.: Persistent items tracking in large data streams based on adaptive sampling. In: Proceedings of the IEEE Conference on Computer Communications, pp. 1948–1957. IEEE (2022)

19. Zhang, Y., et al.: On-off sketch: a fast and accurate sketch on persistence. Proc. VLDB Endow. **14**(2), 128–140 (2020)

20. Dai, H., Shahzad, M., Liu, A.X., Li, M., Zhong, Y., Chen, G.: Identifying and estimating persistent items in data streams. IEEE/ACM Trans. Netw. **26**(6), 2429–2442 (2018)

21. Cheng, S., Yang, D., Yang, T., Zhang, H., Cui, B.: LTC: a fast algorithm to accurately find significant items in data streams. IEEE Trans. Knowl. Data Eng. **34**(9), 4342–4356 (2020)

22. Fan, Z., et al.: Pisketch: finding persistent and infrequent flows. In: Proceedings of the ACM SIGCOMM Workshop on Formal Foundations and Security of Programmable Network Infrastructures, pp. 8–14 (2022)

23. Xiao, Q., Qiao, Y., Zhen, M., Chen, S.: Estimating the persistent spreads in high-speed networks. In: Proceedings of the International Conference on Network Protocols, pp. 131–142. IEEE (2014)

24. Du, Y., et al.: Short-term memory sampling for spread measurement in high-speed networks. In: Proceedings of the IEEE Conference on Computer Communications, pp. 470–479. IEEE (2022)

25. Bloom, B.H.: Space/time trade-offs in hash coding with allowable errors. Commun. ACM **13**(7), 422–426 (1970)

26. Cormode, G., Muthukrishnan, S.: An improved data stream summary: the count-min sketch and its applications. J. Algorithms **55**(1), 58–75 (2005)

27. CAIDA: Anonymized internet traces 2016. https://catalog.caida.org/details/dataset/passive_2016_pcap. Accessed 27 Dec 2021

Joint Optimization of System Bandwidth and Transmitting Power in Space-Air-Ground Integrated Mobile Edge Computing

Yuan Qiu[1,2], Jianwei Niu[1], Yiming Yao[1], Yuxuan Zhao[2], Tao Ren[3(✉)],
Xinzhong Zhu[2], and Kuntuo Zhu[2]

[1] State Key Laboratory of Virtual Reality Technology and Systems, School of
Computer Science and Engineering, Beihang University, Beijing 100191, China
[2] Shanghai Aerospace Electronic Technology Institute, Shanghai 201109, China
[3] Laboratory for Internet Software Technologies, Institute of Software Chinese
Academy of Sciences, Beijing 100190, China
taotao_1982@126.com

Abstract. Thanks to the rapid development of wireless communication
technology, i.e., B5G, 6G, mobile edge computing (MEC) has emerged
as a promising paradigm to facilitate various mobile applications, such
as intelligent connected vehicles, internet of remote things (IoRT), etc.
However, IoRT deployed in remote areas, e.g., oceans and deserts where
terrestrial communication infrastructures are scarce or even unavailable,
still suffers from poor quality of service due to unreliable connectivity. Facing this issue, this paper proposes a space-air-ground integrated
MEC framework with heterogeneous space, air, and ground communication resources to provide seamless and high-throughput traffic offloading
for IoRT. For the intractable traffic offloading problem of task-intensive
IoRT devices in dynamic SAGIN environments due to high-speed satellite movement, we propose a joint optimization method for system bandwidth and transmitting power to minimize total traffic offloading delay
in SAGIN. In view of the sequential decision-making property of the
problem, we further transform it into a Markov decision problem, which
is solved using the popular soft actor-critic reinforcement learning algorithm (SAC) with carefully designed reward functions. Extensive numerical results show that the RL-based traffic offloading policy can sub-
stantially reduce the delay of IoRT tasks, compared to baseline traffic
offloading methods.

Keywords: space-air-ground integrated network · traffic offloading ·
transmitting power · mobile edge computing

This work was supported in part by Zhejiang Provincial Natural Science Foundation of
China under Grant No. LY22F020006, and Beijing Municipal Science and Technology
Program under Grant No. Z221100007722001.

ⓒ The Author(s), under exclusive license to Springer Nature Singapore Pte Ltd. 2024
Z. Tari et al. (Eds.): ICA3PP 2023, LNCS 14492, pp. 133–152, 2024.
https://doi.org/10.1007/978-981-97-0811-6_8

1   Introduction

With the booming development of emerging mobile communication technologies, i.e., beyond 5G (B5G) and 6G, the wide application of internet of things (IoTs) with intelligent sensing capability makes the amount of traffic data increase massively. High quality of service (QoS) requirements such as low delay, high bandwidth, data reliability and network security [1], become more and more essential for IoT applications. In view of this, mobile edge computing (MEC) has been proposed as a promising technology to provide IoT devices with satisfactory computing resources and task delay, primarily by deploying computing and storage resources on the edge of mobile networks close IoT users [2]. Recent years have seen a large number of efforts made by researchers to develop efficient traffic loading approaches in MEC. However, for internet of remote things (IoRT) [3] deployed in areas with scarce or unavailable terrestrial ground communication structures, such as oceans, deserts, and disaster environments, how to provide efficient MEC services is still a challenge.

In recent years, the rapid development of space-air-ground integrated network (SAGIN), which can provide seamless coverage and flexible connection for IoRT devices worldwide, has attracted more and more attention from academic and becomes a key technology of the next generation communication [4]. MEC-assisted SAGIN can provide computing, communication, or storage resources to IoRT devices nearby. Furthermore, it makes use of edge nodes in the air and space segments to provide data sharing and collection, traffic offloading, and data-intensive task processing for IoRT devices (when the ground communication infrastructure is unavailable), thus achieving efficient collaboration between heterogeneous network nodes. For example, continuous situational awareness of the ocean can be gained by collecting information from various types of aquatic devices over a large surface area [5].

In the air segment of SAGIN, data collection, transmission and task processing are mainly carried out by unmanned aerial vehicles (UAVs). The advantages of UAVs lie in their high mobility, easy deployment, line-of-sight (LOS) transmission and low cost. However, the limitation of carrying capacity, antenna size, bandwidth and small-scale fading are constrained by the UAV platform, thus reducing the offloading and processing performance [6].

Satellites can be divided into low earth orbit (LEO), medium earth orbit (MEO) and geostationary earth orbit (GEO) according to orbital altitudes. Among them, low-orbit satellites are more widely used because of their lower delay and less launch cost [7]. Currently, LEO satellite constellations, such as SpaceX and OneWeb [8,9], are under development and deployment, and partly provide application services to IoRT devices [10]. LEO satellites can be used as a useful type of supplement in SAGIN due to their large coverage area and high transmission rate.

Combining the easy deployment of UAVs and wide coverage of satellites, this paper proposes a framework for space/air-based traffic offloading to provide support for data-intensive task requirements of IoRT devices, e.g., data sharing and collection. We investigate how to achieve the optimization objective of minimizing total IoRT task delay by jointly optimizing the system bandwidth, transmitting power, and offloading decisions. The main contributions of this article are summarized as follows:

- We propose a data traffic offloading framework in SAGIN to provide satisfactory data collection services for IoRT devices deployed in remote areas, by jointly utilizing UAVs and LEO satellites.
- We formulate the traffic-delay minimization problem by optimizing system bandwidth, transmitting power and offloading decisions under the constraints of IoRT energy budget and LEO satellite visible periods.
- We transform the minimization problem into a Markov decision problem according to the sequential decision-making property, which are further solved based on the popular soft actor-critic reinforcement learning algorithm.
- Extensive experiments are conducted via numerical simulation, and the experimental results show that the proposed approach could achieve more desirable traffic offloading performances than baseline methods.

The rest of this paper is organized as follows. In Sect. 2, the related works most relevant to this article are discussed. In Sect. 3, we present the network architecture, system model and corresponding problem formulation, followed by a description of our proposed algorithm in Sect. 4. Simulation results are provided in Sect. 5. Finally, the paper is concluded in Sect. 6.

2 Related Works

To develop efficient traffic loading approaches in MEC, researchers have been exploring various innovations, especially in the context of communications among UAV, space and ground.

2.1 Research on Air-Ground Communication

The authors in [11] proposed a UAV-based MEC network framework for 6G wireless communication applications. In their model, a UAV functions as an MEC server, collecting data from various ground users. This framework is enhanced by the use of the intelligent reflection surface (IRS) to boost both wireless data and energy transmissions. Their focus was on minimizing processing time by optimizing the phases of IRS, the UAV's trajectory, and resource allocation of computation capabilities.

Notably, efficient data collection and transmission by UAVs in this heterogeneous and dynamic network environment is challenging due to limited spectrum. Li X et al., in [12], tackled the aforementioned challenges by introducing a joint optimization policy. They employ a model-free deep reinforcement learning algorithm, enabling cognitive UAVs to optimize energy efficiency. Despite UAVs being excellent tools for remote device data collection and situation acquisition, they face difficulties in satisfying stringent requirements like transmission rates and user delays. This is primarily because of their limited coverage capabilities and signal constraints, including phenomena like small-scale fading.

2.2 Research on Space-Ground Communication

Shifting the focus to space-terrestrial communications, spectrum resources become critical, particularly when a vast number of ground users aim to offload data to satellites. The authors in [13] take this into account, introducing a novel multiple access method named pattern division multiple access (PDMA) to optimize satellite-terrestrial cooperative networks. Their resource allocation algorithm addresses spectrum splitting and sub-carrier scheduling problems, aiming for high scalability and numerous connections in future wireless networks. This method also facilitates large range connectivity, utilizing extensive coverage of space-terrestrial communication and inter-satellite link relay.

Due to the high communication power resulting from the high frequency and high rate of space-terrestrial communication, how to achieve the optimal offloading performance remains a challenge. Recognizing this, Peng C et al. in [14] set out to increase the system's uplink rate while reducing energy consumption. Their approach focuses on the joint energy efficiency maximum problem.

2.3 Research on SAGIN Communication

When the SAGIN are considered and applied, some researchers in [15,16] proposed an innovative network framework that utilizes both GEO satellites and LEO satellites. Their approach aims to service remote ground users, where HAP first gathers user device data and forwards it to satellites via a LEO-UAV link. The satellite then relays the data either directly or indirectly to the ground data center.

All of the above studies deal with the problems and solutions encountered in the offloading process of edge computing networks. It can be seen that as the interaction of communication networks becomes complex, the problem of how to optimize the offloading strategy of edge computing to reduce latency and energy consumption is becoming more and more difficult to model and solve mathematically. Some studies have solved the problem mathematically by decomposing the problem, and some have improved the efficiency by adding auxiliary nodes to the model. However, the deep reinforcement learning method has only been used in the space-ground communication research. The authors believe that deep reinforcement learning methods have great potential and feasibility in solving this type of problems with high complexity and stochasticity.

3 System Model and Problem Formulation

In this section, the architecture for space/air-based traffic offloading in SAGIN is presented, and the LEO satellite coverage time model, channel model and time delay model are discussed. Moreover, we formulate the joint optimization problem.

3.1   Network Architecture

The network architecture is shown in Fig. 1. The SAGIN network described in this paper consists of space segment, air segment and ground segment. In order to realize continuous connection for multiple ground terminals in a wide area, the space segment consists of LEO satellite grid constellation, expressed as $\mathbb{L} = \{1, 2, 3, ..., L\}$. The air segment consists of a number of UAVs, denoted as $\mathbb{U} = \{1, 2, 3, ..., U\}$, equipped with inter-UAV communication and radar devices to optimize the flight path while carrying out collision detection to maintain safe flight spacing. There are IoRT devices distributed in the ground segment, e.g., remote area. Due to the requirements of data collection, situational awareness or information coordination, the IoRT devices need to offload the collected data to the space/air-based edge nodes in SAGIN.

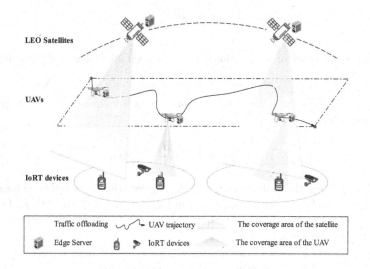

Fig. 1. Network architecture for space-air-based traffic offloading in SAGIN.

3.2   Satellite Coverage Time

The satellite constellation orbits can be divided into polar orbits and inclined orbits. Different from the fixed-position base stations on the ground, LEO satellites fly in a pre-planned orbit, which is assumed to be a circular orbit in these papers [17,18]. Hence, the location of LEO satellites is time-varying and dynamic, and at the same time, the satellite cannot communicate with the IoRT devices at any time unless their geometric relationship satisfies the specific conditions [19]. Space geometry relationship between the LEO satellite and the IoRT devices is shown in Fig. 2.

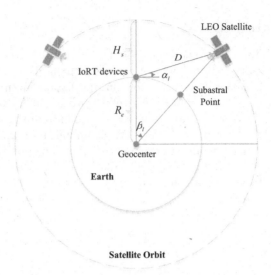

Fig. 2. Space geometry of the relationship between satellite orbit and IoRT devices.

Herein, α_l represents the elevation angle between the IoRT device and the l-th LEO satellite ($l \in \mathbb{L}$), which can be denoted as [17]:

$$\alpha_l = \arccos(\frac{R_e + H_s}{D} \cdot \sin\beta_l),\tag{1}$$

where R_e expresses the radius of the earth, H_s denotes the satellite orbital altitude, D indicates the distance between the IoRT device and the l-th LEO satellite, β_l is the geocentric angle corresponding to the LEO satellite coverage area and can be calculated as:

$$\beta_l = \arccos(\frac{R_e}{R_e + H_s} \cdot \cos\alpha_l) - \alpha_l.\tag{2}$$

L_l is the arc length that the IoRT device can communicate with the LEO satellite which can be obtained by:

$$L_l = 2 \cdot (R_e + H_s) \cdot \beta_l.\tag{3}$$

Thus, the longest communication time between the IoRT device and the LEO, called satellite coverage time, can be denoted as:

$$T_l = \frac{L_l}{v_s},\tag{4}$$

where v_s indicates the speed of the LEO satellite.

3.3   IoRT Devices-Satellite/UAV Channel

IoRT-UAV: Firstly, the air-ground communication channel adopting UHF frequency band is considered. Similar to the existing work [20,21], the channel fading model from the i-th IoRT device to the u-th UAV ($u \in \mathbb{U}$) at the t-th time slot is:

$$h_{i,u}^t = \delta_{i,u}(t)H, \tag{5}$$

where $\delta_{i,u}$ denotes air-ground pass-loss coefficient which can be represented as

$$\delta_{i,u}(t) = \beta_0 d_{i,u}^{-\varepsilon}(t), \tag{6}$$

where β_0 denotes the channel power at the reference distance $d_0 = 1$m, variable ε is the exponent of the path loss function for the air-ground channel, and its value varies with the flight height [22]. Without loss of generality, ε is set as 2 in this paper. In addition, $d_{i,u}$ indicates the distance between the IoRT device and the UAV at the t-th time slot which can be expressed as:

$$d_{i,u}(t) = \sqrt{H_{uav}^2 + ||w_u[t] - w_i||^2}, \tag{7}$$

where H_{uav} denotes the fight altitude of the UAV. Similar to existing work [3], the UAV selects a fixed-wing type and is assumed to fly at a fixed altitude in this paper. Moreover, $||w_u[t] - w_i||$ denotes the horizontal distance between the $i-$th IoRT device and the $u-$th UAV. Herein, the small-scale fading of the IoRT-UAV communication link is expressed as [20,23]:

$$H = \sqrt{\frac{\varsigma}{\varsigma+1}}H_0 + \sqrt{\frac{1}{\varsigma+1}}H_\Delta, \tag{8}$$

where ς indicates Rician fading factor, H_0 is the line-of-sight (LoS) exponent that satisfies $|H_0| = 1$ and H_Δ indicates the none-line-of-sight (NLoS) Rayleigh fading exponent that conforms to $H_\Delta \sim \mathcal{C}, \mathcal{N}(0, I)$. Considering the adjacent interference of multiple IoRT devices, the uplink offloading data rate of the IoRT-UAV link at the $t-$th time slot can be expressed as:

$$R_{i,u}^t = u_u(i)B_u \log_2(1 + \frac{p_i^t h_{i,u}^t}{\sum_{j=1, j \neq i}^N p_j^t h_{j,u}^t + \sigma^2}), \tag{9}$$

where $u_u(i)$, B_u, p_i^t and σ^2 denote the spectrum bandwidth allocated to the i-th IoRT device from the u-th UAV, the maximum communication bandwidth of the u-th UAV, the uplink transmitting power of the i-th IoRT device at the t-th time slot and the additive white Gaussian noise (AWGN) power, respectively.

IoRT-Satellite: Considering the limitation of IoRT device's antenna size and the demand of transmission rate, similar to the existing work [24,25], the space-ground data communication adopts Ku or Ka frequency band in this paper. Therefore, the frequency band is not subject to frequency interference from terrestrial wireless networks. Based on the large-scale fading and free space pass-loss

model in most existing studies, the space-ground communication channel model
from i-th IoRT device to the l-th LEO satellite at the t-th time slot can be represented in terms of the variable. In units, $h_{i,l}^t$ can be calculated as [22, 25, 26]:

$$h_{i,l}^t = 32.44 + 20 \log(f_l) + 20 \log(D^t) + F_{\text{rain}}, \tag{10}$$

where f_l denotes the communication frequency with the unit of MHz. In addition,
the unit of D^t is km, which denotes the distance between the i-th IoRT device
and the l-th LEO satellite at the t-th time slot. It can be seen that the fading
model of the IoRT-Satellite communication link is significantly worsened with
the increase of the distance. Moreover, the channel model considers the effect
of rain attenuation, which will further bring attenuation to the communication
channel (as we know, the higher frequency band will bring more severe rain
attenuation effect). Herein, we're going to represent this variable in terms of a
constant F_{rain}.

Therefore, the uplink offloading data rate of the IoRT-Satellite at the t-th
time slot can be expressed as:

$$R_{i,l}^t = u_l(i) B_l \log_2(1 + \frac{p_i^t h_{i,l}^t}{\sum_{j=1, j \neq i}^N p_j^t h_{j,l}^t + \sigma^2}), \tag{11}$$

where $u_l(i)$ and B_l denote the spectrum bandwidth allocated to the i-th IoRT
device from the l-th LEO and the maximum communication bandwidth of the l-
th LEO, respectively. Due to the high bandwidth and frequency band of satellite-
based communication, it is easier to obtain a higher data rate compared with
the air-ground link, but it also brings greater energy consumption at the same
time. Therefore, the transmitting power allocation of IoRT devices should be
carefully adjusted.

3.4 Traffic Offloading Delay

The i-th IoRT device in the remote area continuously obtains data and information through its own sensing ability, and timely offloads the n bits collected data,
which is denoted as D_i^n, to the edge server to satisfy the requirement of network
node collaboration. The air-ground link adopts the omnidirectional antenna to
realize real-time and stable connectivity. However, the unstable link between
IoRT devices and the LEO satellite is influenced by the directivity of antenna
and time slot allocation, etc. To simplify the description, random accessing time
will not be considered in the traffic offloading delay model. In the scenario of
data traffic offloading task based on edge node in SAGIN, the IoRT device can
choose UAV-based edge node or satellite-based edge node for interaction, while
the data transmission from one IoRT device can only select one edge node once.
Moreover, the time delay from the i-th IoRT device to the u-th UAV and the
time delay from the i-th IoRT device to the l-th LEO satellite can be expressed
as $T_{\text{uav}}^t(i, u) = \frac{D_i^n}{R_{i,u}^t}$ and $T_{\text{leo}}^t(i, l) = \frac{D_i^n}{R_{i,l}^t}$, at the t-th time slot, respectively.

3.5   Problem Formulation

The problem is formulated to minimize the total uplink traffic offloading time from the IoRT devices as follows:

$$
\begin{aligned}
&\min_{u,p,\eta} \ \sum_{t\in TS} F_t \\
&\text{s.t.} \ \ \eta \in [0,1], \forall i \in N, t \in TS, \\
&\quad \sum_{i=1}^{N} (1-\eta)u_u(i) = 1, \forall\, u \in U, \\
&\quad \sum_{i=1}^{N} \eta u_l(i) = 1, \forall\, l \in L, t \in TS, \\
&\quad \sum_{t\in TS} \sum_{l=1}^{L} \sum_{u=1}^{U} ((1-\eta)T_{uav}^{t}(i,u) + \eta\, T_{leo}^{t}(i,l))p_i^t \le E_{i,max}, \\
&\quad \sum_{t\in TS} \sum_{l=1}^{L} \sum_{u=1}^{U} T_{leo}^{t}(i,l) \le T_l,
\end{aligned}
\tag{12}
$$

where $\mathbb{F}_t$ is the objective function in our proposal which can be calculated as $\mathbb{F}_t = \sum_{l=1}^{L} \sum_{u=1}^{U} \sum_{i=1}^{N} ((1-\eta)T_{uav}^{t}(i,u) + \eta T_{leo}^{t}(i,l))$, standing for the total time consumption for all IoRT devices to offload collected data at the t-th time slot. Moreover, binary variable η indicates the offloading decision. When $\eta=0$, it indicates that the current IoRT device will offload the data up to the UAV; Otherwise, the LEO satellite. TS denotes the total number of time slots in the entire offloading process. The bandwidth of IoRT devices adopts subcarrier allocation, which is segmented by a ratio factor, named $u_u(i)$ for IoRT-UAV link and $u_l(i)$ for IoRT-Satellite link, respectively. The sum of the allocation ratio of IoRT devices is 1 in practice. The constraint variable $E_{i,max}$ indicates the maximal energy of the IoRT device for traffic offloading which is calculated by multiplying the offloading time and the transmitting power. The transmitting power needs to be carefully adjusted under different scenarios to achieve the objective of minimizing the total traffic offloading time. Moreover, the sum of the space-based offloading time should be less than the satellite coverage time T_l.

4   Algorithm Design

The target of this paper is to obtain the best offloading decision by minimizing the total latency of the IoRT devices in SAGIN to offload data to the edge nodes throughout the process. By analyzing the related work and the model proposed in this paper, the authors concluded that deep reinforcement learning (SAC in this paper) would be an effective and feasible solution for this type of model with more variables and the presence of parameters that constrain each other. The conclusions obtained in this way are more flexible and expandable than mathematical approaches. In addition, deep reinforcement learning methods are useful to later change the optimization objective based on this mode, for example, to include optimization of energy consumption or to complicate the model by adding other functionalities in addition to task uploading. We give a specific description of Markov decision process (MDP) in the scheme of our proposal firstly, and then the detailed algorithm will be discussed.

4.1 DRL-Oriented Problem Transformation

Our research objective will be achieved in the form of MDP, which is commonly described by a quintuple $M = \langle S, A, P_{\text{state}}, r, \gamma \rangle$ with each parameter representing state, action, state transition probability, immediate reward, and discount factor, respectively [27]. Since minimizing $\sum_{t \in TS} \mathbb{F}_t$ is equivalent to maximizing $\sum_{t \in TS} r(t)$ if $r(t)$ is inversely proportional to $\mathbb{F}_t$, the original objective will be transformed into an MDP problem $\tilde{\mathcal{P}}$ which can be solved from the Deep Reinforcement Learning (DRL) method. In fact, the reward decreases with the discount factor γ, indicating the declining consequence of subsequent rewards. Thus, the transformed problem $\tilde{\mathcal{P}}$ can be express as:

$$\tilde{\mathcal{P}} : \max_{u,p,\eta} \sum_{t=1}^{TS} \gamma_{t-1} r(t). \tag{13}$$

The three significant components of our proposed MDP problem based on the DRL algorithm, i.e., S, A and r, are defined as follows.

- State space: IoRT device makes traffic offloading decisions according to the DRL state $s(t)$ observed in the time-varying environment in each time slot. In this paper, the DRL state is defined as a quintuple consisting of five variables, i.e., task data size of each IoRT device, time tolerance delay of each task, residual energy of IoRT device, residual energy of UAV and residual visible time for LEO satellite. Therefore, the state at the $t-$th time slot is defined as

$$s(t) = \{D^n(t), T^{\text{tol}}(t), E_i^{\text{res}}(t), E_u^{\text{res}}(t), T_l^{\text{visble}}(t)\}. \tag{14}$$
$$\forall i \in N$$

- Action space: In order to maximize the accumulated rewards, IoRT devices make choices of four variables, i.e., IoRT-UAV bandwidth allocation $u_u(t)$, IoRT-Satellite bandwidth allocation $u_l(t)$, transmitting power scheduling of the i-th IoRT device p_i^t, offloading decision $\eta(t)$. Hence, the action at the t-th time slot is defined as

$$a(t) = \{u_u(t), u_l(t), p_i^t, \eta(t)\}. \tag{15}$$

- Reward: After accomplishing an action $a(t)$ in state $s(t)$, a combined reward consisting of the immediate reward $c(t)$ which is the negative of $\mathbb{F}_t$ when all constraints in Eq. (12) are satisfied, a penalty $\hat{p}(t)$ corresponding to the violated constraint and a persistent reward $\hat{r}(t)$ encouraging further exploration step of the simulation episode is awarded. Thus, the reward function at the t-th time slot can be represented as

$$r(t) = c(t) - \tilde{p}(t) + \tilde{r}(t). \tag{16}$$

Herein, the reward function needs to be prudently designed to avoid unpredictable simulation results, e.g., premature stopping due to unexpected situations or failed convergence with continuous exploration.

4.2   SAC-Based Offloading Algorithm

DRL algorithm has been proved to be able to solve the offloading decision problem in MEC environment, based on the existing researches [12, 26–29]. However, the performance of these methods is influenced by two major challenges: high sample complexity and their vulnerability in terms of hyperparameters. To improve the data sample efficiency, a popular off-policy actor-critic (AC) algorithm, named DDPG, which can be viewed both as a deterministic AC method and an approximate Q-learning algorithm, alleviates the challenge of high sampling complexity significantly. Unfortunately, these algorithms, such as DDPG, are still difficult to stabilize and brittle to hyperparameter selection [30, 31].

To solve the above multi-agent MDP, considering the high sample complexity and their vulnerability in terms of hyperparameters, soft actor-critic (SAC) algorithm introduces an off-policy actor-critic framework with a stochastic actor aiming for the objective of entropy maximization, resulting in a drastic improvement in stability and exploration, described in detail in Fig. 3.

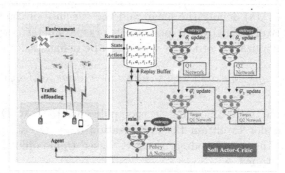

Fig. 3. The SAC-based offloading algorithm for SAGIN.

The final optimization objective is to obtain a policy for continuous action space decision to maximize both the expected return and the expected entropy of the policy, which can be expressed as:

$$\pi^* = arg\max_{\pi} \sum_t \mathbb{E}_{(s_t, a_t) \sim p_\pi}[r(s_t, a_t) + \alpha\mathcal{H}(\pi(\cdot|s_t))], \tag{17}$$

where α denotes the temperature parameter and $\mathcal{H}$ denotes the entropy function. Entropy is the measure of the randomness of a variable, and the entropy function of a variable x following the probability distribution P can be calculated as $\mathcal{H}(P) = \mathbb{E}_{x \sim P}[-logP(x)]$.

The network structure of the proposed algorithm consists of five parts: two critic Q networks, two critic target Q networks, and one actor policy network. The parameters of these networks are $\theta_1, \theta_2, \tilde{\psi}_1, \tilde{\psi}_2, \phi$. These five parameters are also mentioned in the pseudocode below to demonstrate how the SAC algorithm work in the formulation. Similar to other off-policy algorithms, SAC also stores the quadruple $(s_t, a_t, r(s_t, a_t), s_{t+1})$ from each episode into replay buffer $\mathcal{B}$, which is sampled at each step to update the network parameters. Moreover, θ_i is the soft Q-function parameter which can be trained based on the following loss function:

$$J_Q(\theta_\kappa) = \mathbb{E}_{(s_t,a_t)\sim\mathcal{B}}\frac{1}{2}(Q_{\theta_\kappa}(s_t, a_t) - (r(s_t, a_t) + \gamma\mathbb{E}_{s_{t+1}\sim p}[V_{\tilde{\psi}_\kappa}(s_{t+1})]))^2], \quad (18)$$

where $\kappa \in [1,2]$, $\tilde{\psi}_i$ denote the number of Q network parameters, the parameter of the target soft Q-function which can be obtained as a moving average of the soft Q weights, respectively. In addition, the soft state value function can be expressed as:

$$V(s_t) = \mathbb{E}_{a_t\sim\pi}[Q(s_t, a_t) - \alpha log\pi(a_t|s_t)]. \quad (19)$$

Moreover, in order to ensure the randomness of the policy and the differentiability of the process, the reparameterization trick is applied to sample actions from the policy distribution which can be further expressed as:

$$\tilde{a}_t(s, \zeta) = tanh(\mu_\phi(s) + \sigma_\phi(s) \odot \zeta), \zeta \sim \mathcal{N}(0, I), \quad (20)$$

where $\mu_\phi(s)$ and $\sigma_\phi(s)$ denote the mean and the variance of the Gaussian distribution, respectively. Finally, the policy network parameter ϕ will be trained to pursue the maximization objective which can be defined as:

$$\max_\phi\mathbb{E}_{(s_t\sim\mathcal{B},\zeta\sim\mathcal{N})}[\min_{\kappa\in[1,2]} Q_\theta(s_t, \tilde{a}_t(s,\zeta)) - \alpha log\pi_\phi(\tilde{a}_t(s,\zeta)|s_t)]. \quad (21)$$

We optimize the traffic offloading decision policy via the SAC DRL algorithm [30] in each time slot to obtain the optimization objective of minimizing the total time delay. The pseudocode of the SAC-based traffic offloading algorithm is shown in Algorithm 1.

Algorithm 1. SAC-Based Offloading Algorithm for SAGIN

Input: The initial location of the UAV and the LEO satellite, the initial battery capacity of the UAV and IoRT devices, and the initial task.
for all *episode* **do**
    Reset environment
    for all each time slot $t \in TS$ **do**
        Observe state $s(t)$ and select action $a(t)$ based on policy network $\pi(\phi)$
        for all each IoRT device i **do**
            if offloading decision $\eta=0$ **then**
                Calculate the channel gain $h_{i,u}^t$ to u-th UAV via Eq. (5)
                Calculate the uplink transmission delay $T_{uav}^t(i, u) = \frac{D_i^n}{R_{i,u}^t}$
                Calculate the IoRT device's energy consumption via $T_{uav}^t(i, u)p_i^t$
                Calculate the UAV's energy consumption via $T_{uav}^t(i, u)p_{uav}$
            else
                Calculate the channel gain $h_{i,l}^t$ to l-th satellite via Eq. (10)
                Calculate the uplink transmission delay $T_{leo}^t(i, l) = \frac{D_i^n}{R_{i,l}^t}$

Calculate the IoRT device's energy consumption via $T_{\text{leo}}^t(i,l)p_i^t$
end if
Calculate the reward $r(t)$
Update environment to get the next state $s(t+1)$
end for
Push experience $\{s_t, a_t, r_t, s_{t+1}\}$ to experience pool $\mathcal{B}$
if experience pool $\mathcal{B}$ is full **then**
Update policy and critic parameter $(\theta_1, \theta_2, \tilde{\psi}_1, \tilde{\psi}_2, \phi)$ via SAC algorithm
end if
end for
end for

5   Performance Evaluation

In this section, we conduct extensive simulations for traffic offloading in the SAGIN composed of LEO satellite, UAV and IoRT devices to verify the performance of the proposed algorithm. In our proposal, the corresponding algorithms and simulations are implemented Based on Python and evaluated on a workstation with an Intel Core i9-9900K 8-core CPU @3.6GHz, NVIDIA GeForce RTX 2080 Ti and 32 GB RAM.

Table 1. Simulation parameters

Parameters	Values
IoRT-Satellite communication frequency f_{leo}	20 GHz
IoRT-Satellite bandwidth B_l	50 MHz
Rain attenuation F_{rain}	12.2 dB
LEO coverage radius D_{rad}	100 Km
IoRT-UAV communication frequency f_{uav}	400 MHz
IoRT-UAV bandwidth B_u	10 MHz
Rician fading factor ς	6
Max Energy of UAV $E_{u,max}$	100 KJ
Communication noise σ^2	−70 dBm
Number of IoRT devices N	[5,9]
Task data size of i-th IoRT device D_i^n	[100 Kb, 2 Mb]
Max transmitting power of IoRT $p_{i,max}$	6.3 W
Max Energy of IoRT $E_{i,max}$	30 J
UAV moving speed v_{uav}	0–50 m/s
Number of slots TS	60
IoRT moving speed v_{IoRT}	0–5 m/s

5.1   Parameter Setting

Simulation parameters are listed in Table 1. The IoRT devices are randomly distributed in remote area within a geographical size 500 m × 500 m. The IoRT data are randomly generated from 100 Kb to 2 Mb. The UAV flies at a fixed altitude of 500 m during the entire traffic offloading phase. The orbit of the LEO satellite is established by the Satellite Tool Kit (STK), which adopts an altitude of 900 km.

As mentioned earlier, the edge servers with data collection capability in our proposal will be deployed on UAVs and LEO satellites. Hence, offloading decisions of the IoRT devices need to be adjusted. Thus, we consider the following baseline schemes.

- (UAV-Only Scheme) only executes data offloading tasks on the UAV without further offloading to any other MEC servers.
- (Satellite-Only Scheme) only executes data offloading tasks on the satellite without further offloading to any other MEC servers.
- (Random Scheme) randomly executes data offloading tasks on the UAV or satellite.

5.2   Performance Evaluation

In this section, a couple of numerical simulation results are shown as follows. To prove the efficiency and performance of the offloading policy optimization, we have compared our proposed method with other baseline schemes regarding the total time delay of IoRT devices.

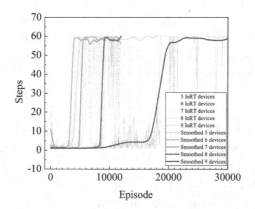

Fig. 4. The training process of SAC-based traffic offloading algorithm.

Firstly, the training process of the SAC algorithm for this scenario is illustrated in Fig. 4. Because of the limited training time and insufficient training episodes, there is still a probability that the curve will dip after convergence.

However, due to the clear convergence trend observed and the fact that the agent performance had already reached the set maximum, there was no need to wait until the curve was fully stabilized. The authors therefore smoothed the actual data curve and fitted a graph that is more convenient for analysis. To facilitate observation, the smoothed curves are depicted in bold. As mentioned, offloading action and successful task completion of an IoRT device are considered as one step. In this environment, the best performance an IoRT device can achieve is to complete all steps during the entire episode without interruption. It can be observed that the algorithm can adapt to the increasing num-ber of IoRT devices in the environment and successfully complete all tasks.

With the increasing of IoRT devices, the complexity of the state space and action space for the agent also grows, making the task more challenging and requiring a larger number of episodes for convergence. In the initial stages when the number of IoRT devices is not large (5–6), the training speed of the algorithm is relatively similar, converging after around 3000–5000 episodes. Interestingly, the training speed even becomes faster when the number of IoRT devices changes from 5 to 6. However, when the number of IoRT devices reaches 8, the complexity increases, and the required number of episodes noticeably increases to around 9000. In the environment with 9 IoRT devices, it can be observed that the algorithm keeps exploring without getting trapped in local optima. It is only after approximately 17000 episodes that the algorithm discovers an effective resource allocation policy and gradually shows signs of convergence. In this environment, the exploration and experimentation of the agent become more challenging, as reflected by the decreased slope of the curve and significantly slower convergence speed when there are 9 IoRT devices. Similar training processes can be observed when changing the maximum data rate and transmitting power in the environment.

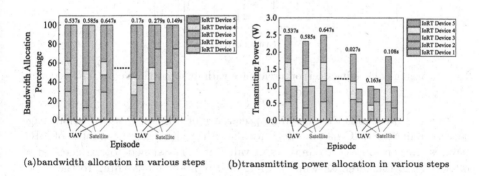

(a)bandwidth allocation in various steps   (b)transmitting power allocation in various steps

Fig. 5. Bandwidth and transmitting power allocation of IoRT devices.

Figure 5(a) depicts the bandwidth allocation proportions when offloading to different edge nodes at different stages of training. The time delay values at the top of the bar chart represent the average time taken by five IoRT devices to offload in that particular step, while the left side shows the performance of three consecutive steps randomly selected around 100 episodes. It can be observed that, at this stage, the traffic offloading exhibits significant randomness and even a tendency towards offloading to the UAV server. This leads to average latency exceeding 0.5 s for the IoRT devices in these three groups. However, after a period of training and optimization, a more reasonable bandwidth allocation becomes evident. The IoRT devices gradually adopt satellite offloading decisions and allocate bandwidth accordingly. Each IoRT device occupies a different portion of the bandwidth from the edge servers in each step. Consequently, the average offloading latency significantly decreases, sometimes reaching as low as 0.149 s.

The power allocation Fig. 5(b) follows a similar trend. In the initial stages of training, power allocation is highly random, resulting in elevated communication latency even after consuming a significant amount of transmitting power. However, as the training progresses, it can be observed that IoRT device exhibit a stronger preference for offloading to the satellite. Due to the superior service quality provided by the satellite, the five IoRT devices achieve latency reduction while maintaining a lower total transmitting power allocation.

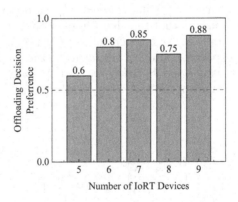

Fig. 6. Average offloading decision preference with different number of IoRT devices.

Figure 6 illustrates the offloading preference of IoRT devices to various edge servers under the SAC-based offloading algorithm with different number of IoRT devices. In convenience to analysis, a value of 1 is assigned to the decision of offloading to the satellite, while a value of 0 represents offloading to the UAV. The average offloading decisions of IoRT devices across each episode were computed to demonstrate the inclination towards a specific offloading option, with 0.5 serving as the threshold. Overall, the offloading choices were consistently greater than 0.5, indicating a ten-dency to offload to the satellite. This can be rationally explained as the satellite's larger bandwidth and superior service stability. However, even the number of IoRT devices varies from 5–9, the offloading preference never reached 1, indicating that in this scenario, UAV edge servers effectively complement and relieve the satellite edge servers. Regardless of communication congestion, the offloading services can offer superior QoS and data rate in this particular scenario.

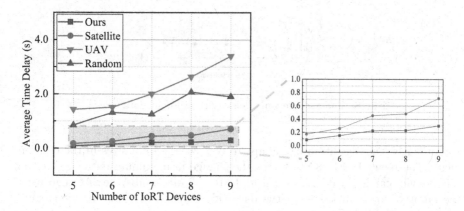

Fig. 7. Average offloading decision preference with different number of IoRT devices.

The traffic offloading delays per step for IoRT devices as the number of IoRT devices increases under four different offloading schemes is shown in Fig. 7. It is evident that the UAV-satellite cooperative offloading policy trained under the SAC algorithm has a significant advantage, consistently achieving faster completion of offloading tasks as the number of IoRT devices increases. Taking the scenario with 9 IoRT devices as an example, under the training of SAC algorithm, the average time taken for each step in the double edge offloading scenario is 0.299 s. This is more than 80% improvement compared to the time of 0.714 s for individual offloading to the satellite. Considering the maximum tolerated delay for tasks set at 0.8–0.85 s, it can be observed that for 9 IoRT devices, individual offloading to the satellite is already close to the upper limit of tolerance. However, the double edge offloading scenario under the SAC algorithm still has significant available capacity, supporting task offloading for more than 9 IoRT devices. It is noteworthy that completing the offloading task within the specified time is achievable by solely uploading to the satellite, likely due to the relatively large communication bandwidth of the satellite. However, offloading to the UAV or random allocation cannot meet the task completion deadline.

With a fixed bandwidth of the communication link, as the number of IoRT devices increases, there is an evident demand for better bandwidth allocation to avoid congestion and competition leading to much co-channel interference. Overall, more IoRT devices will result in higher average offloading time delay, however, our proposal shows a more gradual upward trend compared to the other three policies. The increase in delay is the least pronounced, maximizing the advantages of the different edge servers and finding a more suitable offloading and bandwidth allocation policy for this scenario.

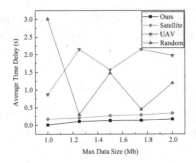

Fig. 8. The average time delay with different data sizes.

Similarly, the algorithm can adapt to the varying data sizes in the environment. The result in Fig. 8 illustrates the change in maximum data size that an IoRT device can generate, ranging from 1 Mb to 2 Mb. Under the SAC algorithm, the latency slowly increases as the data size grows. Unlike the previous simulation result where there is overlap between different numbers of IoRT devices, the data size is generated following a window-shifting method, and there is a significant overlap in the randomly generated task volumes between different intervals. As a result, the latency performance under the scheme of offloading to the UAV and random offloading be-comes irregular. On the other hand, the SAC algorithm demonstrates good adaptability, as the offloading latency does not significantly increase and consistently outperforms the scenario of offloading solely to the satellite.

6   Conclusion

Considering data traffic offloading task generated by IoRT devices deployed in remote areas, the maximum IoRT energy budgets and satellite coverage time, we investigate the traffic offloading problem in space/air-based MEC environments with multiple IoRT devices. We propose a method for joint optimization of bandwidth and transmitting power to minimize total traffic offloading delay of IoRT tasks. The soft actor-critic reinforcement learning algorithm is adopted to optimize traffic offloading and resource allocation. Further, we consider three baseline schemes for comparison in the simulation, i.e., UAV-only scheme, LEO-only scheme and random scheme, respectively. The extensive numerical simulations results demonstrate that the traffic offloading policy generated by our proposed SAC-based offloading algorithm can effectively reduce the delay of IoRT tasks, proving the promising efficiency and superiority. There still exist some specific assumptions and limitations in the research approach. For example, the trajectory of the UAV is set to be random under a certain speed and the UAV energy is set to be large enough to emphasize the time delay rather than energy consumption. In the following works, authors would focus on optimizing the trajectory and energy consumption along with time delay.

References

1. Liu, J., Du, X., Cui, J., et al.: Task-oriented intelligent networking architecture for the space-air-ground-aqua integrated network. IEEE Internet Things J. **7**(6), 5345–5358 (2020)
2. Mach, P., Becvar, Z.: Mobile edge computing: a survey on architecture and computation offloading. IEEE Commun. Surv. Tutor. **19**(3), 1628–1656 (2017)
3. Jia, Z., Sheng, M., Li, J., et al.: LEO-satellite-assisted UAV: joint trajectory and data collection for internet of remote things in 6G aerial access networks. IEEE Internet Things J. **8**(12), 9814–9826 (2020)
4. Qiu, Y., Niu, J., Zhu, X., et al.: Mobile edge computing in space-air-ground integrated networks: architectures, key technologies and challenges. J. Sens. Actuator Netw. **11**(4), 57 (2022)
5. Zhu, X., Jiang, C.: Integrated satellite-terrestrial networks toward 6G: architectures, applications, and challenges. IEEE Internet Things J. **9**(1), 437–461 (2021)
6. Zhan, C., Zeng, Y.: Energy-efficient data uploading for cellular-connected UAV systems. IEEE Trans. Wireless Commun. **19**(11), 7279–7292 (2020)
7. Di, B., Zhang, H., Song, L., et al.: Ultra-dense LEO: integrating terrestrial-satellite networks into 5G and beyond for data offloading. IEEE Trans. Wireless Commun. **18**(1), 47–62 (2018)
8. SpaceX Non-Geostationary Satellite System, Federal Communications Commissions, Washington, DC, USA (2016)
9. OneWeb Non-Geostationary Satellite System, Federal Communications Commissions, Washington, DC, USA (2016)
10. Al-Hourani, A., Guvenc, I.: On modeling satellite-to-ground path-loss in urban environments. IEEE Commun. Lett. **25**(3), 696–700 (2020)
11. Wang, F., Zhang, X.: IRS/UAV-based edge-computing/traffic-offloading over RF-powered 6G mobile wireless networks. In: 2022 IEEE Wireless Communications and Networking Conference (WCNC), pp. 1272–1277. IEEE (2022)
12. Li, X., Cheng, S., Ding, H., et al.: When UAVs meet cognitive radio: offloading traffic under uncertain spectrum environment via deep reinforcement learning. IEEE Trans. Wireless Commun. **22**(2), 824–838 (2022)
13. Wang, J., Li, D., Zhang, Z., et al.: Traffic offloading and resource allocation for PDMA-based integrated satellite/terrestrial networks. In: 2022 IEEE 4th International Conference on Power, Intelligent Computing and Systems (ICPICS), pp. 259–262. IEEE (2022)
14. Peng, C., He, Y., Zhao, S., et al.: Energy efficiency optimization for uplink traffic offloading in the integrated satellite-terrestrial network. Wireless Netw. **28**(3), 1147–1161 (2022)
15. Jia, Z., Sheng, M., Li, J., et al.: Toward data collection and transmission in 6G space-air-ground integrated networks: cooperative HAP and LEO satellite schemes. IEEE Internet Things J. **9**(13), 10516–10528 (2021)
16. Tang, F., Wen, C., Luo, L., et al.: Blockchain-based trusted traffic offloading in space-air-ground integrated networks (SAGIN): a federated reinforcement learning approach. IEEE J. Sel. Areas Commun. **40**(12), 3501–3516 (2022)
17. Tang, Q., Fei, Z., Li, B., et al.: Computation offloading in LEO satellite networks with hybrid cloud and edge computing. IEEE Internet Things J. **8**(11), 9164–9176 (2021)
18. Fu, S., Gao, J., Zhao, L.: Integrated resource management for terrestrial-satellite systems. IEEE Trans. Veh. Technol. **69**(3), 3256–3266 (2020)

19. Elbert, B.R.: Introduction to satellite communication. Artech House (2008)
20. Mao, S., He, S., Wu, J.: Joint UAV position optimization and resource scheduling in space-air-ground integrated networks with mixed cloud-edge computing. IEEE Syst. J. **15**(3), 3992–4002 (2020)
21. Wu, Q., Zeng, Y., Zhang, R.: Joint trajectory and communication design for multi-UAV enabled wireless networks. IEEE Trans. Wireless Commun. **17**(3), 2109–2121 (2018)
22. Song, Z., Hao, Y., Liu, Y., et al.: Energy-efficient multiaccess edge computing for terrestrial-satellite internet of things. IEEE Internet Things J. **8**(18), 14202–14218 (2021)
23. Ding, C., Wang, J.B., Zhang, H., et al.: Joint optimization of radio and computation resources for satellite-aerial assisted edge computing. In: ICC 2021-IEEE International Conference on Communications, pp. 1–6. IEEE (2021)
24. Yu, S., Gong, X., Shi, Q., et al.: EC-SAGINs: edge-computing-enhanced space-air-ground-integrated networks for internet of vehicles. IEEE Internet Things J. **9**(8), 5742–5754 (2021)
25. Pervez, F., Zhao, L., Yang, C.: Joint user association, power optimization and trajectory control in an integrated satellite-aerial-terrestrial network. IEEE Trans. Wireless Commun. **21**(5), 3279–3290 (2021)
26. Tang, F., Hofner, H., Kato, N., et al.: A deep reinforcement learning-based dynamic traffic offloading in space-air-ground integrated networks (SAGIN). IEEE J. Sel. Areas Commun. **40**(1), 276–289 (2021)
27. Li, S., Hu, X., Du, Y.: Deep reinforcement learning for computation offloading and resource allocation in unmanned-aerial-vehicle assisted edge computing. Sensors **21**(19), 6499 (2021)
28. Zhang, D., Cao, L., Zhu, H., et al.: Task offloading method of edge computing in internet of vehicles based on deep reinforcement learning. Clust. Comput. **25**(2), 1175–1187 (2022)
29. Yao, Y., Ren, T., Qiu, Y., et al.: Computation offloading and resource allocation based on multi-agent federated learning. In: Qiu, M., Gai, K., Qiu, H. (eds.) SmartCom 2021, pp. 404–415. Springer, Cham (2022). https://doi.org/10.1007/978-3-030-97774-0_37
30. Haarnoja, T., Zhou, A., Hartikainen, K., et al.: Soft actor-critic algorithms and applications. arXiv preprint arXiv:1812.05905 (2018)
31. Duan, Y., Chen, X., Houthooft, R., et al.: Benchmarking deep reinforcement learning for continuous control. In: International Conference on Machine Learning, pp. 1329–1338. PMLR (2016)

A Novel Sensor Method for Dietary Detection

Long Tan[1], Dengao Li[1], Shuang Xu[1(✉)], Xiuzhen Guo[2], and Shufeng Hao[1]

[1] Taiyuan University of Technology, Taiyuan, China
{lidengao,xushuang,haoshufeng}@tyut.edu.cn
[2] Tsinghua University, Beijing, China

Abstract. A regular diet is closely related to human physical and mental health. Detecting dietary behavior offers a chance to help individuals comprehend their eating habits, promptly recognize any health issues, and adopt suitable measures to enhance their well-being. In this paper, a novel hardware-software sensing method is proposed to enable more people to care about their healthy dietary habits. In this method, the user only needs a speaker-only headphone even without a microphone, typically found in people's lives to achieve diet detection. Besides, the feasibility of utilizing speaker-only headphones as dietary sensors is investigated through mathematical models and experimental validation and this paper focuses on using this sensor in mastication rate detection. We implement our hardware structure on a 1-layer PCB board and follow the IRB protocol to evaluate its performance, and we achieve MAE (0.865), RMSE (1.26), and average error rate (0.11).

Keywords: IoT · Diet detection · Hardware design · Wheatstone bridge · Machine learning

1 Introduction

Currently, there is a growing interest in maintaining a healthy diet, and research has shown that dietary habits play a critical role in the onset of various health conditions, such as depression, obesity, and some types of cancer [1]. Additionally, chewing is a fundamental motion during eating that has a significant impact on the health of individuals. For instance, Kokkinos *et al.* in [2] discovered that eating slowly influenced gut hormones. Otsuka *et al.* in [3] found that eating fast led to obesity. Robinson *et al.* in [4] supported that eating rate affected energy intake. Moreover, an atypical chewing pattern can be indicative of certain medical conditions such as anorexia or tooth decay. Extensive research underscores the significance of both the act of consuming food and the act of chewing in relation to an individual's well-being. As a result, in the pursuit of a healthy life, accurately detecting diet is of significant importance.

In recent years, an increasing number of methods have been proposed for detecting diets, and current works on diet detection can be classified into three

© The Author(s), under exclusive license to Springer Nature Singapore Pte Ltd. 2024
Z. Tari et al. (Eds.): ICA3PP 2023, LNCS 14492, pp. 153–171, 2024.
https://doi.org/10.1007/978-981-97-0811-6_9

distinct categories. The first category adopts image or video information to classify food types [5–7] or count chewing [8,9]. The second method uses audio signals from the microphone to monitor diet behavior [10–15]. The third approach suggests brand-new sensor techniques identify dietary behavior such as accelerator, EMG sensor and piezoelectric sensor [16–18]. These methods necessitate specialized sensors, including cameras, Inertial Measurement Units (IMUs), and in-ear microphones, which are not widely accessible and expensive. As such, there exists a requirement for a novel wearable sensor that can be commonly integrated into daily routines. This will aid a larger population of individuals afflicted with diet-related illnesses in detecting their dietary health using behavioral data.

With the trend of incorporating smart technology, headphones have gained immense popularity worldwide and are attracting increasing attention. Additionally, headphones are commonly used while eating. For instance, a lot of people watch videos or listen to music while wearing headphones, leading to the question of whether headphones could be used as a dietary sensor. Recent works [12,13,19] have focused on using the in-ear microphone in headphones for the purpose of diet detection. However, many commercial headphones don't have inbuilt sensors, and 47% of them don't even have microphones [20]. Bringing intelligence to ordinary headphones at an affordable price remains a meaningful task for the public. The outstanding study conducted by HeadFi [21] demonstrates, through numerous experiments, that regular headphones possess sensing capabilities. However, the question remains: can ordinary headphones be utilized as a dietary sensor?

In this paper, we address the theoretical and experimental evidence that most common headphones can function as dietary sensors and presents a method for detecting diet and chewing counting. The main contributions are summarized as follows:

1. A dual-arm Wheatstone bridge and differential amplifier hardware structure is utilized to transform ordinary headphones with speaker-only into high-sensitivity diet sensors. To our knowledge, this is the first system to propose applying this hardware architecture for dietary detection.
2. The feasibility of utilizing speaker-only headphones as dietary sensors are investigated through mathematical models and experimental validation.
3. According to the characteristics of the signals, we propose a software solution to implement chewing counting detection in a real experiment, and we achieve MAE (0.865), RMSE (1.26), and an average error rate (0.11).

The remainder of this paper is structured as follows. Section 2 presents the related work of diet detection using various sensors. Section 3 presents the hardware-software solution of using ordinary headphones for diet detection. Section 4 describes the dataset, the conducted experiments, the evaluation results, and the discussion of our approach. Finally, the paper concludes in Sect. 5.

2 Related Work

In this section, we review research topics relevant to our work.

Audio Signal Based Dietary Detection Systems. Sound emission is essential in eating, and discriminating food information from sound signals is an important research direction for diet detection. Amft et al. in [22] investigates the relationship between food and acoustics by capturing chewing vibration information through acoustic transducers and applying it for feature extraction and classification. Päßler et al. in [13] records the sound of chewing and swallowing with dual microphones and proposes a daily food intake classification model. Papapanagiotou et al. in [19] proposes a system for detecting mastication that combines in-ear microphones with a photoplethysmography (PPG) sensor, extracts data features from both sensors, fuses the data, and employs Support Vector Machines (SVM) to classify snacking. Earbit [23] proposes a multi-sensor fusion approach with IMUs, proximity sensors, and microphones for mastication behavior detection and chewing data collection and classification utilizing Hidden Markov Models (HMMs). Nyamukuru et al. in [24] proposes a lightweight GRU-based improved algorithm for a diet detection system on a microprocessor. iHearken [12] utilizes outstanding hardware circuits with in-ear microphones to collect audio information of chewing, feature extraction of chewing objects by Bi-LSTM networks, and classification by SoftMax.

However, these methods necessitate the use of specialized sensors for sensing and employ multi-sensor data fusion techniques to enhance accuracy, thereby resulting in a complex system structure and expensive cost.

Image Based Dietary Detection Systems. Image classification is a straightforward method for food classification, and some works have focused on it. Qiu et al. in [5] proposes a technique to record dietary information with a 360 camera and detect dietary behavior utilizing a deep learning approach. AIM-2 [25] reduces privacy leakage by embedding a wide-angle camera, a tri-axis accelerometer, and temporalis muscle sensors into the glasses, using the tri-axis accelerometer and temporalis muscle sensors as triggers to trigger the camera to take pictures of food, thus reducing privacy leakage. Qiu et al. in [26] uses the state-of-the-art networks TSM and SlowFast in video recognition to count the number of chews in the eating video while recognizing the visible food.

Nevertheless, utilizing the image approach escalates the vulnerability of privacy breaches and significantly augments the computational data load. Mounting the camera in a fixed position hinders ease of wear while aiming to avert information leakage.

Other Sensor Based Dietary Detection Systems. In addition to diet detection employing chewing sounds and food image classification, some work focuses on other indirect monitoring methods by sensors. Dong et al. in [27] utilizes a watch-like configuration of sensors to monitor wrist activity continuously and proposed segmentation algorithms to classify eating and non-eating activities. AIM [28] adopts multi-sensor fusion to record human behavioral features with jaw motion sensors, acceleration sensors, and gesture recognition sensors and then employs pattern recognition methods to identify eating behaviors and achieves an average accuracy of 89.9%. Kyritsis *et al.* in [29] records hand motion information with a smart watch's inertial sensor, evaluates the probability distribution of micro-motions utilizing a convolutional neural network (CNN), and implemented behavior recognition by a Long-Short Term Memory (LSTM) network. Ghosh *et al.* in [30] realizes real-time eating behavior monitoring using a decision tree approach by recording chewing and head movement signals with accelerometer sensors. Nicholls *et al.* in [31] automates the detection of eating behaviors such as chewing and swallowing employing wearable electromyography (EMG) sensors.

However, these works require specialized sensors and utilize indirect methods to achieve eating behavior detection as well. Indirect sensing methods have certain limitations when it comes to classifying food. Hence, there is no additional potential to explore this sensing method.

3   Methodology

In this section, the sensor principle of speaker-only headphones is introduced in a lucid manner. The hardware solution is designed to address the low perceptual sensitivity of headphones. In order to ascertain the potential of speaker-only headphones as dietary sensors capable of distinguishing received signals, a classification method based on SVM is proposed. Besides, a chewing counting method is proposed based on the feature of receiving signals.

3.1   Coupling Effect

As shown in Fig. 1, the eardrum, ear canal, and headphones form a circuit system due to the coupling effect, which provides the theoretical support for enabling ordinary headphones with sensor ability. In the equivalent circuit, the headphone is considered the signal source, with its impedance equaling the internal resistance of the signal source. At the same time, the eardrum is seen as a load with

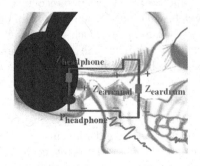

Fig. 1. Coupling effect equivalent circuit

a specific impedance. Besides, when the user wears the headphone, the relationship between the pressure field in the ear and the headphone impedance can be modeled according to Thevenin's equivalent circuit [21].

$$\frac{P_{earcanal}}{P_{\text{headphone}}} = \frac{Z_{\text{earcanal}}}{Z_{\text{earcanal}} + Z_{\text{headphone}}} \tag{1}$$

where $P_{earcanal}$ and $P_{\text{headphone}}$ are ear canal pressure and headphone pressure respectively, Z_{earcanal} and $Z_{\text{headphone}}$ are ear canal impedance and headphone impedance, respectively.

Equation (1) demonstrates the correlation between headphone impedance and ear canal pressure. The impedance of headphones can be affected by three factors ($P_{\text{headphone}}$, P_{earcanal} and Z_{earcanal}). For instance, when the user makes contact with the headphones, it can induce a fluctuation in the pressure of the headphones ($P_{\text{headphone}}$). Likewise, the heartbeat prompts the dilation and constriction of blood vessels in the ear, and this delicate vibrational signal can affect the pressure within the ear canal (P_{earcanal}). In our scene, the process of mastication during eating leads to changes in headphone impedance as well. The process of chewing squeezes the air in the ear, resulting in changes of P_{earcanal}. Besides, the chewing sounds produced when teeth collide with foods are detected through bone conduction and transmitted to the headphones, resulting in vibrations that cause changes in headphone pressure ($P_{\text{headphone}}$). Furthermore, the closed chamber consisting of headphones and ear canal enables signal amplification, which makes them highly suitable for detecting chewing sounds [32].

In conclusion, headphones can be regarded as a diet sensor due to their impedance changing with human chewing behavior. However, the impedance change caused by chewing is so weak.

3.2   Hardware Design

A hardware structure combining a two-arm Wheatstone bridge with a differential amplifier [21] is utilized to amplify changes in headphone impedance due to chewing behavior, as shown in Fig. 3. In this method, the headphone plugs into the Wheatstone bridge as a two-arm, and then the output of the Wheatstone bridge is transmitted to a differential amplifier. The output signal is described as

$$V_o(t) = A_u V_i \Big(\frac{R_1}{R_1 + Z_l + \Delta z_l} - \frac{R_2}{R_2 + Z_r + \Delta z_r} \Big) \tag{2}$$

where V_o is the output voltage of the differential amplifier, V_i is the input voltage, A_u is the gain of the differential amplifier, R_1 and R_2 are the fixed impedances of the left and right bridges, respectively, Z_l and Z_r are the left and right headphone fixed impedance respectively, whose value are determined by the production process of manufacturer. Δz_l and Δz_r are the impedance changes of the left and right headphones, respectively, determined by the vibrations generated during chewing.

If no vibration signal is detected, the output will be zero for the hardware structure. Conversely, the presence of vibration can result in an imbalance in the

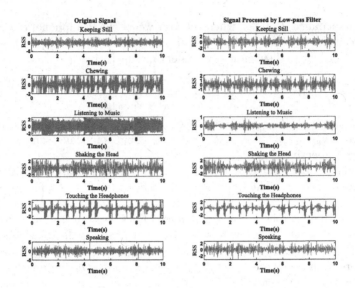

Fig. 2. Received signals of events detection

bridge, and by increasing the gain of the differential amplifier, the degree of this imbalance (represented by $V_o(t)$) can be amplified. As a result, we can analyze the output signals of the differential amplifier to detect dietary behavior.

3.3 Event Detection Method Based on SVM

In this subsection, the event detection algorithm is proposed to verify the feasibility of utilizing speaker-only headphones as dietary sensors. In the context of indoor dining, various behaviors commonly observed among individuals wearing headphones are considered, such as staying still, chewing, listening to music, shaking their heads, touching the headphones, and speaking. These behavior events, all have an effect on the three factors above discussed, resulting in changing the headphone impedance. To facilitate visual observation of the signal, the experiment data of Participant 1-1 is selected and visualized the original signal and the signal after low-pass filtering through 15 Hz, as shown in Fig. 2. When the participant stays still, the heartbeat prompts the constriction of blood vessels in the ear, resulting in a change of the ear canal (P_{earcanal}). The typical range of human heart rates is between 0.9 Hz and 1.7 Hz. Therefore, the noise signal is effectively removed following the application of a low-pass filter with a cutoff frequency of 15 Hz, enabling the emergence of distinct and accurate heart rate information. When the participant chews, the P_{earcanal} and $P_{\text{headphone}}$ will change as discussed before. According to the Eq. (2), the input signal will also influence the receiving signals. Besides, when shaking the head, touching the headphones, and speaking, $P_{\text{headphone}}$ will change as well, but the strength and mode between the three events are different. Importantly, all received signals demonstrate the coupling characteristics of heart rate.

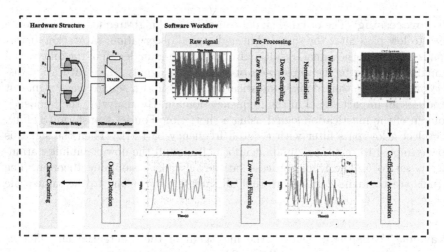

Fig. 3. System workflow

After analyzing the impact of various behavioral events on headphone impedance, it is necessary to differentiate these events from the received signal. In essence, the behavioral event detection task involves binary classification. Therefore, standard methods can be utilized for binary classification, such as SVM and random forest. In our method, the SVM is adopted. To specify our approach, first, the received signals are processed by z-score normalization. Then the signals are passed through a 2nd order Butterworth bandpass filter with passband frequency from 2 Hz to 30 Hz, thus removing the heart rate coupling feature and high-frequency noise. In total, 24 features are extracted, inspired by the work of Saphala *et al.* [33]. Table 1 shows the detailed feature from the time domain (TD), frequency domain (FD), and power spectral density (PSD).

3.4   Chewing Counting Algorithm

In this subsection, we introduce how to extract the chewing feature from the raw signal and implement high-accuracy chew counting. The system workflow is shown in Fig. 3.

Table 1. List of features used for the event detection method.

Frame	Features
TD	Max, min, max-min, rms, median, variance, std, skew, kurtosis, interquartile range
FD	Mean, band power, median, kurtosis, skew
PSD	Max, min, mean, std, spectral entropy, spectral kurtosis, kurtosis, skew, median

Pre-processing: The original signal contains some high-frequency noise, so we need to low-pass filter the signal. Through signal observation, saturation distortion is evident in the signal, resulting from the high-gain design of the differential amplifier, leading to the perception of attenuated signals. Since saturation distortion turns the time domain mastication signal into a square wave signal, it produces an impact signal in the frequency domain. By analyzing the frequency domain of the mastication signal using a short-time Fourier transform (STFT), we select a low-pass filter with a cutoff frequency of 4 kHz. Then the signal is downsampled to remove redundant information, and the downsampling parameter is set to 5. The whole process satisfies Nyquist's sampling theorem. The signals are normalized to ensure that all sample data belong to the same scale range.

Wavelet Transform: Due to the high gain of the differential amplifier as described above, a significant vibrating signal will create output saturation distortion when the participant chews. To extract the chewing features of saturation distortion, the wavelet transform (WT) is utilized to analyze the time and frequency characteristics of the signal. The saturation distortion signal due to mastication is similar to that of the Haar wavelet. Therefore, employing the Haar wavelet as the base wavelet in WT aids in the acquisition of chewing features [34]. We obtain the wavelet coefficient matrix $H = [h_1, h_2, ..., h_m] \in R^{k \times m}$, where k is the biggest scale factor. Scale vector $h_a = [|h(0, a)|, ..., |h(F, a)|]^T \in R^{k \times 1}$, and m is the length of the sequence.

Coefficient Accumulation: To extract the chewing impact signal in the WT spectrogram, we can calculate the vector sum of the scale vector h_a, and obtain the accumulation scale factor $v = [v_1, ..., v_m] = [h_1{}^T h_1, ..., h_m{}^T h_m] \in R^{1 \times m}$.

Low-Pass Filtering: Although the accumulation scale factor has finer-grained chewing information, counting the chewing is difficult due to the noise signals. A low-pass filter can extract the low-pass chewing envelope to remove the vibration noise signal. The signal is considered to be downsampled, so the low-pass filter's cutoff frequency is 10 Hz. After low-pass filtering, we can get the vector v'.

Outlier Detection: Some samples have false positive points after low-pass filtering in the actual signal processing, as shown in Fig. 4. A dynamic threshold clustering algorithm is proposed to reduce the influence of false-positive points on the results. Our basic idea is to ascertain if the current chew is valid by evaluating the intensity of each chew and establishing a dynamic threshold determined by the collective intensity of chews within the sliding window. In Fig. 4, the vertical distance between the ith valley and the two adjacent peaks are d_l^i and d_r^i, respectively. $cs_i = d_l^i + d_r^i$ represents the chewing strength of ith valley, and the *factor* is the weight variable. The number of chews can be obtained by implementing Algorithm 1.

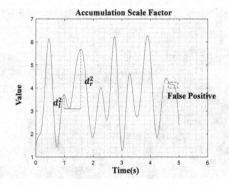

Fig. 4. False positive points in the sample analysis

Algorithm 1: Outlier Detection

Input: v', $factor$
Output: $count$
1 Compute m (the set of all valleys)
2 $n \leftarrow$ length of m
3 **for** $i = 1; i \leq n$ **do**
4 $\quad | \quad cs_i = d_l^i + d_r^i$
5 **end**
6 Compute $\bar{x}$ (the mean value of cs)
7 $gate = factor \times \bar{x}$
8 **for** $i = 1; i \leq n$ **do**
9 $\quad |$ **if** $cs_i < gate$ **then**
10 $\quad | \quad |$ smooth ith valley
11 $\quad |$ **end**
12 **end**
13 Compute $count$ (the number of peaks after smoothing)

---

4   Evaluation and Discussion

4.1   Experimental Methodology

Hardware Implementation: A 42 cm × 48 cm, one-layer printed circuit board (PCB) is used to implement the hardware structure. The signal can be immediately sampled using the sound card of a laptop without additional acquisition hardware. To provide stable power support, the DCDC (direct current to direct current) module is designed to convert the 5 V 1 A power supply from the power bank to a stable voltage of −5 V, 0 V, and 5 V. In the experiment, the RealTek Sound Card ALC256 on the laptop is used, which drives the hardware with a sampling rate of 48 kHz and 16-bit ADC (Analog-to-Digital Converter) precision. Besides, the HIFIMAN HE400SE is the headphone of choice. The experimental equipment is shown in Fig. 5. All data processing by modules of dietary detection implemented in MATLAB.

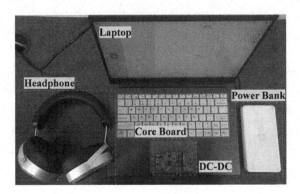

Fig. 5. Experimental equipment

Dataset Collection: Eight (including three females) of the sixteen participants were chosen to participate in the data gathering. All participants signed a written consent form prior to their participation in the experiment and followed the approved ethical inclusion and exclusion criteria:

(i) Participants aged between 23 and 28 years, all of whom were above 18 years old.
(ii) Body Mass Indices (BMI) of the participants ranged from 17.4 kg/m^2 to 26.4 kg/m^2.
(iii) Participants who had a history of any chewing or swallowing disorders were excluded from the study.
(iv) The study included only volunteer participants who had signed an appropriate consent form.

The participants were divided into two groups of five individuals each, with two individuals common to both groups.

- The first group is designed to confirm the distinguishable ability of the received signal. As previously discussed, six events are considered including staying still (reference experiment), chewing, listening to music, shaking the head, touching the headphones, and speaking. In each experimental event, participants were suggested to continue engaging in the activity for a minimum duration of 5 min. A total of 154 min' worth of event signals are gathered from the participants. It needs to be noted that, to generate a more realistic chewing signal, chewing gum is utilized as food throughout the chewing signal acquisition.
- The second group is designed to verify the validity of the chewing counting algorithm. In this experiment, 13 common items, including vegetables, fruits, and staple foods with various regional characteristics, are chosen for the experiments. The food is divided into three categories according to hardness. The first category is the soft category, consisting of bananas, bread, oranges, chewing gum, rice, and soft cookies. Apple, chips, crispy noodles,

cucumber, and potato make up the second crispy category. The last category is a hard group that includes peanuts. A total of 337 min' worth of chewing signals are gathered from the volunteers as they are allowed to chew each food for 4 to 6 min (Table 2).

Table 2. Experimental Food Description

Mark Number	Food Name	Category
F1	Apple	Crispy
F2	Banana	Soft
F3	Bread	Soft
F4	Chips	Crispy
F5	Crispy Cookies	Crispy
F6	Crispy Noddles	Crispy
F7	Cucumber	Crispy
F8	Orange	Soft
F9	Peanut	Hard
F10	Chewing gums	Soft
F11	Rice	Soft
F12	Soft Cookies	Soft
F13	Potato	Crispy

In the experiments, participants are required to wear headphones throughout the experiment that are extremely user-friendly. Participants have the opportunity to engage in experiments that align with their lifestyle habits, such as reading newspapers or watching videos on their cell phones. Given that the majority of eating events with headphones occur indoors, we conducted the experiments in indoor settings. The ambient noise levels varied from 24.3 dB to 81.4 dB, with an average of 50.2 dB.

Evaluation Metrics: The accuracy, precision, recall, and F1-Score are utilized as evaluation metrics for the classification problem.

$$Accuracy = \frac{TP + TN}{TP + TN + FN + FP} \tag{3}$$

$$Precision = \frac{TP}{TP + FP} \tag{4}$$

$$Recall = \frac{TP}{TP + FN} \tag{5}$$

$$F1 - Score = 2 \times \frac{Precision \times Recall}{Precision + Recall} \tag{6}$$

where TP is true positive, TN is true negative, FP is false positive, and FN is false negative.

To measure the difference between the estimated and real values of the chew counting method, we use MAE (Mean Absolute Error), RMSE (Root Mean Squared Error), and Error Rate.

$$MAE = \frac{1}{n} \sum_n |f_t - p_t| \tag{7}$$

$$RMSE = \sqrt{\frac{1}{n} \sum_n (f_t - p_t)^2} \tag{8}$$

$$Error\ Rate = \frac{\sum_n |f_t - p_t|}{\sum_n p_t} \tag{9}$$

where n is the number of samples, f_t and p_t are the estimated and actual values of the tth sample, respectively.

4.2   The Evaluation of Chewing Event Detection

To evaluate the method's performance in chewing event detection, the data set is first divided into a training set and a testing machine according to 6:4. The 6-fold cross-validation strategy is employed to assess the performance of the method. In this strategy, the training set is randomly divided into six folds of data, five folds of data are selected as the training set in each training, and the remaining one fold of data is used as the validation set.

The Impact of Features: When using SVM for classification, it needs to pay attention to the importance of feature selection, inappropriate feature selection can lead to a curse of dimensionality, overfitting, and other problems. All features are listed in Table 1, and different combinations of features will be considered for their impact on the results. In order to control the effects of other variables, in this experiment, the radial basis function (RBF) kernel is set to the kernel, and the length of the sliding window is set to 5 s. To consider the effect of γ, the parameter of RBF kernel, and c, the penalty parameter of regularization, on the results, we evaluate the parameters using the grid search method, where the c take values in the range [0.1, 10, 100, 1000, 10000] and γ take values in the range [0.01, 0.1, 1, 10, 100]. The results are shown in Table 3. It is observed that the features in different are valid to improve the performance of SVM.

The Impact of Kernels: In the SVM classifier, the different kernels, including the linear kernel, polynomial kernel, RBF kernel, and sigmoid kernel, can be chosen to transform a linearly indistinguishable problem in a low-dimensional feature space into a linearly divisible problem in a high-dimensional feature space. In the experiment, the polynomial kernel, RBF kernel, and sigmoid kernel

Table 3. The result of the domain of the different features

Features	No. of Features	Acc. (%)	F1. (%)
TD	10	87.6	82.0
FD	5	73.7	66.9
PSD	9	84.4	77.4
TFD	15	89.7	84.7
PSD and TFD	24	90.9	86.3

are in our consideration due to the problem being a non-linear problem. In this experiment, all features are in consideration, and the length of the sliding window is set to 5 s. In Table 4, it is observed that the RBF kernel has the best performance than others, due to considering global features and the boundaries are smoother and more continuous.

Table 4. The result of different kernels

Kernel	Acc. (%)	F1. (%)
Polynomial kernel	87.2	81.5
RBF kernel	90.9	86.3
Sigmoid kernel	17.6	-

The Impact of the Length of Sliding Window: The length of the sliding window decides the system response speed and the resolution of recognition. In the experiment, sliding windows of different lengths were used to evaluate. The results are shown in Table 5. As the length of the sliding window increases, the model's performance improves because it considers more features. However, it is also important to consider the trade-off between performance and response speed.

Table 5. The result of the length of sliding window

Length (s)	Acc. (%)	F1. (%)
3	86.9	81.1
4	88.9	83.7
5	90.9	86.3
6	92.2	87.5
7	93.0	88.4

The Best Performance Analysis: After the evaluation and discussion above, the detailed model parameters and results are shown in Table 6. These results confirm that the signals received by the hardware can be distinguished due to the different features caused by different behavior events. Besides, the results also prove that the hardware structure can be utilized to collect signals and detect dietary.

Table 6. The parameters setting and results of the models

Kernel	Length(s)	c	γ	Acc. (%)	Prec. (%)	Rec. (%)	F1. (%)
RBF kernel	7	10000	0.1	93.0	86.7	90.5	88.4

4.3   The Evaluation of Chewing Counting

We divide all 337 min of data into 4330 audio files according to a sliding window of 5 s. To ensure the general applicability of the algorithm, we randomly select ten samples from the data set of 13 foods per person according to the ratio for evaluation, and a total of 650 samples are used for algorithm evaluation. Some samples have the phenomenon of chewing interruption, i.e., the volunteer stopped chewing while chewing for a while, and the phenomenon of chewing speeding up and slowing down. To assess the feasibility of the system, the cross-validation method is employed, and partition the dataset into the training and test sets in a 1:1 ratio. The training set is utilized to determine the values of $factor$, while the test set is employed to evaluate the outcomes.

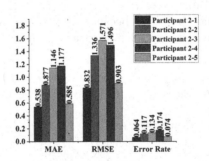

Fig. 6. The results of different individual diet habits

The Impact of Parameter Setting: The $factor$ is the weight variable, and to discuss the effect of $factor$ on the result, we choose the different values of the $factor$ to estimate the results. When the $factor$ is set to 0.4, it performs best in the training set. We utilize the optimal parameters obtained from the training set to evaluate the independent test set. MAE, RMSE, and the average error rate are 0.865, 1.26, and 0.11, respectively.

The Impact of Individual Diet Habit: This part covers the impact of various foods and individuals on the effects of chewing individually. The *factor* is set to 0.4 and evaluate the results. It discovers that various people's chewing patterns impact the outcomes. As shown in Fig. 6, the best performance comes from Participant 2-1. The chewing signal can be easily recovered from the accumulation wavelet coefficients since this volunteer chewed with a strong and large chewing amplitude. Participant 2-4 chewed, in comparison, with a small amplitude, making it difficult to extract the chewing feature.

The Impact of Food Type: From the results of different foods, as shown in Fig. 7, many food-breaking sounds are included in the chewing signals of crunchy food like chips, crispy cookies, and crispy noodles. These sound signals will influence the outcomes.

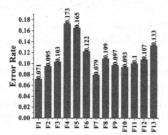
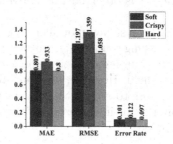

(a) The results of different food   (b) The results of different category

Fig. 7. The results of different foods

Compare with State-of-the-Art: In this part, we contrast with some systems that also count the number of chews using sensors. After testing chewing counts on five participants, the overall ground truth number of chewing in our samples is 5095. Wang *et al.* in [17] performed chewing counts on four individuals using an acceleration sensor, and the total number of actual values of the sample reached 1772. Farooq *et al.* in [18] performed chewing counts using a piezoelectric strain sensor. The comparison result is shown in Fig. 8. The proposed method has a lower overall mean error rate than similar methods using sensors.

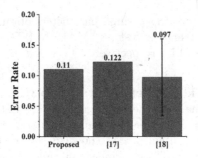

Fig. 8. Comparison of proposed system with [17] and [18]

The Results Analysis: To illustrate the results in more detail, in essence, the signal acquired by our hardware structure is a composite of the change in headphone pressure ($P_{headphone}$) due to the sound signal of mastication and the change in pressure in the ear ($P_{earcanal}$) due to the act of chewing. When our headphones are large enough to cover the entire ear, the jaw movement caused by chewing also causes the headphone pressure ($P_{headphone}$) to change. In our experiments, it is observed that different people's chewing habits, such as the intensity of chewing, can have an impact on the results, and this is a common problem with most jaw movement-based detection methods. Additionally, the signals of chewing sounds transmitted via bone conduction can partially disrupt the accurate counting of chews, particularly when the sounds of food surpass a specific threshold. However, different from the works in [17,18], which utilized different sensors to record the movement of the lower jaw, the act of chewing and the alterations in ear pressure can yield certain benefits to a certain degree in our method.

Computational Complexity Analysis: The whole chewing counting algorithm consists of signal pre-processing, wavelet transform, coefficient accumulation, low-pass filtering, and outlier detection. In order to obtain the computational complexity of the entire algorithm, we have the following definition of data parameters. The data is processed for a length of N. The sequence lengths of the two digital filters are M_1 and M_2, respectively. The length of the downsampled sequence becomes N'. k is the biggest scale factor of WT as defined before. The computation complexity can be obtained as Table 7. The estimation representation is utilized in some complexity calculations. In the subsequent optimization of the system complexity, analog filters can be used to replace the front-end digital signal processing, thus reducing the system complexity and improving the real-time performance of the system.

Table 7. The computational complexity analysis of chewing counting algorithm

Algorithm	Time Complexity	Space Complexity
Low-pass Filtering I	$O(NM_1)$	$O(N + M_1)$
Down Sampling	$O(N')$	$O(N')$
Normalization	$O(N')$	$O(N')$
WT	$O(N'^2)$	$O(kN')$
Coefficient Accumulation	$O(kN')$	$O(N')$
Low-pass Filtering II	$O(N'M_2)$	$O(N' + M_2)$
Outlier Detection	$O(N')$	$O(N')$

5 Conclusion

We have presented the design, implementation, and evaluation of a novel hardware-software diet sensor method detecting dietary events. First, we utilize the Wheatstone bridge and differential amplifier to transform speaker-only headphones into a diet sensor. To investigate whether this hardware structure can be used as a diet sensor, we collected six common behaviors of wearing headphones and explored them using an SVM-based approach, which theoretically and experimentally proved to be very effective in diet detection. The chew counting algorithm is proposed based on the features of the chewing signals acquired by the hardware. Finally, we estimate our method on 13 food items with a total of 337 min of data and achieve MAE (0.865), RMSE (1.26), and an average error rate (0.11). The result shows that our method can implement high-accuracy chew counting.

Acknowledgement. This work is supported by the National Natural Science Foundation of China (No. 62102280), Fundamental Research Program of Shanxi Province (No. 202103 02124167), Key Research and Development Program of Shanxi Province (No. 20210202 0101001), Key Research and Development Program in Shanxi Province (Grant No. 20210202 0101004), the Fundamental Research Program of Shanxi Province (Grant No. 2021030 2124168), the Fund Program for the Scientific Activities of Selected Returned Overseas Professionals in Shanxi Province (Grant No. 20220009), Scientific Research Fund Project of Taiyuan University of Technology (Grant No. 2022QN128).

References

1. Firth, J., et al.: What is the role of dietary inflammation in severe mental illness? A review of observational and experimental findings. Frontiers Psychiatry, 350 (2019)
2. Kokkinos, A., et al.: Eating slowly increases the postprandial response of the anorexigenic gut hormones, peptide YY and glucagon-like peptide-1. J. Clin. Endocrinol. Metab. **95**(1), 333–337 (2010)
3. Otsuka, R., et al.: Eating fast leads to obesity: findings based on self-administered questionnaires among middle-aged Japanese men and women. J. Epidemiol. **16**(3), 117–124 (2006)

4. Robinson, E.: A systematic review and meta-analysis examining the effect of eating rate on energy intake and hunger. Am. J. Clin. Nutr. **100**(1), 123–151 (2014)

5. Qiu, J., Lo, F.P.-W., Lo, B.: Assessing individual dietary intake in food sharing scenarios with a 360 camera and deep learning. In: 2019 IEEE 16th International Conference on Wearable and Implantable Body Sensor Networks (BSN), pp. 1–4. IEEE (2019)

6. Joshua, S.R., Shin, S., Lee, J.-H., Kim, S.K.: Health to eat: a smart plate with food recognition, classification, and weight measurement for type-2 diabetic mellitus patients' nutrition control. Sensors **23**(3), 1656 (2023)

7. Schiboni, G., Wasner, F., Amft, O.: A privacy-preserving wearable camera setup for dietary event spotting in free-living. In: 2018 IEEE International Conference on Pervasive Computing and Communications Workshops (PerCom Workshops), pp. 872–877. IEEE (2018)

8. Alshboul, S., Fraiwan, M.: Determination of chewing count from video recordings using discrete wavelet decomposition and low pass filtration. Sensors **21**(20), 6806 (2021)

9. Hossain, D., Ghosh, T., Sazonov, E.: Automatic count of bites and chews from videos of eating episodes. IEEE Access **8**, 101, 934–101, 945 (2020)

10. Kalantarian, H., Alshurafa, N., Pourhomayoun, M., Sarin, S., Le, T., Sarrafzadeh, M.: Spectrogram-based audio classification of nutrition intake. In: 2014 IEEE Healthcare Innovation Conference (HIC), pp. 161–164. IEEE (2014)

11. Gao, Y.: iHear food: eating detection using commodity Bluetooth headsets. In: 2016 IEEE First International Conference on Connected Health: Applications, Systems and Engineering Technologies (CHASE), pp. 163–172. IEEE (2016)

12. Khan, M.I., Acharya, B., Chaurasiya, R.K.: iHearken: chewing sound signal analysis based food intake recognition system using Bi-LSTM SoftMax network. Comput. Methods Programs Biomed. **221**, 106843 (2022)

13. Päßler, S., Fischer, W.-J.: Acoustical method for objective food intake monitoring using a wearable sensor system. In: 2011 5th International Conference on Pervasive Computing Technologies for Healthcare (PervasiveHealth) and Workshops, pp. 266–269. IEEE (2011)

14. Khan, M.I., Acharya, B., Chaurasiya, R.K.: Hybrid BiLSTM-HMM based event detection and classification system for food intake recognition. In: 2022 First International Conference on Electrical, Electronics, Information and Communication Technologies (ICEEICT), pp. 1–5. IEEE (2022)

15. Kondo, T., Kamachi, H., Ishii, S., Yokokubo, A., Lopez, G.: Robust classification of eating sound collected in natural meal environment. In: Adjunct Proceedings of the 2019 ACM International Joint Conference on Pervasive and Ubiquitous Computing and Proceedings of the 2019 ACM International Symposium on Wearable Computers, pp. 105–108 (2019)

16. Farooq, M., Sazonov, E.: Accelerometer-based detection of food intake in free-living individuals. IEEE Sens. J. **18**(9), 3752–3758 (2018)

17. Wang, S., et al.: Eating detection and chews counting through sensing mastication muscle contraction. Smart Health **9**, 179–191 (2018)

18. Farooq, M., Sazonov, E.: Linear regression models for chew count estimation from piezoelectric sensor signals. In: 2016 10th International Conference on Sensing Technology (ICST), pp. 1–5. IEEE (2016)

19. Papapanagiotou, V., Diou, C., Zhou, L., van den Boer, J., Mars, M., Delopoulos, A.: A novel chewing detection system based on PPG, audio, and accelerometry. IEEE J. Biomed. Health Inform. **21**(3), 607–618 (2016)

20. Olive, S., Khonsaripour, O., Welti, T.: A survey and analysis of consumer and professional headphones based on their objective and subjective performances. In: Audio Engineering Society Convention, vol. 145. Audio Engineering Society (2018)
21. Fan, X.: HeadFi: bringing intelligence to all headphones. In: Proceedings of the 27th Annual International Conference on Mobile Computing and Networking, pp. 147–159 (2021)
22. Amft, O.: A wearable earpad sensor for chewing monitoring. In: Sensors 2010, pp. 222–227. IEEE (2010)
23. Bedri, A.: EarBit: using wearable sensors to detect eating episodes in unconstrained environments. Proc. ACM Interact. Mob. Wearable Ubiquitous Technol. **1**(3), 1–20 (2017)
24. Nyamukuru, M.T., Odame, K.M.: Tiny eats: eating detection on a microcontroller. In: 2020 IEEE Second Workshop on Machine Learning on Edge in Sensor Systems (SenSys-ML), pp. 19–23. IEEE (2020)
25. Doulah, A., Ghosh, T., Hossain, D., Imtiaz, M.H., Sazonov, E.: "Automatic ingestion monitor version 2"-a novel wearable device for automatic food intake detection and passive capture of food images. IEEE J. Biomed. Health Inform. **25**(2), 568–576 (2020)
26. Qiu, J., Lo, F.P.-W., Jiang, S., Tsai, Y.-Y., Sun, Y., Lo, B.: Counting bites and recognizing consumed food from videos for passive dietary monitoring. IEEE J. Biomed. Health Inform. **25**(5), 1471–1482 (2020)
27. Dong, Y., Scisco, J., Wilson, M., Muth, E., Hoover, A.: Detecting periods of eating during free-living by tracking wrist motion. IEEE J. Biomed. Health Inform. **18**(4), 1253–1260 (2013)
28. Fontana, J.M., Farooq, M., Sazonov, E.: Automatic ingestion monitor: a novel wearable device for monitoring of ingestive behavior. IEEE Trans. Biomed. Eng. **61**(6), 1772–1779 (2014)
29. Kyritsis, K., Diou, C., Delopoulos, A.: Modeling wrist micromovements to measure in-meal eating behavior from inertial sensor data. IEEE J. Biomed. Health Inform. **23**(6), 2325–2334 (2019)
30. Ghosh, T., Hossain, D., Imtiaz, M., McCrory, M.A., Sazonov, E.: Implementing real-time food intake detection in a wearable system using accelerometer. In: 2020 IEEE-EMBS Conference on Biomedical Engineering and Sciences (IECBES), pp. 439–443. IEEE (2021)
31. Nicholls, B., et al.: An EMG-based eating behaviour monitoring system with haptic feedback to promote mindful eating. Comput. Biol. Med. **149**, 106068 (2022)
32. Asady, H., Fuente, A., Pourabdian, S., Forouharmajd, F., Shokrolahi, I.: Acoustical role of ear canal in exposure to the typical occupational noise levels. Med. J. Islam Repub. Iran **35**, 58 (2021)
33. Saphala, A., Zhang, R., Amft, O.: Proximity-based eating event detection in smart eyeglasses with expert and data models. In: Proceedings of the 2022 ACM International Symposium on Wearable Computers, pp. 59–63 (2022)
34. Ngui, W.K., Leong, M.S., Hee, L.M., Abdelrhman, A.M.: Wavelet analysis: mother wavelet selection methods. Appl. Mech. Mater. **393**, 953–958 (2013)

Accelerated Optimization for Simulation of Brain Spiking Neural Network on GPGPUs

Fangzhou Zhang[1], Mingyue Cui[1], Jiakang Zhang[1], Yehua Ling[2], Han Liu[1], and Kai Huang[1(✉)]

[1] Sun Yat-Sen University, Guangzhou 510006, China
huangk36@mail.sysu.edu.cn
[2] Guangxi Transportation Science and Technology Group Co., Ltd., Nanning 530007, China

Abstract. As the application scenarios for large-scale spiking neural networks (SNN) increase, efficient SNN simulation becomes more essential. However, simulating such a large-scale network faces expensive overhead in terms of computation and communication, especially for high firing rates. To address this problem, we propose an effective accelerated optimization method for simulating SNN on GPGPUs, which simultaneously takes into account workload balancing and communication overhead. We design a workload-oriented network partition algorithm to minimize the number of external synapses and ensure workload balance. Additionally, we propose spike synchronization optimization by incorporating fine-grained scale, data compression, and full-duplex communication. This optimization aims to achieve lower communication overhead and better performance improvement. Furthermore, to avoid thread warp divergence, we assign an entire thread block for each neuron without collecting information on fired neurons in the spike propagation phase, which simplifies the execution flow and enhances performance. Experimental results demonstrate that our simulator can achieve up to $1.31\times\sim6.74\times$ speedup for SNN with different configurations, and the efficiency is improved by $40.21\%\sim51.11\%$ compared with the state-of-the-art methods.

Keywords: SNN simulation · Accelerated optimization · Load balance · Spiking neural network · High performance · GPGPUs

1 Introduction

Compared with traditional deep neural networks (DNN), spiking neural networks (SNN) are considered to be a more accurate representation of the human brain in biology, offering stronger interpretability and adaptability [23,24]. Simulating SNN serves as a pivotal bridge between neuroscience and artificial intelligence, offering a unique opportunity to unravel the intricacies of neural information

© The Author(s), under exclusive license to Springer Nature Singapore Pte Ltd. 2024
Z. Tari et al. (Eds.): ICA3PP 2023, LNCS 14492, pp. 172–189, 2024.
https://doi.org/10.1007/978-981-97-0811-6_10

processing. SNN incorporates time as an additional input dimension, and their unique spike-driven mechanism provides natural advantages in the field of bionic robots. IBM specifically designs an ultra-low-power programmable neurosynaptic chip called TrueNorth [2]. This chip integrates 5.4 billion transistors, including 1 million neurons and 256 million synapses. Intel also develops Loihi chip [19] for realizing brain-inspired simulation functions, which has a total area of 3840 square millimeters, 8 million neurons, and 8 billion synapses. However, due to selling price or policy restrictions, this has greatly limited the widespread use of SNN.

SNN simulation deployment on traditional commercial semiconductor platforms is one of the solutions. Balaji *et al.* [4] propose a SNN-accelerated GeNN framework, using technologies such as sparse representation of synaptic connections and block size determination based on occupancy to deploy SNN. Ahmad *et al.* [1] further propose an activation synaptic grouping strategy to solve the unbalanced workload of CUDA thread warps. Hazan *et al.* [14] present Bindsnet SNN simulation framework which uses PyTorch library for scheduling heterogeneous resources to achieve better scalability and computation acceleration. These methods realize the simulation deployment of SNN by adjusting the representation of neurons and synapses, but the acceleration effect is often very limited.

The sparsity of data in both time and space adds complexity to achieve a high degree of parallelism in SNN simulations. GPGPU-powered SNN simulators usually introduce synapse-based thread mapping methods that ignore the inherent sparsity property of SNN, which leads to load imbalance and severe limitations on parallelism. Compared with DNN, the topology of SNN tends to be more randomized. This characteristic poses a considerable challenge in achieving an even distribution of resources. Existing simulators focus on the number of synapses to design load balancing, which ignores the impact of the topology of SNN. Maintaining the connectivity and coherence of distributed networks means frequent data exchange during the spike synchronization phase, which brings significant communication overhead. Existing SNN simulators are difficult to make full use of the resources of GPGPUs.

This paper proposes an effective multi-GPU SNN simulator with accelerated optimization. Firstly, we introduce a novel workload-oriented network partition model, which modifies the partition balance condition of Metis [15] algorithm based on the workload model. The partition model reduces the number of external synapses which achieves lower communication overhead in the simulation stage while maintaining workload balance. Subsequently, to boost the computing of the simulation, we simplify execution flow without collecting fired neuron information and propose a branch-matched thread organization strategy to improve the spike propagation performance. This strategy assigns an entire thread block for each neuron to avoid thread warp divergence efficiently for lower computation overhead. Finally, considering the impact of communication overhead on multi-GPU simulation performance, we design a spike synchronization optimization with fine-grained scales, data compression, and full-duplex communication to achieve fewer data transfer sizes and better transfer parallelism during spike synchronization. We conduct extensive performance evaluations for

SNN with a set of different configurations. Experimental results on GPGPUs show that our implementation can achieve better performance, compared with the state-of-the-art methods. The contributions of the work mentioned above can be summarized as follows:

- We propose an accelerated SNN simulation framework on GPGPUs, which can efficiently simulate large-scale SNN networks.
- By introducing the workload-oriented network partition algorithm, our method reduces the number of external synapses, which achieves lower communication overhead during the simulation stage.
- To avoid thread warp divergence, we present a thread organization strategy that allocates thread blocks in units of neurons.
- We design a spike synchronization algorithm with optimization in fine-grained scale, data compression, and full-duplex communication, which further reduces the overhead of communication.

The rest of the paper is organized as follows. Section 2 provides background information and related work. Section 3 describes our method of SNN-specific optimization on GPGPUs. Section 4 presents the experimental details and analyses the simulation results. Finally, we conclude the paper in Sect. 5.

2   Related Work

2.1   Spiking Neural Networks

As the most widely used model in the field of neuroscience and neuromorphic computing, SNN can be described as a directed network, in which neurons connect with each other arbitrarily through synapses. A standard SNN mainly consists of neurons, synapses, and spikes, as shown in Fig. 1.

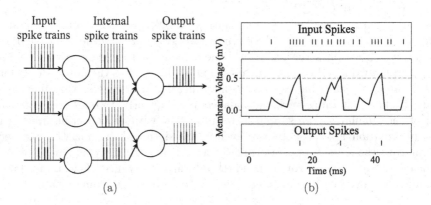

Fig. 1. An example of SNN. (a) The composition of a simple SNN. (b) Membrane voltage, input spike arrival time, and spike generation time for a neuron.

Neurons correspond to signal processing units. Different mathematical expressions of the states distinguish different neuron models. Taking LIF model [13] as an example, its differential equation indicates the iterative formula for neural membrane potential v over time as follows:

$$C_m \frac{dv}{dt} = \frac{C_m(v_{rest} - v)}{\tau_m} + (I_{synE} + I_{synI}) \tag{1}$$

$$\begin{cases} I_{synE} = g_e \times (v_{exc} - v) \\ I_{synI} = g_i \times (v_{inh} - v) \end{cases}, \begin{cases} dg_e/dt = -g_e/\tau_{exc} \\ dg_i/dt = -g_i/\tau_{inh} \end{cases} \tag{2}$$

The iterative update of the membrane potential v depends on the membrane capacitance C_m, membrane time constant τ_m, excitatory and inhibitory synaptic input currents I_{synE} (I_{synI}). The synaptic currents are calculated from the corresponding potentials v_{exc} (v_{inh}) and synaptic conductance g_i (g_e), in which g_i and g_e are also updated iteratively over time based on their respective time constants τ_{exc} (τ_{inh}).

Spikes correspond to the time sequence signal. In SNN, the signals generated by the neurons can be modeled as a binary value (0/1) with a timestamp. Each value in the sequence corresponds to a spike and represents the arrival time of the spike. In general, spikes are used for information transfer between neurons. Spikes can be generated by external input or by neurons during simulation.

Synapses connect neurons and transmit spikes. According to the direction of synapses, neurons are divided into source neurons and target neurons. Spikes sent by the source neurons travel along the synapses to the target neurons. When the source neuron is activated and sends out the spike, it needs to delay d before reaching the target neuron. Besides, if the weight of a synapse is negative, it means that the spikes inhibit the target neuron.

2.2 SNN Simulation Tools

With the growing interest in building large-scale SNN models, several SNN simulators [3,7,10,16,18,22,27] have been proposed to help the research communities. NEST [11] is the first neural network simulator on the traditional computer, which uses multi-core computers and computing cluster resources based on OpenMP. Brain [12] and Brain2 [28] provide equation-oriented methods to define new models, which support a variety of neuron/synapse models. Furthermore, GenEHH [21] lumps some anatomical features into a single compartment to simplify analysis and reduce computational load.

Compare with the CPU-based simulator, using GPU [6,9,17,20,25] has more advantages in simulation speed. SPIKE [1] uses time slice grouping and insensitive delay to improve the simulation speed. CARLsim4 [9] proposes a SNN simulation framework based on heterogeneous resources, which allows multiple GPUs and CPU cores to be used simultaneously. Different from the simulators mentioned above, SNAVA [26] is implemented in modern Field-Programmable Gate Arrays (FPGAs) devices to improve performance execution and flexibility to support large-scale SNN models.

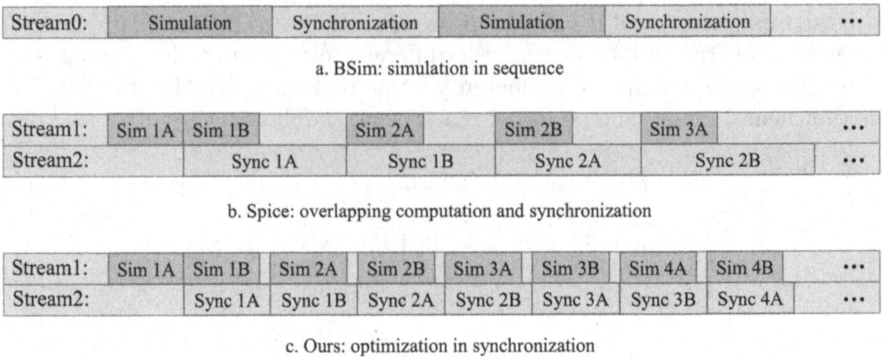

Stream0:	Simulation	Synchronization	Simulation	Synchronization	•••

a. BSim: simulation in sequence

Stream1:	Sim 1A	Sim 1B	Sim 2A	Sim 2B	Sim 3A	•••
Stream2:		Sync 1A	Sync 1B	Sync 2A	Sync 2B	•••

b. Spice: overlapping computation and synchronization

Stream1:	Sim 1A	Sim 1B	Sim 2A	Sim 2B	Sim 3A	Sim 3B	Sim 4A	Sim 4B	•••
Stream2:		Sync 1A	Sync 1B	Sync 2A	Sync 2B	Sync 3A	Sync 3B	Sync 4A	•••

c. Ours: optimization in synchronization

Fig. 2. Comparison of spike synchronization for different methods.

Recently, BSim [22] is a code generation SNN simulation framework, which implements sparsity aware load balance on CUDA threads by distinguishing neuron state in advance. It optimizes transmission by synchronizing neuron states instead of spikes, but the absence of time information leads to no space for further performance improvements. Spice simulator [6] implements a load balancing strategy for synapses and a cache-aware spike transmission algorithm which overlaps the computation and transmission by bisecting the time slice group. Figure 2 shows the process of simulation for different simulators. Different from the serial simulation execution of BSim, the Spice simulator further bisects the time slice group and uses two CUDA streams to realize the parallelism between the synchronization phase and the simulation phase. However, Spice transfers spikes of all devices directly, which ignores the size of data transferred and introduces large communication overhead. Besides, neither BSim nor Spice consider optimizing the number of external synapses in the pre-simulation stage. By contrast, we propose a workload model that guides the Metis algorithm to more evenly distribute neurons and synapses to different devices and reduce the ratio of edge-cut of partition results, which means less communication overhead. We also design a spike synchronization optimization with fine-grained scales, data compression, and full-duplex communication which achieves fewer data transfer sizes and better transfer parallelism. In this way, our method greatly reduces communication overhead and improves simulation performance, compared with others.

2.3   CUDA Programming on Multiple GPUs

CUDA is a parallel computing platform and programming model developed by NVIDIA. It allows developers to harness the power of NVIDIA GPUs (Graphics Processing Units) for general-purpose computing tasks. CUDA programming enables efficient parallel execution of code on multiple GPUs for high-performance computing. To improve performance when using multiple GPUs, it is essential to employ load balancing techniques to evenly distribute the workload across GPUs, preventing one GPU from becoming a bottleneck while others

Single Stream:

Stream0:	Memcpy A	Kernel B	Memcpy C	Memcpy A	•••

Multiple Streams:

Stream1:		A1	B1	C1	A1	B1	C1	A1	B1	•••
Stream2:			A2	B2	C2	A2	B2	C2	A2	•••
Stream3:				A3	B3	C3	A3	B3	C3	•••

Fig. 3. An example of the overlapping using multiple streams.

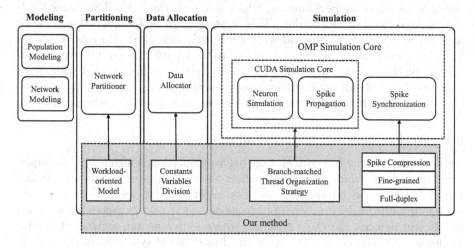

Fig. 4. The framework of accelerated optimization for simulation.

remain idle. Moreover, multi-stream parallelism enables the simultaneous execution of multiple kernel launches and memory operations. This approach effectively reduces the impact of data transfer latency by overlapping transmission and computation through asynchronous memory operation and CUDA streams. By initiating data transfers while the GPUs are still processing previous computations, the overall performance can be greatly enhanced, resulting in improved parallelism and better GPU utilization. Figure 3 shows an example of overlapping transmission and computation. By decomposing tasks and putting them into multiple streams, the computing resources of the GPU are more fully utilized. Incorporating these techniques can optimize the performance of multi-GPU CUDA applications.

3   Method

As is shown in Fig. 4, we propose an effective multi-GPU SNN simulation framework, where the grey area describes the focus of our optimization. The entrance of the framework is a modeling module that defines the model and specifies the network architecture. We provide a series of easy-to-use modeling tools for users in this module. Our method mainly works on the following three fields: partitioning, data allocation, and simulation. Firstly, in the partitioning module, we propose a workload-oriented network partition algorithm combining the workload model and Metis algorithm. Our workload-oriented partition algorithm can achieve a better partition effect by reducing the number of external synapses and taking into account workload balancing. Secondly, at the data allocation module, we take the memory access optimization into the module by extracting invariants in the simulation flow and allocating them to the constant memory of GPGPUs. Besides, we propose a branch-matched thread organization strategy to improve the spike propagation performance, which simplifies execution flow without collecting fired neuron information. This strategy assigns an entire thread block for each neuron to avoid thread warp divergence efficiently which is of benefit for lower computation overhead. Finally, at the simulation module, we present a spike synchronization algorithm with optimization in fine-grained scale, data compression, and full-duplex communication to achieve lower communication overhead and better performance improvement during spike synchronization.

3.1 Workload-Oriented Partitioning

Network partitioning is a critical stage before simulation, which is used to implement static load balancing and may affect the communication overhead during the synchronization phase. The partition algorithm divides a large-scale network into multiple devices, where neurons and their post-synapses are assigned to different devices. For those synapses whose source neuron and target neuron are located in different devices are defined as external synapses. And the ratio of edge-cut indicates the proportion of external synapses to the number of all synapses. The number of neurons and synapses directly affects the workload of the device, while the number of external synapses affects the communication overhead during simulation. In this section, we focus on reducing the number of external synapses while maintaining the load balance. Specifically, we propose a workload model that guides the Metis algorithm to more evenly distribute neurons and synapses to different devices. In this way, our method reduces the ratio of edge-cut, which means less communication overhead of different devices in the simulation stage.

In the workload model, we use piecewise functions wgt_n and wgt_s to specify the simulation overhead of different kinds of neurons and synapses. The equations are as follows:

$$wgt_n(n) = \begin{cases} C_{lif}, & type(n) = lif \\ C_{poisson}, & type(n) = poisson \end{cases} \tag{3}$$

$$wgt_s(s) = \begin{cases} C_{static}, & type(s) = static \\ C_{stdp}, & type(s) = stdp \end{cases} \tag{4}$$

where C_{lif}, $C_{poisson}$, C_{static}, and C_{stdp} define the computational overhead of different types of neurons and synapses. We further define the number of devices as N_d and the number of external synapses is defined as N_{outsyn}. For a device d_i, the number of neurons and synapses are N_{neu}^i and N_{syn}^i. neu_m^i is the m'th neuron in the device d_i. Similarly, syn_n^i is the n'th synapse in the device d_i. Then we can describe the workload V_i of device d_i as follows:

$$V_i = \sum_m^{N_{neu}^i} (wgt_n(neu_m^i)) + \sum_n^{N_{syn}^i} (wgt_s(syn_n^i)) \tag{5}$$

For static load balancing, the goal of our method is to seek the minimum value of the standard deviation of different device workloads:

$$min \quad SD_{load} = \sqrt{\frac{\sum_{N_d}^i (V_i - \bar{V})^2}{N_d}} \tag{6}$$

where $\bar{V}$ is the average workload of all the devices. Finally, we combine the workload model with the Metis algorithm to ensure the workload balance among devices while obtaining partition results with fewer external synapses.

3.2   Memory Access Optimization

Constant memory is a cache that shares data across multiple thread blocks and is read-only for each thread block. It offers lower latency and higher bandwidth compared with global memory, allowing for faster memory access. Based on GPU's constant memory, we divide simulation data into variables and constants and store constant data in constant memory to reduce memory access overheads. As shown in Eq. 1 and Eq. 2, some of elements do not change during the simulation, such as C_m, v_{rest}, τ_m, v_{reset}, τ_{exc}, τ_{inh}, v_{exc}, and v_{inh}. So we divide them from others and copy them into GPU constant memory. Compared with other simulators that store all these quantities in the global memory, it can be shared efficiently among threads without requiring expensive global memory accesses.

3.3   Branch-Matched Thread Organization

The spike propagation phase only performs the spike propagation process on the fired neurons. This indicates using GPU to perform spike propagation in a normal way leads to warp divergence and decreases the performance. When thread wrap diverges, the GPU needs to serialize the execution of different instructions, which leads to high latency and low parallelism. Different from the BSim-based spike propagation algorithm, our method does not need to collect the fired neuron, which simplifies the execution flow. Besides, we introduce a branch-matched thread organization strategy by allocating thread blocks in units of neurons to avoid thread warp divergence for higher performance.

The pseudocode of the spike propagation kernel function is shown in Algorithm 1. Firstly, we launch the spike propagation kernel function by allocating thread blocks consistent with the number of neurons. The dimension of the thread block is set to 32, 64, or 128, according to the scale of the network. Then, we use the index of the thread block as the index of the neuron of the spike to be propagated (line 1). If the neuron is not fired, we terminate the entire thread block to avoid warp divergence (lines 2–3). For a fired neuron, we find all the post-synapses of the neuron (lines 4–5). Then the algorithm assigns all spike propagation tasks of this neuron to threads of the block evenly and efficiently (lines 6–7). Before propagating, the method further judges whether the synapse is an external synapse (8–9). For an internal synapse, the spike is pushed to the corresponding neuron (line 10). For an external synapse, the spike is pushed into the buffer pool of the corresponding device (line 12). By using this thread organization strategy, our method obtains lower overhead in both the neuron simulation phase and the spike propagation phase.

Algorithm 1: Spike propagation algorithm

Input: netId, inBuffers, outBuffers

```
1  nid = blockIdx.x;
2  if fird (nid) then
3  |  return;
4  startloc = get_axon_loc (nid);
5  synsize = get_axon_size (nid);
6  for i = threadIdx.x; i < synsize ; i += blockDim.x do
7  |    synidx = startloc + i;
8  |    zone = get_zone (synidx);
9  |    if zone == netId then
10 |    |  push_inner_spike (synidx, inBuffers);
11 |    else
12 |    |_ push_outter_spike (synidx, zone, outBuffers);
```

3.4   Spike Synchronization Optimization

The spike synchronization which mainly propagates the spikes generated by neurons to neurons in other devices is critical for multi-GPU simulation. This is because the communication overhead in this process has a huge impact on multi-GPU simulation performance. To further reduce simulation overhead, our method overlaps transmission and computation by bisecting time slice group [5]. As shown in Fig. 5, we propose a spike synchronization optimization with fine-grained scales, data compression, and full-duplex communication. We design a compressed spike buffer data structure for spike synchronization, which merges the spikes that have the same target neuron and same arrival time (see Fig. 5(b)). This organization structure can reduce the communication overhead between devices during spike synchronization, especially for the high firing rate.

Different from synchronizing all spike information of the device, our method performs a finer-grained division of spike data to be synchronized for reducing

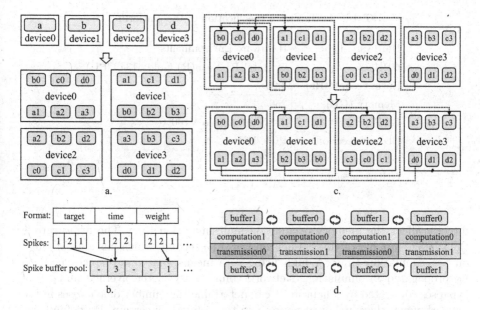

Fig. 5. Schematic diagram of spike synchronization optimization. (a) Finer-grained spike buffer design. (b) Spike data compression. (c) The order of synchronizing spike buffer pools. (d) Double buffering for avoiding conflicts.

redundant information. As is shown in Fig. 5(a), the finer-grained optimization divides the spike buffer of a device into multiple pools according to the device to which the spike propagates, in which green areas are the corresponding receive buffer pool for the spike buffer pool. Considering that computation and transmission to the same data area may lead to conflicts, our method uses double buffering for each buffer pool, as shown in Fig. 5(d). Furthermore, we also optimize the process of spike synchronization of full-duplex communication between GPUs by rearranging the synchronization order of spike buffer pools. From Fig. 5(c), we can see that in the case of no rearrangement, neither device 2 nor device 3 has any download tasks, while device 1 has three download tasks. They do not take full advantage of full-duplex communication at the same time period. However, when using rearrangement, all devices have upload and download tasks at the same period, which means that the data transmission using rearrangement has higher parallelism than that without rearrangement.

4 Experiments

4.1 Setup

In order to verify the effectiveness of our simulator, we consider the following various factors including network scales, firing rate, and network connectivity. Network scales mainly rely on the number of neurons and synapses, which can

Table 1. Hardware specification of the platform

Hardware Type	Hardware Environments
CPU	2× Intel(R) Xeon(R) Gold 6134 CPU
Basic Freq. (GHz)	3.20
Memory (GB)	256
GPU	8× Nvidia Tesla V100
Host connection	PCIe P2P
GPU Memory (GB)	8× 32
CUDA version	10.0

directly affect the memory occupation and simulation time. The firing rate is the average number of times a neuron fires in a second, which has a great effect on the simulation time. A high firing rate means that many spikes need to be propagated in simulation. Network connectivity is the average number of synapses connected by a neuron. It can determine the number of synapses in the network when the number of neurons in the network is certain. By default, we set the network scale and firing rate to 0.75 B and about 150 Hz, respectively. We use the Brunel [8] model as the default network structure, which has Poisson neurons that can manually set the firing rate. For network connectivity, we set the connection probability between populations as 0.1.

As shown in Table 1, we list the hardware specifications of the platform for comparison. All the experiments are deployed on commercial off-the-shelf platforms with Nvidia Tesla V100 GPU and Intel(R) Xeon(R) Gold 6134 CPU. For SNN benchmarks, we choose state-of-the-art methods for comparison, BSim [22] and Spice [5]. For the Spice simulator, in order to provide a corresponding compatible environment, we use docker as a container to deploy in our experimental environment. We use the LIF model to update the internal states of the neurons, which provides a certain degree of biological authenticity while maintaining limited computational complexity. Note that in order to present the results more intuitively, the values of the ordinates of the figures are not consistent. We use the ratio of edge-cut and the standard deviation of workload to evaluate the partition effect of different methods. The ratio of edge-cut reflects the number of external synapses, which affects the communication overhead of simulation. The standard deviation of workload between regions of partition results describes the load balance of the simulation workload on GPGPUs.

4.2   Simulation Results

As shown in Fig. 6, we represent the simulation overhead for different simulators on 8 GPGPUs. We can see that our method is almost the best compared with other methods, especially under high firing rates. When the firing rate is greater than 700 Hz, the efficiency of our simulator is improved by 40.21%~51.11%

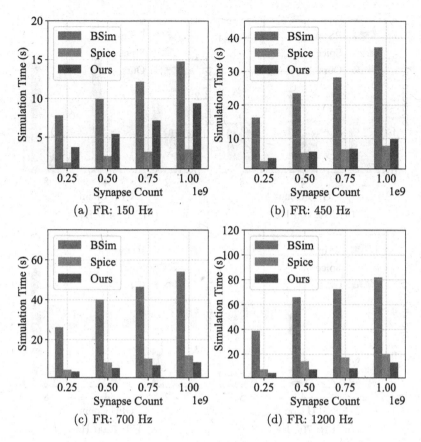

Fig. 6. Simulation time of different simulators with different network scales on 8 GPG-PUs.

compared with the Spice simulator, respectively. Benefiting from the compression of spike information, fine-grained spike buffer division, and optimization of spike synchronization order, our method achieves lower communication overhead, especially for high firing rates. Another observation is that the Spice simulator slightly outperforms our method on low firing rate. The reason is that the Spice simulator calculates the synchronized data transfer size according to the number of spikes, which makes a lower communication overhead at lower firing rates. Our method uses a compressed spike data structure that introduces an inherent overhead, but the data transfer size is kept low at high firing rates. The experimental results also confirmed this point in Fig. 6. In addition, our simulator outperforms the BSim simulator for all conditions. Experimental results demonstrate that our method achieves better accelerated optimization than other methods.

We further verify the speedup performance under different GPGPUs when the number of synapses is 0.75B, as shown in Fig. 7. We can observe that our method achieves significant speedup improvements compared with other meth-

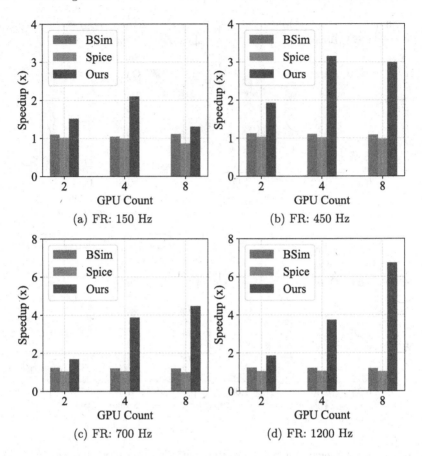

Fig. 7. Speedup of different simulators under different GPGPUs.

ods. The maximum speedup of our simulator can reach 6.74 for a high firing rate on 8 GPGPUs. By contrast, the multi-GPU speedup effect of the BSim and Spice simulator is not obvious. Besides, we notice that the speedup of 4 GPGPUs is better than 8 GPGPUs at low firing rates for each method. Although using 8 GPGPUs provides computation acceleration, it introduces more communication overhead compared with using 4 GPGPUs.

In Fig. 8, we evaluate the computation performance improvement of our method at different stages of the simulation. We extract the time overhead of the CUDA kernel function for neuron simulation and spike propagation and make comparisons on a single GPU to exclude the impact of communication overhead. From Fig. 8, we can see that the time overhead of our simulator for simulating neurons and spike propagation phases on a single GPU is much lower than that of the BSim simulator. Our method simplifies the simulation process through different thread organization methods, which reduces unnecessary computational work while avoiding warp divergence. Experimental results demonstrate that

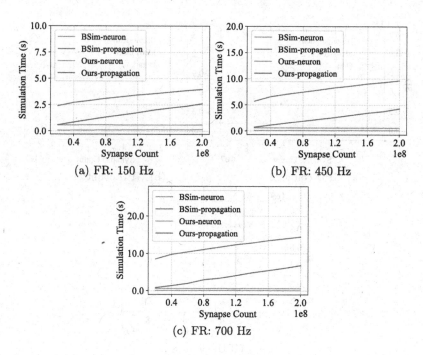

Fig. 8. Time overhead of different simulation stages.

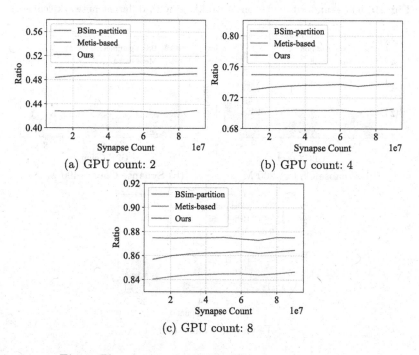

Fig. 9. The edge-cut ratio with different network scales.

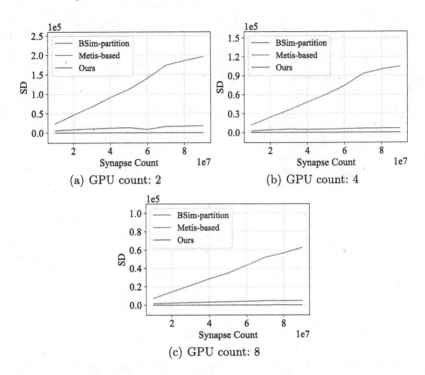

Fig. 10. The standard deviation of workload with different network scales.

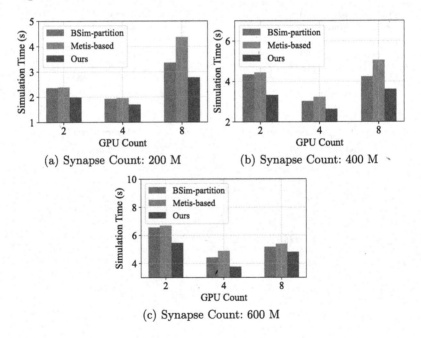

Fig. 11. Simulation overhead with different partition ways on GPGPUs.

our method achieves performance improvements at neuron simulation and spike propagation stage. Note that since the Spice simulator does not perform some necessary computational processes, such as counting the number of neuron fires, for the sake of fairness, we do not show the results of Spice.

The edge-cut ratio and the standard deviation of workload using different partition methods with different network scales are shown in Fig. 9 and Fig. 10. We can observe that the partition results of our method have a lower edge-cut ratio than BSim, which means fewer external synapses in Fig. 9. Our partition results have a better balanced workload effect than others by introducing a workload model, as shown in Fig. 10. Figure 11 further shows the simulation results when using different partition algorithms. Obviously, our partition algorithm is better than directly using the Metis algorithm, and also better than the BSim. Although the edge-cut ratio of the Metis is lower than our method, it ignores the workload balancing that greatly affects the simulation performance.

5 Conclusion

In this paper, we propose an effective accelerated optimization for the simulation of SNN on GPGPUs, simultaneously taking into account the workload balancing and communication overhead. Our method consists of a workload-oriented network partition, spike synchronization optimization with fine-grained scale, data compression, and full-duplex communication, and the spike propagation process with branch-matched thread organization. Experimental results show that our method achieves effective simulation performance and is insensitive to the firing rate variety. We believe that our method provides a novel perspective for this study. In the future, we will further analyze the impact of the optimization algorithm on accuracy performance. More schemes also need to be considered in terms of dynamic load balancing.

Acknowledgements. This work is supported in part by the Open Project Program for the Engineering Research Center of Software/Hardware Co-design Technology and Application, Ministry of Education (East China Normal University), Grant No. 67000-42990016, and in part by Fundamental Research Funds for the Central Universities, Sun Yat-sen University, Grant No. 23qnpy30/67000-31610023.

References

1. Ahmad, N., Isbister, J.B., Smithe, T., Stringer, S.M.: Spike: a GPU optimised spiking neural network simulator. Cold Spring Harbor Laboratory (2018). https://doi.org/10.1101/461160
2. Akopyan, F., Sawada, J., Cassidy, A., Alvarez-Icaza, R., Modha, D.S.: TrueNorth: design and tool flow of a 65 mw 1 million neuron programmable neurosynaptic chip. IEEE Trans. Comput. Aided Des. Integr. Circuits Syst. **34**(10), 1537–1557 (2015). https://doi.org/10.1109/TCAD.2015.2474396
3. Balaji, A., et al.: PyCARL: a PyNN interface for hardware-software co-simulation of spiking neural network (2020). https://doi.org/10.48550/arXiv.2003.09696

4. Balaji, N., Yavuz, E., Nowotny, T.: Scalability and optimization strategies for GPU enhanced neural networks (GENN). Comput. Sci. (2014). https://doi.org/10.48550/arXiv.1412.0595

5. Bautembach, D., Oikonomidis, I., Kyriazis, N., Argyros, A.: Faster and simpler SNN simulation with work queues. In: 2020 International Joint Conference on Neural Networks (IJCNN) (2020). https://doi.org/10.1109/IJCNN48605.2020.9206752

6. Bautembach, D., Oikonomidis, I., Argyros, A.: Multi-GPU SNN simulation with static load balancing. In: 2021 International Joint Conference on Neural Networks (IJCNN), pp. 1–8. IEEE (2021). https://doi.org/10.1109/IJCNN52387.2021.9533921

7. Bekolay, T., et al.: Nengo: a Python tool for building large-scale functional brain models. Front. Neuroinform. **7**, 48 (2014). https://doi.org/10.3389/fninf.2013.00048

8. Brunel, N.: Dynamics of sparsely connected networks of excitatory and inhibitory spiking neurons. J. Comput. Neurosci. **8**, 183–208 (2000). https://doi.org/10.1023/A:1008925309027

9. Chou, T., Kashyap, H., Xing, J., Listopad, S., Rounds, E.L.: CARLsim 4: an open source library for large scale, biologically detailed spiking neural network simulation using heterogeneous clusters. In: IEEE International Joint Conference on Neural Networks (2018). https://doi.org/10.1109/IJCNN.2018.8489326

10. Eppler, J., Helias, M., Muller, E., Diesmann, M., Gewaltig, M.O.: PyNEST: a convenient interface to the nest simulator. Front. Neuroinf. **2** (2009). https://doi.org/10.3389/neuro.11.012.2008

11. Gewaltig, M.O., Diesmann, M.: NEST (neural simulation tool). Scholarpedia **2**(4), 1430 (2007). https://doi.org/10.4249/scholarpedia.1430

12. Goodman, D., Brette, R.: Brian: a simulator for spiking neural networks in Python. Front. Neuroinf. **2** (2008). https://doi.org/10.3389/neuro.11.005.2008

13. Goodman, D.F.M., Brette, R.: The brian simulator. Front. Neurosci. **3**(2) (2009). https://doi.org/10.3389/neuro.01.026.2009

14. Hazan, H., et al.: BindsNET: a machine learning-oriented spiking neural networks library in Python. Front. Neuroinf. **12** (2018). https://doi.org/10.3389/fninf.2018.00089

15. Karypis, G., Kumar, V.: A fast and high quality multilevel scheme for partitioning irregular graphs. SIAM J. Sci. Comput. **20**(1), 359–392 (1998). https://doi.org/10.1137/S1064827595287997

16. Kasap, B., Opstal, A.V.: Dynamic parallelism for synaptic updating in GPU-accelerated spiking neural network simulations. Neurocomputing, S0925231218304168 (2018). https://doi.org/10.1016/j.neucom.2018.04.007

17. Knight, J.C., Nowotny, T.: GPUs outperform current HPC and neuromorphic solutions in terms of speed and energy when simulating a highly-connected cortical model. Front. Neurosci. **12** (2018). https://doi.org/10.3389/fnins.2018.00941

18. Lee, H., Kim, C., Kim, M., Chung, Y., Kim, J.: NeuroSync: a scalable and accurate brain simulator using safe and efficient speculation. In: 2022 IEEE International Symposium on High-Performance Computer Architecture (HPCA), pp. 633–647. IEEE (2022). https://doi.org/10.1109/HPCA53966.2022.00053

19. Lin, C.K., et al.: Programming spiking neural networks on Intel's Loihi. Computer **51**, 52–61 (2018). https://doi.org/10.1109/MC.2018.157113521

20. Mozafari, M., Ganjtabesh, M., Nowzari-Dalini, A., Masquelier, T.: SpykeTorch: efficient simulation of convolutional spiking neural networks with at most one spike per neuron. Front. Neurosci. **13**, 625 (2019). https://doi.org/10.3389/fnins.2019.00625

21. Panagiotou, S., Miedema, R., Sidiropoulos, H., Smaragdos, G., Soudris, D.: A novel simulator for extended Hodgkin-Huxley neural networks. In: 2020 IEEE 20th International Conference on Bioinformatics and Bioengineering (2020). https://doi.org/10.1109/BIBE50027.2020.00071
22. Qu, P., Zhang, Y., Fei, X., Zheng, W.: High performance simulation of spiking neural network on GPGPUs. IEEE Trans. Parallel Distrib. Syst. **31**, 2510–2523(2020). https://doi.org/10.1109/TPDS.2020.2994123
23. Sakemi, Y., Morino, K., Morie, T., Aihara, K.: A supervised learning algorithm for multilayer spiking neural networks based on temporal coding toward energy-efficient VLSI processor design. IEEE Trans. Neural Netw. Learn. Syst. (2021). https://doi.org/10.1109/TNNLS.2021.3095068
24. Shang, Y., Li, Y., You, F., Zhao, R.L.: Conversion-based approach to obtain an SNN construction. Int. J. Software Eng. Knowl. Eng. (2021). https://doi.org/10.1142/S0218194020400318
25. Smaragdos, G., et al.: BrainFrame: a node-level heterogeneous accelerator platform for neuron simulations. J. Neural Eng. **14**(6), 066008.1–066008.15 (2017). https://doi.org/10.1088/1741-2552/aa7fc5
26. Sripad, A., Sanchez, G., Zapata, M., Pirrone, V., Madrenas, J.: SNAVA-a real-time multi-FPGA multi-model spiking neural network simulation architecture. Neural Netw. **97**, 28–45 (2018). https://doi.org/10.1016/j.neunet.2017.09.011
27. Stewart, T.C., Tripp, B., Eliasmith, C.: Python scripting in the Nengo simulator. Front. Neuroinf., 7 (2009). https://doi.org/10.3389/neuro.11.007.2009
28. Stimberg, M., Brette, R., Dan, F.: Brian 2: an intuitive and efficient neural simulator. Cold Spring Harbor Laboratory (2019). https://doi.org/10.7554/eLife.47314

An Improved GPU Acceleration Framework for Smoothed Particle Hydrodynamics

Yuejin Cai[1], Jianguo Wei[1], Jiyou Duan[1], and Qingzhi Hou[2(✉)]

[1] College of Intelligence and Computing, Tianjin University, Tianjin 300350, China
[2] School of Civil Engineering, Tianjin University, Tianjin 300350, China
qhou@tju.edu.cn

Abstract. GPU has drawn much attention on accelerating SPH applications, which need high computational requirements. To eliminate the performance bottlenecks, this paper proposes an efficient GPU-accelerated framework for SPH computation on high-performance computing systems. To this end, several performance acceleration tools are developed to speed up the GPU implementation. The first one puts forward an efficient block size for GPU kernels, the second one determines the optimal workload for each CUDA thread and the third one uses reduction of global memory accesses to optimize data layout. Finally, an improved method makes use of simple GPU kernels instead of heavy ones when computing particle interactions. Comparison and analysis are made among the parallel results by the GPU implementations. As a result, with hundreds of thousands of particles run on advanced GPUs, the performance by the improved GPU implementation is at least 700 times higher than that by the serial CPU code. Compared to multi-thread CPU implementation, the performance increases by a factor over 200 on different GPUs.

Keywords: SPH simulation · GPU · CUDA · Performance improvements · HPC

1 Introduction

Smoothed particle hydrodynamics (SPH) firstly proposed by Lucy [1] and Gingold and Monaghan [2] is a meshless method in which fluid volume is discretized into a series of Lagrangian particles. More recently, SPH has a wide range of applications in many fields [3–5]. However, high computational requirements are needed due to its two time-consuming procedures including neighbor search and particle interaction. To reduce the search time, an improved parallel sorting method was proposed by Sun et al. [6]. Band et al. [7] put forward a compression scheme to store the neighbors by reducing memory consumption. Regretfully, the computing efficiency of the particle interaction in SPH is not improved by these efforts.

© The Author(s), under exclusive license to Springer Nature Singapore Pte Ltd. 2024
Z. Tari et al. (Eds.): ICA3PP 2023, LNCS 14492, pp. 190–201, 2024.
https://doi.org/10.1007/978-981-97-0811-6_11

To minimize the computational cost for SPH interactions, many parallel solutions have been presented to take advantage of GPU computing to complete the computationally intensive tasks. DualSPHysics [8] was the most common GPU solver for implementing SPH on GPUs. Compared to their CPU versions, the computing efficiency obtained by the GPU codes was greatly increased on the targeted NVIDIA graphics devices. Different from depending on CUDA-capable devices, AQUAgpusph [9] was proposed to accelerate SPH simulations with OpenCL, in which accelerator cards from vendors such as IBM, AMD and Intel can also be used. Similarly, Muta et al. [10] developed SISPH by PyOpenCL, which can work together seamlessly with OpenMP and MPI. As expected, these SPH implementations on GPUs markedly decrease computational complexity and save much runtimes compared to serial and parallel implementations achieved on CPUs.

However, due to the complexity and difficulty of the GPU implementation, only a few researches are concerned about GPU optimizations for further speeding up the parallel algorithm. For example, Dominguez et al. [11] developed optimization strategies for the GPU implementation of the SPH method, such as reducing cell size and accessing cells by row in the neighbor search. However, these improvements on performance came at the cost of large memory. In the paper of Winkler et al., A shared memory caching algorithm was presented to accelerate 2D GPU simulations [12]. A major performance penalty of this algorithm was the significant reduction of access efficiency of the shared memory if ill-conditioned access occurs. Wang et al. [13] introduced a series of kernel optimization strategies, such as reducing particle properties, utilizing particle reusability, which got 20x speedups on their peer2peer model.

Nevertheless, the studies about the executive GPU block size, thread workload, GPU data structure and kernel volume are rarely discussed. The block size is very important for the GPU implementation, which directly affects the whole GPU calculations. The thread computing fashion also exerts a strong influence on computational time of the GPU simulator and it is defined by the workload. It is vital to choose array layout to represent particle data. A poor data representation can cause heavy accesses to global memory, leading to reducing the system performance. The overload calculations of GPU kernels can also have similar side effects. In this paper, we will focus on the highly productive solutions for these problems that affect GPU performance.

The rest of this paper is organized as follows. In Sect. 2, the basic SPH equations are reviewed briefly. The parallel SPH scheme by CUDA programming executed on GPU is described in Sect. 3. In Sect. 4, detailed analyses and comparisons are made on the performances of the GPU implementations. Section 5 is conclusion.

2   SPH Method

In SPH [14], the particle properties at an arbitrary point are interpolated from neighbors. For the physical property f_i at particle i, the particle-average based approximation is given as

$$f_i = \sum_j V_j f_j W_{ij}(r_{ij}, h) \tag{1}$$

where $V_j = m_j/\rho_j$ and f_j represent volume and physical property at particle j, respectively. Here, m and ρ are particle mass and density, respectively. $r_{ij} = |r_{ij}| = |\mathbf{r}_i - \mathbf{r}_j|$ is particle distance between particles i and j. $W_{ij}(r_{ij}, h)$ is compact-support kernel function and h is the smoothing length with respect to $\mathbf{r}_i$, position of particle i.

Similarly, the original SPH approximation of the derivative of the physical property at particle i is formulated as

$$\nabla f_i = -\sum_j V_j f_j \nabla_i W_{ij}(r_{ij}, h) \tag{2}$$

where $\nabla_i W_{ij}(r_{ij}, h)$ is the gradient of the kernel function $W_{ij}(r_{ij}, h)$.

Following [15,16], the discretization of the WCSPH method based on a low-dissipation Riemann solution applied for the continuity and momentum equations is

$$\frac{d\rho_i}{dt} = 2\rho_i \sum_j V_j(\mathbf{v}_i - \mathbf{v}^*) \cdot \nabla_i W_{ij}(r_{ij}, h) \tag{3}$$

$$\frac{d\mathbf{v}_i}{dt} = -\frac{2}{m_i} \sum_j V_i V_j P^* \nabla_i W_{ij}(r_{ij}, h) \tag{4}$$

where $\mathbf{v}$ is the velocity. $\mathbf{v}^*$ and P^* are the solutions of the Riemann problem [16].

To close the system Eqs. (3)–(4), the artificial state equation is applied to relate pressure to density [17]. According to [15,18], the dummy particle method is used to impose the wall boundary condition and the Riemann solution is introduced to computed fluid-boundary interaction. Besides, a kick-drift-kick [16,19] scheme with a variable time-step size is used to advance the densities, velocities and positions of all fluid particles in the SPH method.

3   GPU Computation Using CUDA

3.1   Data Layout

In general, structure of arrays (SoA) and array of structures (AoS) are two common data structures that are used for storing SPH particle data. In AoS, a structure is created for each of particles and it is composed of different physical components. Each component represents a kind of property of one particle. For example, a typical AoS is shown as follows:

Struct(
    float velocity[3], position[3], density, pressure, $\cdots$;
)particle[N];

where N is the total number of particles. On the contrary, SoA creates a new property array to store a kind of property of all particles. The following data structure is an example of SoA:

Struct(
       float velocity[4 × N], position[4 × N], density[N], pressure[N], · · · ;
) particle;

As described in [20, 21], by comparing both it can be found that it is better to store particle properties data in SoA instead of AoS to obtain a higher GPU memory bandwidth. This is because that SoA results in a better coalesced memory access while the AoS-based scattered memory access often occurs in GPU implementation. For CUDA-enabled GPU devices, the memory coalescing and successive data manner can satisfy the coalesced rules of global memory access, leading to higher global memory bandwidth and cache-hit rate. Thus, SoA is used to store particle data in our GPU simulation. Besides, an improved strategy for optimizing the SoA model to achieve a higher performance will be described later.

3.2 CUDA Implementation

In SPH, particle computation is independent of each other. In this way, it is very suitable to be fully implemented in parallel on the GPU using CUDA programming language [22]. According to Fig. 1, we map main computational tasks on the modern GPU for the SPH method to fully exploit the parallel capabilities of GPUs. That is, the different computational equations for mass and momentum on SPH fluids are abstracted as CUDA computing kernels and executed on GPU with CUDA compute cores, which including force calculations, density and position renewal.

Before force computation, the searching CUDA kernel is used to obtain SPH neighbors. Then, the first kick stage is activated by launching force CUDA kernels to compute half-step acceleration. With the effort of position kernel, the drift stage for advecting particles is finished. Finally, based on the updated information, the next-time-step force is completed by the CUDA code, which is followed by the calculation of the minimum time-step size according to time integration conditions.

3.3 CUDA Acceleration

Although the CUDA implementation is achieved on the GPU hardware, the higher computing performance is not obtained. There are many barries for GPU acceleration, including thread-grid layout, particle workload, kernel volume and so on. Therefore, the following CUDA suggestions are developed to remove these performance obstacles to accelerate the GPU code.

Using an Efficient Block Size. According to the syntax and style of CUDA programming language, the CUDA block size can be freely determined by the programmer. However, different block sizes seriously affect the computing efficiency of the GPU code. Figure 2 visualizes the GPU runtimes for using different block sizes when implementing the SPH simulation.

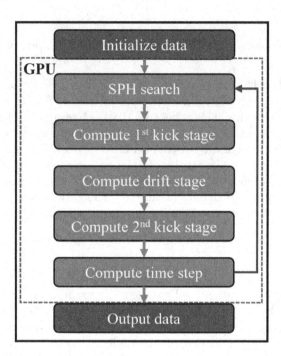

Fig. 1. Flow chart of the GPU-accelerated SPH simulation.

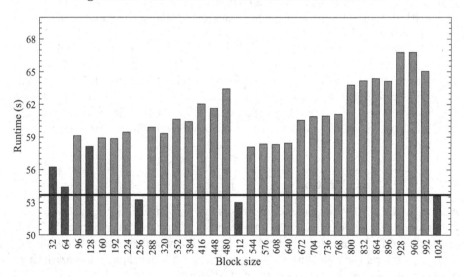

Fig. 2. Runtimes for different block sizes when implementing SPH on a GPU card (black solid line presents the result obtained by the built-in CUDA function, red bars the results of block size with integral power of 2 while blue bars opposite). (Color figure online)

As shown in Fig. 2, it is clearly seen that the results with integral power of 2 take less runtime than those of non-integral power of 2, leading to a maximum speed-up factor of up to 1.3. Among them, two block sizes of 256 and 512 spend the least time and are superior to the result obtained by the built-in *cudaOccupancyMaxPotentialBlockSize* function. Of course, a very slight difference between them is observed but can be neglected. Thus, it is strongly recommended to use the block sizes of 256 or 512 to execute CUDA kernels, which realizing a higher computational performance. In our GPU-SPH framework, the block size of 256 is used.

Optimizing Thread Workload. When running the GPU implementation for SPH, a lot of discrete particles are involved. The CUDA threads are configured to complete particle calculations. Obviously, the computational performance is also determined by the thread workload. For example (see Fig. 3), 16 particle interactions need to be accomplished by GPU threads. The straightforward solution is solved by assigning all particles for a CUDA thread or using one-to-one approach. The workload for computing particles is completely determined by a single CUDA thread.

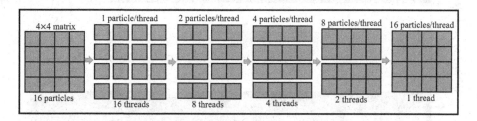

Fig. 3. Example of different number of particles computed by a CUDA thread.

Figure 4 shows GPU runtimes obtained by using different workloads for CUDA threads. As expected, the different solutions have a great difference in GPU performance. The fastest computational speed is obtained by a one-by-one fashion. Moreover, the performance decreases gradually with the workload increasing. To obtain better performance, this present work suggests the biunique approach to assign particle workload for a GPU thread.

Reducing Global Memory Accesses. In this subsection, the issue on optimizing data structure is discussed. In this GPU framework, the CUDA code needs to access data frequently from SoA arrays stored in global memory, which has the maximum access period. To increase global memory access efficiency, the number of accesses to the global memory is decreased by grouping part of GPU arrays. For example, the velocity and density arrays are combined into a new array of 16 bytes using float4 type. Figure 5 shows the improved speedup obtained by the reduction of memory accesses. Consequently, a better performance can be achieved by simplifying GPU arrays.

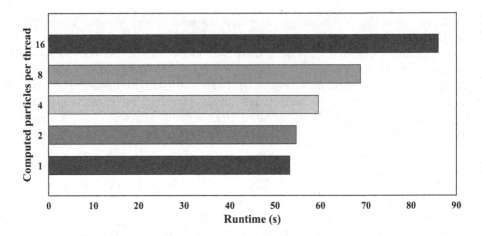

Fig. 4. Runtimes for different workload cases on a CUDA thread.

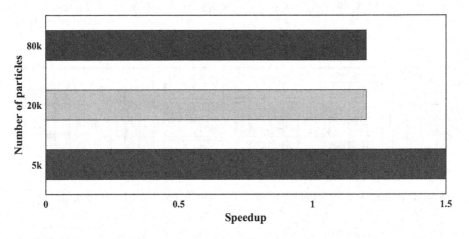

Fig. 5. Speedup achieved on a GPU card for different number of particles using the optimized SoA method.

Reducing GPU Kernel Volume. The complex SPH computations is easy to cause a bloated GPU code. For example, different forces needs to be calculated in a same time-stepping stage by a single GPU kernel. This leads to a increased volume of the GPU function, which holds little register resources. To reduce the kernel complexity, this paper decomposes big GPU kernels into small functions. Note that the achieved performance improvement varies with the CUDA applications.

4   Results

In this section, a dam-break benchmark is calculated to validate the proposed GPU framework. For comparison, the GPU implementations are achieved on two

different workstations. Table 1 shows the GPU configurations. We use CUDA
Toolkit 9.2/GCC 7.5.0/CLion as the development environment.

Table 1. Parameters of two different GPU workstations.

Workstation	W1	W2
CPU	Intel Core Xeon(R) Silver 4210 (2.2 GHz, 10 cores)	Intel Core Xeon(R) E5-1603 v3 (2.8 GHz, 4 cores)
Operating system	CentOS7.0 64-bit	Ubuntu18.0 64-bit
GPU	Titan V	K2200
CUDA cores	5120 Text follows	640
Memory size	96 GB	32 GB
Compute capability	7.0	5.0
GPU architecture	Volta	Maxwell

4.1   GPU Simulation

The dam-break benchmark [23] outlined previously is widely used to validate the
SPH method. The initial water column is supported by a vertical plate, whose
size is $H \times 2H$ ($H = 1$) as shown in Fig. 6 ($T = 0$). Upon sudden removal of the
plate, the water column collapses under gravity ($g = 9.81$ m/s^2) and turns into
a violent flow. In SPH simulations, 12,800 fluid particles in total ($dx = 0.0125$)
are used. All the simulation results in this test case are shown in terms of the
dimensionless time defined by $T = t\sqrt{g/H}$.

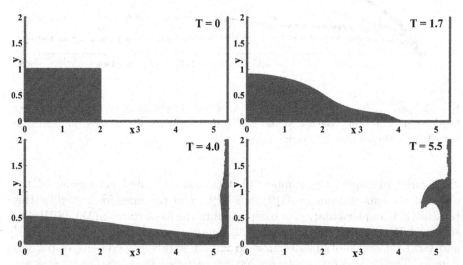

Fig. 6. Snapshots of the simulation results of the dam break case at different time
instants $T = 0, 1.7, 4.0, 5.5$ with $dx = 0.0125$.

Figure 6 shows several different snapshots at different time instants. It can be seen that the main phenomena of the flow including a narrow jet and high roll-up along the right wall are well captured by this GPU framework and are in good agreement with experimental [23] and previous numerical results [15,16].

4.2 Performance Evaluation

The section compares the computational performance of our improved GPU implementations on different GPU architectures with the CPU implementations and the state-of-the-art GPU solver–DualSPHysics [8].

The performance comparison between GPU and CPU can be obtained in Fig. 7. This figure exhibits the speedup achieved with the most efficient GPU implementations in comparison to the single-core and multi-core implementations on CPU. Thus, for example, when simulating for a run involving 200,000 particles, the speedups obtained with Titan V and K2200 compared with the single-threaded CPU implementation are over 3000x and 600x, respectively. When the multi-core implementation on CPU is compared, the speedup obtained with these two GPUs is about 40% of that of single-core CPU version.

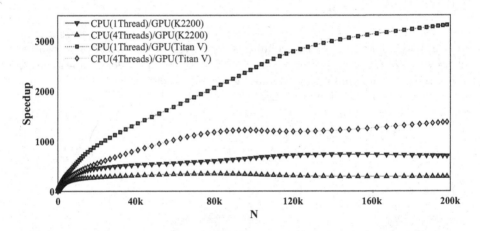

Fig. 7. Speedup for different numbers of particles (N) obtained by the most efficient GPU implementation on Titan V and K2200 in comparison to the CPU implementations (single thread and 4 threads).

Figure 8 highlights the numbers of iterations computed per second by the different implementations on GPU and CPU, and runtimes for a step by this present GPU implementation in comparison to the most common DualSPHysics solver [8]. We can see that the number of steps per second computed by the GPU and CPU implementations decreases gradually with the increasing of particle number. Therefore, for example, for 200,000 particles, the performance achieved on the GPU is 290 steps per second using a Titan V and 50 using a K2200, while

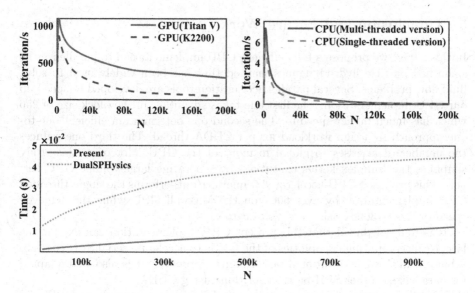

Fig. 8. Iterations per second computed by GPU (top left) and CPU (top right) codes, and runtimes for one iteration (bottom) by the improved GPU solver in comparison to DualSPHysics [8].

only 0.1 steps per second can be computed by the multi-threaded CPU version. For the single-core CPU, this value is barely measurable, which means very low computational efficiency. On the other hand, it can be concluded that the computational performance achieved by this present model outperforms DualSPHysics, leading to a speedup of over 3x with the increasing of particle number.

Table 2 summarizes the runtimes for this present GPU code and CPU implementations. It can be noticed that our GPU-accelerated code works well and achieves a higher performance in comparison to the basic CPU implementations. For example, when running a simulation involving 200,000 particles, the whole simulation takes 12 days on the single-core CPU and only 320 s on the GPU, leading to a speedup of over 3000x. Moreover, since our optimized GPU framework will benefit from more CUDA cores, we can expect higher computational performance on best Nvidia Hopper and future architectures.

Table 2. Runtimes of the GPU and CPU simulations.

Number of particles (N/k)	Runtime (t/s)		
	Volta-GPU	1-thread CPU	4-threads CPU
90	108	251640	155880
140	180	574560	222840
220	324	1055880	440640

5   Conclusion and Future Work

In this paper, we propose a fully efficient GPU implementation to parallelize the smoothed particle hydrodynamics method that has been widely used to solve fluid flow problems. Several parallel acceleration tools are developed to speed up our GPU implementation. The first one takes advantage of the block size of 256 which has much higher speedup. The second one determines a efficient one-to-one approach to assign workload for per CUDA thread. The third one reduces the number of accesses to global memory of the GPU. Moreover, this study simplifies the complex kernel executions by separating large-volume GPU kernels. This proposed GPU-accelerated framework outperforms the single-threaded CPU implementation by over 600x on the Maxwell GPU while the achieved speedup is over 3000x using the Volta card.

It is well concluded that this present GPU implementation can exploit the large-scale computing capabilities of the GPUs to accelerate serial SPH codes to achieve a higher computational performance. Moreover, it is also well adapted for parallelizing other SPH-based algorithms using CUDA.

We will further optimize the improved GPU code and extend our GPU simulator to support multi-GPU system in the future work.

Acknowledgements. This work is supported by the National Key Research and Development Program of China [grant no. 2020YFC1807905] and National Natural Science Foundation of China [grant no. 52079090].

References

1. Lucy, L.B.: A numerical approach to the testing of the fission hypothesis. Astrophys. J. **82**, 1013–1024 (1977)
2. Monaghan, J.J.: Smoothed particle hydrodynamics. Annu. Rev. Astron. Astrophys. **30**, 543–574 (1992)
3. Zhang, A.M., Sun, P.N., Ming, F.R., Colagrossi, A.: Smoothed particle hydrodynamics and its applications in fluid-structure interactions. J. Hydrodyn. **29**, 187–216 (2017)
4. Dominguez, J.M., et al.: DualSPHysics: from fluid dynamics to multiphysics problems. Comput. Part. Mech. (2021)
5. Zhang, F.Q., Wei, Q.M., Xu, L.Q.: An fast simulation tool for fluid animation in VR application based on GPUs. Multimed. Tools Appl. **79**, 16683–16706 (2019)
6. Sun, H.Y., et al.: A special sorting method for neighbor search procedure in smoothed particle hydrodynamics on GPUs. In: 44th International Conference on Parallel Processing Workshops, Beijing, pp. 81–85 (2015)
7. Band, S., Gissler, C., Teschner, M.: Compressed neighbour lists for SPH. Comput. Graph. Forum **39**, 531–542 (2020)
8. Crespo, A., Dominguez, J., Rogers, B., Gomez-Gesteira, M., Longshaw, S., Canelas, R., et al.: DualSPHysics: open-source parallel CFD solver based on smoothed particle hydrodynamics (SPH). Comput. Phys. Commun. **187**, 204–216 (2015)
9. Cercos-Pita, J.L.: AQUAgpusph, a new free 3D SPH solver accelerated with OpenCL. Comput. Phys. Commun. **192**, 295–312 (2015)

10. Muta, A., Ramachandran, P., Negi, P.: An efficient, open source, iterative ISPH scheme. Comput. Phys. Commun. **255** (2020)
11. Dominguez, J.M., Crespo, A.J.C., Gesteira, M.G.: Optimization strategies for CPU and GPU implementations of a smoothed particle hydrodynamics method. Comput. Phys. Commun. **184**, 617–627 (2013)
12. Winkler, D., Meister, M., Rezavand, M., Rauch, W.: gpuSPHASE-a shared memory caching implementation for 2D SPH using CUDA. Comput. Phys. Commun. **235**, 514–516 (2017)
13. Wang, Y.R., Li, L.S., Wang, J.T., Tian, R.: Acceleration of smoothed particle hydrodynamics method on CPU-GPU heterogeneous platform. J. Comput. **40**, 2040–2056 (2017)
14. Liu, M.B., Liu, G.R.: Smoothed particle hydrodynamics (SPH): an overview and recent developments. Arch. Comput. Method Eng. **17**, 25–76 (2010)
15. Zhang, C., Hu, X.Y., Adams, N.A.: A weakly compressible SPH method based on a low-dissipation Riemann solver. J. Comput. Phys. **335**, 605–620 (2017)
16. Rezavand, M., Zhang, C., Hu, X.Y.: A weakly compressible SPH method for violent multi-phase flows with high density ratio. J. Comput. Phys. **402**, 092–109 (2020)
17. Monaghan, J.J.: Simulating free surface flows with SPH. J. Comput. Phys. **110**, 399–406 (1994)
18. Adami, S., Hu, X., Adams, N.: A generalized wall boundary condition for smoothed particle hydrodynamics. J. Comput. Phys. **231**, 7057–7075 (2012)
19. Monaghan, J.J.: Smoothed particle hydrodynamics. Rep. Prog. Phys. **68**, 1703 (2005)
20. Wei, F., Jin, L., Liu, J., Ding, F., Zheng, X.P.: GPU acceleration of a 2D compressible Euler solver on CUDA-based block-structured Cartesian meshes. J. Braz. Soc. Mech. Sci. Eng. **42**, 250 (2020)
21. Wang, X.L., Qiu, Y.X., Slattery, S.R., Fang, Y., Li, M.C., Zhu, S.C., et al.: A massively parallel and scalable multi-GPU material point method. ACM Trans. Graph. **39**, 1–15 (2020)
22. CUDA Toolkit Documentation (v11.4.1). https://docs.nvidia.com/cuda/cuda-toolkit-release-notes/index.html. Accessed 9 June 2023
23. Zhou, Z.Q., De Kat, J.O., Buchner, B.: A nonlinear 3D approach to simulate green water dynamics on deck. In: 7th International Conference on Numerical Ship Hydrody-Namics, Nantes, France, pp. 1–15 (1999)

MDCF: Multiple Dynamic Cuckoo Filters for LSM-Tree

Xingfei Yao[1], Taotao Xie[2], Xiaowei Chen[2], Zhaoyan Shen[1], and Xiaojun Cai[1(✉)]

[1] School of Computer Science and Technology, Shandong University, Qingdao 266237, China
ap0l1o@mail.sdu.edu.cn, {shenzhaoyan,xj_cai}@sdu.edu.cn
[2] Cloud Inspur Information Technology Co., Ltd., Jinan 250000, China
{xie-taotao,chenxiaowei}@inspur.com

Abstract. As a write-optimized data structure, the Log-Structured Merge-tree (LSM-tree) based storage engine, which maintains data in a leveled structure on disk, is widely used in Key-Value (KV) storage systems. Meanwhile, the leveled design also makes it suffer from heavy read amplification since one query may incur multiple file search operations across several levels. To reduce I/O overhead, Bloom filters are adopted to accelerate the query process. Nevertheless, caching Bloom filters in memory incurs substantial memory overhead, and they need to be rebuilt during compaction. In this paper, we analyze the factors causing read amplification in LSM-trees and the challenges for the designing of Bloom filters. Based on our observation and analysis, we propose MDCF, an innovative solution that replaces Bloom filters with Multiple Dynamic Cuckoo Filters (MDCF) and implements it on LevelDB. The basic idea of MDCF is to construct a DCF for each level of the LSM-tree, except for the lowest level. KV pairs in level L_0 are mapped to a combination of a fingerprint and an SSTable identifier in DCF_0, while KV pairs in other levels are mapped to a fingerprint in the corresponding DCF. We demonstrate that this design not only significantly reduces read amplification by directly locating the target SSTable for each read request, but also consumes much less memory space. Experimental results based on YCSB demonstrate that MDCF outperforms LevelDB by 20–89% in read throughput without sacrificing write latency.

Keywords: LSM-tree · Key-Value store · Bloom filter · Read amplification

1 Introduction

Persistent Key-Value (KV) stores play a critical role in a variety of modern data-intensive applications, including cloud systems [16], advertising [5], social networks [2,26], search indexing [23], and online gaming [11]. Based on the used index structures, KV stores can be categorized into hash index-based design [10,

© The Author(s), under exclusive license to Springer Nature Singapore Pte Ltd. 2024
Z. Tari et al. (Eds.): ICA3PP 2023, LNCS 14492, pp. 202–218, 2024.
https://doi.org/10.1007/978-981-97-0811-6_12

11,18], B-tree-based design [15,20], and LSM-tree-based design [19,21,25]. Since hash index-based design requires a large amount of memory and cannot support range queries, and B-tree-based design involves an abundance of random writes, most modern KV stores use LSM-tree, such as BigTable [5] and LevelDB [23] at Google, RocksDB [13] at Facebook, Dynamo at Amazon [12], and Cassandra [1] at Apache.

A typical LSM-tree based KV store is organized into two components: a memory component and a disk component. The memory component seeks to absorb updates. When the memory component becomes full, it is flushed to persistent storage, and a new one is installed. The disk component is organized into levels, with each level containing a number of sorted files called SSTables (sorted string table). The levels close to the memory component hold the fresher information. When level L_i is full, one or more selected files from level L_i are compacted into files at level L_{i+1}, discarding stale values. This compaction operation occurs in the background.

The LSM-tree is designed for high write efficiency on persistent storage devices. On the one hand, it converts small, random writes into sequential append-only writes in large chunks, which optimizes the I/O speed on the disk. On the other hand, LSM-tree based KV stores also suffer from severe read amplification. This is because when lookup a KV pair, the KV store needs to check all possible SSTables of different levels until the key is found or all levels have been checked, leading to a serious read amplification. Furthermore, it is necessary to read multiple metadata blocks to really check whether a KV pair exists in one SSTable. Therefore, the read amplification in LSM-tree KV stores becomes more severe as the number of levels increases and may even reach over $300\times$ [19]. To reduce disk I/O, LSM-trees often use Bloom filters(BF) to eliminate SSTables that do not contain the target KV pair [24]. However, Bloom filters are immutable and need to be rebuilt from scratch after each compaction. In addition, caching Bloom filters in memory incurs a significant memory overhead, and frequent compaction operations not only result in the frequent rebuilding of Bloom filters but also invalidate cached Bloom filters.

In this paper, we introduce MDCF, a novel design that employs **M**ultiple **D**ynamic **C**uckoo **F**ilters to replace Bloom filters used in LSM-trees. The basic idea is to construct a Dynamic Cuckoo Filter (DCF) that supports reliable delete operations and elastic capacity for each level in the LSM-tree, except for the lowest level. For level L_0, where the SSTables are unordered, each KV pair is mapped to a fingerprint and an SSTable identifier in the corresponding DCF_0. As for the lowest level, which contains the most SSTables but receives the fewest requests, no DCF is constructed for it. For the remaining levels, each KV pair is mapped solely to a fingerprint in the corresponding DCF. This mapping operation is carried out during the compaction process and does not incur extra I/O overhead. With this approach, we can achieve a more compact and efficient filter implementation, while also supporting deletion and elastic capacity. Moreover, for a point query request, we can directly get the SSTable number that contains the request KV pair. As a result, unnecessary disk I/Os and memory read operations in the LSM-tree are minimized, leading to reduced read amplification and improved read performance. To demonstrate its efficiency, we carefully built MDCF atop the state-of-the-art KV store LevelDB. Experiments show that the MDCF outperforms LevelDB by 20–89% on read throughput, without sacrificing write latency. The contributions of this work can be summarized as follows:

- We carefully analyzed the causes of the read amplification problem and the worst-case read amplification ratio by tracking the read process of LevelDB.
- We designed and implemented MDCF, which constructs a Dynamic Cuckoo Filter for each level of the LSM-tree, excluding the lowest level. KV pairs in these levels are mapped to either a fingerprint or a combination of a fingerprint and a compressed SSTable identifier in their corresponding DCF.
- We evaluated the performance of MDCF using the YCSB benchmark. The results demonstrate that MDCF significantly enhances read throughput, minimizes memory space overhead, and does not compromise write latency.

2   Background and Motivation

In this section, we will first briefly introduce the LSM-tree and describe how data is organized in it, then analyze the read amplification in LSM-tree based KV stores, and finally explain the problem that motivates this paper.

2.1   LSM-Tree Based KV Store

An LSM-tree is a persistent data structure used in KV stores to efficiently support inserts and updates. LevelDB is a widely used KV store built using the LSM-tree, which we will take as an example to illustrate the working process of the LSM-tree. Figure 1 shows the high-level view of LevelDB, which mainly consists of two components. One is in memory, which includes a *MemTable* (memory table) and an *Immutable MemTable*. The other is in secondary storage, which is divided into multiple levels, with L_0 being the highest level and L_m the lowest, where m depends on the KV store size (the maximum value in LevelDB is 6). Besides, the capacity of level L_i is a multiple (10 in LevelDB by default) of the previous one L_{i-1}, which allows a huge KV store to be organized within a few levels (typically 5 to 7).

Now, we will illustrate how data is stored. Specifically, KV pairs are first written to the *MemTable* which is used to temporarily absorb the updates performed on the KV store. The size of the memory component is typically small, ranging from a few MBs to tens of MBs. When the *MemTable* is filled up, it will be converted into an *Immutable MemTable*, which cannot be written any more. Later, the *Immutable MemTable* will be packed into an SSTable (sorted string table) and appended into L_0 in the secondary storage through a process called *Minor Compaction* (also called *Flush*). *Minor Compaction* is write-efficient because updates are written sequentially in batches without merging with existing data in the store. Thus, the key ranges of different SSTables in L_0 may overlap. If the size of L_0 exceeds a preset threshold, a heavy-weight *Major Compaction*

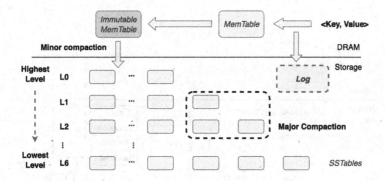

Fig. 1. Structure of an LSM-tree based KV store (LevelDB).

process is triggered to compact all overlapping SSTables into L_1. The *Major Compaction* process performs *Merge Sort* operations to merge the L_0 SSTables with the SSTables in L_1 that have overlapping key ranges. The sorted KV pairs are written back to L_1 in newly formed SSTables. Similarly, if the size of L_1 or any other level exceeds its size limit, the *Major Compaction* process is triggered to select one or multiple SSTables for compaction into the adjacent lower level. In this way, the KV pair updates are rolled down from the top to the bottom, level by level.

2.2   Serious Read Amplification of LSM-Tree

Read amplification is one of the major problems in LSM-trees, such as LevelDB. It is defined as the ratio between the amount of data read from the underlying storage device and the amount of data requested by the user. In this section, we will take LevelDB as an example to analyze the issue of read amplification in LSM-tree.

In LevelDB, each level consists of a single sorted run (except for L_0) where the SSTables have disjoint key ranges. The sorted run of KV pairs in an SSTable is divided into multiple data blocks, and the boundary keys between every two adjacent data blocks are stored in an index block along with the corresponding data block offset within the SSTable. Additionally, each SSTable also contains a bloom filter block to determine the existence of a key in the SSTable, thereby saving unnecessary storage I/Os. A block is the basic storage I/O unit in LSM-tree based KV stores.

The *Get* operation is used to retrieve the value of a specific key in the following sequence: *MemTable* and *Immutable MemTable*, followed by every SSTable in L_0 from the youngest to the oldest, and then L_1 to L_6. To avoid a large Get latency, LevelDB slows down the foreground write traffic if the number of SSTables at L_0 is bigger than eight, in order to wait for the compaction thread to compact some SSTables from L_0 to L_1. As shown in Fig. 2, LevelDB performs the *Get* operation by following the steps outlined below:

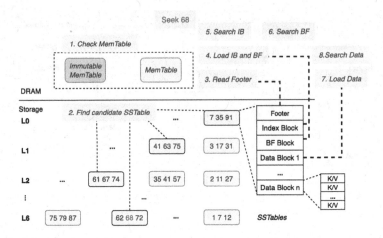

Fig. 2. How a lookup is processed in LevelDB.

1. **Check the *MemTable*.** If the key is found in either the *MemTable* or *Immutable MemTable*, return its value without accessing storage.
2. **Find SSTable Files.** If the key is not found in the *MemTable*, LevelDB identifies the set of candidate SSTable files that may contain the target key. In the worst case, a key could be present in all L_0 SSTables (due to overlapping key ranges) and within one file at each subsequent level.
3. **Read the SSTable Footer.** In each candidate SSTable, an SSTable footer is initially loaded from storage. This footer contains the position and size of the index block and bloom filter block.
4. **Load the Index Block and Bloom Filter Block.** Following that, in accordance with the read SSTable's footer, the index block (IB) and bloom filter block (BF) are fetched from the storage.
5. **Search the Index Block.** LevelDB employs a block-based bloom filter that generates a filter for every 2 KB of contiguous data. To determine the appropriate filter for querying, it is necessary to first query the index block and obtain the data block that may contain the target KV pair.
6. **Search the Bloom Filter Block.** The relevant filter is queried to verify whether the target key is present in the data block.
7. **Load Data Block.** If the filter indicates presence, the data block is loaded.
8. **Search Data Block.** Binary search is performed on the data block.
9. **Read Value.** If the key is found in the data block, the corresponding value is read, and the *Get* operation is complete. If the filter indicates the absence of the key or if the key is not found in the data block, the search will continue to the next candidate SSTable file.

As previously mentioned, LSM-tree based KV stores face a heavy challenge in terms of read amplification. By analyzing the *Get* operation in LevelDB, we have identified two sources of read amplification in LSM-tree based KV stores. Firstly, to locate a specific KV pair, LevelDB may need to examine multiple levels. In the worst-case scenario, LevelDB needs to search through eight SSTables in L_0 and one SSTable in each of the remaining six levels, resulting in a total of 14 SSTables. Secondly, to locate a KV pair within an SSTable file, LevelDB needs to read multiple metadata blocks within the file. The actual amount of data read is determined by the sum of the index block, bloom filter block, and data block. For example, when searching for a 1 KB KV pair, LevelDB must read a 16 KB index block, a 4 KB bloom filter block, and a 4 KB data block, totaling 24 KB. Consequently, in the worst-case scenario with 14 SSTable files, the read amplification of LevelDB amounts to $24 \times 14 = 336$. Smaller KV pairs will result in even higher read amplification.

2.3   Motivation

To reduce extra I/Os induced by checking multiple SSTables, modern KV store designs utilize Bloom filters, which are space-efficient probabilistic data structures. A Bloom filter consists of an array of bits with k hash functions, where each key is mapped to k random bits. By examining these bits, we can determine whether a key exists. If any bit is set to 0, it indicates a negative result, while all bits set to 1 can indicate either a true or false positive outcome. Inserting or querying an existing key in a Bloom filter requires k memory I/Os.

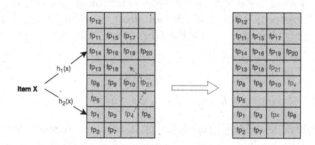

Fig. 3. Insert elements to Cuckoo Filter (each bucket consists of four slots).

The purpose of using Bloom filter in LSM-trees is to avoid the need to further read index blocks and data blocks when reading an SSTable to check whether the target KV pair exists in this SSTable file. LSM-trees often cache the bloom filter block in memory for performance reasons, but this can result in a significant increase in memory usage. If an LSM-tree utilizes a block-based Bloom filter, it becomes necessary to load the index block into memory and query it to identify the specific Bloom filter that needs to be queried within the cached bloom filter block. This is true even if the bloom filter block is already cached in memory.

Additionally, a Bloom filter does not support deletion operations. On the other hand, the *Compaction* operation involves the removal of old SSTable files and the creation of new ones, which may lead to a range of issues. Firstly, it is necessary to rebuild the Bloom filter from scratch for the newly created SSTable file.

In particular, a *Compaction* operation typically involves merging one SSTable from level L_i with multiple SSTables from level L_{i+1}. This implies that when one SSTable is merged into the next level, it results in multiple SSTables having to rebuild their Bloom filters. Secondly, this will render the cached Bloom filter invalid. However, the expired Bloom filter cannot be promptly removed from the cache to free up cache space.

Cuckoo filter (CF) [14] is one of several data structures that have emerged as alternatives to Bloom filters. At their core, these structures employ a compact hash table that stores the fingerprints of keys. A fingerprint is a string of f-bits derived by hashing a key. CF comprises an array of buckets, where each bucket contains s slots for storing fingerprints. As shown in Fig. 3, during insertion, an entry with a key x is hashed to two bucket addresses $h_1(x)$ and $h_2(x)$ using Eq. 2.3. A fingerprint of key x is inserted into the bucket that has available space. If both buckets are full, a fingerprint from one of the two buckets is randomly selected and swapped with its alternative bucket to clear space. By virtue of using the XOR operator, the CF always allows the computation of an entry's alternative bucket using the fingerprint and current bucket address, without requiring the original key. The swapping process continues recursively until a free slot is found for all fingerprints or until a swapping threshold is reached, at which point the insertion fails. Compared to a Bloom filter, the greatest advantage of a CF is the support of the delete operation. A CF achieves deletion by removing the monopolistic fingerprint for an item x_i. It is clear that removing the fingerprint of item x_i will not affect the membership testing of any other elements x_j where $x_j \neq x_i$ in the CF.

$$h_1(x) = hash(x)$$
$$h_2(x) = h_1(x) \oplus hash(x's\ fingerprint)$$

One basic idea to address the issue of insufficient support for deletion in Bloom filters is to replace Bloom filters with CF in LSM-tree. For example, a CF can be constructed for the entire LSM-tree or for each level within the LSM-tree. However, this approach can introduce new issues, such as the significant memory overhead involved in storing CF in memory. Additionally, while a standard Cuckoo filter partially fulfills the deletion requirement for representing dynamic sets, it does not have the capability to easily expand or shrink to accommodate changes in set size. This means that once the capacity limit is reached, a larger CF must be rebuilt to store more KV pairs. This not only suspends the service of the CF, but also incurs a certain cost. Given the limitations of the traditional LSM-tree Bloom filter and Cuckoo Filter, we will introduce a new method in the next section to address these issues. This method aims to improve the space efficiency and scalability of LSM-tree filters, reduce the number of disk I/Os for read operations, and mitigate the read amplification issue of LSM-tree to enhance read performance.

3 Proposed Approach

To address the limitations of the Cuckoo Filter, a Dynamic Cuckoo Filter (DCF) [6] leverages the CF as a building block and consists of a set of n linked homogeneous CFs $\{CF_1, ..., CF_n\}$. Initially, a DCF consists of a single CF (CF_1), but it can extend its capacity by appending new CFs. The DCF provides a compact operation to move fingerprints from sparse CFs to their corresponding buckets in other denser CFs, while also eliminating empty CFs to improve space efficiency. Building on this analysis, we designed MDCF, which replaces the Bloom filter with multiple different DCF variants. MDCF further innovates in two directions.

Dynamic Adjustment. MDCF utilizes multiple DCFs to replace the Bloom filter in the conventional LSM-tree. It has the ability to dynamically adjust based on variations in the data volume within the LSM-tree. If the amount of data in the LSM-tree suddenly increases, MDCF can be dynamically expanded without incurring reconstruction overhead. Similarly, if the amount of data in the LSM-tree suddenly decreases, MDCF can shrink in time to reduce memory space overhead.

Memory Space Overhead. MDCF only adds KV pairs in levels other than the lowest level to the corresponding DCF. For the KV pairs in level L_0, their corresponding SSTable identifiers and fingerprints need to be stored, while for the remaining KV pairs, only the fingerprint needs to be stored. This approach leads to significant savings in memory space.

3.1 The Design of MDCF

The main idea of MDCF is to construct multiple Dynamic Cuckoo Filters for different levels in memory except the lowest level. As shown in Fig. 4, each KV pair in L_0 is mapped to a fingerprint and an SSTable identifier (SID) in the corresponding DCF_0. And for the intermediate levels, such as L_5, each KV pair is only mapped to a fingerprint in DCF_5. The specific design approach is as follows.

Map the KV Pairs in L_0 to DCF_0. The SSTables in L_0 are not sorted, which means that one SSTable may have overlapping ranges with another SSTable. Therefore, even if we know that a KV pair is in L_0, we cannot determine in which SSTable it resides. We have to check each possible SSTable one by one. Thus, for each KV pair in L_0, there is one DCF_0 entry consisting of a fingerprint and an SSTable identifier that maps its current SSTable number in L_0.

Map the KV Pairs in the Intermediate Levels to Their Corresponding DCF_i. For the other levels of the LSM-tree, the SSTables are strictly ordered. Thus, if we can determine that a KV pair exists in one of these levels, we can also determine in which SSTable it is located. For each KV pair in intermediate level L_i, there is a corresponding DCF_i entry that contains only a fingerprint used to verify whether the KV pair is a member of L_i. There is no additional overhead in terms of memory space.

No DCF for the Lowest Level in the LSM-Tree. In the LSM-tree, the lowest level contains approximately 70% of the SSTables, but it only receives about 9% of the requests [32]. If we cannot find the target KV pair in the DCF of the upper levels, then it is either in the lowest level or does not exist in the LSM-tree. Therefore, we only construct a DCF for each of the upper levels. When the target KV pair is not found in the upper levels, we go directly to the lowest level to locate it.

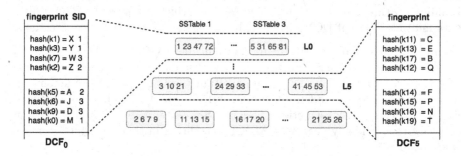

Fig. 4. Map the KV pairs to the corresponding DCF.

3.2 Maintenance of MDCF

We create a variant of DCF called DCF_0 for L_0. For each KV pair in L_0, there is a corresponding DCF entry that includes a fingerprint and an SSTable identifier. This entry maps the KV pair to the SSTable in which it is stored. Storing the complete SSTable number in the DCF would consume an excessive amount of space. Therefore, we encode the SSTable number instead. In LevelDB, L_0 can hold a maximum of 12 SSTable files. When the number of files reaches 12, a write pause is triggered. Therefore, we only need four bits to identify all SSTable numbers in L_0.

As data is continuously written to the LSM-tree, new SSTables are written in L_0, while the old SSTables are merged into the next level. Consequently, the KV pairs contained in the SSTable in L_0 and the SSTable number will change. To address this issue, we maintain a dynamic mapping table called SMAP in memory. This table enables a bidirectional mapping between SSTable numbers and a 4-bit fixed-length encoding. When the *MemTable* stored in memory requires flushing to storage, it is written to an SSTable file located in L_0. MDCF first obtains the number of the new SSTable file and then queries SMAP to obtain an available 4-bit SSTable identifier (SID). Finally, the fingerprint of each KV pair in the *MemTable* and its corresponding 4-bit SID are combined to create a DCF entry. This entry is then inserted into DCF_0.

During a *Compaction* operation from L_0 to L_1, MDCF determines whether to create a new DCF_0 for L_0 or update the existing DCF_0 based on the number of SSTable files involved in the compaction. Since the SSTables in L_0 are unordered, the compaction operation from L_0 to L_1 typically includes all SSTable files in

L_0. Therefore, in such cases, it is sufficient to create a new empty DCF_0 for L_0. If only a subset of SSTables from L_0 are involved in the compaction, the existing DCF_0 is updated by removing the corresponding mapping entries associated with those SSTables.

When a KV pair is moved from one level to the next during the *Compaction* process (L_i to L_{i+1} and $i > 0$), and the target level is an intermediate level L_{i+1} rather than the lowest level in the LSM-tree, MDCF calculates the fingerprint of the KV pair and removes it from DCF_i. Subsequently, the fingerprint is added as a DCF entry to the corresponding standard DCF_{i+1} of L_{i+1}. As mentioned earlier, the lowest level of the LSM-tree contains the highest number of SSTables but has the lowest access frequency. Therefore, when the *Compaction* operation outputs to the lowest level in the current LSM-tree, MDCF does not generate DCF entries for the KV pairs involved in this compaction and avoids including them in a DCF.

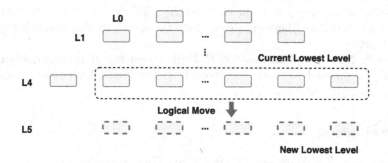

Fig. 5. New logical Move operation.

An additional concern arises as data is continuously written to the LSM-tree, eventually causing the current lowest level to transition into an intermediate level. When dealing with an intermediate level, it is necessary to construct a DCF and establish mappings for all KV pairs within this level. However, constructing a DCF for an intermediate level by reading all of its KV pairs introduces significant overhead.

Through observation, we have noticed that the first SSTable file in the lowest level of the LSM-tree is transferred from the previous level to this level through a logical *Move* operation, without any physical read or write operations involved. As shown in Fig. 5, when a *Move* operation is triggered to generate a new lowest level from the current lowest level of the LSM-tree, only one SSTable file is kept in the current lowest level. The other SSTable files in this level are logically moved to the new lowest level. A DCF is then created for the level that changes from the lowest level of an LSM-tree to the intermediate level through this *Move* operation. At this point, it is only necessary to read the single SSTable in this level to complete the construction of the DCF.

3.3   The Flow of Read Operations

Now, we will introduce the reading process of the MDCF. Denote K as the key to be searched. First, the *MemTable* is searched for K as it potentially contains the latest version of the corresponding value. If K is not found, then each DCF is queried sequentially. One of the following cases will occur:

- **Case 1: Found in** DCF_0. If the SID of the target key K is obtained from DCF_0, the corresponding SSTable number can be obtained through SMAP. Once the SSTable number has been obtained, the corresponding SSTable can be queried. If the KV pair is found in the SSTable, it will be returned directly. Otherwise, MDCF will continue to check other DCFs sequentially.
- **Case 2: Found in other DCFs.** If the fingerprint of the target key K is found in DCFs other than DCF_0, we can directly determine in the SSTable where K is located. This is because the SSTables in the level are all strictly ordered, except for L_0. Once we have determined the SSTable where K is located, we can query it. If the entry for K is found in the SSTable, it will be returned directly. Otherwise, we continue the search by checking other DCFs in sequence.
- **Case 3: Not found in any DCF.** If the target key K is not found in any DCF, we proceed to check the SSTable at the lowest level that covers the range of keys containing K.

4   Evaluation

We implemented a prototype of MDCF based on LevelDB v1.23 in C++, by adding or modifying approximately 2K lines of code. Most of the changes are related to building and maintaining the Dynamic Cuckoo Filter. To evaluate the performance of our approach, we measured the average read latency and throughput for four distinct workloads, each with different read-to-write ratios.

4.1   Experiment Setup

We conducted all experiments on a machine equipped with an Intel Core i7-8700 3.2 GHz processor, 64 GB of RAM, and a 500 GB SSD. For the software, we use Ubuntu 18.04 LTS with Linux Kernel 5.4.0 and the Ext4 file system. We used the YCSB-cpp [28] benchmark, which is the C++ version of YCSB [7] and has low overhead.

To alleviate memory pressure while creating SSTables, the original LevelDB generates a separate Bloom filter for each 2 KB block of key values. During the *Get* operation, the index block is accessed first to obtain the offset of the block that may contain the KV pair. The corresponding Bloom filter is loaded based on the offset. If the filter indicates that the key might exist, the actual data block is searched for the key. However, even if the filter produces negative responses, the index block still needs to be loaded and examined. Furthermore, the index and bloom filter blocks are stored in the `table_cache` and are subject to caching. Unfortunately, users have no control over the amount of memory allocated for caching these blocks, except by adjusting the `max_open_files` setting. To address these issues and ensure fair comparisons, we have implemented a modified version of LevelDB. This version creates a single Bloom filter for the entire SSTable and caches it in a filter cache. In this work, we will refer to the original LevelDB as LevelDB and the enhanced version as Full-Filter-LevelDB.

4.2 Workloads

We evaluated the performance of MDCF using YCSB benchmarks, which provide a range of workloads with different combinations of KV operations. Specifically, Workload A consisted of 50% reads and 50% updates, Workload B consisted of 95% reads and 5% updates, Workload C consisted of 100% reads, and Workload D consisted of 95% reads and 5% inserts. Note that Workload D uses the Latest distribution, while the others follow a Zipfian distribution. In this set of experiments, we set the size of each KV pair to 1 KB, randomly loaded 20 million KV items and then issued 30 million operations. For LevelDB and Full-Filter-LevelDB, we used Bloom filters with 12 `bits-per-key`. Similarly, we allocate 12 bits for the fingerprint in MDCF.

4.3 Performance Evaluation

The cache space used by DCFs in MDCF does not exceed 16 MB for any of the four workloads. Therefore, we have set the `max_open_files` parameter of LevelDB to 1100, which is expected to occupy approximately 16 MB of memory. Additionally, we have set the filter cache size of Full-Filter-LevelDB to 16 MB.

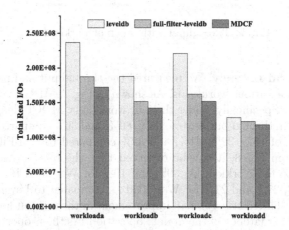

Fig. 6. Read I/Os with different workloads.

Read I/Os. We evaluated read I/Os under four workloads with mixed read and write ratios. The number of reading I/Os is an essential indicator for evaluating our design because a lower number of reading I/Os means that fewer blocks need to be read and accessed when executing read requests. As shown in Fig. 6, compared to other KV stores, our design significantly reduces the number of reading I/Os. Specifically, it decreases by 28% and 9% under workload A, 25% and 7% under workload B, and 32% and 7% under workload C for the two KV stores, respectively. Finally, under workload D, the number of reading I/Os is decreased by 9% and 4%, respectively. Under workload D, the reduction of reading I/Os is limited. This is because LevelDB and Full-Filter-LevelDB can maintain a high cache hit rate for this workload. It is important to note that locating a KV pair from an SSTable involves reading various types of information, including footers, meta blocks, index blocks, bloom filter blocks, and data blocks. These blocks have varying sizes, which can result in variations in the time required to read them. Therefore, the number of read I/Os alone cannot accurately reflect changes in reading performance.

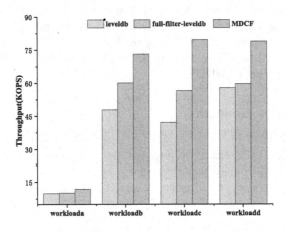

Fig. 7. Throughput with different workloads.

Throughput and Latency. We measured the throughput and read latency of each KV store for various workloads. As shown in Fig. 7, MDCF outperforms all other KV stores. Specifically, under the four workloads, MDCF demonstrated an overall increase in throughput of 20%, 53%, 89%, and 36% compared to LevelDB, and an increase of 18%, 22%, 41%, and 32% compared to Full-Filter-LevelDB, respectively. From Fig. 8, it can be observed that the read latency is reduced by 43% and 21% for Workload A, 39% and 19% for Workload B, 48% and 39% for Workload C, 27% and 21% for Workload D, compared to LevelDB and Full-Filter-LevelDB. The relatively small improvement in throughput for Workload A is due to the higher latency of the update operation. As the proportion of update operations increases, they contribute more latency than the read operations.

Write Performance. To study the impact on write performance, we first randomly loaded 20 million KV pairs for different KV stores and then evaluated the performance of 30 million mixed operations. We compared the load throughput

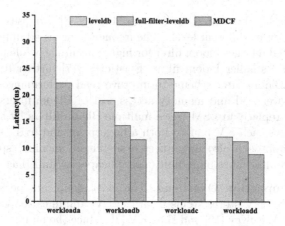

Fig. 8. Read latency with different workloads.

in order to evaluate the write performance, and the results are shown in Fig. 9. As we can see, the write performance remains almost the same, with MDCF demonstrating slightly better performance, even when the Dynamic Cuckoo Filter is integrated into the KV stores. The primary reason for this is that Bloom filters are organized into blocks.

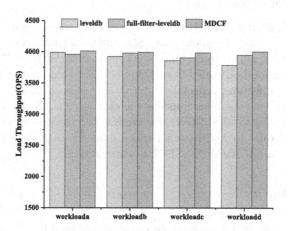

Fig. 9. Throughput in load data phase.

5   Related Work

In recent years, many studies have proposed new designs based on the LSM-tree.

Improving Read Performance with Filters. Several studies have aimed to optimize the Bloom filter for improved read performance. The accuracy of a bloom filter is highly related to its size, Monkey [8] proposes differentiating Bloom filters between different levels. The frequently accessed upper levels of the LSM-tree adopted a larger bloom filter for higher accuracy, while the lower levels only demanded a smaller bloom filter. ElasticBF [17] further develops a fine-grained elastic Bloom filter scheme to improve read performance. To avoid the reconstruction overhead and memory access cost of Bloom filters, SlimDB [22] and Chucky [9] replace an LSM-tree's multiple Bloom filters by a Cuckoo filter variant that maps each KV pair to both a fingerprint and to a *Level ID* (LID). However, this approach also introduces a significant memory space overhead, and the Cuckoo filter must be rebuilt once its capacity limit has been reached.

Reducing Compaction Overhead. The most significant performance challenge of LSM-tree KV stores stems from the I/O amplification caused by compaction actions. WiscKey [19] and HashKV [4] reduce the number of compaction I/Os by employing key-value separation techniques to manage keys and metadata in the LSM-tree, while storing values in an append-only log. Skip-tree [31] allows KV items to be written to a deeper level without going through the level-by-level approach. VT-tree [25] reduces disk writes by reusing existing data in the old tables. TRIAD [3] keeps hot data in memory, avoids duplicate writes of the log component, and compacts an SSTable only when there is sufficient overlap with lower-level SSTables. PebblesDB [21] reorganizes the storage layout inspired by skip lists, thereby avoiding data rewriting within the same level to reduce the compaction overhead.

Co-design of Software and Hardware. LOCS [27] was designed to exploit the parallelism of customized open-channel SSDs and to optimize the scheduling and dispatching policies, which could improve performance. GearDB [29] introduced a new compaction method called Gear Compaction to eliminate the overhead of on-disk garbage collection, especially for the new host-managed shingled magnetic recording drives, and thus improve the compaction efficiency. To solve the problem of write stalls in LSM-trees, MatrixKV [30] designs a matrix container to receive MemTables in NVMs and performs fine-grained data compaction to reduce the overhead of NVM-to-SSD data compaction.

6   Conclusion

In this paper, we conducted an analysis of the severe read amplification problem and identified some shortcomings of Bloom filters in LevelDB during the read operation process. Based on our findings, we propose MDCF, which uses multiple variants of Dynamic Cuckoo Filter to replace the Bloom filter used in traditional LSM-tree based KV stores. With MDCF, certain KV pairs in the LSM-tree are selectively chosen and mapped to their corresponding level or SSTable file. As a result, for a *Get* operation, we can directly locate the SSTable that contains the target KV pair. Our experiments demonstrate that the MDCF significantly improves reading performance.

Acknowledgements. This work is supported by Key Research and Development Program of Shandong Province of China (Grant No. 2022CXGC020107).

References

1. Apache: Cassandra. https://cassandra.apache.org
2. Armstrong, T.G., Ponnekanti, V., Borthakur, D., Callaghan, M.: LinkBench: a database benchmark based on the Facebook social graph. In: Proceedings of the 2013 ACM SIGMOD International Conference on Management of Data, pp. 1185–1196 (2013)
3. Balmau, O., et al.: TRIAD: creating synergies between memory, disk and log in log structured key-value stores. In: 2017 USENIX Annual Technical Conference (USENIX ATC 2017), pp. 363–375 (2017)
4. Chan, H.H., Li, Y., Lee, P.P., Xu, Y.: HashKV: enabling efficient updates in KV storage via hashing. In: 2018 USENIX Annual Technical Conference (USENIX ATC 2018), pp. 1007–1019 (2018)
5. Chang, F., et al.: Bigtable: a distributed storage system for structured data. ACM Trans. Comput. Syst. (TOCS) **26**(2), 1–26 (2008)
6. Chen, H., Liao, L., Jin, H., Wu, J.: The dynamic cuckoo filter. In: 2017 IEEE 25th International Conference on Network Protocols (ICNP), pp. 1–10. IEEE (2017)
7. Cooper, B.F., Silberstein, A., Tam, E., Ramakrishnan, R., Sears, R.: Benchmarking cloud serving systems with YCSB. In: Proceedings of the 1st ACM Symposium on Cloud Computing, pp. 143–154 (2010)
8. Dayan, N., Athanassoulis, M., Idreos, S.: Monkey: optimal navigable key-value store. In: Proceedings of the 2017 ACM International Conference on Management of Data, pp. 79–94 (2017)
9. Dayan, N., Twitto, M.: Chucky: a succinct Cuckoo filter for LSM-tree. In: Proceedings of the 2021 International Conference on Management of Data, pp. 365–378 (2021)
10. Debnath, B., Sengupta, S., Li, J.: FlashStore: high throughput persistent key-value store. Proc. VLDB Endowment **3**(1–2), 1414–1425 (2010)
11. Debnath, B., Sengupta, S., Li, J.: SkimpyStash: RAM space skimpy key-value store on flash-based storage. In: Proceedings of the 2011 ACM SIGMOD International Conference on Management of Data, pp. 25–36 (2011)
12. DeCandia, G., et al.: Dynamo: Amazon's highly available key-value store. ACM SIGOPS Oper. Syst. Rev. **41**(6), 205–220 (2007)
13. Facebook: RocksDB. https://rocksdb.org/
14. Fan, B., Andersen, D.G., Kaminsky, M., Mitzenmacher, M.D.: Cuckoo filter: practically better than Bloom. In: Proceedings of the 10th ACM International on Conference on Emerging Networking Experiments and Technologies, pp. 75–88 (2014)
15. Frühwirt, P., Huber, M., Mulazzani, M., Weippl, E.R.: InnoDB database forensics. In: 2010 24th IEEE International Conference on Advanced Information Networking and Applications, pp. 1028–1036. IEEE (2010)
16. Lai, C., et al.: Atlas: Baidu's key-value storage system for cloud data. In: 2015 31st Symposium on Mass Storage Systems and Technologies (MSST), pp. 1–14. IEEE (2015)
17. Li, Y., Tian, C., Guo, F., Li, C., Xu, Y.: ElasticBF: elastic bloom filter with hotness awareness for boosting read performance in large key-value stores. In: USENIX Annual Technical Conference, pp. 739–752 (2019)

18. Lu, G., Nam, Y.J., Du, D.H.: BloomStore: bloom-filter based memory-efficient key-value store for indexing of data deduplication on flash. In: 2012 IEEE 28th Symposium on Mass Storage Systems and Technologies (MSST), pp. 1–11. IEEE (2012)

19. Lu, L., Pillai, T.S., Gopalakrishnan, H., Arpaci-Dusseau, A.C., Arpaći-Dusseau, R.H.: WiscKey: separating keys from values in SSD-conscious storage. ACM Trans. Storage (TOS) **13**(1), 1–28 (2017)

20. Papagiannis, A., Saloustros, G., González-Férez, P., Bilas, A.: Tucana: design and implementation of a fast and efficient scale-up key-value store. In: 2016 USENIX Annual Technical Conference (USENIX ATC 2016), pp. 537–550 (2016)

21. Raju, P., Kadekodi, R., Chidambaram, V., Abraham, I.: PebblesDB: building key-value stores using fragmented log-structured merge trees. In: Proceedings of the 26th Symposium on Operating Systems Principles, pp. 497–514 (2017)

22. Ren, K., Zheng, Q., Arulraj, J., Gibson, G.: SlimDB: a space-efficient key-value storage engine for semi-sorted data. Proc. VLDB Endowment **10**(13), 2037–2048 (2017)

23. Sanjay Ghemawat, J.D.: LevelDB. https://github.com/google/leveldb

24. Sears, R., Ramakrishnan, R.: bLSM: a general purpose log structured merge tree. In: Proceedings of the 2012 ACM SIGMOD International Conference on Management of Data, pp. 217–228 (2012)

25. Shetty, P.J., Spillane, R.P., Malpani, R.R., Andrews, B., Seyster, J., Zadok, E.: Building workload-independent storage with VT-trees. In: Presented as Part of the 11th USENIX Conference on File and Storage Technologies (FAST 2013), pp. 17–30 (2013)

26. Sumbaly, R., Kreps, J., Gao, L., Feinberg, A., Soman, C., Shah, S.: Serving large-scale batch computed data with project Voldemort. In: FAST, vol. 12, pp. 18–18 (2012)

27. Wang, P., et al.: An efficient design and implementation of LSM-tree based key-value store on open-channel SSD. In: Proceedings of the Ninth European Conference on Computer Systems, pp. 1–14 (2014)

28. Lee, Y., Ren, J.: RocksDB. https://github.com/ls4154/YCSB-cpp

29. Yao, T., et al.: GearDB: a GC-free key-value store on HM-SMR drives with gear compaction. In: 19th USENIX Conference on File and Storage Technologies (FAST) (2019)

30. Yao, T., et al.: MatrixKV: reducing write stalls and write amplification in LSM-tree based KV stores with a matrix container in NVM. In: Proceedings of the 2020 USENIX Conference on Usenix Annual Technical Conference, pp. 17–31 (2020)

31. Yue, Y., He, B., Li, Y., Wang, W.: Building an efficient put-intensive key-value store with skip-tree. IEEE Trans. Parallel Distrib. Syst. **28**(4), 961–973 (2016)

32. Zhang, Q., Li, Y., Lee, P.P., Xu, Y., Cui, Q., Tang, L.: UniKV: toward high-performance and scalable KV storage in mixed workloads via unified indexing. In: 2020 IEEE 36th International Conference on Data Engineering (ICDE), pp. 313–324. IEEE (2020)

A Game Theory Based Task Offloading Scheme for Maximizing Social Welfare in Edge Computing

Chen Sheng[1], Liu Yang[2], Chen Baochao[1], Hong Tu[1], Tan Renrui[1],
and Tao Xiaoyi[3(✉)]

[1] Tianjin University, Tianjin, China
{chensheng,cbcchenbaochao,fs50ht,trr}@tju.edu.cn
[2] Beijing Institute of Computer Technology and Applications, Beijing, China
[3] Dalian Maritime University, Dalian, China
xytao@dlmu.edu.cn

Abstract. Edge computing, as a computing paradigm that enables the decentralization of cloud computing services to the edge of the network, effectively addresses the issue of service unavailability caused by power constraints on end devices when handling user application requests. End users offload computational tasks and associated data to the infrastructures at the network edge. Even if executing tasks at the edge can reduce energy consumption and computational latency compared to local execution on end devices, offloading a large number of tasks consumes wireless channel resources and computational resources of the edge infrastructures, resulting in additional transmission costs and energy consumption. Moreover, competition among multiple users for limited resources at the edge nodes leads to a situation where it is challenging to balance the utilities of all the users and the Edge Service Provider (ESP).

In this paper, we address the scenarios where both the battery capacity of end devices and the resource capacity at the edge are limited. We propose a computation offloading scheme based on a master-slave Stackelberg game. We provide theoretical proof of the existence of a unique Nash equilibrium in the proposed game and optimize the energy consumption and user benefits during the offloading process. Furthermore, ESP improves its revenue by servicing more user requests. The simulation results show that the proposed algorithm performs well in terms of energy consumption and user utility.

Keywords: Mobile Edge Computing · Stackelberg Game · Task Offloading

1 Introduction

As a mature and widely-used general technical solution, cloud computing provides flexible and reliable computing, storage, and bandwidth resources for various application services. With the increasing popularity of 5G and IoT technology, various terminal applications with low latency and high bandwidth

© The Author(s), under exclusive license to Springer Nature Singapore Pte Ltd. 2024
Z. Tari et al. (Eds.): ICA3PP 2023, LNCS 14492, pp. 219–239, 2024.
https://doi.org/10.1007/978-981-97-0811-6_13

requirements have emerged, such as virtual reality, high-definition video, and autonomous driving. If the services are hosted in the cloud far away from the users, it will inevitably bring huge communication costs and unbearable latency experiences. Edge computing, compared to the traditional cloud computing model, can deploy its service capabilities closer to user requests. Specifically, edge computing can migrate the computation and storage resources from the cloud to the network edge, such as base stations, WiFi, etc., which greatly reduces the transmission latency of user request [24].

Some edge devices, such as cameras, can be equipped with hardware that provides a certain amount of computation capacity. At the same time, they are also places where a large amount of data is generated, making them the best location for collecting, filtering, analyzing, and extracting big data. However, as the functionality of terminal applications increases and strengthens, the limitations of computation capacity and energy consumption become bottlenecks for local processing. To address this issue, energy-intensive tasks can be offloaded to edge servers for processing [15]. However, the first key issue in this process is how to choose the appropriate edge server for task offloading.

Generally speaking, task offloading involves two processes: communication and computation. The communication process includes two parts: the edge device transfers the local task to the edge server and returns the computation result from the edge server. This process involves the energy consumption of the edge device. The computation process of task offloading is performed on the edge server, which involves the revenue of ESP, that is, to ensure the reliable service quality of end users and obtain more users. Specifically, ESP hopes to obtain the maximum revenue by providing computation resources to end users, while the end devices hope to reduce the consumption of local computation resources and energy consumption. Existing work on computation offloading has almost failed to simultaneously improve the utility of both ESP and end users, especially when considering the performance guarantee of users [1,5,7,11,17,19]. Therefore, the goal of this paper is to optimize the utility of both users and ESP while minimizing the overall energy consumption of user devices as much as possible. However, the key challenge of this process is how end users select the appropriate edge server from multiple candidate of ESP.

In this paper, energy consumption is taken as one of the indicators, and we propose an offloading strategy that maximizes user request utility. This paper defines the problem as a master-slave Stackelberg game. The first stage is to determine the upper and lower bounds of the payment of end users to edge servers, and the second stage is to decide which edge server to offload the task and the corresponding payment based on the goal of minimizing user energy consumption. Specifically, the end user has an initial bid for each task request, and based on their bids and task requirements we determine the end users. Finally, the task offloading is determined by optimizing the target of minimizing the energy consumption of all users. Through theoretical analysis, we prove that we can obtain a unique Nash equilibrium. In addition, we design a multi-round computation task offloading algorithm, which determines the payment and offloading strategies by considering the trade-off between request energy consumption and user utility. On the one hand, it aims to reduce the users' request payment and energy consumption, and on the other hand, it aims to improve the offloading rate of users, thereby increasing the overall revenue of ESP. Simulation results show that the proposed algorithm performs well in terms of energy consumption and user utility, and effectively motivates users to offload tasks to non-nearby edge servers, increasing the offloading rate of user requests and increasing the revenue of ESP.

The rest of this paper is organized as follows. Section 2 introduces the motivation and related works. Section 3 presents the system model in this paper and mathematically formulates the user utility maximzation problem. In Sect. 4, we propose a Stackerblg game based offloading algorithm. Section 5 demonstrates the simulation results and analysis. Finally, we conclude our work in Sect. 6.

2 Motivation and Related Works

2.1 Motivation

Typically, ESP places one or more edge servers at different network edges. For end users, due to the limitations of device size, their resources are relatively limited, and they need to offload computation tasks that are difficult to process locally to edge servers. Similarly, due to the limited computing and channel resources of edge servers, it is hard to satisfy offloading energy consumption requirements and utility of all users in the face of concentrated competitive offloading requests. If we only consider optimizing the revenue of ESP, it leads to a situation where edge users make malicious bids for requests, which will reduce the offloading rate and resource utilization rate, seriously damaging the Quality of Experience (QoE) of users. Therefore, this paper starts from the perspective of users in the task offloading, using limited resources to maximize user utility and energy consumption requirements, thereby improving the offloading rate of user requests and further increasing the revenue of ESP, achieving a win-win situation. We first establish the energy consumption model and utility model of users, and based on the fact that the user's payment strategy has a decisive impact on the offloading strategy, we propose a computation offloading scheme based on Stackelberg game, optimizing the overall utility and energy consumption of user requests, improving the resource utilization rate of ESP, and achieving the optimization of social welfare.

2.2 Related Works

Computation offloading refers to drive the computationally intensive tasks from edge devices to edge servers with relatively abundant computational resources. Energy consumption and system utility are significant indicators in computation offloading. Based on the divisibility of tasks, computation offloading can be divided into two different granularity modes: complete offloading and partial offloading [13].

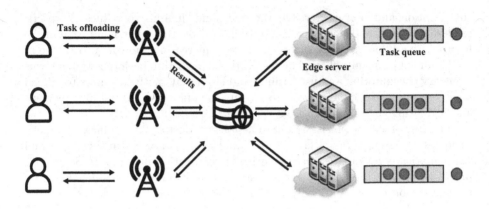

Fig. 1. User offloading Scene

Complete offloading means that the computational tasks are indivisible, and they are either executed by the edge devices themselves or offloaded to edge servers for processing. In terms of minimizing energy consumption, Li et al. [8] explored the energy minimization problem in edge offloading systems under the dual-connectivity and Non-Orthogonal Multiple Access (NOMA) modes. They designed an iterative optimization algorithm to obtain suboptimal time scheduling and task allocation scheme and introduced an intelligent offloading algorithm based on deep learning, making full use of the data and computational capabilities at the network edge. Chen et al. [3] proposed a task offloading and resource allocation problem for Augmented Reality (AR) scenarios, aiming to minimize the energy consumption of each end user under the delay requirements of AR tasks and resource capacity constraints. The authors used the Deep Deterministic Policy Gradient (DDPG) method of multi-agent reinforcement learning [12] to solve this mixed competition and cooperation problem among multiple users. Pliatsios et al. [16] studied the computation offloading problem in a multi-access edge computing vehicular network environment. They jointly considered the energy and bandwidth capacity constraints to minimize the overall energy consumption of the system and achieved an optimal solution using the block coordinate descent method [22]. Tang et al. [21] investigated the computation offloading problem in a near-earth orbit satellite network. They proposed a solution to minimize the energy consumption of ground users considering the coverage time of near-earth satellites. By using binary variable relaxation, the original non-convex problem was transformed into a linear programming problem, and a distributed algorithm based on the alternating direction method of multipliers [2] was proposed to obtain a suboptimal solution. Liu et al. [10] aimed to provide high-reliability and low-latency services for task-intensive applications in multi-user edge scenarios. They used extreme value theory [4] to set a threshold for the length of the task queue. The authors defined the problem as a network-wide energy minimization problem and proposed a two time scale strategy for edge server selection and task offloading. In the long time scale, matching theory [18] was used to associate edge devices with edge servers, while in the short time scale, decisions were made for task offloading and resource allocation on a given server. Finally, the Lyapunov technique [14] was employed to handle the stochasticity of task arrivals, wireless channels, and task queue lengths. In terms of optimizing system utility, Yang et al. [25] found that multiple edge devices use the same task code and studied cooperation among devices to eliminate redundant data transmission. They employed a cooperative offloading algorithm based on coalition game theory to maximize cost savings. Jovsilo et al. [7] studied the collaboration problem among wireless devices that periodically generate computationally intensive tasks to minimize the total cost of devices and determine whether to offload tasks to the edge for execution. Qiu et al. [17] pursued the goal of minimizing the system cost in dynamic edge environments and investigated the applicability of deep reinforcement learning in the multi-user computation offloading problem from a practical perspective.

Partial offloading allows for the offloading of a portion of computationally intensive tasks to be executed at the edge, where both the end devices and edge servers contribute to the execution of tasks. In terms of minimizing energy consumption, Huang et al. [6] combined small cell networks with mobile edge computing to investigate task offloading and interference coordination between small cell networks. They proposed a distributed multi-agent deep reinforcement learning approach to minimize overall energy consumption. The approach deeply explores the collaboration among small cell base stations to adaptively adjust strategies for task offloading, channel and computation resource allocation, and transmit power control. Tang et al. [20] studied the problem of task distribution among local devices, unmanned aerial vehicles (UAVs), and base stations. They optimized the task offloading ratio, resource allocation, and UAV trajectory to minimize the energy consumption of all players in the system. In terms of optimizing system utility, Bi et al. [1] considered the problem of wireless terminal task offloading in a multi-user mobile edge network scenario. They jointly optimized the terminal offloading selection mode and system transmission time allocation to maximize the weighted sum of the compute rates of all wireless terminals in the network. Due to the strong coupling between offloading mode selection and time allocation, the authors assumed that the offloading selection strategy was given and used a simple binary search algorithm to obtain conditionally optimal time allocation strategies. Additionally, they designed a coordinate descent method to optimize the offloading mode selection and applied the alternating direction method of multipliers decomposition technique [2] to reduce the computational complexity resulting from the increasing network scale. Li et al. [9] proposed a novel utility function to guide task offloading decisions in the mobile edge environment. Unlike previous resource pricing approaches that relied solely on abstract utility functions independent of physical layer parameters, their utility function measured the cost reduction brought about by task offloading compared to local execution. Based on this function, the authors designed two game-theoretic approaches for task offloading to either maximize individual utility or maximize global system interests. They also proposed a pricing-based optimal scheme and proved that the interactive decision process of self-interested users can converge to a Nash equilibrium.

Most of these works focus on maximizing utility in specific objectives, such as minimizing user energy consumption or minimizing system costs. However, there is often a conflict of interest between edge operators and users, making it challenging to improve the utility for both simultaneously. Therefore, in this paper, we take a user-centric approach and consider both the optimization of users' economic utility and the minimization of their energy consumption. By increasing the task offloading rate and thereby increasing the number of paying users, we aim to improve the economic revenue for ESP while taking into account the users' energy consumption minimization objective.

3   System Model

The considered edge system in this paper involves multiple mobile users and multiple user edge nodes. Each edge node is capable of serving the users in its respective area. Each mobile device has the option to offload task requests to an edge server, as shown in Fig. 1. Assuming that the edge system can collect the status of all edge servers and requests. Specifically, the scenario of the edge system consists of n users, denoted as $U = \{i \mid 1 \leq i \leq n\}$, and a set of m edge servers denoted as $M = \{j \mid 1 \leq j \leq m\}$. The edge server j contains computational resources $Q = \{Q_j \mid 1 \leq j \leq m\}$ and bandwidth B_j allocated to user requests $UR = \{r \mid 1 \leq r \leq k\}$. It is assumed that each user generates multiple requests, and each individual request r has multiple requirements, including computational demand C_r and bandwidth resource demand W_r. The system determines the offloading strategy of a computational task based on the bidding price and resource requirements of user requests. The variable x_{rj} is used to represent whether request r is offloaded to server j. The key parameter symbols for this paper are shown in Table 1.

3.1   User Energy Consumption Model

Battery life has become an important consideration for the development of mobile devices. Currently, the demand for computational performance in devices such as smartphones is growing much faster than the technological advancements in battery capacity. Therefore, in order to meet the increasing task requirements of mobile devices, it is crucial to focus on the energy consumption levels of devices during task processing.

In this paper, we are considering the offloading of large-scale data tasks from user terminals to edge servers. The energy consumption generated during the task offloading process needs to be taken into account. We assume that the communication between users and edge servers is carried out using the Non-Orthogonal Multiple Access technique, which ensures that there is no interference between different channels at the same time. When a user sends an offloading request, data communication incurs a certain amount of energy consumption,

Table 1. Notation

Notation	Definition
U	User Set
g_j	Channel gain of edge node j
UR	User request
σ^2	Channel Gaussian noise of edge node j
M	Edge node set
b_{rj}	Rate of request r offloading to node j
Q_j	Computing capacity of node j
u_i	Utility of user i
C_r	Required computing resource of request r
$pmax_i$	Max power consumption of user i
s_r	Data size of request r
p_{rj}	Transmission power of request r offloading to node j
B_j	Channel bandwidth of j
e_{rj}	Consumption of request r offloading to node j
W_r	Required bandwidth of request r
$e_{r,loc}$	Local computing consumption of request r
pa_{rj}	Bidding price of request r for edge node j
x_{rj}	Request r offloading to node j
pa_r	Bidding set of request r
y_{ir}	Whether Request r belongs to user i

which is dependent on the data transmission power and transmission time. The communication for the offloading request occupies the bandwidth of the edge node. The data transmission rate of computation request r at the server can be represented as

$$b_{rj} = B_j log_2 \left(1 + \frac{p_{rj}g_j^2}{\sigma^2 N_0} \right). \tag{1}$$

B_j represents the channel bandwidth of edge server j, σ^2 represents the Gaussian noise power of the channel, g_j represents the channel gain, and N_0 represents the channel noise power density.

The energy consumption of task offloading includes both the data communication process and the computational processing phase. However, in this paper, the energy consumption generated during the computation phase at the edge server is neglected and not considered part of the user's energy consumption. Therefore, the energy consumption of offloading request r to edge server j is defined in terms of the request size s_r, transmission rate b_{rj}, and power p_{rj}.

$$e_{rj} = p_{rj}t_j. \tag{2}$$

Additionally, for edge server j, there exists a bandwidth capacity constraint. The sum of bandwidth requirements W_r for all requests offloaded to edge server j should be less than the channel bandwidth of edge server j. Here, x_{rj} represents whether request r is offloaded to edge server j, $x_{rj} \in \{0,1\}$. This constraint can be expressed as follows:

$$\sum_{r \in UR} W_r x_{rj} \leq B_j, \forall j. \tag{3}$$

Additionally, each edge server has a fixed capacity for computational resources, and there is a limit on the allocatable quota of computational resources at any given moment. This means that the sum of computational resource demands on that server should be less than the computational resource capacity of that node. Mathematically, it can be expressed as:

$$\sum_{r \in UR} C_r x_{rj} \leq Q_j, \forall j. \tag{4}$$

Assuming that each terminal user is sensitive to energy consumption, the energy consumption of local computation is greater than that of offloading tasks to edge servers. In other words, there exists at least one server j among all edge servers such that $e_{rj} \leq e_{r,loc}$, user's request to be offloaded to the server for computation. In this case, the objective of minimizing the request energy consumption can be represented as minimizing $\sum_{r \in UR} \left(\sum_{j \in M} x_{rj} e_{rj} + f(pa_{rj}) \right)$, where $f(pa_{rj})$ represents the energy consumption incurred during the bargaining process, defined as a second-order function with respect to the payment strategy. Energy consumption is shared between the user and the edge system. Let $pmax_i$ denote the maximum energy consumption value that each user can tolerate, which is only used to limit the energy consumption generated during the transmission process. Thus, for each user, the sum of all request energy consumption should be less than or equal to the maximum energy consumption value $pmax_i$.

The objective of minimizing the request energy consumption can be formulated as follows:

$$\begin{aligned}
\max \quad & \sum_{i \in U} pmax_i - \sum_{r \in UR} \left(\sum_{j \in M} x_{rj} e_{rj} y_{ir} + f(pa_{rj}) \right) \\
s.t. \quad & \sum_{r \in UR} e_{rj} x_{rj} y_{ir} \leq pmax_i, \forall i \\
& x_{rj} \in \{0,1\} \\
& Eqs.(3),(4)
\end{aligned} \tag{5}$$

3.2 User Utility Model

The objective of the system considered in this paper is to enhance the utility for both users and ESP. Due to the limited resources of edge devices, ESP prices the computation resources based on user task demands and its own service capabilities to maximize its own utility. However, conducting auctions for limited resources solely from the perspective of ESP can lead to malicious bidding behavior from users. Some high-value user requests may possess stronger bargaining power, resulting in a priority decision-making power for offloading positions. In such a scenario, the entire system fails to efficiently coordinate the limited resources, resulting in that low-value requests are unable to obtain optimal offloading positions. Consequently, it impairs users' QoE and offloading acceptance rates. Therefore, in this paper, the focus is on enhancing the overall utility for users while reducing their energy consumption and increasing their overall utility. The goal is to increase the offloading ratio of user requests, promote ESP's revenue, and optimize social welfare.

We define a reliable resource trading system, which can collect user request information and ESP edge server resource status. It provides pricing and resource deployment services for all players. The user's value of offloading request r to server j is denoted as b_{rj}. The payment of user i for request r on edge server j is represented as pa_{rj}. Therefore, the total revenue for all requests on edge server j can be expressed as:

$$u_j(x_{rj}) = \sum_{r \in UR} x_{rj}(b_{rj} - pa_{rj}). \tag{6}$$

In order to better enhance users' QoE, we define user satisfaction using a function that relates the consumption associated with request offloading to the maximum energy consumption of the user. User satisfaction is an important component of social welfare, as it expresses user utility from another perspective, shown as follows:

$$g_r(x_{rj}) = e^{1 - \frac{e_{rj} x_{rj} y_{ir}}{pmax_i}}, \tag{7}$$

where y_{ir} represents the relationship between the user and the request, which is a constant value during the interaction between the user and the ESP. To simplify the formula, we can use $pmax$ to represent $\frac{y_{ir}}{pmax_i}$. It is evident that if the energy consumption associated with fulfilling a user's request is smaller, the ratio $\frac{e_{rj} x_{rj}}{pmax}$ will be smaller as well. As $g_r(x_{rj})$ is a decreasing function, a smaller value of $\frac{e_{rj} x_{rj}}{pmax}$ corresponds to higher user satisfaction. In this section, the utility of a user is defined as the sum of the two most important indicators: energy consumption and benefits. Therefore, the offloading problem can be transformed into a maximization problem of user utility:

$$\begin{aligned} \max \sum_{j \in M} \sum_{r \in UR} (u(x_{rj}) + g(x_{rj})) \\ s.t. \quad x_{rj} \in \{0, 1\} \\ Eqs.(3), (4) \end{aligned} \tag{8}$$

In this model, it is evident that the user's payment strategy directly affects the specific offloading position. Requests with higher bids can prioritize selecting computing nodes that are closer in proximity, while users with lower bids may have to choose to offload to distant nodes or even give up offloading. Therefore, in the next section, the problem of maximizing user utility in task offloading is modeled as a Stackelberg game to obtain the optimal payment strategy for computational resources and offloading strategy. Specifically, the range for determining the request payment strategy pa_{rj} is defined as the leader, the set of payment strategies is denoted as **pa**, and the process of calculating the request offloading strategy based on the specified payment price range is defined as the follower. The offloading strategy is denoted as x_{rj}, and the set is denoted as **x**.

4   Stackelberg Game Based Offloading Algorithm

To address the problem of maximizing user utility, we propose a computation offloading algorithm based on a leader-follower Stackelberg game (IMO). The algorithm begins by determining the upper and lower bounds of the payment set pa_r for each user request r, and then determining the offloading position for the request. It is demonstrated that the proposed game problem can achieve a Nash equilibrium, where user utility is maximized. Let x_r^* represent the offloading strategy for request r in Eq. (8), with $x_r^* = [x_{r1}^*, \ldots, x_{rm}^*]$. Additionally, let pa_{rj}^* denote the payment for request r to be offloaded to server j, and pa_r^* represent the payment strategy for request r, with $pa_r^* = [pa_{r1}^*, \ldots, pa_{rm}^*]$. At the Nash equilibrium, the offloading strategy and payment strategy are mutually constrained, such that requests cannot change their offloading strategies by simply increasing the payment.

If the strategy $\{x_r, pa_r\}$ is the solution that maximizes $H = u + g$, and it satisfies the following conditions, then $\{x_r^*, pa_r^*\}$ is a Nash equilibrium solution for the Stackelberg problem. The conditions are as follows:

$$\{x_r^*, pa_r^*\} = \arg \max H, \tag{9}$$

while pa_{rj}^* satisfy

$$pa_{rj}^* \in pa_r. \tag{10}$$

Let's define the lower limit of payment pa_{rj} for request r to server j as pa_{rj}^0, and the upper limit of payment as pa_{rj}^1. The offloading results can be denoted as

$$x_{rj}^* \in \mathbf{x}, \quad \forall r \in UR. \tag{11}$$

$\{x_r^*, pa_r^*\}$ is the solution to problem (5) and (8). Next, we prove that this problem has a unique solution and can achieve Nash equilibrium.

Theorem 1. *The strategy of the user utility model is a non-empty finite set.*

Proof. The first and second partial derivatives of $H = u + g$ are as follows:

$$\frac{\partial H}{\partial x} = (b_{rj} - pa_{rj}) - \frac{e_{rj}}{pmax} e^{\left(1 - \frac{e_{rj}x_{rj}}{pmax}\right)}. \tag{12}$$

$$\frac{\partial^2 H}{\partial x^2} = \frac{e_{rj}^2}{pmax} e^{\left(1 - \frac{e_{rj}x_{rj}}{pmax}\right)}. \tag{13}$$

By setting Eq. (12) equal to 0, we can obtain:

$$x_{rj}^* = \frac{pmax}{e_{rj}} \left(1 - \frac{pmax}{e_{rj}} ln(b_{rj} - pa_{rj})\right), \tag{14}$$

where x_{rj} equals 0 and 1 respectively, the corresponding upper and lower bounds of the payment are:

$$pa_{rj}^0 = b_{rj} - e^{\frac{e_{rj}}{pmax}}. \tag{15}$$

$$pa_{rj}^1 = b_{rj} - e^{\frac{e_{rj}}{pmax}(1-\frac{e_{rj}}{pmax})}. \tag{16}$$

Therefore, the payment strategy has well-defined upper and lower bounds, and $\{x_r^*, pa_r^*\}$ can be searched within a finite space, specifically within the range of the upper and lower bounds. Hence, the payment strategy is non-empty and finite, and the theorem holds.

For the solution x_{rj}^*, the offloading strategy satisfies the following conditions within the effective range of the payment strategy pa_{rj}:

$$x_{rj}^* = \begin{cases} 0 & pa_{rj} \leq pa_{rj}^0 \\ \frac{pmax}{e_{rj}}\left(1 - \frac{pmax}{e_{rj}}ln(b_{rj} - pa_{rj})\right) & pa_{rj}^0 < pa_{rj} < pa_{rj}^1 \\ 1 & pa_{rj} \geq pa_{rj}^1 \end{cases} \tag{17}$$

The offloading strategy of the request is based on the payment strategy. When the payment strategy pa_{rj} falls within the range of $pa_{rj}^0 \leq pa_{rj} \leq pa_{rj}^1$ in the user utility model, the offloading strategy x_{rj} is related to the transmit power of the edge device and the energy consumption of offloading the task to the edge server. Therefore, the request offloading can be determined through the payment strategy, and the final offloading strategy can be obtained using Eq. (14). The solution process can be divided into two stages: First, determining the range of the payment strategy in the user utility model, and second obtaining the request offloading strategy within the payment strategy range.

Theorem 2. *The user utility model has a unique optimal solution.*

Proof. Intuitively, if a sufficiently large payment $pa_{rj} \geq pa_{rj}^1$ is set, the request can be offloaded to the nearest edge server. Conversely, if $pa_{rj} \leq pa_{rj}^0$, the task cannot be offloaded. Therefore, when $pa_{rj}^0 \leq pa_{rj} \leq pa_{rj}^1$, we can conclude that:

$$\frac{\partial x_{rj}^*}{\partial pa_{rj}} = \frac{pmax}{e_{rj}(b_{rj} - pa_{rj})}. \tag{18}$$

$$\frac{\partial^2 x_{rj}^*}{\partial pa_{rj}^2} = \frac{-pmax}{e_{rj}(b_{rj} - pa_{rj})^2}. \tag{19}$$

It can be observed that $\frac{\partial x_{rj}^*}{\partial pa_{rj}} > 0$, indicating that as the generated benefits from the request increase, the user is able to afford higher costs. Clearly, the value of Eq. (19) is less than 0. Therefore, considering Eqs. (18) and (19), it can be concluded that the optimal strategy in Eq. (14) is unique.

Since the payment strategy pa_{rj} in the user utility model determines the offloading strategy x_{rj} in the energy consumption model, we will now prove that within the solution of the user utility model, the energy consumption model also achieves its optimal solution.

Theorem 3. *In the optimization solution of the user utility model, the energy consumption model has a unique optimal solution.*

Proof.

$$G = pmax - \left(x_{rj}^* e_{rj} + f(pa_{rj})\right). \tag{20}$$

Taking the second partial derivative of $G(pa_{rj}^*)$:

$$\frac{\partial^2 G}{\partial pa_{rj}^2} = -e_{rj} \frac{\partial^2 x_{rj}^*}{\partial pa_{rj}^2} - f''(pa_{rj}). \tag{21}$$

$$\frac{\partial^2 G}{\partial pa_{rj} \partial pa_{rj1}} = -e_{rj} \frac{\partial x_{rj}^*}{\partial pa_{rj}} \frac{\partial x_{rj}^*}{\partial pa_{rj1}}. \tag{22}$$

Given that $\frac{\partial x_{rj}^*}{\partial pa_{rj}} \geq 0$, and $\frac{\partial^2 x_{rj}^*}{\partial pa_{rj}^2}$ is related to $\frac{\partial^2 G}{\partial pa_{rj} \partial pa_{rj1}}$, and $\frac{\partial^2 x_{rj}^*}{\partial pa_{rj}^2} < 0$, we can infer that $G(x_{rj}, pa_{rj})$ is a convex function. Therefore, pa_{rj}^* corresponds to a unique x_{rj}. Since G is a transformed version of the objective function in problem (5), the unique solution x_{rj} corresponding to G is also the unique solution of the original problem. Thus, the proof is complete.

Based on the aforementioned methodology, we propose a computational offloading algorithm based on multi-round Stackelberg game (IMO). Due to the decisive impact of the bidding strategy on the offloading decisions, the algorithm first determines the upper and lower bounds for the user request payments. Subsequently, within this range, it searches for the optimal offloading strategy and its corresponding payment strategy. Algorithm 1 provides a specific description of the overall procedure of the Stackelberg game approach.

To begin, the algorithm establishes the initial values for the payment strategy as $\underline{pa_{rj}} = \frac{pa_{rj}^0 + pa_{rj}^1}{2}$ and $\overline{pa_{rj}} = \frac{pa_{rj} + pa_{rj}^1}{2}$. By comparing the values of $G(x, pa_{rj})$, the values of $pa_{rj}^0 = \underline{pa_{rj}}$ and $pa_{rj}^1 = \overline{pa_{rj}}$ are used to gradually narrow down the range of $\overline{pa} - \underline{pa}$. The entire process can be iterated until $\overline{pa} - \underline{pa} < \epsilon$, where ϵ represents an extremely small value. As a result, the algorithm obtains the payment strategy pa_{rj} and the corresponding offloading strategy x_{rj}.

Theorem 4. *The solution obtained by the algorithm can achieve a unique Nash equilibrium.*

Proof. Since G is a convex function, the algorithm guarantees finding pa_{rj}^* that maximizes G, and accordingly, the offloading strategy x_{rj}^* that maximizes the utility H. Consequently, the corresponding pa_{rj}^* in the user utility is determined, and the algorithm achieves a unique Nash equilibrium.

5   Evaluation

In this section, we evaluate the performance of the IMO algorithm through simulation experiments.

Algorithm 1: Maximizing User Utility Task Offloading Algorithm

---

1 **Input:** Set of requests UR, request capacity C_r;
2 **Output:** Payment strategy **pa**, offloading strategy **x**;
3 Initialization: $\underline{pa_{rj}} = \frac{pa_{rj}^0 + pa_{rj}^1}{2}$, $\overline{pa_{rj}} = \frac{pa_{rj} + pa_{rj}^1}{2}$;
4 Initialization: $\underline{pa} = \frac{1}{N}\sum_{UR} pa_r^0$, $\overline{pa} = \frac{1}{N}\sum_{UR} pa_r^1$;
5 **while** $\overline{pa} - \underline{pa} > \epsilon$ **do**
6   Calculate $x_{rj}(\overline{pa_{rj}})$ using Eq. (14);
7   Calculate $G(x, \overline{pa_{rj}})$ using Eq. (20);
8   Calculate $x_{rj}(\underline{pa_{rj}})$ using Eq. (14);
9   Calculate $G(x, \underline{pa_{rj}})$ using Eq. (20);
10   **if** $G(x, \underline{pa_{rj}}) < G(x, \overline{pa_{rj}})$ **then**
11 | Update $pa_{rj}^0 = \underline{pa_{rj}}$;
12   **end**
13   **if** $G(x, \underline{pa_{rj}}) > G(x, \overline{pa_{rj}})$ **then**
14 | . Update $pa_{rj}^1 = \overline{pa_{rj}}$;
15   **end**
16   **if** $G(x, \overline{pa_{rj}}) = G(x, \underline{pa_{rj}})$ **then**
17 | Update $\underline{pa_{rj}} = \frac{pa_{rj}^0 + pa_{rj}^1}{2}$, $\overline{pa_{rj}} = \frac{pa_{rj} + pa_{rj}^1}{2}$;
18   **end**
19 **end**
20 Update $pa_{rj} = \overline{pa_{rj}}$;
21 $\underline{pa} = \frac{1}{N}\sum_{UR} pa_r^0$, $\overline{pa} = \frac{1}{N}\sum_{UR} pa_r^1$.

5.1 Evaluation Settings

To evaluate the effectiveness of the proposed algorithm, the IMO algorithm was implemented using MATLAB R2022a in this section. The simulations were conducted on an Intel i7 2.1 GHz host machine with 32 GB RAM. The scenario studied in this section is an edge computing system consisting of multiple users and edge servers. In the simulation, eight edge servers were simulated. Each edge server can serve a subset of the terminal users. To optimize user utility, the revenue of users and the energy consumption of offloading tasks to servers were considered in selecting suitable servers for task offloading. The servers have different locations, capacities, and energy consumption levels, while each request has different sizes and workloads. Unless otherwise specified, the specific values of the parameters used in the comparative experiments are shown in Table 2.

To demonstrate the effectiveness of the proposed method, the following comparative methods are considered:

– **Simulated Annealing Algorithm (SAA)** [1]: SAA is a classical heuristic algorithm that can effectively solve NP-hard problems. The SAA algorithm employs random perturbations and accepts suboptimal solutions with a certain probability to avoid getting trapped in local optima.

Table 2. Settings

Parameter	Value
User request number	32
Edge node M	8
Workload C_r	1500 Mhz
Computing capacity Q	30 GHz
Request size S_r	700 MB
Gaussian noise σ^2	0.01 dB/Hz
Bandwidth B	40 MHz

– **Particle Swarm Optimization (PSO)** [23]: PSO algorithm searches for
the global optimum based on the current best value using a swarm of particles.
Each particle has a position and velocity, and it moves based on its own
historical experience and the experience of its neighbors. PSO algorithm is
known for its high precision and fast convergence.

These comparative methods will be used to evaluate the performance of the
proposed approach.

The performance evaluation metrics for the experimental study include user
energy consumption, user utility, and the revenue of ESP. These metrics respectively reflect the algorithm's performance in ensuring the energy consumption
level of edge users, the overall utility of edge users, and the revenue of ESP.

5.2   Results

The proposed algorithm aims to improve the utility of edge users while simultaneously reducing the energy consumption of end devices. Therefore, the performance analysis primarily focuses on two metrics: energy consumption and utility.
The effectiveness of the proposed method is validated based on these metrics.

Additionally, the ultimate goal of optimizing user requests is to increase the
request offloading rate, thereby enhancing the revenue of ESP. Therefore, the
performance analysis also considers the task offloading rate and the overall revenue of ESP under different numbers of user requests. By analyzing these metrics,
the performance evaluation provides insights into the effectiveness of the proposed algorithm in terms of energy consumption reduction, utility improvement,
task offloading rate, and ESP revenue.

First, we discuss the impact of algorithm iterations on user energy consumption. As shown in Fig. 2(a), when the maximum power p_{max} is set to 10W and
the number of requests is 32, with a total of 5000 iterations, it can be observed
that the energy consumption of end users gradually decreases with increasing
the number of iterations. After around 400 iterations, it starts to converge, and
after 500 iterations, the energy consumption no longer decreases significantly.
This indicates that the proposed IMO algorithm effectively reduces the energy

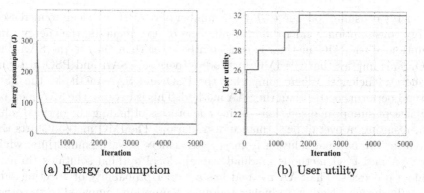

(a) Energy consumption   (b) User utility

Fig. 2. The impact of number of iterations

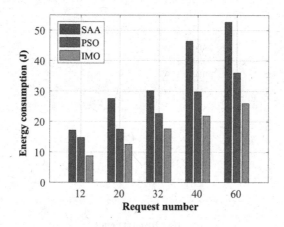

Fig. 3. User request vs consumption

consumption of end devices, achieving the optimization goal of minimizing user energy consumption. Since the energy consumption no longer decreases significantly after 500 iterations, it can be concluded that the IMO algorithm proposed in this paper converges well.

Figure 2(b) illustrates the impact of iteration on user utility, using the same settings as in Fig. 2(a): $p_{max} = 10$ W, 32 requests, and a total of 5000 iterations. It can be observed that user utility increases with the number of iterations and reaches its peak at around 1650 iterations. Between iterations 450 and 1650, there are two instances of improvement in user utility. If a lower number of iterations is chosen, there would be a loss in user utility performance.

Figure 3 presents the relationship between energy consumption and the number of user requests for the three methods. The maximum power p_{max} of user devices is set to 10 W. As shown, it can be observed that as the number of user requests increases, the energy consumption of all three methods also increases. This indicates a positive correlation between the number of user requests and the energy consumption. Increasing the number of requests will lead to increased energy consumption. Under different numbers of user requests, the energy consumption of the IMO method is consistently lower than that of the SAA and PSO. This implies that the IMO method outperforms SAA and PSO in terms of energy efficiency. When comparing the PSO and SAA methods, the PSO method performs better than the SAA method. This is because the SAA method may accept suboptimal solutions during the process of finding the optimal solution, resulting in overall performance degradation. The IMO method starts the search for the optimal solution from pa_{rj} to achieve a Nash equilibrium, while the SAA and PSO methods gradually obtain local optimal solutions through random processes. The SAA method has a higher probability of selecting suboptimal solutions, leading to higher energy consumption compared to the other two methods.

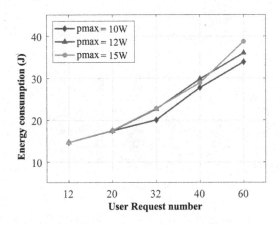

Fig. 4. The impact of maximum power on energy consumption

Figure 4 illustrates the impact of varying the maximum power p_{max} on energy consumption and further demonstrates the relationship between p_{max} and the number of user requests. p_{max} is set to three different values: 10 W, 12 W, and 15 W. It can be observed that energy consumption is positively correlated with p_{max}, indicating that a higher p_{max} allows for a higher acceptable energy consumption for user devices, resulting in higher energy consumption during the request offloading process in the IMO algorithm. This is because with a higher p_{max}, users have a wider range of edge servers to choose from, leading to higher average transmission power and consequently more energy consumption. In the figure, when the number of requests is 40, there is a situation where the energy consumption with $p_{max} = 12$ W is greater than that with $p_{max} = 15$ W. This is because, under certain conditions, as the number of user requests increases, the maximum power is no longer the primary limiting factor for energy consumption. Both p_{max} values cover the range of edge servers that can satisfy user requests, and the request size becomes a factor that influences energy consumption. Therefore, further experimental studies investigate the impact of request size on energy consumption.

Table 3. The impact of request size

Request size(MB)	500	600	700	800	900
Consumption(J)	17.9989	23.9952	30.9894	36.0156	43.0382

Table 3 presents the relationship between energy consumption and request size when the maximum power p_{max} of user devices is set to 12 W, and the number of requests is fixed at 40. This table further illustrates the impact of adjusting request size on energy consumption when other variables are held constant. As the request size increases, the energy consumption of users also increases, which explains the occurrence of higher energy consumption with $p_{max} = 12$ W compared to $p_{max} = 15$ W when the number of requests is 40, as observed in Fig. 4. Hence, user energy consumption is influenced not only by the maximum power but also by the size of the requests. This table provides a clear understanding that as the request size increases, the energy consumption of users also increases. It further emphasizes that user energy consumption is not only related to the maximum power but also influenced by the size of the requests.

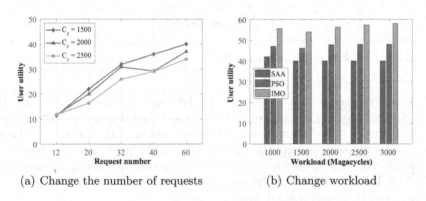

(a) Change the number of requests   (b) Change workload

Fig. 5. The impact of request load on user utility

Figure 5 presents the impact of request workload on user utility. In Fig. 5(a), the performance of the IMO algorithm is evaluated under different request workloads, specifically $C_r = 1500$, 2000, and 2500. As shown, user utility exhibits a negative correlation with request workload. Particularly, when the request workload exceeds 2000, the user utility value significantly decreases. This is because the computational resources of the edge servers are insufficient to handle the offloading of additional computational request workloads, resulting in limited improvement in user utility. Additionally, when the bidding of the same user does not increase with an increase in computational workload, it implies that with lower request workloads, the servers can accommodate more requests, thereby achieving higher user utility. Setting the request workload to 1500, Fig. 5(b) demonstrates the impact of request workload on user utility for the three methods. Consistent with the SAA and PSO methods, increasing computational demands does not proportionally increase system benefits. This is due to the capacity constraints of the edge servers, which cannot meet the high demand for computational workloads and thus cannot achieve higher revenue despite the larger workload.

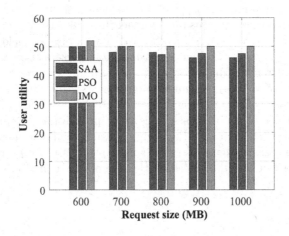

Fig. 6. The impact of request size on user utility

Figure 6 illustrates that as the request data size increases, the user utility value does not significantly improve. It can be observed that as the request data size S_r increases, the system revenue decreases slowly. This is because a large amount of input data reduces the number of requests that the edge servers can handle, thus decreasing user utility. Despite the overall trend of decreasing user utility, the IMO algorithm outperforms other methods in terms of revenue under different workloads and request sizes.

The revenue of ESP is a key research metric emphasized in this paper. Figure 7(a) shows the impact of the number of requests on the user request offloading ratio. The maximum power p_{max} of user devices is set to 12W, and the request size is fixed at 600 MB. Due to the limited resources of the edge servers, as the number of requests increases, the user offloading ratio decreases. The average offloading ratios for the three methods are 0.975, 0.9302, and 0.903, respectively. The offloading ratio represents the request acceptance rate of the edge servers and reflects their utilization to some extent. A higher offloading ratio indicates higher resource utilization. As shown, it can be observed that

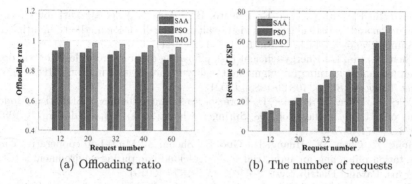

(a) Offloading ratio   (b) The number of requests

Fig. 7. Performance comparison from ESP perspective

the IMO method achieves the best performance in terms of the offloading ratio. To further demonstrate the optimization of social welfare in the edge system, Fig. 7(b) illustrates the relationship between ESP revenue and the number of user requests. Here, ESP revenue is represented by the product of the number of requests, the ratio of accepted requests, and the payment value. The IMO algorithm proposed outperforms the SAA and PSO methods in terms of ESP revenue.

6   Conclusion

This paper investigated the task offloading problem in edge systems and proposed a two-stage task offloading approach based on game theory to minimize user energy consumption and maximize social welfare. Firstly, a payment strategy that aligns with system revenue was proposed based on the relationship between user bidding for request tasks and system pricing. Then, a task offloading method with the minimum energy consumption was selected based on the payment strategy. Simulation results demonstrated that the IMO method effectively improves the revenue of edge systems by obtaining optimal task offloading payment strategies and finding the optimal energy offloading strategy based on high revenue. This study did not consider the latency requirements and deadline constraints in user requests. Future research directions could focus on incorporating latency requirements and deadline constraints in user requests and improving the performance of the algorithm.

References

1. Bi, J., Yuan, H., Duanmu, S., Zhou, M., Abusorrah, A.: Energy-optimized partial computation offloading in mobile-edge computing with genetic simulated-annealing-based particle swarm optimization. IEEE Internet Things J. **8**(5), 3774–3785 (2020)

2. Boyd, S., Parikh, N., Chu, E., Peleato, B., Eckstein, J., et al.: Distributed optimization and statistical learning via the alternating direction method of multipliers. Found. Trends® Mach. Learn. **3**(1), 1–122 (2011)
3. Chen, X., Liu, G.: Energy-efficient task offloading and resource allocation via deep reinforcement learning for augmented reality in mobile edge networks. IEEE Internet Things J. **8**(13), 10843–10856 (2021)
4. Coles, S., Bawa, J., Trenner, L., Dorazio, P.: An Introduction to Statistical Modeling of Extreme Values, vol. 208. Springer, Cham (2001). https://doi.org/10.1007/978-1-4471-3675-0
5. Hong, Z., Chen, W., Huang, H., Guo, S., Zheng, Z.: Multi-hop cooperative computation offloading for industrial IoT-edge-cloud computing environments. IEEE Trans. Parallel Distrib. Syst. **30**(12), 2759–2774 (2019)
6. Huang, X., Leng, S., Maharjan, S., Zhang, Y.: Multi-agent deep reinforcement learning for computation offloading and interference coordination in small cell networks. IEEE Trans. Veh. Technol. **70**(9), 9282–9293 (2021)
7. Jošilo, S., Dán, G.: Computation offloading scheduling for periodic tasks in mobile edge computing. IEEE/ACM Trans. Networking **28**(2), 667–680 (2020)
8. Li, C., Wang, H., Song, R.: Intelligent offloading for NOMA-assisted MEC via dual connectivity. IEEE Internet Things J. **8**(4), 2802–2813 (2020)
9. Li, L., Quek, T.Q., Ren, J., Yang, H.H., Chen, Z., Zhang, Y.: An incentive-aware job offloading control framework for multi-access edge computing. IEEE Trans. Mob. Comput. **20**(1), 63–75 (2019)
10. Liu, C.F., Bennis, M., Debbah, M., Poor, H.V.: Dynamic task offloading and resource allocation for ultra-reliable low-latency edge computing. IEEE Trans. Commun. **67**(6), 4132–4150 (2019)
11. Liu, Y., Xu, C., Zhan, Y., Liu, Z., Guan, J., Zhang, H.: Incentive mechanism for computation offloading using edge computing: a stackelberg game approach. Comput. Netw. **129**, 399–409 (2017)
12. Lowe, R., Wu, Y.I., Tamar, A., Harb, J., Pieter Abbeel, O., Mordatch, I.: Multi-agent actor-critic for mixed cooperative-competitive environments. In: Advances in Neural Information Processing Systems, vol. 30 (2017)
13. Luo, Q., Hu, S., Li, C., Li, G., Shi, W.: Resource scheduling in edge computing: a survey. IEEE Commun. Surv. Tutorials **23**(4), 2131–2165 (2021)
14. Neely, M.J.: Stochastic network optimization with application to communication and queueing systems. Synthesis Lectures on Communication Networks **3**(1), 1–211 (2010)
15. Ning, Z., Zhang, K., Wang, X., Guo, L., Hu, X., Huang, J., Hu, B., Kwok, R.Y.: Intelligent edge computing in internet of vehicles: a joint computation offloading and caching solution. IEEE Trans. Intell. Transp. Syst. **22**(4), 2212–2225 (2020)
16. Pliatsios, D., Sarigiannidis, P., Lagkas, T.D., Argyriou, V., Boulogeorgos, A.A.A., Baziana, P.: Joint wireless resource and computation offloading optimization for energy efficient internet of vehicles. IEEE Trans. Green Commun. Networking **6**(3), 1468–1480 (2022)
17. Qiu, X., Zhang, W., Chen, W., Zheng, Z.: Distributed and collective deep reinforcement learning for computation offloading: a practical perspective. IEEE Trans. Parallel Distrib. Syst. **32**(5), 1085–1101 (2020)
18. Roth, A.E., Sotomayor, M.: Two-sided matching. Handbook Game Theor. Econ. Appl. **1**, 485–541 (1992)
19. Sahni, Y., Cao, J., Yang, L., Ji, Y.: Multi-hop multi-task partial computation offloading in collaborative edge computing. IEEE Trans. Parallel Distrib. Syst. **32**(5), 1133–1145 (2020)

20. Tang, Q., Liu, L., Jin, C., Wang, J., Liao, Z., Luo, Y.: An UAV-assisted mobile edge computing offloading strategy for minimizing energy consumption. Comput. Netw. **207**, 108857 (2022)
21. Tang, Q., Fei, Z., Li, B., Han, Z.: Computation offloading in LEO satellite networks with hybrid cloud and edge computing. IEEE Internet Things J. **8**(11), 9164–9176 (2021)
22. Tseng, P.: Convergence of a block coordinate descent method for nondifferentiable minimization. J. Optim. Theory Appl. **109**(3), 475 (2001)
23. Uguz, S., Sahin, U., Sahin, F.: Edge detection with fuzzy cellular automata transition function optimized by PSO. Comput. Electr. Eng. **43**, 180–192 (2015)
24. Wang, X., Han, Y., Leung, V.C., Niyato, D., Yan, X., Chen, X.: Convergence of edge computing and deep learning: a comprehensive survey. IEEE Commun. Surv. Tutorials **22**(2), 869–904 (2020)
25. Yang, X., Luo, H., Sun, Y., Zou, J., Guizani, M.: Coalitional game-based cooperative computation offloading in MEC for reusable tasks. IEEE Internet Things J. **8**(16), 12968–12982 (2021)

Research on Dos Attack Simulation and Detection in Low-Orbit Satellite Network

Nannan Xie[1,2], Lijia Xie[1,2]([✉]), Qizhao Yuan[1,2], and Dongbo Zhao[1,2]

[1] School of Computer Science and Technology Changchun University of Science and Technology, Changchun, Jilin 130022, China
[2] Jilin Province Key Laboratory of Network and Information Security, Changchun University of Science and Technology, Changchun, Jilin 130022, China
lijiaxie2023@163.com

Abstract. Low-Earth Orbit (LEO) satellite networks are increasingly becoming the communication infrastructure of critical sectors because of their extensive coverage and global connectivity. However, due to the unique nature of satellite networks, they face the threat of DoS attacks from malicious actors. In this paper, we propose SDN-based framework and detection method for simulating and detecting DoS attacks in LEO satellite network. In terms of attack simulation, we design dynamic single-path and multi-path forwarding strategies to simulate various DoS attack scenarios encountered in the real network, and construct the marked attack data set. For attack detection, we propose an ensemble learning detection method based on the stacking framework. We incorporate six base classifiers to train and classify the dataset, realize the detection of DoS attacks in LEO satellite networks. Experimental results demonstrate the effectiveness of the attack simulation and detection process. We compare our proposed method with other classifiers and show that the attack simulation framework accurately simulates different types of DoS attack scenarios in LEO satellite networks. Furthermore, our detection method achieves a detection rate of 0.997, validating the effectiveness and feasibility of our proposed approach.

Keywords: Low-orbit Satellite · Dos Attack Simulate · Stacking

1 Introduction

1.1 A Subsection Sample

Satellite networks possess unique advantages with global coverage and rapid deployment, particularly suitable for remote areas and emergency communication needs. However, due to their dynamic nature and extensive coverage,

This work was supported in part by the Science and Technology Research Project of the Education Department of Jilin Province under Grant No. JJKH20230850KJ, and the Science and Technology Development Plan Project of Jilin Province under Grant No. 20230508096RC.

© The Author(s), under exclusive license to Springer Nature Singapore Pte Ltd. 2024
Z. Tari et al. (Eds.): ICA3PP 2023, LNCS 14492, pp. 240–251, 2024.
https://doi.org/10.1007/978-981-97-0811-6_14

they are susceptible to various attacks and threats. Attackers exploit the global coverage of satellite networks, creating vulnerabilities to launch DDoS attacks from multiple locations [20]. For instance, in October 2019 [2], the satellite TV company DISH Network experienced a DDoS attack resulting in website and application unavailability, negatively impacting subscribers. Similarly, in March 2021, European satellite operator SES experienced a large-scale DDoS attack, disrupting its services and affecting various sectors. Therefore, detecting and defending against attacks in low Earth orbit satellite networks is essential.

Intrusion detection technology, which primarily focuses on monitoring and detecting malicious activities within computer networks, faces unique challenges when applied to satellite networks, especially those operating in low Earth orbit (LEO). Moreover, the scarcity of attack datasets tailored to satellite networks presents a significant hurdle for research in this domain. This article addresses these issues by providing a feasible solution for attacks, detection, and defense within satellite networks. The main contributions are as follows:

(1) Construct a Software-Defined Networking (SDN)-based Low Earth Orbit (LEO) simulated satellite network. Define single-path and multi-path forwarding rules and simulate six types of DoS attacks separately in single-path and multi-path forwarding: SYN, ACK, ICMP, UDP, CC, and MYSQL. In the multi-path scenario, utilize both ground nodes and satellite nodes to launch the attacks.
(2) Create a dataset for low Earth orbit satellite network attacks. Collect traffic data from both normal communication and attack scenarios within the network. Extract features and preprocess the data. The final dataset should encompass over 300,000 traffic records.
(3) Utilizing an enhanced stacked ensemble machine learning approach to process a constructed attack dataset and achieve the recognition of attack data from normal data.

2   Related Work

LEO is a multifunctional satellite network for tasks like communication, navigation, and Earth observation, requires efficient management. Traditional satellite network approaches rely on fixed resource allocation and routing, lacking flexibility. SDN offers vital advantages, including unified resource management, minimal device requirements, and configurable policies [11]. In recent years, researchers have proposed innovative SDN-based satellite network architectures [9,12,19]. These studies explore multi-layered satellite networks, integrating various orbital levels, such as GEO, MEO, and LEO, into the network framework.

Due to the unique characteristics of LEO satellite networks, such as transmission latency, limited bandwidth resources, link instability, and hardware/software constraints, they are relatively susceptible to the threat of DoS attacks. In LEO DoS attacks, attackers aim to disrupt system availability by flooding the victim's bandwidth or targeting available resources, rendering legitimate users unable to access services. DoS attacks can be categorized into two main types: bandwidth attacks and resource exhaustion attacks. Bandwidth attacks, particularly botnet-driven floods, are common. Attackers infect victim hosts with malicious software, creating zombie networks known as Bot/Zombies [5,6,16]. These networks generate massive traffic, overwhelming the victim's resources and reducing their ability to serve legitimate users. Common flooding attack types include UDP, HTTP, and ICMP flooding [18].

Research on DoS attack detection in satellite networks is emerging, with machine learning methods playing a prominent role. Previous studies [14,21] have demonstrated the effectiveness of machine learning in satellite network attack detection. Ensemble algorithms, as intrusion detection methods, offer significant advantages in satellite network intrusion detection, addressing issues like data imbalance and gaining attention. Literature [7] utilizes ensemble intrusion detection system algorithms to implement lightweight detection models, addressing resource limitations in satellite networks.

However, existing intrusion detection methods for satellite networks have not genuinely considered the impact of the network's real environment on network traffic. Factors such as the dynamics and high latency of satellite networks can alter the characteristics of satellite network traffic data. This paper addresses this gap by simulating real satellite network conditions, including LEO attacks, to collect more authentic satellite network traffic data for subsequent research on intrusion detection algorithms in satellite networks.

3   Simulating DoS Attacks in Low Earth Orbit Satellite Networks

In order to closely mimic the actual topology of a satellite network, this paper employs the Iridium satellite constellation for satellite transmission network simulation [2]. As illustrated in Fig. 1, the simulation of LEO attacks consists of four main modules: simulating satellite network physical links using Systems Tool Kit (STK), transferring the physical link connectivity information of LEO to an SDN controller based on RYU for traffic control, dynamically managing and controlling OpenvSwitch (OVS) connections using the SDN controller, and finally, deploying network attacks using Docker on simulated dynamic LEO networks.

3.1   Satellite Physical Link Simulation Based on STK

As a satellite physical link simulator, STK finds extensive applications in the field of satellite simulation research [13]. It can provide satellite position and attitude data, simulate satellite orbits, and analyze satellite coverage and visibility. In specific instances, STK can calculate link budgets for each link using user-defined parameters and models. Furthermore, it can simulate physical parameters of certain links, such as Bit Error Rate (BER) and transmission delay, to serve as link weights for establishing paths between source and destination nodes [8]. Subsequently, it provides orbit information and physical parameters to the network simulator to build a topology and offers the link's physical parameters as

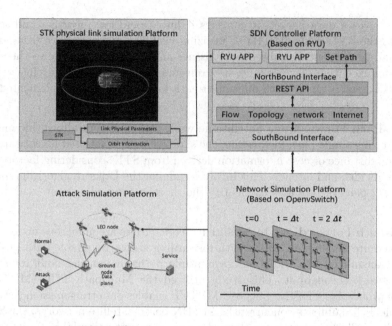

Fig. 1. Attack Simulation Platform for LSCN

weights to the SDN controller. The physical link simulation also includes connections between satellites and the ground, as well as inter-satellite connections. According to previous work [15], this paper have set the capacity of both the uplink and downlink to 4 Gbps. Additionally, to ensure signal transmission, we have configured a minimum elevation angle of 40° and set the ISL for 10 Gbps simplex transmission.

3.2 Network Simulation Platform Based on OVS

OVS serves as the network simulator, an open-source virtual switch with support for various protocols and advanced network functions. In satellite network simulation, OVS emulates satellite nodes by incorporating physical link data generated using STK. This data encompasses satellite orbits, communication satellite positions, and link quality. OVS with its API facilitates dynamic communication and data transmission among network satellites. Its virtualization capabilities enable thorough testing, evaluation, and optimization of network performance and satellite communication quality.

3.3 SDN Control Platform Based on RYU

The RYU controller was utilized for SDN management in this study. RYU interfaces with conventional network elements such as switches and routers through REST API, employing the OpenFlow protocol to manage OVS switches. When

OVS receives a data packet, it converts it into OpenFlow messages and sends them to the RYU controller. RYU, utilizing visibility matrices for routing decisions and traffic control, subsequently issues an OpenFlow message to the OVS switch for execution. Additionally, two conventional routing strategies in satellite networks, namely single-path and multi-path forwarding, were implemented.

Single-Path Forwarding. Satellite network paths are typically stable in seconds. During single-path forwarding, the shortest path is selected for forwarding based on distance or cost information derived from STK, considering factors like link bandwidth and latency. This involves running the Dijkstra algorithm for all $N(N-1)/2$ pairs of satellites to identify the shortest forwarding route.

Multi-path Forwarding. In satellite networks, there are numerous nodes, and the structure is complex. Multi-path forwarding strategies enable data packets to be transmitted in parallel across multiple paths, enhancing fault tolerance and network throughput. This study utilized the Multipath TCP (MPTCP) control protocol [38] for this purpose. MPTCP enables data transmission through multiple TCP subflows concurrently. In SDN-based satellite networks, the SDN controller collects network topology data and computes available paths using shortest path algorithms. It then communicates this information to relevant devices within the satellite network, allowing them to configure rules for multipath forwarding. At the receiving end, MPTCP terminal nodes receive data from various subflows and reassemble it into a complete data stream, offering improved bandwidth utilization and fault tolerance.

3.4   Network Attack Deployment Based on Docker Containers

Docker was linked to the OVS on the satellite nodes, serving as satellite terminals. In this configuration, normal and malicious traffic packets were routed to the ground station server via distinct paths, each traversing different satellite nodes. As detailed in Sect. 3.3, in both the single-path and multi-path transmission modes, malicious traffic is transmitted from the client node to the server via one or two routes. Additionally, we accounted for scenarios in which satellite nodes were compromised, enabling potential attacks to originate not only from ground nodes but also concurrently from satellite nodes towards the server.

The simulation encompassed prevalent DDoS attack types that have emerged in recent years, including User Datagram Protocol (UDP) flooding, Hypertext Transfer Protocol (HTTP) flooding, and Internet Control Message Protocol (ICMP) flooding [10]. Additionally, we examined SYN flooding, ACK flooding, and MySQL database exploitation attacks.

4 Building and Detecting Datasets for Satellite Network Attacks

4.1 Overall Framework

The Satellite Network Attack Detection Framework distinguishes normal and attack traffic in network data using a dataset generated from simulating six attacks. This framework includes data collection, feature extraction, and detection using a stacking algorithm. Data collection gathers network traffic data, while feature extraction extracts relevant features. These features are used to classify data as malicious or legitimate using statistical or machine learning methods. For a visual representation, refer to Fig. 2.

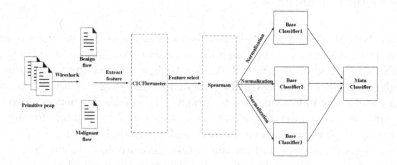

Fig. 2. Attack Simulation Platform for LSCN

4.2 Dataset Construction

Data Collection. The dataset is designed to collect various types of attacks simulated in a dynamic satellite network, based on the satellite network attack scenarios explained in Sect. 3.3. In scenarios corresponding to the three transmission modes, we implemented the transmission of both normal traffic and six types of attack traffic. In order to accurately simulate a real threat environment, approximately one hour of traffic data was collected for each type. Tcpdump was employed at the server end to collect various traffic data. The specific composition of the dataset is detailed in Table 1, comprising a total of over 300,000 entries of normal and attack traffic data. Additionally, we utilized CICFlowMeter to extract traffic features [1]. CICFlowMeter is a flow-based feature extraction tool capable of extracting 80 features from pcap files.

Data Preprocessing. The majority of machine learning algorithms primarily accommodate numerical features. However, when data includes both categorical values and corresponding numerical attributes, This study employ one-hot encoding [17] for feature transformation. This approach proves advantageous for

Table 1. Data composition

Simulation scene	SYN	ICMP	UDP	ACK	HTTP	Mysql	Total
Single path	17452	14364	14462	13159	16556	13884	102413
Multipath	12363	13452	12323	16452	17582	15673	101490
Multinode	15625	14512	16322	12365	16222	12322	100823

linear model performance. One-hot encoding translates categorical variables into binary representations, thereby mitigating issues related to correlation and magnitude, particularly suitable for discrete features. Additionally, our data preprocessing involves normalization and standardization to ensure that feature values are within the appropriate range.

Feature Selection. Feature selection is an essential step before constructing the dataset. Furthermore, most algorithms benefit from feature selection, such as KNN, MLP, random forests and decision trees, can also improve performance through scaling. Spearman correlation coefficient and Pearson correlation coefficient are statistical indicators used to measure the correlation between two variables, but they have different calculation methods and are suitable for different types of data. Spearman correlation coefficient is suitable for ordered categorical variables or ranking data and measures their monotonic relationship by comparing the rankings of variables. Calculate the Spearman correlation coefficient according to the formula below

$$\rho = 1 - \frac{6 \sum d_i^2}{n(n^2 - 1)} \tag{1}$$

where d_i represents the difference between the rank values of the i-th data pair, and n is the total number of observations. Compared with Pearson correlation coefficient, Spearman correlation coefficient is less affected by outliers and does not require a linear relationship between variables [4]. The range of Spearman correlation coefficient is also between −1 and 1, but it is usually used for analyzing non-normal distribution or ordered categorical data. We use Spearman correlation coefficient (rs) ranging from −1 to 1, where rs > 0 represents positive correlation, and rs < 0 represents negative correlation. The larger the absolute value of rs, the stronger the correlation between the feature and the label. Table 2 below shows the feature set we selected.

4.3 Stacking-Based Attack Detection

Stacking-based intrusion detection is a method that combines multiple classifiers under the stacking framework. In this framework, each classifier can classify intrusion behaviors, and decision making is made through ensemble of a meta-classifier. This framework consists of two main parts: base classifiers and meta-classifiers, as shown in Fig. 3.

Table 2. Selected Features with Absolute Correlation Values≥0.4

feature	Correlation
Total Fwd Packet	−0.623648
Bwd Packet Length Min	−0.560408
Flow IAT Mean	0.435638
SYN Flag Count	0.497625
Bwd Bytes/Bulk Avg	−0.671878
Bwd Packet/Bulk Avg	−0.671906
Bwd Bulk Rate Avg	−0.671684

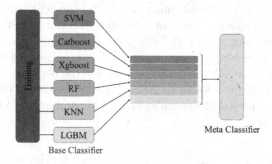

Fig. 3. Stacking Framework Structure

Due to the limited computing resources on the satellite network, lighter-weight algorithms should be used for the combination of base classifiers and meta-classifiers. We used six relatively lightweight algorithms as base classifiers, namely SVM, CatBoost, XGBoost, RF, KNN, and LightGBM. The selected base classifiers were trained on the training set using K-fold cross-validation. Each base model generated a set of prediction results on the validation set.

This study focused on configuring the base classifiers and the meta-classifier. Initially, a new training dataset was created by merging the predictions of the first-layer models on the validation set. The predictions from the base classifiers, either class labels or probabilities, were used for training and decision-making by the meta-classifier.

In the meta-classifier phase, the meta-classifier made classification decisions based on the outputs of the individual base classifiers, learning from their performance to generate a collective classification decision. A straightforward logistic regression (LR) algorithm was chosen for the meta-classifier to prevent overfitting.

Grid search was employed to meticulously fine-tune hyperparameters and pinpoint the optimal combination for achieving the highest F1-score. When fine-tuning these hyperparameters, particular emphasis was placed on minimizing both false positives and false negatives, with the ultimate goal of enhancing accuracy and reducing false alarms. This aspect is of paramount importance in practical intrusion detection systems. Combinations that exhibited lower false positive rates were prioritized to ensure the model's robustness and reliability in real-world applications.

5   Experimental Results

The model was implemented in Python using scikit-learn and the classifiers were trained on a Windows system equipped with a 2.59 GHz 6-core Intel(R) Core(TM) i7 processor. In the following sections, we discuss bagging, boosting, and stacking ensemble learners, as well as compare them with other classifiers.

The process of path selection implemented by the RYU controller and the procedures for conducting attacks are detailed in Table 3. Initially, we established a network topology using an LEO network constructed with OVS, where in Docker terminals were designated as client and server ground stations. Subsequently, we initiated the RYU controller to oversee the network, employing both single-path and multi-path forwarding transmission modes. Within the Docker environment, we employed tools such as hping3, ab, and hydra to initiate attacks on the LEO network.

Table 3. LSCN Attack Simulation Process and Terminal Commands

Implementation process	Terminal command
Step1 Set up the network topology:	ovs-docker add-port <LEO_network> eth0 <docker_terminal>
Step2 Start the RYU controller:	ryu-manager <GEO_controller>.py
Step3 Launch attacks using tools:	hping3 −1 -a <source_IP> <LEO_network>

In the context of limited satellite network resources, our use of a stacking algorithm necessitates the selection of lightweight classifiers for the initial layer of base classifiers, complemented by the Logistic Regression algorithm as the second-layer meta-classifier. The second-layer model must possess unique generalization capabilities to handle prior layer outputs and accurately predict unknown data, while also offering attributes such as computational efficiency, lightweight implementation, and ease of deployment. Consequently, Logistic Regression stands out as an appropriate choice for the second layer. Hyperparameter tuning via grid search ensures optimal model performance for each classifier.

To effectively assess the performance of the proposed network intrusion detection model, the following performance evaluation metrics are considered. This paper selected five common metrics to evaluate the classification performance of the proposed stacking-based satellite network intrusion detection method: accuracy, precision, recall, and F1-score.

In the comparative experiment of classifier performance, we initially assessed the performance of each classifier. As shown in Fig. 4, it is evident from the graph that both LGBM and KNN exhibit high accuracy, precision, and F1-score, but their performance in terms of recall is relatively lower. These results represent the best performance achieved for each classifier through grid search.

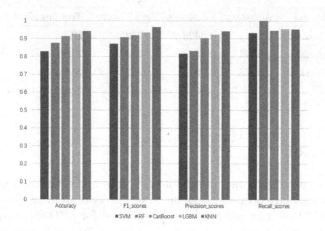

Fig. 4. Stacking Framework Structure

Comparison of Different Layers Stacking. Additionally, this study compared the performance of stacked models utilizing two, three, and four optimal learner models. As shown in Table 4, All models demonstrate very high accuracy, recall, and F1-score, with low false positive rates. From the results, it can be observed that using only the KNN algorithm in the first layer achieves high accuracy and F1-score. The experiment also indicates that having more base learners does not necessarily improve performance, as the best three learner models have lower performance compared to the best four learner models.

Table 4. Effects of different classifiers on results without changing Mata learner

Base Learners	Accuracy	F1-scores	Precision	Recall
KNN	0.9973	0.9978	0.9956	0.9962
KNN, LGBM	0.9937	0.9948	0.9945	0.9957
RF, KNN, CATTREE	0.9937	0.9948	0.9976	0.9950

Finally, a comparison of bagging, boosting, and stacking algorithms was conducted in the experiments. As shown in Table 5, it can be observed that the stacking algorithm has higher Accuracy, F1-score, and Recall-score. However, the training time of the stacking algorithm is twice as long as that of the boosting algorithm.

Table 5. Comparison of stacking with bagging and boosting

Model	Accuracy	F1_scores	Recall_scores	Training_time
Boosting	0.9334	0.8423	0.8732	20.4 s
Bagging	0.9334	0.8423	0.8734	79.8 s
Stacking	0.9973	0.9978	0.9962	20.3 s

6 Conclusion

With the widespread adoption of satellite networks in civilian communications, satellite networks have been facing an increasing array of security threats. In this study, a DoS attack simulation framework tailored for LEO satellite networks was proposed. It placed particular emphasis on the analysis of the three most susceptible transmission methods within satellite networks and simulated six types of satellite network attacks. Data comprising both normal and abnormal traffic were collected during specific time intervals to create a satellite network intrusion detection dataset. In terms of intrusion detection, a stacking-based ensemble learning method was employed to process the data and compared against bagging and boosting ensemble algorithms. The results demonstrated the superiority of the proposed method in terms of accuracy, F1-score, and recall, achieving a detection accuracy of 0.997. Additionally, it was characterized by its lightweight nature, making it suitable for deployment in satellite networks.

However, there are limitations in the types of simulated attacks in this study. Since all six simulated attacks belong to DoS attacks, they may exhibit similarities in terms of traffic, which leads to insufficient robustness verification of the method. In future work, it is recommended to consider simulating more diverse types of attacks in the satellite network environment to further enhance this research.

References

1. Datasets, https://www.unb.ca/cic/datasets/index.html. Accessed 7 Jun 2023
2. Iridium. https://www.iridium.com. Accessed 26 Apr 2023
3. Scmagazine. https://www.scmagazine.com/news/ransomware/dishnetworkconfirmscyberattack. Accessed 10 Feb 2023
4. Spearman. https://statisticsbyjim.com/basics/spearmanscorrelation. Accessed 28 Apr 2023

5. Abbas, N., Nasser, Y., Shehab, M., et al.: Attack-specific feature selection for anomaly detection in software-defined networks. In: 2021 3rd IEEE Middle East and North Africa COMMunications Conference (MENACOMM), pp. 142–146. IEEE (2021)

6. Alomari, E., Manickam, S., Gupta, B., et al.: A survey of botnet-based ddos flooding attacks of application layer: Detection and mitigation approaches. In: Handbook of Research on Modern Cryptographic Solutions for Computer and Cyber Security, pp. 52–79. IGI Global (2016)

7. Ashraf, I., Narra, M., Umer, M., et al.: A deep learning-based smart framework for cyber-physical and satellite system security threats detection. Electronics **11**(4), 667 (2022)

8. Barritt, B., Bhasin, K., Eddy, W., et al.: Unified approach to modeling & simulation of space communication networks and systems. In: 2010 IEEE International Systems Conference, pp. 133–136. IEEE (2010)

9. Cheng, N., Quan, W., Shi, W., et al.: A comprehensive simulation platform for space-air-ground integrated network. IEEE Wirel. Commun. **27**(1), 178–185 (2020)

10. Gaurav, A., Gupta, B.B., Alhalabi, W., et al.: A comprehensive survey on ddos attacks on various intelligent systems and it's defense techniques. Int. J. Intell. Syst. **37**(12), 11407–11431 (2022)

11. Gopal, R., Ravishankar, C.: Software defined satellite networks. In: 32nd AIAA International Communications Satellite Systems Conference, p. 4480 (2014)

12. Hu, F., Hao, Q., Bao, K.: A survey on software-defined network and openflow: From concept to implementation. IEEE Commun. Surv. Tutorials **16**(4), 2181–2206 (2014)

13. Lei, H., Lian, B., He, W., et al.: Analysis of DOP for BeiDou navigation system in Asia-pacific region based on stk. In: The Fourth China Satellite Navigation Academic Annual Symposium-S6 Beidou/GNSS Test Evaluation Technology (2013)

14. Li, Z., Yang, B., Zhang, X., et al.: Ddos defense method in software-defined space-air-ground network from dynamic bayesian game perspective. Secur. Commun. Networks **2022**, 1–13 (2022)

15. del Portillo, I., Cameron, B., Crawley, E.: Ground segment architectures for large LEO constellations with feeder links in EHF-bands. In: 2018 IEEE Aerospace Conference, pp. 1–14. IEEE (2018)

16. Saravanan, A., Bama, S.S., Kadry, S., et al.: A new framework to alleviate DDoS vulnerabilities in cloud computing. Int. J. Electr. Comput. Eng. (2088-8708) **9**(5), 4163–4175 (2019)

17. Shafieian, S., Zulkernine, M.: Multi-layer stacking ensemble learners for low footprint network intrusion detection. Complex Intell. Syst. **9**, 3787–3799 (2022). https://doi.org/10.1007/s40747-022-00809-3

18. Specht, S.M., Lee, R.B.: Distributed Denial of Service: Taxonomies of Attacks, Tools and Countermeasures, Princeton Architecture Laboratory for Multimedia and Security. ISCA, Princeton, NJ (2003)

19. Wang, Y., Pan, C., Automation, S.: Architecture design and performance analysis of double-layer satellite networks. Comput. Eng. Des. (2016)

20. Zhang, Y., Wang, Y., Hu, Y., et al.: Security performance analysis of Leo satellite constellation networks under DDoS attack. Sensors **22**(19), 7286 (2022)

21. Zhu, J., Wang, C.F.: Satellite networking intrusion detection system design based on deep learning method. In: Liang, Q., Mu, J., Jia, M., Wang, W., Feng, X., Zhang, B. (eds.) CSPS 2017. LNEE, vol. 463, pp. 2295–2304. Springer, Singapore (2019). https://doi.org/10.1007/978-981-10-6571-2_280

Malware Detection Method Based on Visualization

Nannan Xie[1,2], Haoxiang Liang[1,2(✉)], Linyang Mu[1,2], and Chuanxue Zhang[1,2]

[1] School of Computer Science and Technology, Changchun University of Science and Technology, Changchun, Jilin 130022, China
lianghaoxiang147@163.com
[2] Jilin Province Key Laboratory of Network and Information Security, Changchun University of Science and Technology, Changchun, Jilin 130022, China

Abstract. The rapid development of information technology and computer networks has led to the emergence of various new applications on both PC platforms and mobile devices. Malware continues to evolve and update, which often developing new variants or changing existing features to evade detection. Traditional feature based malware detection methods are limited in their ability to detect variants, and are computationally resource-intensive. Considering these issues, a new visualization-based and integrated malware detection method, Mal_Vis, is introduced. It decompiles the application software and applies PCA to reduce the feature dimension, then visualises the decompiled data to greyscale and RGB image. A Stacking-based ensemble machine learning algorithm is used to classify the visualized images to detect malware. Experiments show the method achievs detection accuracy of 98.19% and 93.03% in the Windows and Android application software datasets.

Keywords: Malware Detection · Visualization · Grayscale Image · RGB Image · Feature Dimensionality Reduction · Stacking

1 Introduction

The development of computer network has led to the integration of multiple PC and mobile applications into daily life. Windows and Android as the leading global operating systems, have become top targets for these attacks. Based on AV-TEST analysis [2], as of May 2023, Windows operating systems are host to an estimated 41.55 million malware applications, with Trojan, Worm, Downloader, and Backdoor accounting for about 80% of the top attack vectors. Similarly, Android systems have been infested by a majority of 33.66 million ransomware, dropper, Trojan, and other malware. Malware creators continuously modify their attack methods and obfuscate their work using techniques such as packing and

This work was supported in part by the Science and Technology Research Project of the Education Department of Jilin Province under Grant No. JJKH20230850KJ, and the Science and Technology Development Plan Project of Jilin Province under Grant No. 20230508096RC.

© The Author(s), under exclusive license to Springer Nature Singapore Pte Ltd. 2024
Z. Tari et al. (Eds.): ICA3PP 2023, LNCS 14492, pp. 252–264, 2024.
https://doi.org/10.1007/978-981-97-0811-6_15

deformation. Therefore, researching new malware detection methods is a very urgent matter.

Traditional malware detection mainly extracts static or dynamic features and employs machine learning techniques to classify or construct malicious behaviours to achieve classification-based or match-based detection. However, these methods are not very effective in detecting malware variants. Nataraj et al. [8] has first proposed a method to visualize application software binaries as grey scale images and classify malware using image processing techniques. Visualisation methods have gradually attracted the attention of researchers, since they can identify the characteristics of malware in a new way.

Emerging malware use advanced techniques like packing and obfuscation to bypass conventional detection measures. To address this challenge, decompiling malware samples into binary data can be advantageous. This process enables the circumvention of the impact of malware variants. We characterizes malware features using data visualization techniques and deploys an integrated machine learning method to detect these features. The main contributions are as follows.

(1) The proposed method, Mal_Vis, is a framework for detecting malware that combines data visualization techniques and integrated machine learning. It includes three modules, namely data preprocessing, visualization, and detection and classification, working together to achieve efficient detection and categorization of malicious software.

(2) This study accomplishes the visualization of application software by generating grayscale and RGB images. The first step involves decompiling the binary files of the application software dataset samples. The next step applies principal component analysis (PCA) to mitigate the computational demands and reduce the dimensionality of the data. Finally, the reduced data is used to create grayscale and RGB images for all application software samples.

(3) This study implements an integrated approach to malicious software detection based on Stacking ensemble. By combining three distinct detection models via the Stacking integration algorithm and implementing this approach for image datasets, the efficacy of malicious software detection is significantly improved.

2   Related Work

2.1   Malware and Detection Methods

Windows and Android systems, which dominate PC and mobile respectively, are particularly favoured by malware attackers. Common types of malware mainly include worms, backdoors, Trojans, etc. The emergence of technologies such as obfuscation and shelling has reduced the producing cost of malware. Obfuscation makes the malware code complex and difficult to understand to mask the true purpose, enhance resistance to static analysis and detection, and render traditional signature matching ineffective. Shelling, on the other hand, protects and hides malware by modifying the binary file, encapsulating it in a custom shell, and automatically decrypting and loading it into computer memory for execution at runtime.

Static method analyze the software by decompiling the underlying code to extract features without runing the program. Common reverse engineering tools utilized for the process include IDA Pro, Hopper, and OllyDbg. Static features, including the permissions, API calling sequence, and operation code sequence, are extracted by conducting reverse engineering analysis. The static method is easy to implement and not need a high-performance system environment, but it's vulnerable to code obfuscation and performs poorly against targeted attacks.

Dynamic detection requires executing the program in a monitored environment to capture vital features such as registry modifications, API calls, and memory writing effectively. Dynamic approaches focus on the behaviour of the malware rather than on specific code or signatures, and also perform detection in a virtual environment where samples are run without endangering the real system, but such environments lead to a certain performance overhead and need to be updated in a timely manner to adapt to new threats as the malware evolves. The related reserach include Shaid S Z M et al. [11], Akinori Fujino et al. [4], and Gianni D'Angelo et al. [3].

2.2 Visualization

Visualization is technology that uses computer graphics and image processing techniques to convert data into graphics or images to be displayed on the screen and interactively processed. Depending on the principles of visualisation, they can be classified as geometry-based techniques, pixel-based techniques, icon-based techniques, hierarchy-based techniques, image-based techniques, and distributed techniques [5].

Most malware detection research use pixel based techniques to convert malware into grey scale images or colour RGB images. Sang Ni et al. [9] converted decomposed malware code into grey scale image based on SimHash and then detected it by Convolutional Neural Networks with multiple hashing, master block selection and bilinear differencing to improve the performance. Kamran Shaukat et al. [12] proposed a deep learning based malware detection method by first visualising the executable file as a colour images, then extract features from the images using deep learning models and finally use SVM to detect malware.

Visualisation techniques provide an intuitive and easy-to-analyse methodology that helps researchers better understand and analyse malware operations, leading to more effective strategies. Visualisation is particularly useful in analysing and detecting obfuscated and shelled malware. Usually obfuscation, shelling and other techniques make the malware code complex and obscure. For example, it dynamically decrypting and decompressing the code at runtime, thus complicating detection. Visualisation can better demonstrate the hidden logic and code execution paths, making it easier for researchers to understand the malware operation and reveal evasion. However, it is found through investigation that the visualisation process in research consumes more computational resources and the computation time is relatively slow.

2.3   Ensemble Machine Learning

In order to solve the problem of uneven detection accuracy of a single machine learning model, the integration algorithm is proposed. The main idea of integrated learning is to combine multiple classifiers for joint prediction. Bagging, Boosting and Stacking are the three mainstream algorithms of integrated learning. Integrated learning has better generalisation performance than individual models, Bagging and Boosting are base classifiers that are combined using deterministic algorithms, while Stacking is a multi-layer integrated learning approach that aims to combine the strengths of each base classifier to produce better classification results. There are two stages of models in a general Stacking framework, including the base model (also called level-0 model) and the meta-model (also called level-1 model).

Stacking has gained significant traction in the field of malware detection. Zhu et al. [14] introduced an ensemble framework powered by Stacking which utilizes fused classifiers such as basic Multilayer Perceptron and Support Vector Machine (SVM) for Android malware detection. Xue et al. [13] developed a Stacking-based ensemble learning approach to classify malware by constructing host-based traffic fingerprints through the extraction of network traffic features. Shafin et al. [10] presented a two-layer machine learning-based model for accurately predicting and classifying attacks on Android smartphones. These studies are effective attempts to apply integrated methods to malware detection.

3   Visualization-Based Malware Detection

3.1   Method Framework

A method for applying visualization to malware detection is proposed in this work, as shown in Fig. 1. The malware are transformed into images, which are then classified by ensembled stacking, in order to recognized the malware from the normal applications.

The proposed framework shown in Fig. 1 consists of three modules: data preprocessing, data visualization, and malware detection. This method integrates visualization techniques into the detection process, allowing for the representation of high-dimensional data in lower dimensions and ultimately improving the detection accuracy, with the goal to develop a lightweight visualization-based method for detecting malware.

(1) The data preprocessing module is composed of three steps to prepare the data for analysis. Firstly, it collect applications from open-source datasets and third-party markets, which include normal and malware for both Windows and Android. Secondly, we decompose the applications into binary files, specifically PE files for Windows and APK files for Android, which serve as inputs for PCA. Lastly, the PCA method is applied to reduce feature dimensionality and to make the visualizations more practical.

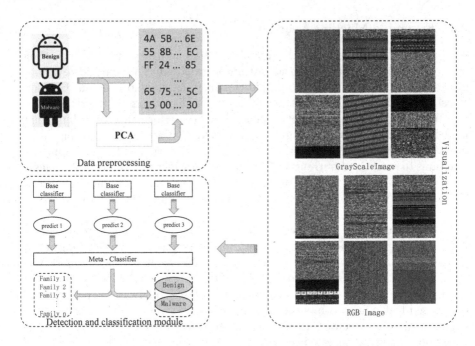

Fig. 1. Visualization-based malware detection model

(2) The application visualization module converts the reduced-dimensional data into grayscale and RGB images and generates image datasets of application samples.

(3) The malware detection module utilizes integrated Stacking to process the images obtained from the previous module. The module combines three high-performing base classifiers, SVM, LGBM, and XGBoost, with LR as the meta-classifier. This integration aims to improve the detection of malware by reducing false positives and achieving high accuracy in a time-efficient way.

3.2 Specific Implementation

(1) Feature dimentionality reduction by PCA

Using PCA to reduce the dimensionality of Hexadecimal files of applications can decrease the visualization time and the space needed. The steps are as follows:

Input: Convert the Hexadecimal files of the software under test into a two-dimensional matrix.

Output: The hexadecimal file obtained after processing.

Step 1: Subtract the mean value of each column of the original data.

Step 2: The covariance matrix identifies the correlation and calculates eigenvectors and eigenvalues to identify the main components. The number of dimensions of the main components is the minimum value of the rows and columns of the original data.

Step 3: Retain the main components, divide each column by its standard deviation (perform whitening processing) to recast the data.

Step 4: Stores data in hexadecimal form in a two-dimensional matrix and converts to a hexadecimal file.

(2) Dataset construction by visualization

Once the dataset is preprocessed, the hexadecimal machine code is converted to a one-dimensional array of decimal values, and the array is converted to grayscale values (0 for black, 255 for white) for generating a corresponding image. The image width is kept constant, while the height varies based on the size of the file. Figure 2 depicts this conversion process.

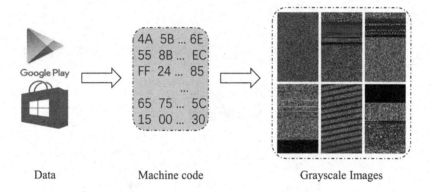

Data   Machine code   Grayscale Images

Fig. 2. Grayscale images generation

The method of generating RGB images through visualization is similar to that of grayscale images. The difference is that an RGB image can represent three chaneels of the values between 0 and 255. For example, yellow can be represented as [R: 255, G: 255, B: 0]. Each pixel can be generated by inputting three values for R, G, and B channels and assigning it to the respective channels. The resulting pixel includes the values of all three channels.

(3) Stacking algorithm

A malware detection framework is presented that utilizes Stacking to enhance the detection accuracy of the generated image samples. Specifically, Stacking's combination function is used to create a classification model that detects malware through the results of multiple classifiers. As shown in Fig. 3, the framework consists of two layers: the first layer consists of base classifiers, while the second layer is the meta-classifier.

(A) SVM, XGBoost, and LightGBM were selected as base classifiers in the first layer. Each base classifier used the training image dataset as input and produced a feature vector. The output f each classifier was a feature matric which combined i feature vectors. The number of columns i in the feature matrix was equal to the number of base classifiers, and the number of rows was equal to the number of samples in the training set.

Using five-fold cross-validation, each base classifier is trained in five sessions, with one-fifth of the samples retained as test data for each session. The model predicts the test data and stacks the results to obtain full sample predictions. Simultaneously, for the test data prediction, each model will produce five prediction results. These five results will be averaged based on the average of each classifier model after the run. This averaging provides the unique prediction of the test sample, serving as new features for the test data. Each base classifier will generate such new features. Finally, the new feature is combined with the original sample prediction to form the second-layer input.

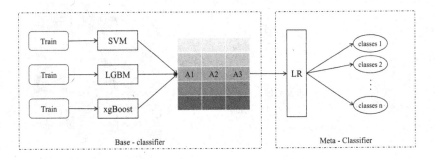

Fig. 3. Stacking detection framework

(B) The meta-classifier aggregates the output of the previous layer's base models to calculate the final classification result. To prevent overfitting, a simple logistic regression (LR) model is used for both training and prediction. The LR model's straightforward nature makes it efficient and easily interpretable.
The steps of the stacking operation are as follows:

Input: The dataset generated by visualizing the binary files as images.

Output: The final prediction results.

Step 1: Design the types and quantity of base classifier models, divide the training data into five non-overlapping parts, and label them as train 1 - train 5.

Step 2: Select a base classification model, use the other four parts to train the model, and predict train 1. Preserve the prediction results.

Step 3: Use train 2 as the prediction set, train on the other four parts, and predict train 2. Repeat this step for all five parts, stack the results vertically, and combine the predictions of the five parts using the first base classification model.

Step 4: Predict the test set separately in the process of steps 2–3, and save the results. Then, average the results, and consider them as the first basic classification model's predicted data for the test set.

Step 5: Repeat the process 2–4 separately for all the base classifiers; this generates several columns of feature expressions that correspond to the number of base classifiers and the corresponding new feature expressions for the test set.

Step 6: Combine the results of each base model into a matrix and train the second-level meta-classifier model to obtain the final prediction result.

4   Experiments and Evaluations

4.1   Data Set and Experimental Environment

(1) Data set

In the Windows dataset, the malware is sourced from the 9th Big Data and Computing Intelligence Competition (CCF BDCI 2021) [7], while benign software is obtained from the Software Applications Platform. The dataset comprises 7,307 benign software and 3,396 malware samples, with the latter belonging to 10 distinct malicious families.

The Android dataset consisted of 2929 randomly selected malware samples from the CIC MalDroid 2020 [6] dataset, and 7366 benign applications downloaded from third-party app markets. To classify the malware samples according to their families, we used the Drebin dataset [1], which consists of 5,560 malware samples from 179 different families collected between August 2010 and October 2012 by the Mobile-Sandbox project.

To obtain accurate labels for the dataset, the aforementioned samples were uploaded onto Virus Total to identify and label each sample.

(2) experimental environment

All experiments were conducted on a Linux operating system with Ubuntu version 22.04.2, using an Intel(R) Xeon(R) Silver 4108 1.80 GHz CPU and 64 GB of RAM. Visual Studio Code version 1.78.2 was used as the programming tool with Python version 3.10.6. The Pandas, Keras, and Matplotlib libraries were employed during the implementation process.

4.2   Experimentation

This section will contain three experiments, including software visualization application, visualization-based detection of malware, and visualization-based classification of malicious families.

Experiment one: Visualization of Application Software

Principal Component Analysis (PCA) is utilized to select essential feature subsets from the decompiled features of the original samples, aimed at reducing computational expenses associated with data visualization. Also, Window's application software processes its Portable Executable file while Android samples extract classes.dex from their APK files. The visualization times for Windows and Android are presented in Tables 1 and 2, respectively.

Table 1. Comparison of visualization times for Windows (Seconds per image)

Windows	Grayscale Images			RGB Images		
	Original	PCA	Percentage	Original	PCA	Percentage
Benign	5.98	0.85	14.21%	22.36	3.86	17.26%
Malware	0.17	0.03	17.64%	0.58	0.08	13.79%

Table 2. Comparison of visualization times for Android (Seconds per image)

Android	Grayscale Images			RGB Images		
	Original	PCA	Percentage	Original	PCA	Percentage
Benign	0.60	0.08	13.33%	1.10	0.15	13.64%
Malware	0.58	0.07	12.07%	1.17	0.15	12.82%

Table 1 shows that the average time to generate a grayscale image from raw data of benign Windows software visualization is 5.98 s, which was reduced to 0.85 s (14.21% of original time) after using PCA for dimensionality reduction purposes. Similarly, Android software visualization resulted in an 80% reduction in visualization time.

Experiment two: Visualization-Based Malware Detection

(A) Detection of Windows malware
Employing the dataset of Windows application software, visualize the dataset as grayscale and RGB images, respectively. Subsequently, apply five classification algorithms to identify malware in both types of images. Figure 4(a) presents a comparison of the grayscale image detection algorithm, while Table 3 illustrates a comparison of the RGB image detection algorithm.

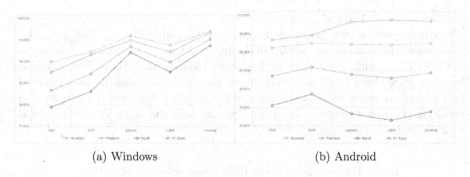

(a) Windows   (b) Android

Fig. 4. Grayscale image detection of malware

Figure 4(a) depicts that when machine learning algorithms were employed to classify and detect the Windows image dataset, the Stacking ensemble learning algorithm attained the highest detection accuracy of 97.11% and outperformed other algorithms in terms of precision, recall, and F1-Score.

When using Stacking ensemble learning to detect RGB images of the original dataset, a detection accuracy rate of 99.37% was achieved, while detecting RGB images of the data generated by PCA technique after dimensionality reduction yielded a detection accuracy rate of 98.19%, as presented in Table 3. Evidently, the use of PCA technique can significantly reduce temporal and spatial complexity without a considerable effect on detection accuracy.

Table 3. Comparison of the RGB image detection algorithm

Aldorithms	Data	Accuracy	Precision	Recall	F1-score
CNN	Original	91.51%	98.76%	85.10%	91.42%
	PCA	90.03%	90.79%	76.66%	83.12%
SVM	Original	96.21%	96.60%	96.11%	96.35%
	PCA	93.18%	93.45%	85.49%	89.30%
xgBoot	Original	99.32%	99.80%	98.90%	99.35%
	PCA	97.91%	98.03%	95.18%	96.58%
LGBM	Original	99.17%	99.80%	98.61%	99.20%
	PCA	96.58%	96.73%	92.06%	94.34%
Stacking	Original	99.37%	99.70%	99.10%	99.40%
	PCA	98.19%	98.54%	95.58%	97.04%

Results from the above experiments suggest that the detection rate of RGB images is higher than that of grayscale images, which implies that the features contained in RGB images are superior to those in grayscale images and are capable of better representing the characteristics of application software.

(B) Detection of Android malware

The comparison of the grayscale image detection algorithm is shown in Fig. 4(b), while the comparison of the RGB image detection algorithm is shown in Table 4. The results illustrated in Fig. 4(b) indicate that the use of machine learning algorithms to classify and detect the Android image dataset resulted in a slightly reduced recall and F1-Score but improved detection accuracy and precision compared to other algorithms. Specifically, the detection accuracy was 92.39%, and the precision was 98.45%.

Table 4. Comparison of the RGB image detection algorithm

Algorithm	Data	Accuracy	Precision	Recall	F1-score
CNN	Original	91.87%	90.96%	78.21%	84.11%
	PCA	91.20%	97.86%	71.24%	82.45%
SVM	Original	92.62%	96.51%	76.50%	85.35%
	PCA	97.03%	96.19%	75.03%	84.29%
xgBoot	Original	92.03%	97.98%	72.97%	83.64%
	PCA	92.97%	97.50%	76.80%	85..92%
LGBM	Original	92.76%	98.90%	75.71%	85.77%
	PCA	91.80%	98.61%	72.28%	83.42%
Stacking	Original	92.13%	98.43%	72.97%	83.81%
	PCA	93.03%	97.65%	76.91%	86.04%

The results presented in Table 4 demonstrate that the detection accuracy rate of the original RGB dataset using Stacking ensemble learning was 92.13%, and that for the RGB dataset generated by PCA technique was 93.03%. Furthermore, the abundance of features in RGB images in comparison to grayscale images validates the practicability of the proposed method.

Experiment three: Visualization-Based Classification Of Malicious Families. Attempted to utilize the proposed method for classifying the malicious families of malware. The Windows dataset remained unchanged as previously mentioned. For the Android dataset, whereas, the authors chose the six classes with the highest frequency from Drebin dataset. To identify the malicious families of malware, employed the Stacking ensemble algorithm. As shown in Table 5.

Table 5. Accuracy of malicious family classification

OS	Grayscale Image	RGB Image
Windows	92.94%	92.35%
Android	87.81%	89.53%

Table 5 shows that the Mal_Vis technique is able to classify malicious families with 92.94% accuracy for the Windows dataset and with 89.53% accuracy for Android. The results suggest that the proposed technique can effectively classify malicious families.

The results of the three experiments suggest that applying PCA technique to visualize application software as images in Experiment 1 can help to reduce computational costs. In Experiments 2 and 3, it was shown that visualizing the reduced-dimension data can effectively detect and classify malware with detection accuracy remaining essentially constant compared to that of the original sample visualization.

5   Conclusion

In malware detection, how to improve the detection rate and reduce system consumption has always been a research focus. We propose a mixed detection approach that combines visualization and stacking alrorithms. Since the visulization process consume most of the computation time, PCA is employed to reduce the feature dimension to be visulized. With SVM, LGBM, xgBoost as base classifiers, we use Stacking to esemble them to classify malware. The experiment's results demonstrated that the detection rate of grayscale images was 97.11% and RGB images was 98.19% of Windows malware. These two numbers are 92.39% and 93.03% in the Android operating system. These results illustrate the practicality and effectiveness of the proposed method.

We are relatively single in the use of feature selection, only PCA algorithm for experimental verification, in future work can explore other feature extraction or selection methods, so that in the detection of the basic accuracy of the case of the same to reduce the generation of images brought about by the large amount of time and space consumption, in the visualisation process only used grey scale images and RGB images for detection, future research will explore new visualisation techniques will be explored in future research to solve the above problem. Finally we will also explore the impact of emerging machine learning models as classifiers for detecting images.

References

1. Arp, D., Spreitzenbarth, M., Hubner, M., et al.: Drebin: Effective and explainable detection of android malware in your pocket. In: Ndss, vol. 14, pp. 23–26 (2014)
2. AV-ATLAS: Av-atlas malware statistics. AV-TEST. https://portal.av-atlas.org/malware. Accessed 09 Apr 2023
3. D'Angelo, G., Ficco, M., Palmieri, F.: Malware detection in mobile environments based on autoencoders and API-images. J. Parallel Distrib. Comput. **137**, 26–33 (2020)
4. Fujino, A., Murakami, J., Mori, T.: Discovering similar malware samples using API call topics. In: 2015 12th Annual IEEE Consumer Communications and Networking Conference (CCNC), pp. 140–147. IEEE (2015)
5. Kan Liu, Xiaozheng Zhou, D.Z.: Data visualization research and development. Comput. Eng. **08**, 1–2+63 (2002)
6. Mahdavifar, S., Kadir, A.F.A., Fatemi, R., et al.: Dynamic android malware category classification using semi-supervised deep learning. In: 2020 IEEE International Conference on Dependable, Autonomic and Secure Computing, International Conference on Pervasive Intelligence and Computing, International Conference on Cloud and Big Data Computing, International Conference on Cyber Science and Technology Congress (DASC/PiCom/CBDCom/CyberSciTech), pp. 515–522. IEEE (2020)
7. Microsoft: Datafountain competition dataset. https://www.datafountain.cn/comp-etitions/507/datasets. Accessed 16 Mar 2023
8. Nataraj, Lakshmanan, K., et al.: Malware images: visualization and automatic classification. In: Proceedings of the 8th International Symposium on Visualization for Cyber Security, pp. 1–7 (2011)
9. Ni, S., Qian, Q., Zhang, R.: Malware identification using visualization images and deep learning. Comput. Secur. **77**, 871–885 (2018)
10. Shafin, S.S., Ahmed, M.M., Pranto, M.A., et al.: Detection of android malware using tree-based ensemble stacking model. In: 2021 IEEE Asia-Pacific Conference on Computer Science and Data Engineering (CSDE), pp. 1–6. IEEE (2021)
11. Shaid, S.Z.M., Maarof, M.A.: Malware behaviour visualization. Jurnal Teknologi **70**(5), 25–33 (2014)
12. Shaukat, K., Luo, S., Varadharajan, V.: A novel deep learning-based approach for malware detection. Eng. Appl. Artif. Intell. **122**, 106030 (2023)
13. Xue, Z., Niu, W., Ren, X., et al.: A stacking-based classification approach to android malware using host-level encrypted traffic. In: Journal of Physics: Conference Series, vol. 2024, p. 012049. IOP Publishing (2021)
14. Zhu, H., Li, Y., Li, R., et al.: SEDMDroid: an enhanced stacking ensemble framework for android malware detection. IEEE Trans. Netw. Sci. Eng. **8**(2), 984–994 (2020)

TOC: Joint Task Offloading and Computation Reuse in Vehicular Edge Computing

Kaiyue Li[1($\boxtimes$)], Shihong Hu[1,2], and Bin Tang[1,2]

[1] School of Computer and Information, Hohai University, Nanjing 211100, China
{kaiyueli,shihonghu,cstb}@hhu.edu.cn
[2] Key Laboratory of Water Big Data Technology of Ministry of Water Resources, Houhai University, Nanjing 211100, China

Abstract. With the proliferation of intelligent vehicles, addressing the demands of computing-intensive and delay-sensitive vehicle tasks has become a formidable challenge. Vehicle edge computing (VEC) has been proposed as an advanced paradigm that leverages edge servers such as road side units (RSUs) to offload tasks, thereby enhancing vehicle services. However, similar computation tasks in the VEC environment result in computational redundancy, imposing additional burden on the limited edge resources. Moreover, the increased interdependency among different tasks of vehicle tasks adds complexity to the offloading strategy. To this end, we propose a collaborative task offloading and computation reuse framework, called *TOC*, which enables RSUs to reuse previous computations and design a task offloading scheme based on a Conflict Graph(CG) model. We also evaluate the efficiency and effectiveness of *TOC* using real-world datasets, and our results show that *TOC* is able to reduce the task completion time by 48.73% compared to baselines.

Keywords: Edge computing · service offloading · computation reuse · task dependency

1 Introduction

VEC has emerged in the Internet of Vehicles (IoV) as a new paradigm that offloads computation tasks to RSUs aiming to reduce the processing delay as well as the resource consumption of vehicles [1]. Unfortunately, these tasks have stringent latency requirements, which are severely constrained by the limited computational and storage capacities of intelligent vehicles to perform computationally intensive tasks [2]. As an effective alternative to cloud computing, VEC

This work was supported in part by the National Natural Science Foundation of China under Grant No. 62202140, and the Natural Science Foundation of Jiangsu Province of China under Grant No. BK20220974, and the Future Network Scientific Research Foundation Project FNSRFP-2021-ZD-7

© The Author(s), under exclusive license to Springer Nature Singapore Pte Ltd. 2024
Z. Tari et al. (Eds.): ICA3PP 2023, LNCS 14492, pp. 265–282, 2024.
https://doi.org/10.1007/978-981-97-0811-6_16

is able to meet ultra-low execution latency requirements and reduce backhaul pressure on cloud data centers by sinking processing and storage resources to a Base Station(BS) or RSU near the intelligent vehicle [3].

In addition, in the vehicle-road collaboration scenario, vehicle tasks and user requirements present a many-to-one relationship between the inputs and outputs imposed on location-aware computing. This phenomenon leads to input data that are correlated in time, space, and even in meaning [4]. To reduce the computation cost, redundant/duplicate computations can be avoided by reusing the output of previous (similar) computations instead of starting the computation from scratch [5]. In a VEC system, the RSUs must be capable of receiving and executing tasks in real-time, despite the uncertainty of task offload time due to the mobility of the vehicles [6]. In addition, the vehicle task consists of multiple sub-tasks that may exhibit dependencies between them, usually represented using Directed Acyclic Graphs (DAGs) [7], each task requires varying levels of computational power and computation time. Sub-tasks may be shareable among multiple tasks, and storing execution results on the RSUs can prevent redundant computations. These stored results can be reused for future sub-tasks executions with the same input.

In this paper, we considered a scenario with different mobile vehicles distributed in an intersection environment, as shown in Fig. 1. This paper proposes a collaborative task offloading and computational reuse framework, called *TOC*. Compared to existing task offloading approaches, this framework addresses the computational reuse problem of RSUs in computing offloading tasks while considering task dependencies, and designs a task offloading scheme based on a conflict graph model. Specifically, the main contributions of this article are summarized as follows.

- We build a computational reuse model on the RSUs side that aims to achieve three things: 1) determine which computations to store for potential future reuse; 2) store as many computations as possible that may be useful in the future; 3) mitigate the cost of searching for computations that can be reused once the RSUs receives incoming tasks.
- We propose a task offloading scheme, which is based on a conflict graph model that dynamically generates offloading policies for vehicle tasks. By combing the computational reuse model and a dynamic offloading scheme, we develop the *TOC* framework.
- We verify the performance of *TOC* using a real-world dataset, combining two comparison algorithms and designing six different experimental scenarios. Finally, the *TOC* is evaluated by comparing the completion time of all tasks and the computation execution time of all RSUs.

The remainder of this paper is structured as follows. In Sect. 2, we provide a brief background and review the related literature. Section 3 outlines our system model and problem formulation, while Sect. 4 elaborates on the RSUs side computational reuse and vehicles side task offloading scheme. The efficiency and effectiveness of *TOC* are evaluated in Sect. 5. Finally, Sect. 6 concludes the paper.

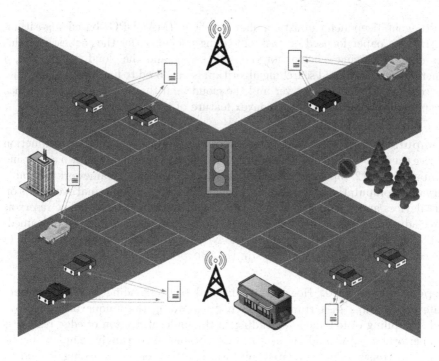

Fig. 1. The architecture of collaborative task offloading and computation reuse in VEC system.

2   Related Work

Task Offloading. Various studies have proposed many offloading methods that enable resource-intensive intelligent vehicles to transfer computation-heavy activities to RSUs [8]. In [9], a cloud-edge collaboration hierarchical intelligent-driven VEC network architecture is proposed, which utilizes the heterogeneous computation capabilities of cloud center, aggregation servers, and MEC servers to achieve comprehensive collaboration and intelligent management of network resources. Lu *et al.* [10] proposed a distributed intelligent task offloading and workload balance (DIOW) framework. In the framework, the base stations, mounted with mobile edge computing (MEC) servers, can execute the tasks from task vehicles. To obtain the optimum design, the framework adopts a multi-agent deep deterministic policy gradient (MADDPG)-based algorithm. In [11], the paper focused on task offloading allocation for the requesting vehicles and the pricing schemes for the edge server and the cloud. Meanwhile, a genetic algorithm-based searching algorithm is proposed to find the optimal pricing schemes for the edge server and the cloud, and the proposed algorithm has a rapid convergence due to the convex feature of the objective problem.

Computational Reuse. The practice of using previously completed function service execution results for upcoming calculations is commonly known as reusing computations [12,13]. Azad *et al.* [14] proposed Reservoir, a framework to enable pervasive computational reuse at the edge, while imposing marginal overheads on user devices and the operation of the edge network infrastructure. The reservoir takes advantage of Locality Sensitive Hashing (LSH) and runs on top of Named-Data Networking (NDN), extending the NDN architecture for the realization of the computational reuse semantics in the network.

Dependency-Aware. However, traditional task offloading solutions concentrate on one-time task transfers, often disregarding the unique task topology and scheduling of IoT devices, leading to the underutilization of edge resources and performance. As modern tasks in VECs become increasingly complex, mobile tasks may consist of many related sub-tasks [15]. Therefore, offloading dependent sub-tasks is necessary for many practical tasks among VECs. Chang *et al.* [16] studied the fine-grained task offloading problem in edge computing for low-power IoT systems, and proposed a lightweight but efficient multi-user edge offloading scheme by fully considering the topology of tasks. In [7], the paper considered the completion time constraint of each task and the execution dependency of multiple sub-tasks belonging to the same task and developed an efficient task scheduling method with the basic idea of prioritizing multiple tasks to ensure task completion.

3   System Model and Problem Formulation

3.1   System Model

System Model. As shown in Fig. 1, there are a number of moving service vehicles(SVs) $s = \{1, 2, ..., |S|\}$ in an ordinary intersection, each of which has several tasks that can be offloaded to the RSUs for execution. There is a set of RSUs that are evenly distributed next to the buildings: $R = \{1, 2, 3, ..., |R|\}$. The considered system operates in a time slot manner, with distinct time periods made up of the timeline: $t = \{1, 2, 3, ..., T\}$. The set of tasks carried by each vehicle is expressed as $A_s = \left\{ a_1^s, a_2^s, ..., a_n^s, .., a_{|A_s|}^s \right\}$. The location of each vehicle is represented using two-dimensional coordinates: $L_s(t) = (L_s^x(t), L_s^y(t))$. In the time interval from t to t', the location $L_s(t') = (L_s^x(t'), L_s^y(t'))$ of SV s at time slot t' is given by

$$L_s^x(t') = L_s^x(t) + d_s cos\theta_s, \tag{1}$$

and

$$L_s^y(t') = L_s^y(t) + d_s sin\theta_s, \tag{2}$$

where d_s is the distance of SV s between time slot t and t', θ_s is the moving angle of SV s.

Task Model. We assume that there are several tasks to be executed on each SV, each task is characterized by input data I_n^s, output data O_n^s, workload W_n^s, and a deadline constraint Δ_n^s. A task is made up of a number of sub-tasks that have particular dependencies on one another are $a_n^s = \{\tau_1, \tau_2, ..., \tau_i, ..., \tau_K\}$. A DAG diagram $G_a(V, l)$, $l = \{(p, q)|p, q \in \tau, p \to q\}$, where V represents the collection of sub-tasks and l as the collection of edges that characterize the interactions between sub-tasks, used to illustrate the dependencies between these K sub-tasks. For instance, the statement (p, q) suggests that sub-task τ_q can only be executed after sub-task τ_p is finished. The workload of each sub-task is expressed as $\omega(\tau_{n,i}^s)$.

Communication Model. Besides, the signal noise ratio between SV s and RSU R is

$$SNR_{s,R}(t) = \frac{P_{s,R}(t) \cdot G_{s,R}(t)}{\epsilon_{s,R}(t) \cdot d_{s,R}(t) \cdot \sigma_{s,R}(t)^2}, \tag{3}$$

where the $d_{s,R}(t)$ is the distance between SV s and RSU R, $P_{s,R}(t)$ is the transmission power, $G_{s,R}(t)$ is the channel gain, $\epsilon_{s,R}(t)$ is the path loss, $\sigma_{s,R}(t)^2$ is the power of the white gaussian noise. Since SVs are moving and their mobility varies with time, the distance between SV s and RSU R is determined as the Euclidean Distance:

$$d_{s,R}^t = \sqrt{(L_s^x - L_R^x)^2 + (L_s^y - L_R^y)^2}. \tag{4}$$

The allocated bandwidth resource between SV s and RSU R is denoted by $b_{s,R}(t)$. Therefore, the data uplink transmission rate and data downlink transmission rate are denoted as

$$V_{s,R}(t) = b_{s,R}(t) \cdot Log_2(1 + SNR_{s,R}(t)), \tag{5}$$

$$V_{R,s}(t) = b_{s,R}(t) \cdot Log_2(1 + SNR_{s,R}(t)). \tag{6}$$

Computation Model. We identify two computation cases, $Y_{n,i}^s(t)$ denotes the sub-task τ_i whether offloading to the RSU (0) or not (1).

Sub-task Local Execution. When a sub-task is selected for local execution, the local execution time is

$$T_{n,i}^{s,L} = \frac{\omega(\tau_{n,i}^s)}{C_s^L}, \tag{7}$$

where C_s^L represents the computation capability at the local.

Offload sub-tasks to the RSUs. A sub-task is offloaded to the RSU for execution and computational reused whenever possible. We utilize the results of a previously executed sub-task that the RSUs has stored rather than starting each sub-task from scratch. Specifically, when a sub-task is offloaded to the RSU, the results of a similar sub-task performed previously can be queried, but if it is not found, the sub-task will be executed from the beginning. Note that each RSU can only process one sub-task at a time. The computation time of RSU is

$$T_{n,i}^{s,R} = \frac{W_{lp} \cdot e_{n,i}^s + (1 - f_{n,i}^s) \cdot \omega(\tau_{n,i}^s)}{C_s^R}, \tag{8}$$

where C_s^R denotes the computation capability at RSU, W_{lp} denotes the lookup workload, $e_{n,i}^s$ indicates whether a lookup operation was conducted before the execution of the sub-task τ_i (1) or not (0). Whether the outcome of sub-task τ_i is attained through computational reuse is indicated by $f_{n,i}^s$ (1) or not (0).

3.2 Problem Formulation

Sub-task Local Execution. If the local execution of the sub-task τ_i occurs, the sub-task τ_i only has the execution time $T_{n,i}^{s,L}$. The earliest finish time of the sub-task τ_i is

$$FT_{s,n,i}^L = max\left\{FT_v^L, FT_v^{dl,R}\right\} + T_{n,i}^{s,L}, v \in pre(i), \tag{9}$$

where v denotes the precursor of sub-task τ_i in the DAG graph. Since the sub-task τ_i can only be executed if all of its predecessor sub-tasks have been finished.

Offload Sub-tasks to the RSU. In the event that the sub-task τ_i is sent from the SV s to the RSU R for execution, its transmit completion time is

$$FT_{s,n,i}^{ul,R} = max\left\{FT_v^L, FT_v^{dl,R}\right\} + T_{s,n,i}^{ul}, v \in pre(i). \tag{10}$$

Execution completion time is

$$FT_{s,n,i}^{exe,R} = FT_{s,n,i}^{ul,R} + T_{s,n,i}^{exe,R}, v \in pre(i). \tag{11}$$

Receive completion time is

$$FT_{s,n,i}^{dl} = FT_{s,n,i}^{exe,R} + T_{s,n,i}^{dl}, v \in pre(i). \tag{12}$$

Meanwhile, we consider the time $T_{s,n,i}^{ul}$ for the data sending from the SV s to the RSU R, the sub-task execution time on the RSU R is $T_{s,n,i}^{exe,R}$, and the time $T_{s,n,i}^{dl}$ represents that the sub-task τ_i execution result is transmitted back to the SV s. The computational execution time of sub-task τ_i executed on the RSU R is

$$C_{s,n,i}^R = (1 - Y_{n,i}^s) \cdot T_{n,i}^{s,R}. \tag{13}$$

Problem Formulation. Our problem is defined as two goals simultaneously, the first is to minimize the completion time of all tasks and the second is to minimize the computational execution time of all RSUs. As the computational reuse rate of RSUs increases, the execution time decreases, and consequently, the completion time of all tasks decreases.

Problem 1:

$$CT = \min \sum_{s=1}^{S} \sum_{n=1}^{|A_s|} (Y_{n,i}^s \cdot FT_{s,n,i}^l + (1 - Y_{n,i}^s) \cdot FT_{s,n,i}^{dl}). \tag{14}$$

Problem 2:

$$ET = \min \sum_{s=1}^{S} \sum_{n=1}^{|A_s|} \sum_{i=1}^{K} (1 - Y_{n,i}^S) \cdot C_{s,n,i}^R, \tag{15}$$

$$s.t. \quad CT \leq \max(\Delta D_n^S), \forall S, \forall n, Y_{n,i}^S \in 0, 1, \tag{16}$$

where CT represents the minimum completion time of all tasks and ET represents the minimum computational execution time of all RSUs, and $max(\Delta D_n^S)$ represents the deadline of the last sub-task for all service vehicles.

4 *TOC* Framework

In the *TOC* framework, we design the service vehicles side and the RSUs side separately. On the RSUs side, we design a computational reuse model that aims to use previously stored computational results as much as possible and reduces the computational execution time of the RSUs. On the service vehicles side, we design a Minimum Weighted Vertex Search Algorithm based on the conflict graph model with the aim of obtaining the best offloading scheme for the SVs.

4.1 Computational Reuse

We present the design of the computational reuse model, which aims to reduce the computational execution time of tasks on the RSUs. This computational reuse model includes mainly *Merged Graph*, *Reusable Graph*, *Reuse Table*, and associated managers.

Merged Graph. In the model of computational reuse, for all received offload sub-tasks, each RSU keeps a *merged graph*, which is generated dynamically based on the received task DAG graphs. We use an adjacency list to represent the *merged graph*. As shown in Fig. 2, each node has two linked lists, namely Edge List and Record List. The *merged graph* is not only used for aggregating the performed sub-tasks at the edge and the dependencies among them, but also for revealing the estimation of reusing the results of these sub-tasks in future computations. When the execution of the SV's last sub-task is finished, the DAG

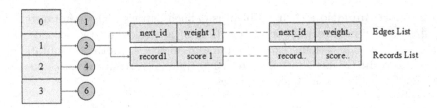

Fig. 2. Merged graph using adjacency list representation.

graph of the task will be sent to the RSUs, and the *worker* adds the DAG graph to the *merged graph*.

Each sub-task in our computational reuse model has a metric called potential reusability, denoted by δ. This metric indicates the likelihood of the computational result of the sub-task being reused in future executions. Furthermore, the edge connecting two sub-tasks (p, q) has a weight that represents the number of times the output of sub-task p is utilized as the input of sub-task q. Each sub-task has a set of records, which are represented as $Record_\tau = \{record_\tau^1, ..., record_\tau^m\}$. A record denotes a previous execution of a sub-task with I_τ^r as the input data and O_τ^r as the output data. Each record has a rate $rd(r_\tau)$ that indicates the number of times that sub-task τ has been used with input I_τ^r. Additionally, each record in the *merged graph* is given a score $score(r)$ that represents the total influence of the record on the other records, represented as $\sqrt{w(\tau) \cdot rd(r_\tau)}$. The potential reusability metric is

$$\delta = \frac{\sum_{r \in Record_\tau} rd(r_\tau)}{|Record_\tau|}, \tag{17}$$

where $\sum_{r \in Record_\tau} rd(r_\tau)$ is the presence of sub-task τ in all the received tasks' DAG, $|Record_\tau|$ represents the number of different input records of sub-task τ.

Reusable Graph. A sub-graph from the *merged graph* is *reusable graph*, which holds the sub-tasks that the RSUs can highly reuse when performing sub-tasks in the future. We use *Gain* to identify the reusability criterion of sub-task τ, expressed as $\sqrt{w(\tau) \cdot \delta_\tau}$, ensuring a proper balance between the potential reusability of a sub-task and its workload. Extracting the *reusable graph* from the *merged graph* depends on $g - thres$ and $\rho - thres$, which represent the minimum gain values of vertices and edges. In order to be part of the *reusable graph*, these thresholds can be dynamically adjusted by each RSU based on its available storage resources.

Reuse Table. The *reuse table* serves as a repository for sub-tasks records that may be reused in the future. In order to achieve computational reuse, RSUs search the *reuse table* for past computation results. The table's size has a strong impact on lookup operation costs. A large table could maximize the potential of computational reuse, but at the same time make lookup operations expensive and lengthy. Regardless of the storage capacity at the edge, the size of the *reuse table* should be limited, and make the best use of this limited size. To address this issue, the *reuse table* is limited based on the storage capacities of the RSUs. In this experiment, we used a fixed size *reuse table*, the specific size will be mentioned in Table 1. The *min_score* is saved to filter out records with high computational reuse rates.

Algorithm 1. *Computational_Reuse_DFS*

Input:
       Incoming task DAG, vertice root

1: **if** $(root.visited = false)$ **then**
2:    index = G[root].index
3:    **if** (index = null) **then**
4:       index = MG.AddVertices(root)
5:    **end if**
6:    v = MG.getVertices(index)
7:    record=MG.FindSameRecord(v)
8:    **if** (record != null) **then**
9:       record.rd ++
10:      UpdateScore(r)
11:      **if** (v.selected = true) **then**
12:         **if** $(record.score \geq min\_score$
              $\&\&record.selected = false)$ **then**
13:           SubmitToTable(record)
14:           v.selectedRecords ++
15:         **end if**
16:      **end if**
17:    **else**
18:       AddRecord(v, record)
19:    **end if**
20:    Check Vertice (v)
21:    **for** $(w = root.Neighbor)$ **do**
22:       **if** w.visited = false **then**
23:          *Computational_Reuse_DFS(G, w)*
24:       **end if**
25:    **end for**
26: **end if**

Because the incoming tasks are dynamic, timely update operations (vertices or edges, and their weights) are required on the RSUs. We use *worker* for the dynamic management of the *merged graph*, as described in Algorithm 1. We use the DFS algorithm for depth-first traversal of the DAG graph, where the *worker* explores all vertices starting from the root of the received DAG graph, using an index to label a sub-task(line 4). Note that two sub-tasks requesting the same service are recorded with the same index, and the same sub-task with different input data is recorded with a different record. After locating the index of a vertex, the *worker* should search the *merged graph* for any records that have the same input as the current sub-task (line 7). If there is no record of the current sub-task's input data in the *merged graph*, the *worker* should update the *merged graph* by adding a new record (line 18). Furthermore, if the record has already been saved, the rate of the record should be increased accordingly (Line 9), and the score of the record needs to be updated (line 10).Then, we call Algorithm 2 to check the vertice.

Algorithm 2. *Check_Vertice*

Input:

vertice v

1: $\delta_v = \frac{\sum_{r \in Record_v} rd(r_\tau)}{|Record_v|}$

2: $g_v = \sqrt{\omega(v) \cdot \delta_v}$

3: $g_{thres} = \frac{g_{thres} + g_v}{2}$

4: **if** $(gv \geq g_{thres} \&\& v.selected = false)$ **then**

5:    **for** record **do**

6:        SubmitToTable(record)

7:        **if** (record.selected = true) **then**

8:            v.selectedRecords ++

9:        **end if**

10:    **end for**

11:    **if** $(v.selectedRecords \geq 0)$ **then**

12:        v.selected = true

13:    **end if**

14: **end if**

In Algorithnm 2, the potential reusability and the gain of the sub-task are calculated (line 1 and line 2). The value of *g_thres* indicates the minimum gain required for the sub-tasks to be submitted to the *reusable graph*. This threshold is dynamically adjusted over time and is calculated as the average between the gain of the received sub-task and the previous threshold (line 3). If the gain of the sub-task exceeds the latest threshold, it will be submitted to the *reuse table* (line 6). After each update, the *reuse table* stores the lowest score in the table, denoted as *min_score*. If a new record arrives with a score greater than *min_score*, it will replace the lowest score in the table and become the new *min_score*. Finally, the *worker* continue to explore the adjacent vertices in the DAG graph (line 23). Figure 3 shows an example of adding an incoming DAG graph to the *merged graph*. Existing merged graph refers to the *merged graph* stored in the RSU recently. When the RSU receives a new DAG, the *worker* will traverse it starting from the root node. For example, in Fig. 3, node 1 is shown to have an input data of 4. The *worker* will then search the existing merged graph to find if it contains a similar sub-task with the same input data record. If such a record is found in the existing merged graph, the score of that record will be incremented by 1, and the new edge nodes will be added to the edge list. During the traversal process, if the value of the node's g_v is greater than g_{thres}

and the node has not been visited before, then all its records will be sent to the *reuse table*. The decision of whether to add these records to the *reuse table* is based on the *min_score* criterion. If all the records of a particular node are not present in the *reuse table*, then that node will be removed from the *merged graph*. Additionally, any edge nodes that have saved connections with this node will also be deleted. Next, the *worker* attempts to visit the second node. If the *worker* cannot find the second node saved in the existing merged graph, it will add a new node to the existing merged graph with an index of 4 to store node 2. Additionally, it will save a record with input data of 7 for this newly added node. Next, the *worker* traverse the other nodes in turn.

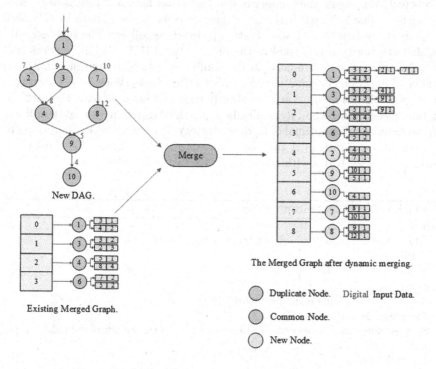

Fig. 3. Dynamically add DAG graph to merged graph.

4.2 Minimum Weighted Vertex Search Algorithm

On the service vehicles side, we propose the conflict graph model that represents the offloading conflict. The conflict graph is an undirected graph with two values for each vertex, representing the association between the RSU and the sub-task. The proposed conflict graph is constructed as follows:

Vertices Set. The size of the vertex set is $|K| \cdot |R|$, and each vertex V_{ij} represents the offloading of sub-task τ_i to RSU j. To select the offloading decision, we give each vertex in conflict graph a weight Wp, which is defined as Wp_{ij} and $Wp_{ij} = T_{s,n,i}^{ul} + T_{s,n,i}^{exe,R} + T_{s,n,i}^{dl} + d_{s,R}^{t}$.

Scheduling Conflict Edges. Furthermore, for any two vertices V_{ij} and V_{mn} in the conflict graph, if i or j is equal to m or n, they will be set adjacent by a conflict edge.

The Algorithm 3 prioritizes each sub-task in descending order by $\Delta_i + F_i$ (line 4), where F_i represents the number of sub-tasks that have antecedent dependencies with τ_i. The sub-task with the highest priority is the first to be offloaded to the most optimal RSU. Each iteration proceeds as follows: The sub-task with the highest priority is offloaded to the most optimal RSU, and all vertices that are adjacent to it are eliminated in the conflict graph. This continues until the number of iterations is equal to the number of sub-tasks (lines 5–9). Given that the service vehicles operate in a constantly changing environment, we track the location of each vehicle and dynamically apply the Algorithm 3 every 10 milliseconds to determine an optimal offloading strategy for the sub-tasks. Additionally, once the last sub-task of a task is executed, the DAG graph for the task is transmitted to all RSUs within the region for future reference.

Algorithm 3. Conflict Graph offloading

Input:
    The Directed Acyclic Graph $\boldsymbol{DAG}$, Conflict Graph $\boldsymbol{CG}$, $wpij$
Output:
    Independent Set Ω

1: Initialization:
2: Set the selected independent set Ω=NULL
3: Calculate $\Delta_i = W_{i1}$ and $\boldsymbol{F_i}$
4: Sort the sub-tasks in descending order according to the weight of data $\Delta_i + F_i$, put in the array $\zeta[K]$
5: **for** p=1 to K **do**
6:     $V_{ij}(p) = min_{j \in R} Wp_{ij}(p)$
7:     add $V_{ij}(p)$ to Ω
8:     $G = G$ without $V_{ij}(p)$
9: **end for**

---

5   Experiment and Evaluation

We evaluate TOC by using the Alibaba cluster dataset in this study. The dataset provides us with the DAG graphs of tasks, the computation capabilities required for sub-tasks, instances of sub-tasks included, and types of services required for sub-tasks, among other things. To implement the TOC framework, we use Java's

JgraphT library, which covers mathematical graph theory along with a collection of sophisticated graph algorithms. Our experiments are conducted on a machine running Windows 10, with an Intel Core i7-7500U processor and 8GB of RAM. Table 1 displays our experiment parameters. We compare the performance of *TOC* with the following algorithms.

- Greedy. This policy adopts a greedy approach to select RSUs for sub-tasks that require immediate execution, with the RSU chosen to be the one in closest proximity. To enhance the efficiency and timeliness of task execution, this algorithm prioritizes proximity over other factors [17].
- Random. This policy dynamically and randomly determines whether a sub-task should be offloaded and which RSU to use.

Moreover, we also consider whether to add the computational reuse model of RSUs and obtain six experimental scenarios based on these three algorithms.

Table 1. Experiment parameters.

Items	Parameters
Number of RSUs	4
Number of SVs	20
Bandwidth of RSUs	20 MHz [18]
Gauss white noise power	-174 dBm/Hz [19]
Computation capacity of SVs and RSUs	0.2,1GHz [19]
Input range	4–16
Lookup workload	5
Reuse table size(RG size)	300 records
Threshold	Dynamic

5.1 Effects of Different Numbers of Tasks on Completion Time and Computational Execution Time on RSUs

In the first experiment, we simulate an intersection with four RSUs evenly distributed, where 20 cars are in motion. Each vehicle continuously generates tasks with varying workloads and dependencies. The objective of this experiment is to investigate how the total completion time of all the tasks and the computational execution time of all RSUs change as the number of tasks carried by service vehicles increases from 20 to 70.

As shown in Fig. 4, as the number of tasks increases, the completion time of the scenarios without the computational reuse model is significantly higher than the scenarios with the computational reuse model. The Algorithm 3, the greedy algorithm and the random algorithm all show steady growth when the

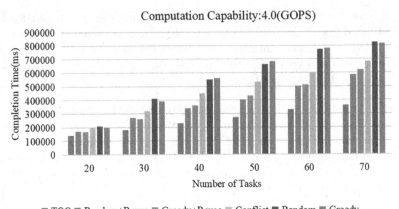

Fig. 4. Effects of different numbers of tasks on completion time.

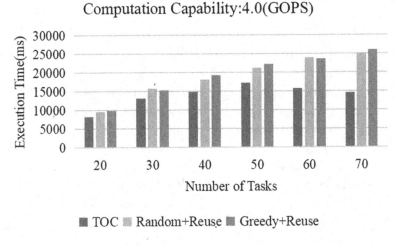

Fig. 5. Effects of different numbers of tasks on computational execution time on RSUs.

computational reuse model is employed. However, the TOC framework grows much slower, and when the number of tasks reaches 70, the completion time decreases by 38.11%, 41.6%, 46.91%, 58.65%, and 58.4% for the TOC compared to the other scenarios, respectively.

The figure displayed in Fig. 5 demonstrates the variation in the computational execution time of the RSUs as service vehicles carry varying numbers of tasks. It is evident that in the case of greedy and random algorithms, the computational execution time of the RSUs gradually increases. On the other hand, in the TOC, the computational execution time gradually increases until the task number reaches 50, after which it decreases. This is due to the fact that the Algorithm 3 is superior to the other two algorithms, allowing for an increased computational reuse rate on the RSUs.

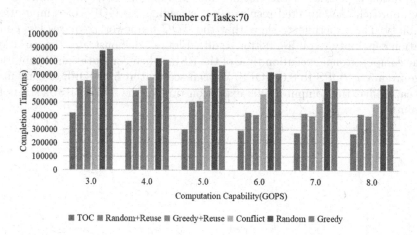

Fig. 6. Effects of different computation capabilities on completion time.

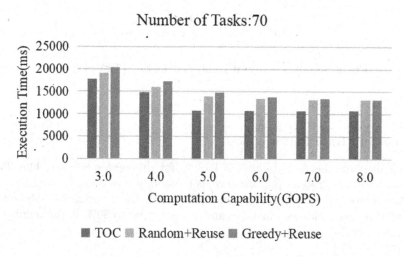

Fig. 7. Effects of different computation capabilities on computational execution time.

5.2 Effects of Different Computation Capabilities on Completion Time and Computational Execution Time on RSUs

In the second experiment, we simulate a set of RSUs with computation capabilities ranging from 3.0 GOPS to 8.0 GOPS [18]. Figure 6 shows how the completion time of the different scenarios varies as the computation capabilities of the RSUs increases, with random and greedy slowly decreasing due to randomness and blindness. The *TOC* framework achieves the expected results when the computation capacities of the RSUs is 5.0 GOPS. The other algorithms gradually approach the expected results after exceeding 7.0 GOPS of computation capability. The experiments show that the *TOC* framework requires less computation capability to obtain better results. Figure 7 demonstrates the variation in the computational execution time of the RSUs when computation capabilities range from 3.0 GOPS to 8.0 GOPS. We can see that at 5.0 GOPS of computation capability, the computational execution time of the *TOC*'s RSUs starts to remain flat.

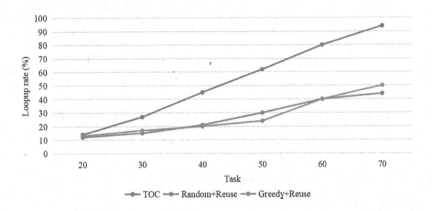

Fig. 8. Effects of different numbers of tasks on lookup rate.

5.3 Lookup Rate Performance

Figure 8 displays the probabilities of lookup for different scenarios. The *TOC* framework shows a robust performance as its lookup rate consistently increases with the number of tasks, ranging from 15% to 92%. On the contrary, the lookup rates of other scenarios are unstable and remain close to 50% as the number of tasks increases.

6 Conclusion

In this paper, we introduce a framework that enables RSUs to reuse past computations when carrying out interdependent tasks. The proposed framework, known as *TOC*, was tested by examining the system model using actual datasets. We are able to obtain optimal performance by using an offloading algorithm based on a conflict graph model to obtain offloading schemes and employing efficient computational caching and lookup mechanisms on the RSUs. Simulation results show that the proposed TOC provides performance close to the optimal solution and reduces the task completion time. Despite the achievements of this work in computational multiplexing of RSUs, the high-speed mobility of vehicles is an unavoidable problem. We will strive to include consideration of vehicle's mobility-aware offloading in our future work, which can be a great challenge.

References

1. Su, M., Cao, C., Dai, M., Li, J., Li, Y.: Towards fast and energy-efficient offloading for vehicular edge computing. In: 2022 IEEE 28th International Conference on Parallel and Distributed Systems (ICPADS), pp. 649–656. IEEE (2023)
2. Shakarami, A., Ghobaei-Arani, M., Shahidinejad, A.: A survey on the computation offloading approaches in mobile edge computing: a machine learning-based perspective. Comput. Netw. **182**, 107496 (2020)
3. Tan, K., Feng, L., Dán, G., Törngren, M.: Decentralized convex optimization for joint task offloading and resource allocation of vehicular edge computing systems. IEEE Trans. Veh. Technol. **71**(12), 13226–13241 (2022)
4. Bellal, Z., Nour, B., Mastorakis, S.: CoxNet: a computation reuse architecture at the edge. IEEE Trans. Green Commun. Networking **5**(2), 765–777 (2021)
5. Al Azad, M.W., Mastorakis, S.: The promise and challenges of computation deduplication and reuse at the network edge. IEEE Wirel. Commun. **29**(6), 112–118 (2022)
6. Tang, J., Li, X., Jin, M., Lu, Y.: A mobility aware task offloading scheme for vehicle edge computing. In: 2021 13th International Conference on Wireless Communications and Signal Processing (WCSP), pp. 1–5 (2021)
7. Shen, Q., Hu, B.J., Xia, E.: Dependency-aware task offloading and service caching in vehicular edge computing. IEEE Trans. Veh. Technol. **71**(12), 13182–13197 (2022)
8. Feng, W., Lin, S., Zhang, N., Wang, G., Ai, B., Cai, L.: C-v2x based offloading strategy in multi-tier vehicular edge computing system. In: GLOBECOM 2022–2022 IEEE Global Communications Conference, pp. 5947–5952. IEEE (2022)
9. Wang, S., Xin, N., Luo, Z., Lin, T.: An efficient computation offloading strategy based on cloud-edge collaboration in vehicular edge computing. In: 2022 International Conference on Computing, Communication, Perception and Quantum Technology (CCPQT), pp. 193–197. IEEE (2022)
10. Lu, Y., Han, D., Wang, X., Gao, Q.: Distributed task offloading for large-scale VEC systems: a multi-agent deep reinforcement learning method. In: 2022 14th International Conference on Communication Software and Networks (ICCSN), pp. 161–165. IEEE (2022)
11. Zhang, Z., Zeng, F.: Efficient task allocation for computation offloading in vehicular edge computing. IEEE Internet Things J. **10**(6), 5595–5606 (2022)
12. Mastorakis, S., Mtibaa, A., Lee, J., Misra, S.: ICedge: when edge computing meets information-centric networking. IEEE Internet Things J. **7**(5), 4203–4217 (2020)
13. Nour, B., Cherkaoui, S.: A network-based compute reuse architecture for IoT applications. arXiv preprint arXiv:2104.03818 (2021)
14. Al Azad, M.W., Mastorakis, S.: Reservoir: named data for pervasive computation reuse at the network edge. In: 2022 IEEE International Conference on Pervasive Computing and Communications (PerCom), pp. 141–151. IEEE (2022)

15. Dass, P., Misra, S.: DeTTO: dependency-aware trustworthy task offloading in vehicular IoT. IEEE Trans. Intell. Transp. Syst. **23**(12), 24369–24378 (2022)
16. Kai, C., Xiao, S., Yi, Y., Peng, M., Huang, W.: Dependency-aware parallel offloading and computation in MEC-enabled networks. IEEE Commun. Lett. **26**(4), 853–857 (2022)
17. Liu, Y., et al.: Dependency-aware task scheduling in vehicular edge computing. IEEE Internet Things J. **7**(6), 4961–4971 (2020)
18. Al-Habob, A.A., Dobre, O.A., Armada, A.G., Muhaidat, S.: Task scheduling for mobile edge computing using genetic algorithm and conflict graphs. IEEE Trans. Veh. Technol. **69**(8), 8805–8819 (2020)
19. Tian, H., et al.: CoPace: edge computation offloading and caching for self-driving with deep reinforcement learning. IEEE Trans. Veh. Technol. **70**(12), 13281–13293 (2021)

Distributed Task Offloading for IoAV Using DDP-DQN

Xiting Peng[1,3] ⓘ, Wenjie Li[1] ⓘ, Xiaoyu Zhang[2,3](✉) ⓘ, Mianxiong Dong[4] ⓘ,
Kaoru Ota[4] ⓘ, and Shun Song[1] ⓘ

[1] School of Information Science and Engineering, Shenyang University of Technology,
Shenyang 110870, China
xt.peng@sut.edu.cn, {liwenjie929,songshun626}@smail.sut.edu.cn
[2] School of Artificial Intelligence, Shenyang University of Technology,
Shenyang 110870, China
xy.zhang@sut.edu.cn
[3] Shenyang Key Laboratory of Information Perception and Edge Computing,
Shenyang, China
[4] The Department of Information and Electronic Engineering,
Muroran Institute of Technology, Muroran 050-8585, Japan
{mx.dong,ota}@csse.muroran-it.ac.jp

Abstract. The swift progress of the Internet of Vehicles (IoV) and
autonomous driving technology has facilitated the emergence of the
Internet of Autonomous Vehicles (IoAV). If delay-sensitive vehicle tasks
are not completed on time, it will lead to bad consequences for IoAV.
Task offloading technology can solve the problem that the vehicle cannot
meet the task requirements. However, highly dynamic vehicle networks
and diverse vehicle applications require more intelligent task offloading
strategies. Therefore, this paper addresses the distributed task offloading problem in the IoAV to meet diverse vehicle task demands. First,
we model the vehicle task offloading problem as a decision problem,
and a deep reinforcement learning (DRL) algorithm named DDP-DQN
(double-dueling-prioritize-DQN) is applied to complete vehicle tasks
more efficiently. Then, we design a reward function to complete the
task within the acceptable maximum delay of the task while reducing
the consumption of resources. Simulations demonstrate the outperforming of the DDP-DQN compared with other three reinforcement learning
algorithms.

Keywords: Internet of autonomous vehicles · Task offloading · Vehicle
edge computing · Deep reinforcement learning

This study was supported by the "Chunhui Plan" Cooperative Research forthe Ministry
of Education (HZKY20220407), the Natural Science Foundationof Liaoning Province
(Grant No. LJKZ0136), part of National Natural ScienceFoundation of China (Grant
No. 62001313), the National Key Research and Development Program of China (Grant
No. 2022YFE011400), the Applied Basic Research Program of Liaoning Province
(Grant No. 2022JH2/101300246).

© The Author(s), under exclusive license to Springer Nature Singapore Pte Ltd. 2024
Z. Tari et al. (Eds.): ICA3PP 2023, LNCS 14492, pp. 283–298, 2024.
https://doi.org/10.1007/978-981-97-0811-6_17

1   Introduction

The Internet of Vehicles (IoV) is a prime example of how the Internet of Things (IoT) technology has been applied to Intelligent Transportation Systems (ITS) [1]. IoV not only enhances the efficiency of transportation but also provides users with excellent driving experience. As autonomous driving technology continues to advance and vehicle communication improves, massive autonomous vehicles have been integrated into IoV, resulting in the emergence of a novel concept known as the Internet of Autonomous Vehicles (IoAV). The applications in IoAV such as voice recognition, navigation services, vehicle control, augmented reality, virtual reality, online multiplayer gaming, etc.

However, these applications are CPU-bound and delay-sensitive [2], failing to complete these vehicle tasks on time can result in a negative driving experience for passengers and have adverse effects on the IoAV. Due to the limited computing resources within the vehicle, tasks are offloaded to devices with greater computational capabilities. Nevertheless, offloading the entire task can result in prolonged transmission times and hinder task completion. Consequently, a potential solution is to divide complex vehicle tasks into smaller sub-tasks and offload them separately. One possible approach is to use mobile cloud computing (MCC) [3], although the cloud has powerful computing capacity, it may not meet the stringent requirements of vehicle applications due to factors such as transmission distance, costs, latency, and possible security issues. In contrast, Mobile edge computing (MEC) [4] is considered to be a promising solution as it can largely reduce latency and expand the computing capabilities of vehicles. Vehicular edge computing (VEC), which is a key application of MEC in IoAV scenarios, has garnered attention from numerous scholars.

In vehicular edge networks, MEC servers are usually deployed in roadside units (RSU) located on the sides of the road. Vehicles can transmit tasks to MEC servers through vehicle-to-infrastructure (V2I) [5] communication, but excessively offloading tasks to RSU can lead to server overload, resulting in a variety of negative consequences. The emergence of autonomous vehicles and the rapid advancement of IoV technology is increasing the processing capacity of vehicles. In addition, nearby vehicles with available computing resources can serve as potential offloading destinations, utilizing vehicle-to-vehicle (V2V) [6] communication. As a supplement to V2I mode, V2V mode can effectively utilize idle resources and largely reduce the load of RSU. In conclusion, task offloading plays a crucial role in reducing the delay of vehicle applications, ensuring driving safety, and enhancing passengers' overall driving experience.

In most cases, the task offloading problem is formulated as a decision problem, and using traditional methods(such as game theory and heuristic algorithms) is not only computationally intensive but also unable to complete vehicle tasks efficiently in the highly dynamic IoAV scenario. Fortunately, deep reinforcement learning (DRL) has tremendous advantages when solving decision problems. DRL can learn the optimal policy through the interaction of agents with the environment without any prior knowledge, and secondly, DRL can make full use of neural networks to adapt to more complex problems and still obtain the optimal policy in the presence of a large action or state information. These reasons position DRL favorably for the IoAV offloading problem.

Our research is focused on addressing the task offloading problem in IoAV. In this context, mobile vehicles, including traditional and autonomous vehicles, serve as mobile edge servers while RSUs act as fixed edge servers. To tackle this challenge, we applied a deep Q-network (DQN) based vehicle task offloading scheme - DDP-DQN (double-dueling-prioritize-DQN). Specifically, DDP-DQN integrates two DQN variants, Double-DQN (DDQN) and Dueling-DQN, and introduces a prioritized experienced replay (PER) mechanism. This integration effectively leverages the resources of both fixed and mobile edge servers and accomplishes tasks more effectively. Additionally, a well-designed reward function is crucial in DRL as it helps the agent select the optimal policy to maximize the reward. In our work, we have formulated a reward function that aims to minimize costs and energy consumption while ensuring timely task completion.

The main contributions of this paper are as follows:

- We delve into the matter of vehicle task offloading within the IoAV scenario. Our objective is to address this problem by formulating it as a decision problem and subsequently employing a DRL algorithm named DDP-DQN to offer a solution.
- Given the constrained computing and communication resources present in the IoAV, we design a reward function that encompasses task delay, energy consumption, and computation cost. The primary objective is to effectively execute the task within the allowable maximum delay while concurrently minimizing the utilization of resources and cost.
- The scheme was evaluated by simulation experiments. We used the VISSIM simulator to collect data, and the potential of the DDP-DQN algorithm is illustrated in the 6-group experiment.

The rest of paper is organized as follows. Section 2 reviews the related works about vehicle task offloading. Section 3 presents the IoAV model. Section 4 proposes the network model and details of algorithm. Section 5 conducts simulation experiments and analysis. Section 6 summarizes the paper.

2   Related Work

Some scholars try to use traditional methods to obtain task offloading strategies. Zeng et al. [7] designed an improved genetic algorithm to find the optimal offloading server. Furthermore, they have put forth an incentive mechanism that aims to compensate vehicles for their valuable contribution of unused computing resources. Zhu et al. [8] proposed to use the PSO algorithm to obtain the offloading strategy. It is noticed that in the immensely dynamic environment of the IoV, the conventional approaches will encounter considerable obstacles. Consequently, it becomes imperative to discover a decision mechanism for task offloading that can effectively adjust to the dynamic environment.

In recent times, reinforcement learning has garnered significant interest among scholars due to its ability to effectively handle the extensive state and action space found within the dynamic environment of the IoV. Moreover, reinforcement learning has proven to be successful in deriving suitable offloading decisions. Several researchers have explored utilize MEC servers for processing vehicle tasks. Zhan et al. [9] studied the distributed task offloading in expressway scenarios. By constructing a queueing model, introducing relevant transition states, and using the PPO algorithm to learn the optimal decision method. Ning et al. [10] also focused on the issue of vehicle resource allocation. The authors considered using base stations and roadside units to process vehicle requests and establish an optimization framework to find the optimal allocation scheme based on the DDPG. Yi et al. [11] designed a distributed computing model to deal with large-scale data requests of vehicles and reduce resource consumption and used the Dueling-DQN method.

In addition to leveraging edge servers to process tasks, nearby vehicles can also be used to process vehicle task requests. Shi et al. [12] proposed a distributed dynamic pricing scheme for vehicles, a DDQN algorithm is used to maximize the cumulative reward. Chen et al. [13] created a distributed computing model in which complex tasks are decomposed, and a sub-task allocation method is designed using the DDQN algorithm. Their findings indicate that this scheme significantly reduces task execution time. Zhao et al. [14] designed a distributed contract scheme to obtain lower resource consumption, reduce allocation, and conflict incentive vehicles to contribute more resources by selecting appropriate offloading targets. In [15], a partial offloading method is used to solve the resource offloading issue in the MEC scenario, and a DRL algorithm was used to obtain the best offloading decision. Liu et al. [16] believe that the computing resources owned by vehicles can be used as edge servers to provide offloading services. The author utilizes the DRL algorithm aims to improve the utility of the VEC network. Distributed also has other applications such as Li et al. [17] researched that multiple coflows in a data center network, the authors designed a novel algorithm to reduce the metric-coflow age. In [18], the author proposed a framework aims to solve endpoint placement and coflow scheduling problems, experiments prove the effectiveness of the framework.

A majority of the extant scholarly literature has focused on the investigation of task offloading within the IoV context. However, there is a scarcity of studies that have examined the scenario of IoAV. It is evident that, shortly, the influence of autonomous vehicles on IoV is expected to be more substantial and far-reaching.

3   System Model

In this section, we propose the system overview, transmission model, delay model, and energy consumption model. At the end of this section we describe the problem to be solved.

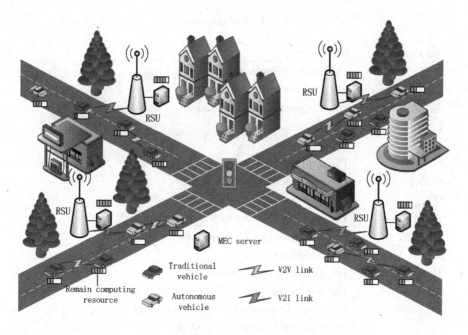

Fig. 1. Task offloading in IoAV.

3.1 System Overview

In the IoAV scenario designed by us in Fig. 1, the RSUs with high-speed communication capability and are equipped with the MEC server. The MEC's rich computing resources can handle the offloading task with plentiful data. The remaining computing resources owned by some vehicles are not enough to meet the needs of the vehicle tasks, and the vehicle must offload part of the tasks to nearby vehicles or RSUs for execution.

The vehicles are divided into two categories, task vehicles, and service vehicles. Some vehicles cannot fit the bill of vehicle tasks owing to insufficient resources, so the tasks need to be offloaded to nearby vehicles, which are called task vehicles (TV). We named the vehicles that can provide offloading service for TVs as service vehicles (SV), which are consisted of traditional vehicles and autonomous vehicles. The difference between traditional and autonomous vehicles is their computing capacity, but both of them can provide offloading services as mobile edge services. Suppose that during task offloading, there are Z service vehicles, among which there are Z_1 traditional vehicles and Z_2 autonomous vehicles, denoted as SV = $\{SV_1^{trad}, SV_2^{trad}, ..., SV_{z_1}^{trad}, SV_1^{auto}, SV_2^{auto}, ..., SV_{z_2}^{auto}\}$. Partial offloading and binary offloading are two ways of vehicle task offloading. Binary offloading means that the task can only be transmitted in one data block or not offloaded, while partial offloading can further split the complex vehicle request into sub-tasks with no intrinsic connection and assign them to VEC devices for execution. The tasks of the vehicle have different attributes.

Table 1. Key notations

Notation	Description
Z	The service vehicles index
Z_1	The traditional vehicles index
Z_2	The autonomous vehicles index
T_n	The properties of tasks
D_n	Data size
C_n	The required CPU cycles
t_n^{max}	The acceptable maximum delay
t_{delay}^n	The task completion time
R_{V2I}, R_{V2V}	The transmission rate of V2I mode and V2V mode
B_{V2I}, B_{V2V}	The bandwidth in V2I modes and V2V modes
p_t	Transmission power
P_{loc}	The local processing power of TV
P_{RSU}, P_{SV}	The processing power of RSU and SV
N_{V2I}, N_{V2V}	The power of noise in V2I mode and V2V mode
$d_{TV,RSU}$	The distance between TV and RSU
$d_{TV,SV}$	The distance between TV and SV
f_{loc}, f_{rsu}, f_{sv}	The computing resources of TV, RSU, and SV
t_{loc}	The total latency for local processing
t_{delay}^{rsu}	The total latency for RSU processing
t_{delay}^{sv}	The total latency for SV processing
E_{loc}, E_{RSU}, E_{SV}	The total energy consumption for local, RSU, SV processing
ι	The offloading ratio

Suppose that there are s tasks need to be processed, and use a triplet $T_s = \{D_s, C_s, t_s^{max}\}$ to represent the properties of tasks, where D_s is the data size, C_s is the CPU cycles, t_s^{max} is the acceptable maximum delay. We have summed up the key notations in Table 1.

3.2 Transmission Model

When a vehicle offloads tasks to an RSU through V2I mode or to a service vehicle through V2V mode, the transmission delay of tasks mainly includes sending latency and return latency. Usually, the results back time can be overlooked. The transmission rate of V2I mode is given by

$$R_{V2I} = B_{V2I} \log_2 \left(1 + \frac{p_t * |h|^2}{N_{V2I} * \left(d_{TV,RSU} \right)^v} \right) \tag{1}$$

The transmission rate of V2V mode is given by

$$R_{V2V} = B_{V2V} \log_2 \left(1 + \frac{p_t * |h|^2}{N_{V2V} * \left(d_{TV,SV} \right)^v} \right) \tag{2}$$

where p_t represents the transmission power, N_{V2I} and N_{V2V} are the power of noise in different modes, B_{V2I} and B_{V2V} are the bandwidth in V2I and V2V modes respectively, $d_{TV,RSU}$ is the distance between TV and RSU, $d_{TV,SV}$ is the distance between TV and SVs, v and h are relevant parameters.

3.3 Task Delay

If task is not offloaded, the delay is given by

$$t_{loc} = \frac{(1 - \iota) * C_s}{f_{loc}} \tag{3}$$

where f_{loc} is the computing ability of task vehicle, ι is the offloading ratio.

When the task selects the VEC server as the execution location, the delay contains transmission and calculation delay.

$$t_{delay}^{rsu} = t_{procrsu} + t_{trans}^{V2I} = \frac{\iota * C_n}{f_{rsu}} + \frac{\iota * d_n}{R_{V2I}} \tag{4}$$

$$t_{delay}^{sv} = t_{procsv} + t_{trans}^{V2V} = \frac{\iota * C_n}{f_{sv}} + \frac{\iota * d_n}{R_{V2V}} \tag{5}$$

where t_{trans}^{V2I} is the transmission time in V2I mode, and t_{trans}^{V2V} is the transmission time in V2V mode, f_{rsu} and f_{sv} represent the computing ability of the RSUs and SVs respectively.

3.4 Energy Consumption Model

When processing vehicle tasks, it needs to consume a certain amount of energy. If the task is not offloaded, there is no transmission energy consumption involved. Local processing energy consumption is represented as follows:

$$E_{loc} = P_{loc} * t_{loc} \tag{6}$$

If the task selects the VEC server as the execution location, the consumption of task offloading is given by

$$E_{RSU} = P_t * t_{trans}^{V2I} + P_{RSU} * t_{procrsu} \tag{7}$$

where P_{loc} is the local processing power.

$$E_{SV} = P_t * t_{trans}^{V2V} + P_{SV} * t_{procsv} \tag{8}$$

where P_{RSU} and P_{SV} represent the RSU processing power and the service vehicle processing power respectively.

3.5   Formulation of Task Offloading Problem

Our objective is complete the task within the acceptable maximum delay, which is expressed as follows:

$$\max \frac{1}{N} \sum_{n=1}^{N} \sum_{O=1}^{O} \iota_o^n * (\log_2(1 + (t_n^{max} - t_{delay}^n)) - E_{sv} - d_n * r_v) - (1 - \iota_o^n)$$
$$* (\log_2(1 + (t_n^{max} - t_{delay}^n)) - E_{loc}) - \iota_o^n * (\log_2(1 + (t_n^{max} - t_{delay}^n)) \qquad (9)$$
$$- E_{rsu} - d_n * r_u)$$

$$s.t. \quad \begin{cases} C1 : \iota_o^n \in [0,1], \\ C2 : t_n^{max} \geq t_{delay}^n, \forall n \in N. \end{cases}$$

C1 represent the offloading ratio. C2 guarantees that the total delay at most the acceptable maximum delay. Using traditional scheme is difficult to solve such problem. So we use a DDP-DQN algorithm to solve it.

4   DDP-DQN Algorithm in IoAV Offloading

4.1   IoAV Environment Model

In our design, the vehicle terminal is the agent to decide where the task is executed, and the task can no offloaded or partial offloaded to VEC servers. At the same time, the VEC servers and the vehicle's processing unit will inform the vehicle terminal of the current remaining computing resources. We can devise the S_t, A_t and R_t.

State. The system state is delineated as

$$S_t = \{D_t, C_t, t_t^{max}, f_{loc}, f_{rsu}, f_{trad1}, f_{trad2}, \qquad (10)$$
$$..., f_{tradz_1}, f_{auto1}, f_{auto2}, ..., f_{autoz_2}\}$$

where D_t, C_t, t_t^{max} are the triplet of tasks, f_{tradz_i} is computing resources of traditional vehicle i, f_{autoz_i} is computing resources of autonomous vehicle i. f_{loc} is the remaining computing resources of task vehicle, f_{rsu} is the computing resources of RSU.

Actions. The system action space for vehicle task offloading is given as

$$A_t = \{(o_{auto_1}, \iota), (o_{auto_2}, \iota), ..., (o_{auto_{z_2}}, \iota), \qquad (11)$$
$$(o_l, 1 - \iota), (o_{rsu}, \iota), (o_{trad_1}, \iota), (o_{trad_2}, \iota), ..., (o_{trad_{z_1}}, \iota)\}$$

Vehicle tasks can be separated into the unrelated sub-tasks, and tasks can choose local processing, or offloading to RSU or SV. Where o_l, o_{rsu}, $o_{trad_{z_i}}$, $o_{auto_{z_i}} \in (0,1)$, $o_l = 1$ means vehicle tasks are processed locally, $o_{rsu} = 1$ means task offloaded to RSU, $o_{trad_{z_i}} = 1$ means task offloaded to traditional vehicle i, $o_{auto_{z_i}} = 1$ means task offloaded to autonomous vehicle i, ι is the offloading ratio.

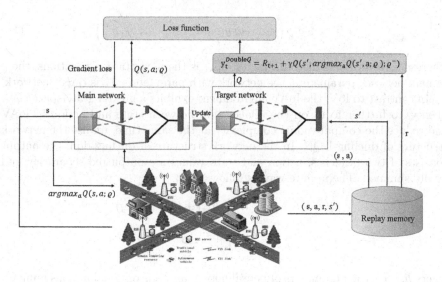

Fig. 2. Architecture of DDP-DQN.

Rewards. Our goal is to jointly optimize the reward function of vehicle task, delay, transmission and computational consumption, and processing cost. The reward R_t is given by
If $t^{max} \geq t_{delay}$,

$$
\begin{aligned}
R_t = K * \log_2(1 + (t^{max} - t_{delay})) - (o_l * E_{loc} + o_{rsu} \\
* E_{rsu} + o_{tradi} * E_{SV} + o_{autoi} * E_{SV}) - d_n * r_n
\end{aligned}
\tag{12}
$$

If $t^{max} < t_{delay}$,

$$
R_t = T^- \tag{13}
$$

where t^{max} is the acceptable maximum delay, t_{delay} is the total delay, K is a positive constant. As shown in action space, o_l, o_{rsu}, $o_{trad_{z_i}}$, $o_{auto_{z_i}}$ are flag bits, represents where the task is offloaded, E_{loc}, E_{rsu}, E_{SV} are energy consumption, r_n means the cost of processing 1MB task, if task choose local processing, $r_n = 0$. $T^- = -1000$, represents the punishment for exceeding the task's maximum tolerable delay. There is also a penalty for exceeding the maximum communication range during offloading.

4.2 DDP-DQN Based Task Offloading Algorithm

The traditional DQN approach tends to use the greedy strategy of choosing the action that maximizes the next reward, which inevitably leads to overestimation. In order to better solve this problem, we use DDQN method to calculate the target Q value so as to overcome the overestimation problem in DQN, the Q-values are updated as follows:

$$y = r_k + \delta Q\left(s_{k+1}, \underset{a}{argmax} Q\left(s_{k+1}, a; \varrho\right), \varrho^-\right) \tag{14}$$

where r is reward, δ is discount factor. s_{t+1} is the next state, a is actions, the ϱ is main network parameters, the network with parameters ϱ^- is target network.

In contrast to IoV, the IoAV environment exhibits a greater state-space complexity, to further focus on the relationship between state and action in IoAV and reduce the computational complexity at the same time, we use the network structure of dueling-DQN. In the network structure of dueling-dqn, the output consists of two branches, action and state values are outputted separately and finally summed. The update way as follows:

$$Q(s, a, \hbar, \imath, \jmath) = W(s, \hbar, \imath) + (T(s, a, \hbar, \jmath)$$
$$- \frac{1}{|\Lambda|} * \sum_{a' \in \Lambda} T(s, a', \hbar, \jmath)) \tag{15}$$

where $\hbar, \imath, \jmath$ is the parameters of neural networks, $|\Lambda|$ is optional actions number, $W(s, \hbar, \imath)$ is only related to s, $T(s, a, \hbar, \jmath)$ is related to s and a.

Finally, we use PER mechanism to improve the utilization of significant experience. In PER, the TD error measures the magnitude of the experience value. The greater the TD error, the higher the correlation, and the greater the impact on back-propagation, indicating the higher sample importance. When the sampling probability is higher for the samples with larger absolute TD error values, the algorithm tends to converge more easily. The TD error Ω is

$$\Omega = r + \gamma Q_{target}\left(s_{k+1}, a_{k+1}, \varrho^-\right) - Q\left(s_k, a_k, \varrho\right) \tag{16}$$

The structure of DDP-DQN algorithm is shown in Fig. 2.

DDP-DQN contains several episodes. First, vehicle terminal obtains the s_1 from IoAV, selects an action in each step using the ε-greedy strategy, and calculates the reward value R. Then, the state changes from s_k to s_{k+1}. The quadruple (s_k, a_k, R_k, s_{k+1}) is saved in the prioritized replay buffer. Then, vehicle terminal samples experiences from the prioritized replay buffer to update the ϱ and ϱ^-. In every N step, the ϱ copies itself to the ϱ^-. Algorithm 1 illustrate the DDP-DQN algorithm.

5   Performance Evaluation

We conduct experiments and verify the DDP-DQN algorithm. First, we describe the simulation scenario and relevant experimental parameters and then evaluate the simulation results.

Algorithm 1. DDP-DQN based Task Offloading Algorithm

Input: States s_k, action a_k.
Output: Reward R_k.
Initialize prioritized experience buffer $\wp$, the weight ϱ, the weight $\varrho^- = \varrho$.

1: **for** each episode **do**
2:　　Collect observation s_1 from IoAV.
3:　　**for** step k=1,2,...K **do**
4:　　　　Choose s with random probability.
5:　　　　**if** s < ε **then**
6:　　　　　　Random select a_k.
7:　　　　**else**
8:　　　　　　Choose action
9:　　　　　　$a_k = argmax\limits_{a} Q\left(s_k, a; \varrho\right).$
10:　　　　The state transits from s_k to s_{k+1} and save
11:　　　　(s_k, a_k, R_k, s_{k+1}) into D.
12:　　　　Sample a batch of experiences $(s_\rho, a_\rho, R_\rho, s_{\rho+1})$
13:　　　　from $\wp$.
14:　　　　Calculate the target Q-value with
15:　　　　$y_\rho = R_\rho + \delta Q\left(s_{\rho+1}, argmax\limits_{a} Q\left(s_{\rho+1}, a; \varrho\right); \varrho^-\right).$
16:　　　　Update the ϱ by
17:　　　　$L_\varrho = \left(y_\rho - Q\left(s_\rho, a_\rho; \varrho\right)\right)^2$
18:　　　　In every N step,update $\varrho^- = \varrho$.
19:　　**end for**
20: **end for**

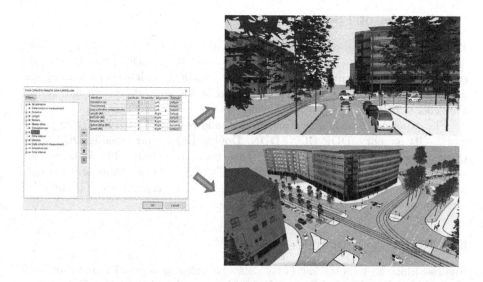

Fig. 3. The scenario of VISSIM simulator

Table 2. Experimental Parameters

Parameter	Value
Data size	10 MB–50 MB
Computation size of task (10^8 cycles)	0.5–1
The acceptable maximum delay	2 s–3 s
The bandwidth in V2I modes and V2V modes	10 MHz, 5 MHz
The h and v	2
Transmission power	1 W
The local processing power	2.74 W
The processing power of RSU	10 W
The processing power of traditional vehicle	2.74 W
The processing power of autonomous vehicle	6.4 W
The power of noise in V2I mode	−120 dbm
The power of noise in V2V mode	−144 dbm
The local computational capability	5 MHz–15 MHz
The RSU computational capability	60 MHz–80 MHz
The traditional vehicle computational capability	25 MHz–40 MHz
The autonomous vehicle computational capability	40 Hz–60 MHz
The basic unit price of traditional vehicle	0.1
The basic unit price of autonomous vehicle	0.2
The basic unit price of RSU	0.5

5.1   Experimental Settings

We use the VISSIM simulator to simulate the vehicle edge calculation scenario, and the simulator can load different road types and vehicle types, we can get real-time information including vehicle type, location, vehicle speed, and power. As shown in Fig. 3, we select an intersection as the simulation area and collect the generated data.

The experiment was carried out on a Windows system with i7-6700HQ CPU and 16GB memory. DDP-DQN, DQN, Double DQN, and Dueling DQN algorithms were implemented using the TensorFlow framework. The learning rate was set to 0.01, the discount factor was set to 0.9, and the value of ε was 0.9. Memory size and batch size are set to 500 and 32, respectively. The detailed experimental parameters are shown in Table 2.

5.2   Results and Analysis

Figure 4 illustrates the average task delay for different Z(total number of vehicles). Based on the findings, the DDP-DQN algorithm demonstrated the lowest average delay. The increase in the number of vehicles causes the state space

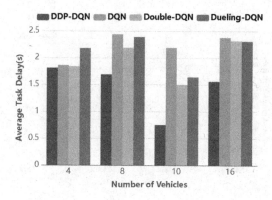

Fig. 4. Average task delay with different Z.

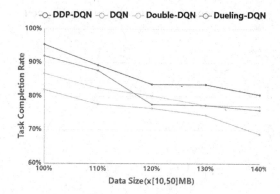

Fig. 5. Task completion rate with different data size.

to become more complex, while our algorithm can still find relatively suitable targets for offloading, thus guaranteeing a low task delay.

In Fig. 5, the task completion rates of various algorithms are compared based on different data sizes(The default data size (100%) is [10,50]MB). Both transmission time and execution time are impacted by the increase in task data. Results indicate that, among the four algorithms tested, the DDP-DQN algorithm has a better performance.

The plot in Fig. 6 displays the task completion rates for varying acceptable maximum delays(The default acceptable maximum delays (100%) is (2–3)s). The acceptable maximum delay size is of utmost importance in ensuring timely completion of tasks. Findings indicate that the DDP-DQN algorithm demonstrates superior performance.

The comparison of task completion rates for different bandwidth sizes is shown in Fig. 7. We set the bandwidth of V2V mode(default is 5MHz) from 50% to 130%. The size of the bandwidth directly affects the transmission rate, and thus has a certain impact on the task completion rate. The results surface that DDP-DQN can maintain a high task completion rate under different bandwidth sizes.

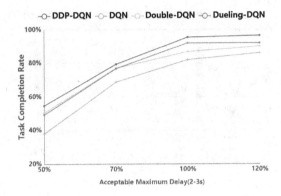

Fig. 6. Task completion rate with different maxdelay.

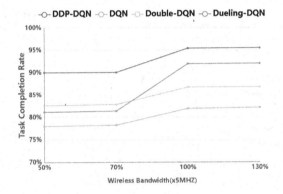

Fig. 7. Task completion rate with different bandwidth.

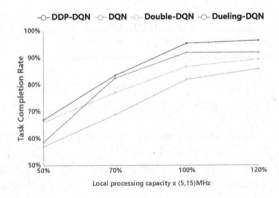

Fig. 8. Task completion rate with different local processing capacity.

In Fig. 8, task completion rates are plotted for different local processing capabilities. Since the paper mainly focuses on partial offloading, local processing capacity becomes a factor in determining the success rate of task completion. As

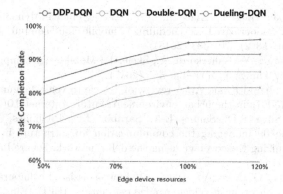

Fig. 9. Task completion rate with different edge devices resources.

local processing capabilities increase, all schemes exhibit an increased completion rate. The DDP-DQN algorithm indicates its superior performance compared to other three algorithms.

In Fig. 9, task completion rates are compared for different edge device capabilities. The processing capacity of the edge equipment is also a critical factor affecting task completion rates. In Figs. 8 and 9, the completion rate of four schemes increases with an increase in the processing capacity of the local and edge devices. Notably, the algorithm employed in this paper achieves a higher task completion rate.

6   Conclusion

In this paper, we focus on distributed vehicle task offloading in the context of IoAV. The problem is formulated as a decision problem and uses the DDP-DQN algorithm to solve it, and we design a reward function that jointly optimizes the delay energy cost under the limited resource constraint. The experiment demonstrates that our algorithm outperforms the other three compared algorithms in terms of delay and task completion rate. In our future work, we will consider more complex IoAV scenarios and reward functions, while we note that in addition to the execution delay of the task, the decision time of the task is equally important. Considering the decision time of the task is also one of our future research directions.

References

1. Wang, J., Jiang, C., Zhang, K., Quek, T.Q., Ren, Y., Hanzo, L.: Vehicular sensing networks in a smart city: principles, technologies and applications. IEEE Wirel. Commun. **25**(1), 122–132 (2017)
2. Uhlemann, E.: Connected-vehicles applications are emerging [connected vehicles]. IEEE Veh. Technol. Mag. **11**(1), 25–96 (2016)

3. Guo, S., Liu, J., Yang, Y., Xiao, B., Li, Z.: Energy-efficient dynamic computation offloading and cooperative task scheduling in mobile cloud computing. IEEE Trans. Mob. Comput. **18**(2), 319–333 (2018)
4. Abbas, N., Zhang, Y., Taherkordi, A., Skeie, T.: Mobile edge computing: a survey. IEEE Internet Things J. **5**(1), 450–465 (2017)
5. Abbasi, I.A., Shahid Khan, A.: A review of vehicle to vehicle communication protocols for VANETs in the urban environment. Future Internet **10**(2), 14 (2018)
6. Silva, C.M., Silva, L.D., Santos, L.A., Sarubbi, J.F., Pitsillides, A.: Broadening understanding on managing the communication infrastructure in vehicular networks: customizing the coverage using the delta network. Future Internet **11**(1), 1 (2018)
7. Zeng, F., Chen, Q., Meng, L., Wu, J.: Volunteer assisted collaborative offloading and resource allocation in vehicular edge computing. IEEE Trans. Intell. Transp. Syst. **22**(6), 3247–3257 (2020)
8. Zhu, C., et al.: Folo: latency and quality optimized task allocation in vehicular fog computing. IEEE Internet Things J. **6**(3), 4150–4161 (2018)
9. Zhan, W., et al.: Deep-reinforcement-learning-based offloading scheduling for vehicular edge computing. IEEE Internet Things J. **7**(6), 5449–5465 (2020)
10. Ning, Z., et al.: Joint computing and caching in 5G-envisioned internet of vehicles: a deep reinforcement learning-based traffic control system. IEEE Trans. Intell. Transp. Syst. **22**(8), 5201–5212 (2020)
11. Ouyang, Y.: Task offloading algorithm of vehicle edge computing environment based on Dueling-DQN. In: Journal of Physics: Conference Series, vol. 1873, p. 012046. IOP Publishing (2021)
12. Shi, J., Du, J., Wang, J., Yuan, J.: Distributed V2V computation offloading based on dynamic pricing using deep reinforcement learning. In: 2020 IEEE Wireless Communications and Networking Conference (WCNC), pp. 1–6. IEEE (2020)
13. Chen, C., Zhang, Y., Wang, Z., Wan, S., Pei, Q.: Distributed computation offloading method based on deep reinforcement learning in ICV. Appl. Soft Comput. **103**, 107108 (2021)
14. Zhao, J., Kong, M., Li, Q., Sun, X.: Contract-based computing resource management via deep reinforcement learning in vehicular fog computing. IEEE Access **8**, 3319–3329 (2019)
15. Truong, T.P., Nguyen, T.V., Noh, W., Cho, S., et al.: Partial computation offloading in NOMA-assisted mobile-edge computing systems using deep reinforcement learning. IEEE Internet Things J. **8**(17), 13196–13208 (2021)
16. Liu, Y., Yu, H., Xie, S., Zhang, Y.: Deep reinforcement learning for offloading and resource allocation in vehicle edge computing and networks. IEEE Trans. Veh. Technol. **68**(11), 11158–11168 (2019)
17. Li, W., et al.: Efficient coflow transmission for distributed stream processing. In: IEEE INFOCOM 2020-IEEE Conference on Computer Communications, pp. 1319–1328. IEEE (2020)
18. Li, W., Yuan, X., Li, K., Qi, H., Zhou, X., Xu, R.: Endpoint-flexible coflow scheduling across geo-distributed datacenters. IEEE Trans. Parallel Distrib. Syst. **31**(10), 2466–2481 (2020)

A Task Offloading and Resource Allocation Optimization Method in End-Edge-Cloud Orchestrated Computing

Bo Peng, Shi Lin Peng, Qiang Li[✉], Cheng Chen, Yu Zhu Zhou,
and Xiang Lei

School of Information Engineering, Southwest University of Science and Technology,
Mianyang, China
liqiangsir@swust.edu.cn

Abstract. The limited resources of mobile devices (MDs) pose an emerging requirement, resulting in its essential to reducing task processing latency and energy consumption on MDs with efficient task offloading and scheduling strategies. In this paper, we aim to minimize the weighted sum of the task processing time and energy consumption of MDs in end-edge-cloud orchestrated computing (EECOC). To solve the non-convex problem caused by joint optimization and multiple constraints, a task offloading and resource allocation method based on deep reinforcement learning (DRL) is proposed. The proposed algorithm adopts a hierarchical structure, where the upper layer employs game theory to determine task offloading strategies through a competitive game among MDs. The lower layer leverages the proximal policy optimization (PPO) approach to optimize the channel bandwidth and computation capability problem of servers. We conducted multiple experiments in diverse EECOC scenarios to evaluate the performance of our proposed approach. Experimental results demonstrate that the proposed method outperforms traditional offloading algorithms and effectively reduces the task processing time and energy consumption of MDs.

Keywords: End-edge-cloud orchestrated computing · Task offloading · Resource allocation · Proximal policy optimization

1 Introduction

The previous decade has been marked by the swift advancement of mobile devices (MDs) [1], along with massive novel applications, e.g., face recognition and virtual reality [2,3]. In general, these applications place significant demands on computation and latency, while MDs are typically resource-constrained in terms of computational capacity and energy resources. Motivated by this, the emergence

---

This work is supported by Heilongjiang Provincial Science and Technology Program (No. 2022ZX01A16) and Sichuan Science and Technology Program (No. 2022ZHCG0001).

© The Author(s), under exclusive license to Springer Nature Singapore Pte Ltd. 2024
Z. Tari et al. (Eds.): ICA3PP 2023, LNCS 14492, pp. 299–310, 2024.
https://doi.org/10.1007/978-981-97-0811-6_18

of mobile edge computing (MEC) has become a promising technology to overcome this dilemma, by offloading MDs' computing tasks to edge servers closer to MDs [4]. Nevertheless, faced with a rapidly increasing number of tasks and a more complex edge environment, edge servers will also be limited by their computing capacity. Therefore, end-edge-cloud orchestrated computing (EECOC) combines the advantages of cloud computing and edge computing [5], enabling tasks to be calculated on edge servers or cloud servers according to their computation and delay requirements to achieve better performance.

Within the EECOC system, the selection of an effective task offloading method holds significance, which is crucial for the task offloading approach to intelligently and efficiently utilize the resources available from MDs, edge servers, and cloud servers. Typically, a task comprises a series of interdependent subtasks, with their dependencies often represented as a directed acyclic graph (DAG) referred to as fine-grained task [6]. By leveraging fine-grained offloading opportunities and effectively minimizing consumption of MD, the key challenges that need to be addressed in this context include:1) How to coordinate the offloading strategy of subtasks in a multi-user and multi-server environment, 2) How to effectively allocate computation and communication resources of servers.

To address the above challenges, A DRL-based task offloading and resource allocation method proposed for EECOC in this paper. Specifically, We first model the task as a DAG. Then we build a multi-user, multi-server EECOC system model and formulate the problem of minimizing task processing time and energy consumption of MDs. Furthermore, we address the problem using the combination of game theory and proximal policy optimization (Game-PPO) algorithm for optimization. The primary contributions of this article can be summarized as follows:

- We built a multi-user, multi-server EECOC system model, defined the task offloading and resource allocation of EECOC as a markov decision process (MDP) problem and further solved it through the Game-PPO algorithm.
- We conduct extensive experiments to evaluate the performances of the proposed approach compared to conventional and baseline methods across diverse EECOC scenarios.

The subsequent sections of this paper are organized as follows. Section 2 reviews related works. Section 3 describes EECOC system model and problem definition. Task offloading and resource allocation algorithm in Sect. 4. Performance evaluation and discussion are given in Sect. 5, and conclusions are presented in Sect. 6.

2   Related Work

With the rapid advancements in MDs and applications, there has been a surge of research in MEC to address the performance demands of various MDs. Among them, An essential area of research involves optimizing the design of task computation offload within the EECOC architecture [7]. Task offloading approaches in EECOC mainly has the following two aspects:1) Traditional algorithms with deterministic rules and mathematical models [8] and 2) Intelligent optimization algorithm represented by heuristic and deep reinforcement learning [9].

Traditional computing offloading algorithms usually use certain mathematical rules or deterministic strategies. A potential game framework for optimizing the offloading strategy is presented [10]. An optimized greedy task offloading method is proposed to effectively address the problem of overloading edge servers during peak hours [11]. As the size of applications for MDs and the EECOC network increases become more complex, traditional optimization approaches may struggle to find feasible solutions.

Researchers had started exploring intelligent optimization algorithms to address the task offloading challenges in EECOC, aimed to optimize the overall performance of EECOC networks. [12] leveraged improved Strength Pareto Evolutionary Algorithm 2 (SPEA2) to minimize energy consumption, computing delay and maximize server resource utilization in EECOC environment. [13] adopted Deep Deterministic Policy Gradient (DDPG) to make optimized decisions of computation offloading, achieving the purpose of minimizing delay.

On the whole, research on EECOC is still limited, with a majority of existing studies primarily focusing on computation offloading optimization while neglecting resource allocation. Besides, the resource-constrained situation for MDs is seldom concerned. On account of that, our paper proposes a task offloading and resource allocation approach based on DRL for resource-constrained MDs to optimize MDs' task processing time and energy consumption.

3   System Model and Problem Formulation

In this section, we describe the EECOC system model with constrained MDs resources first and then formulate the problem of minimizing task processing time and energy consumption of MDs as an optimization problem.

3.1   Network Model

In this paper, the EECOC network model is composed of three layers: end, edge, and cloud, which including multiple MDs $\mathcal{M} = \{1, 2, ...m,..., M\}$, multiple edge servers $\mathcal{N} = \{1, 2, ...,n,...N\}$, and one cloud server. The task of MD is partitioned into some interdependent subtasks, denoted by a set of $\mathcal{V} = \{1, 2, ..., v\}$. We utilize a directed acyclic task graph $G = (V, E)$ to represent the dependency relationships among the subtasks. Furthermore, V denotes the set of subtasks and E represents the edges connecting the tasks. Each edge server's computation capacity is f^{ES}, and the computation capacity of the cloud server is denoted as f^{CS}.

3.2   Communication Model

MDs can execute tasks locally, or they can choose to offload tasks either to nearby edge servers using a wireless local area network (WLAN), or to cloud servers using a wide area network (WAN). In the wireless communication system, the transmission rate is typically determined by the network bandwidth. This work consider the multi-user and multi-server system which may co-exist over a certain area and result in co-channel interference with each other. Therefore, this paper employs Shannon's formula to compute the maximum achievable transmission rate as the task transmission rate of MDs [14]. Specifically, we let γ_m^n denote the data transmission rate between MD m and edge server n, which can be calculated as

$$\gamma_m^n = w_m^n log_2(1 + \frac{p_m g_m^n}{\sigma^2 + N}) \tag{1}$$

where w_m^n denotes the channel bandwidth assigned by edge server n to MD m, p_m is the transmission power of MD and g_m^n is the channel gain between the MD m and the edge server n. σ^2 denotes background noise power and $N = \sum_{m' \in m, m' \neq m} p_{m'} g_{m',n}[D_m^* = D_{m'}]$ means the interference to the wireless channel when other devices are offloaded by the same server as MD m.

3.3 Computation Model

In this section, we discuss time and energy consumption in the calculation process of the task in MD, edge server, and cloud server, respectively.

1) Local computing: When tasks are processed locally, there is a queue delay due to all subtasks being processed sequentially, denoted by $t_{m,i}^{lc,w}$. Let f^m be the computation capability of the MD m, the local execution time of subtask i is given by

$$t_{m,i}^{lc} = \frac{c_{m,i}}{f_m} + t_{m,i}^{lc,w} \tag{2}$$

where $c_{m,i}$ represents the computation frequency required by subtask i of device m. Next, we can calculate the energy consumption for a local subtask as

$$e_{m,i}^{lc} = \xi(f_m)^2 c_{m,i} \tag{3}$$

where ξ represents the chip-dependent computing coefficient of MD m [15].

2) Edge computing: When MDs offload subtask to edge servers for computation, the total latency of subtask in edge server n is achieved by four part: local queue latency, offload latency, server queue latency, and server processing latency. Let f_m^n denote the computation capability allocated by edge server n to device m. The edge computing time $t_{m,i}^n$ of subtask can be expressed as

$$t_{m,i}^n = t_{m,i}^{lc,w} + t_{m,i}^{n,off} + t_{m,i}^{n,w} + \frac{c_{m,i}}{f_m^n} \tag{4}$$

$$t_{m,i}^{n,off} = \frac{d_{m,i}}{\gamma_m^n} \tag{5}$$

where $t_{m,i}^{n,w}$ denotes queue delay of subtasks i at edge server n, $d_{m,i}$ denotes the size of task i for MD m. Therefore, the energy consumption of MD m for offloading subtask i to edge server n can be expressed as

$$e_{n,i}^{n,off} = t_{m,i}^{n,off} p_m \tag{6}$$

3) Cloud computing: Offload task to cloud server for computing, similar to edge server computing. Let f_m^{cloud}, w_m^{cloud} denotes the computation capability and the bandwidth allocated by cloud server to MD m respectively. We can calculate the transmission rate γ_m^{cloud} of MD m using (1).

In particular, there is an additionally fixed transmission latency T due to the long geographical distances between the cloud server and MDs. The cloud computing delay and energy consumption can be expressed as

$$t_{m,i}^{cloud} = t_{m,i}^{lc,w} + \frac{d_{m,i}}{\gamma_m^{cloud}} + t_{m,i}^{cloud,w} + \frac{c_{m,i}}{f_m^{cloud}} + T \tag{7}$$

$$e_{m,i}^{cloud,off} = t_{m,i}^{cloud,off} p_m \tag{8}$$

3.4 Problem Formulation

This work aims to minimize the task processing latency and energy consumption of MDs. In order to handle the task with a series of interdependent subtasks, it is necessary to define the AST (actual start time) and AFT (actual finish time). $AST(m, i)$ is the actual execution start time of subtask i on MD m, And $AFT(m, i)$ is the actual execution finish time of subtask i on MD m. The AFT of subtask i on MD m can be expressed as follows:

$$AFT(m, i) = AST(m, i) + t_{m,i}^{k'} \tag{9}$$

where $t_{m,i}^{k'}$ is the execution time of the subtask i of MD m, and k' indicates the actual execution position of the subtask. Due to limitations of DAG tasks, one subtask can not be executed earlier than its dependent tasks. At the same time, it must be ensured that the subtask has reached the computing server t_{arrive}, and the server must be available $avail[k']$. Based on that, $AST(m, i)$ can be defined as follows:

$$AST(m, i) = max(avail[k'], t_{arrive}, \max_{i' \in pred(i)} (AFT(m, i'))) \tag{10}$$

Consequently, the task offloading and resource allocation scheduling problems in the studied EECOC networks can be formulated as follows:

$$(P1) : \min_{w_k^p, f_k^p} \sum_{j=1}^{m} \sum_{i=1}^{v} (w_1 t_{j,i} + w_2 e_{j,i}) \tag{11}$$

$$s.t. \sum_{j=1}^{k} w_k^n \leq w, 1 \leq k \leq m, \quad n = \{1, 2, ...N\}, \tag{12}$$

$$\sum_{j=1}^{k} f_k^n \leq f, 1 \leq k \leq m, \quad n = \{1, 2, ...N\} \tag{13}$$

In the above equation, w_1, w_2 weighted coefficients for delay and energy consumption respectively. In the following we refer to the consumption of all MDs as the system consumption. Formulas (12), (13) express the same concept, which means the allocated transmission and computing resources for the servers should not exceed their respective maximum values.

4 DRL-Based Task Offloading and Resource Allocation in EECOC

In this section, a task offloading and resource allocation algorithm with a hierarchical structure is proposed according to the optimization problem defined in Eq. (11). Initially, we introduce a task offloading approach based on game theory. Subsequently, we present a resource allocation using proximal policy optimization, building upon the aforementioned task offloading framework.

4.1 Game-Based Task Offloading

In a multi-user, multi-server EECOC model, the task offloading of multiple MDs can be regarded as a process of server resource preemption, and each MD competes for server resources to offload their tasks and achieve better performance. Therefore, to handle the competition among multiple MDs, we employ a game-theoretic approach and consider each MD as a game player. In each iteration, each gamer will make their own optimal offloading strategy according to the current environment. After finite steps, all the gamers can collectively reach a state of mutual Equilibrium: a Nash Equilibrium which implies that no gamer can further reduce its consumption by changing its strategy unilaterally.

4.2 DRL-Based Resource Allocation

Based on this task offloading strategy, PPO is introduced to address the above-mentioned resource allocation problem. PPO is an algorithm that belongs to the family of policy gradient methods in RL. It is designed to optimize and update the policy parameters iteratively in an efficient and stable manner. It achieves this by utilizing a trust region approach, which constrains the policy update to ensure that it doesn't deviate too far from the previous policy.

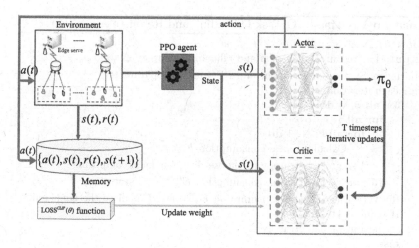

Fig. 1. The Game-PPO framework for offloading strategy and resource allocation

After the task offloading strategy of the upper-level MDs is determined, we transform the resource allocation of the server problem into a MDP, and solved based on PPO, which is defined by the following three elements.

1) State: The state space s is a finite set that captures the essential observable in the environment, which consists of bandwidth and frequency resources of edge servers and cloud servers. So the state can be expressed as

$$s = \{w^{ES}, f^{ES}, w^{CS}, f^{CS}\} \tag{14}$$

2) Action: For the corresponding action, given that the task offloading strategy has been determined, the focus is on allocating specific bandwidth and frequency resources to MDs who have computing tasks on their respective servers, according to the requirements of each server. Thus action can be expressed as

$$a = \{w_k^p, f_k^p\} \tag{15}$$

where w_k^p, f_k^p denotes bandwidth resources and frequency resources allocated by server p to MD k, respectively.

3) Reward: The reward function, denoted as r, is a function that determines the immediate value assigned to the transition of taking action a under the state s. Therefore, the reward should be aligned with the optimization goal. As our objective is to minimize the overall system consumption of the MDs, we set the reward to be

$$r = (\sum_{j=1}^{m} \sum_{i=1}^{v} (t_{j,i} + e_{j,i}))_{local} / (\sum_{j=1}^{m} \sum_{i=1}^{v} (t_{j,i} + e_{j,i}))_t \tag{16}$$

where $(\sum_{j=1}^{m} \sum_{i=1}^{v} (t_{j,i} + e_{j,i}))_{local}$ denotes system consumption when all tasks are executed locally, and $(\sum_{j=1}^{m} \sum_{i=1}^{v} (t_{j,i} + e_{j,i}))_t$ denotes system consumption of current steps.

Algorithm 1. Game-PPO Task Offloading and Resource Allocation

---

Input: $\mathcal{M}, \mathcal{N}, \mathcal{V}, f_m, f_n^{ES}, w_n^{ES}, f^{CS}, w^{CS}$
Output: The resource allocation and offloading strategy Δ to MDs

1: Initialize: System environment and PPO network parameters
2: **for** each iteration t **do**
3:     **for all** $i \in$ M **do**
4:         **for all** $j \in v$ **do**
5:             **for all** $k \in$ N+1 **do**
6:                 Calculate total consumption E_{total}
7:                 Use greedy strategy to choose server or local computing
8:             Record each MD total consumption E_m^{total} and state S
9:         Record each MD total consumption E_m^{total} and Calculate the game gain
10:     **if** Gain=1 **then**
11:         output offload strategy Δ
12:     **else**
13:         Select the E_{min} MD to update the offload strategy Δ
14: **for** $episode = 1$ to $episodes$ **do**
15:     Reset the EECOC environment and observe s
16:     **for** $t = 1$ to $steps$ **do**
17:         Generate an action a_t and compute the reward r_t
18:         Transition to the next system state $s\_$
19:         Store transition($s_{norm}, a_t, r_t, s_{\_norm}$), and update $s = s\_$
20:         **if** repaly buffer is full **then**
21:             Update the PPO network
22:         **if** servers finish allocating resources to MDs **then**
23:             **return** Task offloading and Resource allocation strategy

---

As shown in Fig. 1, the edge orchestrator acts as the agent of the PPO network. It selects the appropriate action by observing the state of the EECOC environment, and calculates the reward based on the current action, and finally records the converted system state. So as to realize the allocation of the resources of MDs by the servers. Details about the Game-PPO task offloading and resource allocation are illustrated in Algorithm 1.

5   Performance Evaluation

In this section, extensive experimental evaluations were conducted to assess the performance of our proposed approach. The experimental setting was thoroughly described at first, and then we compare baseline and traditional algorithm in various EECOC scenarios to demonstrate the effectiveness of our proposed algorithm.

5.1   Experimental Setting

The simulation is conducted using the specified parameter values for EECOC, which are provided in Table 1.

Table 1. Primary parameters of EECOC.

Parameter	Description	Value
p	The transmission power of MD	2 W
f^{MD}	Computing frequency for MDs	1 GHz
f^{ES}	Computing frequency for edge servers	10 GHz
f^{CE}	Computing frequency for cloud server	50 GHz
w^{ES}	Wide area network	100 MHz
w^{CS}	Wireless local area network	200 MHz

Performance comparison is conducted by considering the following baselines and approaches for evaluation.

- **Local Computing (LC)**: The LC means each MD will execute its task locally without offloading tasks to an edge server or a cloud server, which is mainly used as a baseline value.
- **Greedy**: The greedy algorithm is widely employed for task offloading. each step aims to find a locally optimal solution that minimizes the system consumption for task offloading, similar to [16].
- **Game theory**: An algorithm achieves task offloading and resource allocation through games between MDs.
- **Game-PPO**: The task offloading strategy obtained from the upper game is combined with PPO for resource allocation optimization as shown in Algorithm 1.

5.2 Performance Evaluation and Analysis

To validate the versatility and effectiveness of our proposed algorithm, three sets of simulation experiments are conducted in diverse EECOC environments to evaluate the performance of different approaches. The base parameter values for these experiments were taken from the values in Table 1 above, and in each set of experiments, one variable was varied around the control value while keeping the other variables constant.

In the following experiments, the system consumption is the weighting of energy and latency, displayed as Energy-Delay-Product. In the EECOC environment, each MD can offload its subtasks to edge servers or cloud server, and different MDs on the same server will use the servers' bandwidth resources and frequency resources together. Hence, variations in the number of MDs have a significant impact on the experiment. As shown in Fig. 2, we conducted two separate experiments, each with the same number of MDs from 8 to 15, with CCR ratios of 1:10 and 1:5, respectively. It is seen that, increased with the rise in the number of MDs. Comparing the two sets of experiments at the same time, when faced with computationally intensive tasks as shown in Fig. 2 (a), by offloading the tasks to the server, the reduced system is better than the Fig. 2 (b).

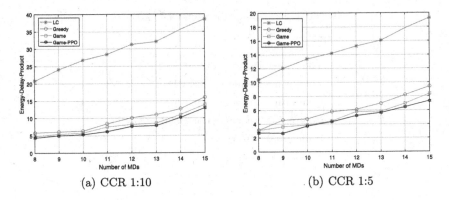

(a) CCR 1:10   (b) CCR 1:5

Fig. 2. System consumption for different numbers of MDs

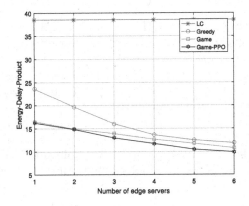

Fig. 3. System consumption for different numbers of edge servers

Figure 3 illustrates the system consumption with different numbers of edge servers ranging from 1 to 6, and fixing one cloud server. At the same time, we fixed the number of MDs to 8 in this and subsequent experiments. When edge servers only have one or two, our proposed algorithm Game and Game-PPO are basically the same, but obviously better than Greedy. As the number of edge servers increases, the difference in system consumption between different algorithms gradually decreases due to the increase in the number of edge servers, transmission and computing resources are relatively redundant.

In the last set of experiments, we take into account two key task characteristics. The first is the CCR of the task, which reflects the relative amount of communication required compared to the computation. The second is the size of task, which refers to the amount of data or computational complexity associated with the task. These two types of features can well reflect different types of tasks in the real world. As the Fig. 4 (a) shown, with the CCR increases, tasks are offloaded to servers with abundant computing resources, resulting in more efficient reduction of system consumption. In Fig. 4 (b), as the

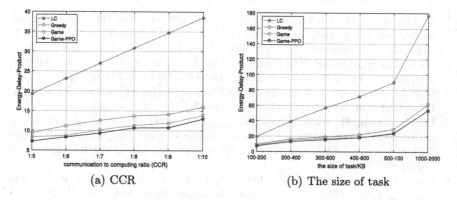

Fig. 4. System consumption for different numbers of MDs

size of tasks increases, it implies that the number and size of subtasks composing them also increase. The two algorithms we proposed exhibit similar performance probably because each MD has multiple subtasks to be computed on a single server, thereby mitigating the resource waste issue highlighted in the game theory approach.

6 Conclusion

In this paper, we optimize the problem of minimizing task processing latency and energy consumption of MDs, by proposing a task offloading and resource allocation approach in a EECOC environment. In this method, we optimize task offload strategy and resource allocation through the utilization of (Game-PPO) algorithm. Finally, through extensive simulation experiments conducted in various EECOC scenarios, the results demonstrate the significant reduction in energy consumption of MDs and task computation delays achieved by our proposed algorithm. The performance of our algorithm outperforms the baseline and traditional algorithms, highlighting its effectiveness.

In our future work, we are considering incorporating real-world data into our algorithm to validate its feasibility. Additionally, we plan to introduce additional constraints to our model that align with the characteristics of real data tasks and MDs. This will enable us to evaluate the performance and effectiveness of our algorithm in practical settings.

References

1. Al-Sarawi, S., Anbar, M., Abdullah, R., Al Hawari, A.B.: Internet of things market analysis forecasts, 2020–2030. In: 2020 Fourth World Conference on Smart Trends in Systems, Security and Sustainability (WorldS4), pp. 449–453. IEEE (2020)
2. Aceto, L., Morichetta, A., Tiezzi, F.: Decision support for mobile cloud computing applications via model checking. In: 2015 3rd IEEE International Conference on Mobile Cloud Computing, Services, and Engineering, pp. 199–204. IEEE (2015)

3. Erol-Kantarci, M., Sukhmani, S.: Caching and computing at the edge for mobile augmented reality and virtual reality (AR/VR) IN 5G. In: Zhou, Y., Kunz, T. (eds.) Ad Hoc Networks. Lecture Notes of the Institute for Computer Sciences, Social Informatics and Telecommunications Engineering, vol. 223, pp. 169–177. Springer, Cham (2018). https://doi.org/10.1007/978-3-319-74439-1_15

4. Sun, X., Ansari, N.: Latency aware workload offloading in the cloudlet network. IEEE Commun. Lett. **21**(7), 1481–1484 (2017)

5. Ren, J., Zhang, D., He, S., Zhang, Y., Li, T.: A survey on end-edge-cloud orchestrated network computing paradigms: transparent computing, mobile edge computing, fog computing, and cloudlet. ACM Comput. Surv. (CSUR) **52**(6), 1–36 (2019)

6. Liang, J., Li, K., Liu, C., Li, K.: Joint offloading and scheduling decisions for DAG applications in mobile edge computing. Neurocomputing **424**, 160–171 (2021)

7. Wang, H., et al.: Architectural design alternatives based on cloud/edge/fog computing for connected vehicles. IEEE Commun. Surv. Tutorials **22**(4), 2349–2377 (2020)

8. Hong, Z., Chen, W., Huang, H., Guo, S., Zheng, Z.: Multi-hop cooperative computation offloading for industrial IoT-edge-cloud computing environments. IEEE Trans. Parallel Distrib. Syst. **30**(12), 2759–2774 (2019)

9. Qu, G., Wu, H., Li, R., Jiao, P.: DMRO: a deep meta reinforcement learning-based task offloading framework for edge-cloud computing. IEEE Trans. Netw. Serv. Manage. **18**(3), 3448–3459 (2021)

10. Ding, Y., Li, K., Liu, C., Li, K.: A potential game theoretic approach to computation offloading strategy optimization in end-edge-cloud computing. IEEE Trans. Parallel Distrib. Syst. **33**(6), 1503–1519 (2021)

11. Zhou, W., Lin, C., Duan, J., Ren, K., Zhang, X., Dou, W.: An optimized greedy-based task offloading method for mobile edge computing. In: Lai, Y., Wang, T., Jiang, M., Xu, G., Liang, W., Castiglione, A. (eds.) Algorithms and Architectures for Parallel Processing. Lecture Notes in Computer Science(), vol. 13155, pp. 494–508. Springer, Cham (2022). https://doi.org/10.1007/978-3-030-95384-3_31

12. Peng, K., Huang, H., Wan, S., Leung, V.C.: End-edge-cloud collaborative computation offloading for multiple mobile users in heterogeneous edge-server environment. Wireless Netw., 1–12 (2020)

13. Li, Y., Qi, F., Wang, Z., Yu, X., Shao, S.: Distributed edge computing offloading algorithm based on deep reinforcement learning. IEEE Access **8**, 85204–85215 (2020)

14. Yang, L., Zhang, H., Li, X., Ji, H., Leung, V.C.: A distributed computation offloading strategy in small-cell networks integrated with mobile edge computing. IEEE/ACM Trans. Networking **26**(6), 2762–2773 (2018)

15. Zhang, W., Wen, Y., Guan, K., Kilper, D., Luo, H., Wu, D.O.: Energy-optimal mobile cloud computing under stochastic wireless channel. IEEE Trans. Wireless Commun. **12**(9), 4569–4581 (2013)

16. Wei, F., Chen, S., Zou, W.: A greedy algorithm for task offloading in mobile edge computing system. China Commun. **15**(11), 149–157 (2018)

Multi-agent Cooperative Intrusion Detection Based on Generative Data Augmentation

Ming Liu, Yungang Jia, Chao Li, Peiguo Fu, and Zhen Zhang[✉]

National Computer Network Emergency Response Technical Team/Coordination Center of China, Beijing 100029, China
{liuming,zhangzhen}@cert.org.cn

Abstract. Existing supervised learning methods are difficult to adapt the rapidly evolving network attacks. They are effective for malicious flows with clear features, but struggle with flows that reveal unclear or sparse characteristics. This is a concern as malicious flows are rare and discrete in real-world situations. To overcome these challenges, this research paper introduces a novel few-shot sample malicious flow detection model that leverages data augmentation techniques. The model's core objective is to train agents to distinguish between normal and malicious flows. On this basis, the model enhances the agents' ability to recognize malicious flows through discrete information interactions. Experimental results confirm that the data augmentation method effectively improves the agents' understanding of network traffic. Additionally, it successfully enhances intrusion detection capabilities in multiple agents, diverse datasets, and varied scenarios. Notably, in few-shot sample scenarios, the method greatly boosts the overall accuracy rate.

Keywords: Generative Adversarial Networks · Multi-agent System · Intrusion detection · Reinforcement Learning

1 Introduction

Currently, intrusion detection models empowered by machine learning methods excel in scenarios with balanced feature distribution and continuously optimize algorithms to enhance overall classification accuracy. Consequently, even if a few samples are misclassified, the detection accuracy remains unaffected [1]. However, in practice, these samples often pose the most serious threat to network security. Therefore, there is an urgent need to study how to improve the accuracy of detecting malicious traffic in few-shot scenarios. Building upon our previous work [10], which demonstrated the benefits of agent interactions in enhancing decision-making efficiency, particularly in detecting known network attacks [11].

This work is financially supported by the National Natural Science Foundation of China under Grant 62106060.

© The Author(s), under exclusive license to Springer Nature Singapore Pte Ltd. 2024
Z. Tari et al. (Eds.): ICA3PP 2023, LNCS 14492, pp. 311–328, 2024.
https://doi.org/10.1007/978-981-97-0811-6_19

Therefore, this paper aims to investigate the problem of detecting few-shot malicious samples through multi-agent collaborative decision making.

The existing line of research in the analysis of few-shot samples primarily revolves around achieving data balance by adjusting data distribution, with a particular emphasis on utilizing undersampling and oversampling techniques [6,8,13,14], Each of these studies has its own unique characteristics, which are not further elaborated on. Nevertheless, it is important to acknowledge that these approaches involve the exclusion of specific samples, which may lead to the removal of crucial data features and subsequently diminish the overall performance of the detection model. The initial conceptualization of the Generative Adversarial Network (GAN) was proposed by Goodfellow [6]. It comprises a generator and a discriminator that engage in an adversarial training process aimed at attaining equilibrium. In subsequent advancements, the Wasserstein GAN-Gradient Penalty (WGAN-GP) [8] technique emerged, which introduced batch normalization and adopted a fully convolutional architecture for both the generator and discriminator models, thereby enhancing the stability of the training procedure. In accordance with previously published studies [3,5,12], our approach encompasses the utilization of multiple generators, however, it distinctively emphasizes the necessity of capturing a wide range of diverse modes.

Deep learning techniques have revolutionized Reinforcement Learning (RL) by enabling agents to learn in high-dimensional or continuous state spaces [4]. However, the nonstationarity of the environment poses a significant challenge for the application of Multi-agent RL (MARL) methods [7]. In this study, we focus on the problem of few-shot malicious samples, which is insufficient for practical multi-agent independent training-collaborative attack detection. To address this, we propose the following innovations: Firstly, we employ the Wasserstein GAN-Gradient Penalty (WGAN-GP) technique to augment the few-shot sample data, and its efficacy is assessed across diverse datasets. Moreover, independent training datasets for individual agents are constructed based on the expanded dataset obtained through the previous step. Afterwards, by leveraging the Actor-Critic algorithm, we develop a framework for multi-agent collaborative attack detection. This framework facilitates the online interactive update of agent decision parameters, leading to improved efficiency in multi-agent collaborative decision-making. Eventually, the effectiveness of our proposed approach is validated through experimental results. In summary, our study broadens the application of multi-agent deep learning in the realm of network security, with a particular focus on validating its efficacy in detecting few-shot malicious samples.

2   Problem Description

Considering in an open network $\mathcal{N}$ with fixed topology, N decentralized IDS (Intrusion Detection System) are deployed on N links, each IDS detects cyber attack on its link, these N independently facilities can be modeled as a N decentralized agent system. In our previous work [10,11], we found that the interaction between agents can effectively enhance the collaborative defense decision-making effects, therefore, for the issue of few-shot sample cyber attack detection, we continue our research on the basis of the above conclusions, and the multi-agent collaborative cyber attack detection structure is shown in Fig. 1.

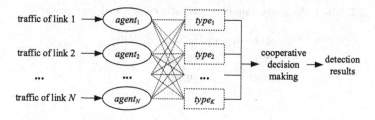

Fig. 1. Multi-agent collaborative cyber attack detection structure.

It can be found that each agent independently classifies network traffic the link it deploys, and obtains the final detection result through collaborative decision making, thus improving the overall accuracy and reliability of malicious traffic detection. However, there has 2 technical challenges here:

– Few-shot malicious samples is not enough to meet the training needs of agents.
– Online collaborative decision-making requires timely updating of agents' decision parameters.

3   Few-Shot Sample Detection Based on Data Augmentation

For the first issue, we generate more realistic sample data based on WGAN-GP to enhance the few-shot samples to meet agent's training requirements. On this basis, a malicious traffic detection training process consisting of two stages for each agent is constructed, which will reduce the complexity of traditional malicious traffic detection model and improve the detection accuracy.

3.1   Data Preprocessing

In order to make the features of samples exist in Euclidean space, we adopt one-hot encoding to numerize the data with string values, which uses binary coding and borrows the idea of register storage. For a feature with k string values, k-bit status registers are used, and only one bit is valid, which represents the unique position of its character value. That is, each character has its unique register position, the result obtained after one-hot encoding for each character is unique and mutually exclusive, indicating that each value is independent.

The numerization characteristic of protocol type features are defined in Table 1. There are two more features with string values, which are processed in the same way as protocol type. Service is transformed into a 70-dimensional vector, and flag is transformed into an 11-dimensional vector. Then, the three features with string values are expanded from 3 dimensions to 84 dimensions, so the original 41-dimensional features are expanded to 122 dimensions.

Table 1. Numerization characteristic of protocol type features.

protocol_type	one-hot encoding
TCP	(0,0,1)
UDP	(0,1,0)
ICMP	(1,0,0)

Normalization is mainly performed to analyze the distribution of the original samples and organize and summarize the different dimensions of features, so that features with different dimensions can also be compared with each other. In this paper, the range transformation method is used to standardize the data to $[0,1]$, which is implemented Eq. 1.

$$y_i = \frac{x_i - \min(x)}{\max(x) - \min(x)} \tag{1}$$

$\min(x)$ and $\max(x)$ indicate the lower limit and upper limit of a feature, respectively. x_i represents the value of the i-th sample, and y_i represents the value obtained after normalizing x_i, that is, the maximum and minimum values of each feature in the training set need to be obtained first, and then the features of the samples in the test set are normalized based on these values, and the value interval is $[0,1]$. To perform feature dimensionality reduction, one feature was removed from the 122 features. The principles of selection is: analyze all the features of each record, calculate the variance corresponding to each feature value, and remove the feature value with the smallest variance. If the variance of a feature value is 0, it means that the value of this feature is the same for all samples. This indicates that no differences can be observed between samples from this feature, that it, no help to the model training, so it can be deleted. After performing feature dimensionality reduction, 121 features remained, and after image processing, an $11 * 11$ two-dimensional image could be obtained.

3.2   WGAN-GP Based Data Augmentation

It is desired to load a dataset with homogeneous feature distribution so that the detection algorithm can achieve a better accuracy and recall rate. Therefore, we adpot WGAN-GP to augment the few-shot samples, and fuses the generated data with the recognized and classified data to form a new training set. The reconstructed input sample x is the real value rather than a binary value, so the distribution of the generated data follows a Gaussian distribution of $N[\mu(z, \ddot{z}, y), \delta]$, where z is random noise, $\ddot{z}$ denotes potential features of the sample learned by agent i during the training, $P(y|x)$ denotes sample x being identified as type y, therefore, we adopts $\mu(z, \ddot{z}, y)$ as the reconstructed value, and δ is a constant and is related to the size of the generated data (Fig. 2).

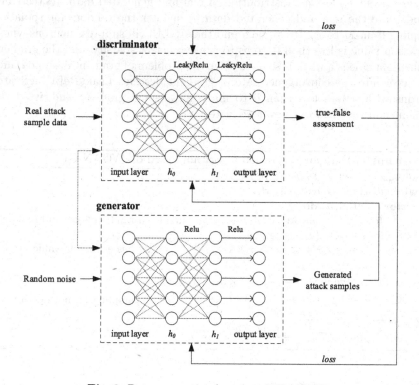

Fig. 2. Data generation based on WGAN-GP.

For agent i, the data-enhanced network attack training process is as follows: Firstly, WGAN-GP is used to expand the few-shot sample data, and then the generated data is disturbed to obtain more realistic sample data with an increased scale. Then, i is supervised trained by combining the generated data with known network attack samples and normal network traffic as inputs.

Specifically, we adopts a fully connected network with multiple hidden layers as the basic network module of the generators and discriminators. The network structure of the generator model consists of one input layer, two hidden layers and one output layer. The Dropout layer is introduced to overcome the problem of long training time and overfitting. The activation function selects the ReLU function to avoid the problem of the reverse computation gradient disappearing when training the network, and also to reduce the operation load, the value range is $(0, x)$. The discriminator adopts the same neural network structure, the difference is it adopts LeakyReLU function and its value range is $(-\infty, +\infty)$. We assume that the discriminator follows the 1-Lipschitz condition hypothesis [15], the loss functions are defined as follows:

$$G : -E_{x \sim P_g}[d(x)]$$
$$D : E_{x \sim P_g}[d(x)] - E_{x \sim P_r}[d(x)] + \lambda E_{x \sim P_{\tilde{x}}}[\nabla_x D(x)_p - 1]^2 \tag{2}$$

where, P_r and P_g are the distribution of real and generated data, respectively. x and y are the real and generated data, $\tilde{x}$ denotes the random interpolation sampling from x_r to x_g. It can be found that ReLU will suppress neurons whose activation value is less than 0, so as to achieve better sparsity, and the gradient on the right side is always 1, so as to solve the problem of gradient disappearance and accelerate the convergence rate of gradient descent. LeakyReLU activates neurons with values less than 0 to maintain a small gradient and avoid the phenomenon that neurons never activate.

Algorithm 1. Data augmentation algorithm based on WGAN-GP

1: initializing w_0 for D and θ_g for G
2: **while** θ_g is not convergence **do**
3:     **for** $t = 0, ..., n_c$ **do**
4:         **for** $i = 1, ..., m$ **do**   ▷ Sampling from random noise and real data
5:             $x \leftarrow G_{\theta_g}(z)$
6:             $x \leftarrow \varepsilon * x + (1 - \varepsilon) * x$ ▷ ε is a random value in $[0, 1]$
7:             $L^i \leftarrow D_w(x) + \lambda[|||\nabla_x D_x(x)||_2 - 1]^2$
8:         $w \leftarrow Adam(\nabla_m \frac{\sum_{i=1}^m L^i}{m}, w, \alpha, \beta_1, \beta_2)$
9:     $\{z^i\}_{i=1}^m \sim P_z$ ▷ Re-sampling from random noise
10:     $g_\theta \leftarrow -\nabla_\theta \frac{\sum_{i=1}^m f_w[g_\theta(z^i)]}{m}$
11:     $\theta \leftarrow Adam\{-\nabla_m \frac{\sum_{i=1}^m D_w[G_\theta(z)]}{m}\}, \theta, \alpha, \beta_1, \beta_2]$

In the above algorithm, λ is the gradient penalty coefficient, n_c is the number of critic iterations of generator, that is, the number of training sessions. α, β_1 and β_2 are the parameters of the Adam algorithm respectively. For the generated network attack samples, Shapley Additive exPlanations (SHAP) [2] is used to calculate the top k features in the data that are more important for detection, then part of the generated data is randomly selected, and the corresponding features are replaced with the features of real network traffic, so as to achieve the disturbance of the overall input traffic and make it more realistic, the disturbance process is shown in Fig. 3 and the details will not be repeated.

Length of the Gaussian distribution random number is defined as 100, and a generated image of $(11, 11, 1)$ will be obtained. In the deconvolution process, the height and width of the feature layer are continually increased. In the generator, a fully connected layer is used to connect the input to a length of 4608 $(3 \times 3 \times 512)$, so that the result of this layer can be reshaped into a corresponding same size of the feature layer. After deconvolution, the feature layer becomes larger with a reduced number of channels, that is, $(3, 3, 512) \rightarrow (5, 5, 256) \rightarrow (7, 7, 128) \rightarrow (9, 9, 64) \rightarrow (11, 11, 1)$.

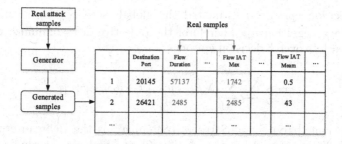

Fig. 3. Generated samples disturbance.

4   Interactive Multi-agent Cooperative Detection

For the second issue, on the basis of our previous work [11], we improve and propose an interactive multi-agent cooperative detection and decision model, that is, agents are first trained independently using the generated data set of disturbance, in the cooperative decision-making stage, the newly identified sample features and type labels, especially the learned potential features, are shared. Then, the detection of few-shot samples is realized by integrating the decision belief of each agent for similar samples' features and labels.

4.1   Multi-agent Cooperative Detection Decision Model

Since agent i can only observe the link it deploys, even with unrestricted interactions within $\mathcal{N}$, it is still difficult to establish the global state of $\mathcal{N}$, in particular, the types and arrival time of cyber attacks are all random. In addition, the interaction between agents actually has a one-step delay for decision-making, that is, the current decision-making adopts the information obtained from the last round of interaction. Therefore, N agents collaborative malicious traffic detection can be model as an Actor-Critic based Delayed Interactive Multi-agent Markov Decision Problems of $\langle \Pi, \mathcal{S}, O, h, r \rangle$, where:

- $\delta_i = \{\pi_0, \pi_1, ..., \pi_K\}$ is the available decision strategies set of agent i, $\pi_i^t = \varsigma^t \langle \phi_i, \psi_i \rangle$ denotes the functional decision strategy at time t, and ϕ_i and ψ_i are the parameterized Q function and V function of i, respectively.
- $\mathcal{S}^t = \{S_0, \widetilde{\mathfrak{S}}_1^t \cap ... \cap \widetilde{\mathfrak{S}}_N^t\}$ denotes the global state of $\mathcal{N}$ at time t, S_0 is the unchangeable parts of $\mathcal{N}$, $\widetilde{\mathfrak{S}}_i^t$ denotes the observed link state by i, specifically, the detected malicious network traffic records.
- $h_i^t = \langle \omega_i^t, \sum_{j \in neighbor} I_{j,i} \rangle \circ h_i^{t-1}$ denotes the cognition update process of i and h_i^t is the current cognition, $\sum_{j \in neighbor} I_{j,i}$ is the valid information set that i obtained from its neighbor agents, $I_{j,i} = \sum_k \langle x, y \rangle$, which represents the newly detected sample features and type labels.
- $O_{\pi_i^t}(\omega^t | \mathcal{S}_{i,b}^t)$ denotes the probability of i observing ω^t on the estimated state $\mathcal{S}_{i,b}^t$ when strategy π_i^t is taken, which $\mathcal{S}_{i,b}^t = \arg\max b(\mathcal{S}_{i,b}^t | \widetilde{\mathfrak{S}}_i^t, h_i^t)$.
- $r_{\pi_i^t}$ denotes the expected reward obtained by taking π_i^t at state $\mathcal{S}_{i,b}^t$.

In each decision episode, i estimated the global state $\mathcal{S}_{i,b}^t$ based on ω_i^t and previous-stage cognition h_i^t. Therefore, the objective is to minimize the joint accumulated loss over k decision episodes:

$$\min_{\langle \phi_i, \psi_i \rangle} \sum_0^t \sum_{i=1}^N (\lambda_i \beta_i - \mu_i r_i) + \aleph[\langle \phi_i, \psi_i \rangle_{i=1}^N] \tag{3}$$

where $\beta_i \leftarrow \max[r(\mathcal{S}) + \gamma \psi_i(\mathcal{S})_0^t] - \psi_i(\mathcal{S})|\langle \omega_i^t, h_i^t \rangle$ is the Bellman error of i's Critic module, γ is the interference coefficient, it simulates the random impact of the unobserved elements on $\mathcal{N}$. λ_i and μ_i are the penalty coefficients of the Actor and Critic modules, respectively. $\aleph$ represents the regularization terms of $\langle \phi, \psi \rangle$. One practical term needs to be considered is that in order to achieve decentralized cooperation and avoid network redundant information congestion, agents prefer sharing partial information instead of all information their held, therefore, the part of the parameters that are not shared are marked as $\langle \phi^\ell, \psi^\ell \rangle$, Eq. 3 can be rewriten as:

$$\min_{\langle \phi_i, \psi_i \rangle} \sum_{i=1}^N \underbrace{[\lambda_i \beta(\phi_i, \phi_{-i}, \psi^\ell) - \mu_i r(\phi_i, \phi_i^\ell)]}_{\tau_i \{\phi_i, \phi_{-i}, \phi^\ell, \psi^\ell\}}$$
$$+ \aleph[\langle \phi_i, \psi_i \rangle_{i=1}^N, \phi^\ell, \psi^\ell] \tag{4}$$

where $\tau(\cdot)$ denotes the interaction process, that is to realize $\langle \phi_i, \psi_i \rangle$ sharing between agents. In addition, all agents tend to be greedy because every link security is important, and this is significantly different from previous studies.

The multi-agent cooperative decision-making framework is based on AC (Actor Critic) algorithm. The core objective is to alternately minimize the Bellman error in the Critic module and maximize the cumulative reward in the Actor module throughout the training process. However, different from the traditional researches, this paper is dedicated to achieve joint optimization based on the interaction between agents, so that the above two objectives can promote each other to improve the training efficiency and the final algorithm performance.

As shown in Fig. 4, for any agent i, the architecture in Fig. 4 can map i's recognition h_i^{t-1}, local observation ω^t and the defense task features to its next policy π^t, which can be denoted as $\Delta_{\sum_{i,j}} \pi(h, h')$, it means that i obtains additional information from its neighbors through interaction to enhance h, and transfer from h to h', which j denotes the neighbor that i can currently interact with, therefore, Δ can be marked as the interactive indicator. During interactions, agents share an embedded recurrent architecture accounting for the value iteration, which maps task parameters to agents' Q and V values, then updates h_i^t. The belief update module is a Bayesian filter that recursively propagates i's belief and outputs values to the decision module for evaluating π_t. Therefore, we named the decision algorithm based on the above framework Interactive Adam, which action strategy optimizer is Adam.

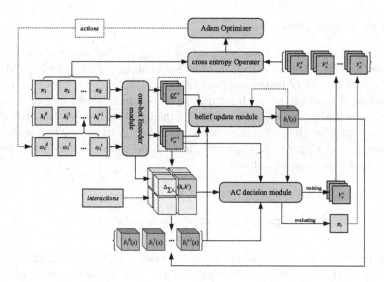

Fig. 4. The general DI-MMDPs architecture, consisting of the MDP planning module and the belief update module, agent is trained by minimizing the cross entropy between the output and the expert's decision results.

Algorithm 2. Interactive Adam Algorithm

1: **while** *training* **do**
2:     $g_t = \nabla f_t(x_t)$
3:     **if** $\max_i g_t \geq \mathcal{L}\varphi_t$ **then**
4:         $v_t = \beta v_{t-1} + (1 - \beta)g_t$
5:     **else**
6:         $v_t = v_{t-1}$
7:     $g'_t = \nabla f_t(x_{t+1})$
8:     $l_t = \frac{\|g'_t - g_t\|_2}{\|x_{t+1} - x_t\|_2}$
9:     $\mathcal{L} = \max(\mathcal{L}, l_t)$

As explained in Algorithm 2, the learning objective function of can be defined as $x \in \mathbb{R}^d$ as $f(x)$, and then the general adaptive method could be denoted as $x_{t+1} = x_t - \varphi_t \frac{\widetilde{m}_t}{\widetilde{v}_t}$, and it assumed that f is $\mathcal{L}$-smooth function, which is a typical and weak assumption for analysis of first-order optimization methods [9]. φ_t is the learning rate, and $\widetilde{m}_t$ is a typical momentum of gradients. The core is evaluating the next momentum $\widetilde{v}_t$. $\widetilde{v}_t$ is a nonnegative d-dimensional vector which adjusts learning rate for each dimension. It should be noted that, based on the above framework, the core process of aAdam algorithm has not been changed, so it would be no further elaboration. In addition, since the value function of traditional MDP provides a linear constraint on a single confidence space, AC can be used for strategy evaluation to obtain the highest expected value. After the value iterative solver outputs the Q value, we weight it with the belief tensor and outputs the softmax policies to the trainer, where the cross entropy between the output and the expert (in h) is the training loss.

4.2 Two Stages Detection

When training agents, the data set used by N agents for independent training is marked as $\{D_1, D_2, ..., D_N\}$, each data set contains of M samples and each sample x_i consists of G features $x_i = \{f_1, f_2, ...F_G\}$, whose corresponding type is labeled y_i. Therefore, for each sample x_i in D_i, agent i will output a unique result. Due to the limited cognitive ability of the agent, this paper the probability $P_{y_i}(x_i)$ to represent that sample x_i is recognized as y_i in the current state by i, and it has $\sum_{I \in M} P(x_i) = 1$. Therefore, in each decision episode, the detection accuracy of agent i can be defined as:

$$f(x_i) = \arg\max P_{y_i}(x_i)$$
$$acc(D_i) = \frac{\sum_{i \in M} \phi[f(x_i), y_i]}{|D_i|} \tag{5}$$

The first formula represents the selection of the category label y_i with the highest probability as the detection result of sample x_i. Then the detection accuracy of all samples in D_i is calculated according to the second formula, that is, the number of samples correctly detected by agent i is divided by the total number of samples. ϕ is a symbolic function, when the detection result is the same as the actual result, $\phi[f(x_i), y_i] = 1$, otherwise is 0. In the training stage, each agent outputs different detection accuracy for independent data D_i, which directly reflects the sensitivity of each agent to different sample data.

In order to maintain the accuracy of classification of known normal and malicious sample data while identifying few-shot malicious sample, the detection training process is divided into two stages. The 1st stage is to classify known attack and normal traffic data, which goal is to minimize the possibility of misclassification of known attack sample data, but the distribution of few-shot sample is not specifically constructed. In the 2nd stage, based on the classification completed in the 1st stage, the discriminant boundary of few-shot malicious sample is characterized in a specific area to achieve further fine-grained division. Therefore, constructing two separate functions to decouple two classification problems can provide enough flexibility to improve the ability of few-shot malicious sample identification, which will possible maintaining the classification accuracy of known attacks, which objective funtions can be defined as:

$$g_1^* = \arg\min_{x \to y \in K} F[g_1(x)]$$
$$g_2^* = \arg\min_{x \to y \in K \bigcup U} F[g_2|y = g_1^*(x)] \tag{6}$$

which purpose is to minimize the error rate F of sample x being identified as type y, K represents the known characteristic space and U represents unknown. Specifically, in the 1st stage, it can adopt standard RF (Random Forest) algorithm, the normal flow and sampling data of known malicious attacks are classified. The key is to maximize the accurate of x is identified as type y, that is, $P(y|x)$ approximation optimal g_1^*, and details are no longer repeated.

In the 2nd stage, the agent further distinguishes the unknown malicious sample data, so the it will produce a large reconstruction error according to the error category label it determined in the 1st stage. In order to ensure the recognition accuracy, it is necessary to maximize the generation probability of input samples. In addition, it should be emphasized that in order to improve the data diversity and ensure the agent's detection performance of unknown sample, the potential feature $\ddot{z}$ learned in 1st stage of known samples should be reconstructed and inputs. The objective is defined as:

$$\arg \max_{(x,y) \in K \bigcup U} E_{\ddot{z} \sim D_c(\ddot{z}|x,y)}[\log P(x|\ddot{z}, y)] \tag{7}$$

It can be found that the discriminant model $D_c(x, y)$ of the 1st stage can be directly used to construct the feature distribution $D_c(\ddot{z}|x,y)$, so it is only need to construct the training model $P(x|\ddot{z}, y)$ in this stage, which reconstructs the input sample x from the potential feature $\ddot{z}$ and label y.

5   Experimental Result and Analysis

In the experimental chapter, we select data from open datasets NSL-KDD, CIC-IDS 2017 and CICDDoS 2019 to verify the proposed method from different perspectives and analyze the effects presented. Experimental tests were conducted on a Windows 10 and utilizing a hardware configuration comprising an Intel Core i7 8700 CPU and 16GB of memory. PyCharm served as the integrated development environment (IDE), while the programming language selected for implementation was Python 3.7. The deep learning framework utilized consisted of TensorFlow and Keras, which provided a robust foundation for model construction and training. Moreover, fundamental data processing libraries, including pandas, numpy, and matplotlib, were employed to facilitate data manipulation, analysis, and visualization throughout the experimental process.

5.1   Data Augmentation Test and Analysis

In this section, we performed an evaluation of the proposed method using a subset of the NSL-KDD dataset. Specifically, we selected 10,000 normal samples and 13,000 attack samples for testing purposes. We conducted 500 rounds of testing, and for each round, the samples from NSL-KDD were randomly divided into an 80% training dataset and a 20% test dataset. This rigorous evaluation allowed us to thoroughly assess the performance of the proposed method. This experimental session focuses on the parameters used in the model and their implementation during training and decision making. Table 2 indicates the optimal parameters of RNN-based benchmark intrusion detection model.

During the testing, the benchmark model is trained in order to compare the detection of few-shot sample data generated by data augmentation using WGAN-GP, and the parameter settings are shown in Table 3. This test is used to verify the effectiveness of the WGAN-GP data augmentation. The trained

Table 2. Parameter setting of benchmark intrusion detection model.

Parameters	Numerical values
Input layer	11*11
Output layer	5
Activation function	ReLU
Dropout	0.5
Output layer activation function	Softmax
Bias	0.1
Backpropagation algorithm	Adam
Batch size	128
Learning rate	0.0002
Epoch	500

intrusion detection model is then saved, and during the evaluation process using the test set, the model parameters are adjusted based on the experimental results in order to achieve the best detection rate.

Table 3. Parameter settings of WGAN-GP based data augmentation.

Parameters	Numerical values
discriminator input layer	11*11
discriminator output layer	[0,1]
discriminator activation function	LeakyReLU
discriminator loss function	Sigmoid
discriminator batch_size	32
discriminator convolutional layers	4
discriminator dropout	0.5
generator input layer	100
generator output layer	11*11
generator activation function	ReLU
generator deconvolutional layers	4
Epoch	500

The main goal is to improve the detection performance of few-shot sample data, the effectiveness of data generation based on WGAN-GP is the focus, so a self-contrast is used to compare the generated few-shot sample data and the intrusion detection models without data enhancement. The optimal result is obtained by adjusting the parameters during the testing process using a test set to evaluate the detection performance of the model. WGAN-GP is used to generate data to balance the training set, and the intrusion detection neural network model is trained again with the same training parameters. The detection performance of the network is evaluated again using the test set. The precision, recall, and F1-Score of each attack type are compared and analyzed, and the resulting confusion matrix obtained from the experiment is in Table 4.

Table 4. Confusion matrix of benchmark model.

True Label	Predicted Label				
	Normal	DoS	Probe	R2L	U2R
Normal	9365	307	26	7	6
DoS	277	7125	54	2	0
Probe	31	121	2256	12	0
R2L	2059	67	163	437	28
DDoS	43	19	83	18	37

Based on the confusion matrix, the precision, recall, F1-Score, and overall accuracy of each sample type can be calculated. The specific results are shown in Table 5. On this basis, the accuracy of the intrusion detection neural network model is 85.26%. However, the recall rate of the R2L and U2R few-shot sample

Table 5. Testing results of benchmark model.

Labels	Test Results			
	Precision	Recall	F1-Score	Accuracy
Normal	79.53%	96.44%	87.17%	85.26%
DoS	93.27%	95.53%	94.39%	
Probe	87.37%	93.18%	90.18%	
R2L	91.61%	15.87%	27.05%	
DDoS	52.11%	18.50%	27.31%	

Table 6. Confusion matrix of WGAN-GP based augmentation.

True Labels	Predicted Labels				
	Normal	DoS	Probe	R2L	U2R
Normal	9364	255	52	28	12
DoS	216	7148	83	11	0
Probe	41	93	2274	13	0
R2L	854	8	61	1768	63
U2R	21	2	45	14	118

types is very low, only 15.87% and 18.50%, and F1-Score scores are also only 27.05% and 27.31%. The reason is, there is little training data for these two attack types, and the intrusion detection model cannot extract sample features well and learn sufficiently. Therefore, the detection effect for R2L and U2R is not good. Then, the performance of the intrusion detection algorithm based on WGAN-GP-generated few-shot sample data was tested, and the confusion matrix obtained from the experiment is shown in Table 6.

Similarly, based on the confusion matrix, the precision, recall, F1-Score, and overall accuracy of each category can be calculated. The specific results are shown in Table 7. In summary, according to Tables 6 and 7, the overall accuracy of the intrusion detection based on WGAN-GP-generated approach is 91.70%, which is 6.44% higher than that of the benchmark model.

Table 7. Testing results of WGAN-GP based augmentation.

Labels	Test Results			
	Precision	Recall	F1-Score	Accuracy
Normal	89.21%	96.43%	92.68%	91.70%
DoS	95.23%	95.84%	95.53%	
Probe	90.42%	93.93%	92.14%	
R2L	96.40%	64.20%	77.07%	
U2R	61.14%	59.00%	60.05%	

5.2   Data Disturbance Testing and SHAP Feature Analysis

The CICDDoS2019 dataset is comprising a total of 50,006,249 attack samples and 56,863 normal samples. To ensure data equilibrium, 130,000 samples of DDoS attack data in CICDDoS2019 and 130,000 samples of normal traffic in CIC-IDS2017 are randomly selected as the base data set. The data are still randomly divided into 80% training dataset and 20% test dataset. First, the training data is used to train 4 different network structures of CNN, and then test their detection performance. The accuracy, recall rate, accuracy and F1 scores obtained are shown in the Table 8.

Table 8. Detection results of 4 different CNN structures on the base data set.

	Accuracy	Recall	Precision	F1_Score
C3P3F3	98.50%	99.04%	98.85%	98.94%
C3P3F2	98.42%	99.17%	98.79%	98.97%
C2P2F3	97.96%	98.56%	98.34%	98.44%
C2P2F2	97.58%	98.64%	98.50%	98.57%

It can be found that the detection performance of CNN with 3-layer convolutional pooling structure is better than that of CNN with 2-layers, because the filter is responsible for learning the features, and more filters will extract deeper features, and the detection performance will be better. In addition, reducing the number of convolutional pooling layers decreases the overall detection performance, while reducing the fully connected layers decreases the detection accuracy, but the F1 score increases. Therefore, C3P3F2 is chosen as the base detection model. On this basis, we improve the model by adding the Dropout layer with parameter 0.5 to the fully connected layer and replacing the fully connected layer with the global average pooling layer. The 3 models were compared and theresults as shown in Table 9.

Table 9. Detection results of 3 models on the base data set.

	Accuracy	Recall	Precision	F1_Score
CNN	97.67%	97.75%	97.69%	97.72%
Dropout	98.31%	98.37%	98.34%	98.35%
improved-CNN	99.67%	99.68%	99.65%	99.66%

On this basis, WGAN-GP was used to generate 20,000 attack samples, and then 20,000 real attack data from the base data set were selected and combined as training data, 80% for training and 20% for testing. 4 detection models are selected, XGBoost, LSTM, CNN and improved-CNN. LSTM is a simple LSTM model with DNN layers, adopts ReLU activation function in all layers and Sigmoid activation function in the last layer, and using binary cross entropy as the loss function and Adam as the optimizer. These models are trained with the base data set as the training set, and the model is evaluated with the real data and generated data as the test set. The F1 score values are output, and the results are shown in Table 10.

Table 10. Detection results of 4 models on real samples and generated samples.

models	F1(real samples)	F1(generated samples)
XGBoost	95.15%	90.31%
LSTM	98.14%	97.91%
CNN	97.22%	96.25%
CNN	99.36%	98.36%

According to the test results, although the detection performance of the four models on the generated samples is inferior to that of the real samples, they can all be correctly classified by the model, and the probability of correct classification is more than 90%, indicating that the samples generated based on

WGAN-GP are effective. However, in order to improve the resistance of the generated DDoS attack data to the detection model, the generated data still needs to be disturbed. Specifically, SHAP analysis was performed on 260,000 samples. The results are shown in Fig. 5 The horizontal axis represents the average value of absolute SHAP, and the vertical axis represents specific features, with the importance decreasing from top to bottom.

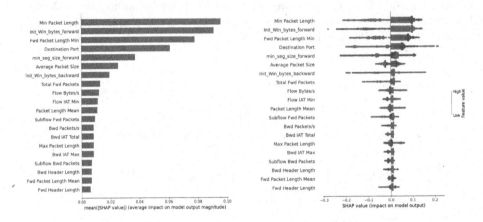

Fig. 5. Feature ordering.   **Fig. 6.** Feature influence ordering.

It can be found that among the 20 features, the feature of packet length type and the feature of byte number type are more important for detecting DDoS attacks. When an attacker sends a request packet to the victim, these packets are intended to establish the communication between the attacked, and do not contain real and valid data. Therefore, the characteristics of the packet length and number of bytes to a large extent indicate the existence of the attack. When subjected to a DDoS attack, packets flood the device in a short period of time, so the characteristics of the time type are also helpful in detecting the attack. However, it still cannot reveal whether the feature has positive or negative impact on the output result, evne to understand the impact of each feature.

As shown in Fig. 6, the horizontal axis represents the SHAP value, the vertical axis represents the feature, a point represents a sample, a wide area represents a large number of samples gathered here, and the color of the point represents the size of the sample's feature value. The darker the red, the larger the feature value, and the lighter the blue, the value smaller. It can be found that the red point in the Min Packet Length feature is located at the right end of the SHAP value, that is, the larger the value of the feature, the larger the SHAP value, and the greater the positive impact on the model. The lower the value of the feature of the total number of bytes sent backward (Init_win_bytes_backward) in the initial, the larger the SHAP value, and the more correct result would be, however, the higher the feature value, the smaller the SHAP value, and the worse result would be. In the feature of min_seg_size_forward, the minimum length of the data segment (min_seg_size_forward) in forward forwarding, there are many dark red points on both sides of the value, that is, when the feature value is large, it may to play a positive role in the detection, or a small SHAP value to play a negative role. Therefore, when the feature value is large, it may cause a large SHAP value to play a negative role, but when the characteristic value is small, it is concentrated on the right side of the SHAP value, which plays a positive role to the result, so the smaller the characteristic value, the better.

6 Conclusions

We conducted a comprehensive analysis and evaluation of multi-agent collaborative intrusion detection for few-shot sample cyber attack scenarios. In this study, we proposed a framework based on WGAN-GP that incorporates few-shot sample data augmentation. To assess the efficacy of the data enhancement, we employed the NSL-KDD, CIC-IDS 2017 and CICDDoS 2019 datasets and constructed networks with varying structures. Subsequently, XGBoost, LSTM, CNN, and an improved-CNN algorithm were selected to evaluate the impact of data enhancement on detection performance. Furthermore, a comparative analysis of algorithm performance was conducted to investigate the collaborative detection effectiveness of multi-agents in a multi-link setting. The experimental results indicate that optimizing the parameter interaction among agents effectively enhances the detection rate of few-shot sample cyber attacks. However, it should be noted that while the utilization of WGAN-GP enables data augmentation, the recognition rate of few-shot sample attack types still requires improvement when compared to large-sample data. In recent years, there has been a limited amount of research conducted on multi-agent based network intrusion detection. Consequently, conducting a comprehensive horizontal comparison of related technical aspects has proven challenging. In future work, we propose exploring more intricate sampling methods to achieve better data distribution. Additionally, techniques such as active learning and cost-sensitive learning can be further explored to enhance the recognition rate of few-shot sample. Moreover, we firmly believe that different technical routes exhibit their unique strengths in addressing various technical challenges. Consequently, there remains a need to continuously explore novel technologies, including the utilization of classical multi-agent systems, remains imperative in order to effectively adapt to the ever-evolving complexity presented by cyber attacks.

References

1. Ahmad, Z., Shahid Khan, A., Wai Shiang, C., Abdullah, J., Ahmad, F.: Network intrusion detection system: a systematic study of machine learning and deep learning approaches. Trans. Emerg. Telecommun. Technol. **32**(1), e4150 (2021)
2. Albini, E., Long, J., Dervovic, D., Magazzeni, D.: Counterfactual shapley additive explanations. In: 2022 ACM Conference on Fairness, Accountability, and Transparency, pp. 1054–1070 (2022)

3. Arora, S., Ge, R., Liang, Y., Ma, T., Zhang, Y.: Generalization and equilibrium in generative adversarial nets (GANs). In: International Conference on Machine Learning, pp. 224–232. PMLR (2017)
4. Arulkumaran, K., Deisenroth, M.P., Brundage, M., Bharath, A.A.: Deep reinforcement learning: a brief survey. IEEE Signal Process. Mag. **34**(6), 26–38 (2017)
5. Ghosh, A., Kulharia, V., Namboodiri, V.P., Torr, P.H., Dokania, P.K.: Multi-agent diverse generative adversarial networks. In: Proceedings of the IEEE Conference on Computer Vision and Pattern Recognition, pp. 8513–8521 (2018)
6. Goodfellow, I., et al.: Generative adversarial networks. Commun. ACM **63**(11), 139–144 (2020)
7. Gronauer, S., Diepold, K.: Multi-agent deep reinforcement learning: a survey. Artif. Intell. Rev., 1–49 (2022)
8. Gulrajani, I., Ahmed, F., Arjovsky, M., Dumoulin, V., Courville, A.C.: Improved training of Wasserstein GANs. In: Advances in Neural Information Processing Systems, vol. 30 (2017)
9. Jais, I.K.M., Ismail, A.R., Nisa, S.Q.: Adam optimization algorithm for wide and deep neural network. Knowl. Eng. Data Sci. **2**(1), 41–46 (2019)
10. Liu, M., Chang, W., Li, C., Ji, Y., Li, R., Feng, M.: Discrete interactions in decentralized multiagent coordination: a probabilistic perspective. IEEE Trans. Cogn. Dev. Syst. **13**(4), 1010–1022 (2021)
11. Liu, M., et al.: Modeling and analysis of the decentralized interactive cyber defense approach. China Commun. **19**(10), 116–128 (2022)
12. Niu, W., Zhou, J., Zhao, Y., Zhang, X., Peng, Y., Huang, C.: Uncovering apt malware traffic using deep learning combined with time sequence and association analysis. Comput. Secur. **120**, 102809 (2022)
13. Palli, A.S., Jaafar, J., Hashmani, M.A., Gomes, H.M., Gilal, A.R.: A hybrid sampling approach for imbalanced binary and multi-class data using clustering analysis. IEEE Access **10**, 118639–118653 (2022)
14. Zhang, J., Wang, T., Ng, W.W., Zhang, S., Nugent, C.D.: Undersampling near decision boundary for imbalance problems. In: 2019 International Conference on Machine Learning And Cybernetics (ICMLC), pp. 1–8. IEEE (2019)
15. Zhang, Z., Zeng, Y., Bai, L., Hu, Y., Wu, M., Wang, S., Hancock, E.R.: Spectral bounding: strictly satisfying the 1-Lipschitz property for generative adversarial networks. Pattern Recogn. **105**, 107179 (2020)

Long Short-Term Deterministic Policy Gradient for Joint Optimization of Computational Offloading and Resource Allocation in MEC

Xiang Lei, Qiang Li$^{(\boxtimes)}$, Peng Bo, Yu Zhu Zhou, Cheng Chen,
and Si Ling Peng

School of Information Engineering, Southwest University of Science and Technology,
Mianyang, China
liqiangsir@swust.edu.cn

Abstract. Mobile Edge Computing (MEC) is regarded as a promising paradigm for reducing service latency in Mobile users data processing by providing computing resources at the network edge. Existing deep reinforcement learning (DRL) algorithms struggle to effectively handle the joint optimization of computational offloading and resource allocation (JCORA). To overcome this challenge, we propose a Long Short-Term Deterministic Policy Gradient (LSTDPG) approach to tackle JCORA. Building upon the Deep Deterministic Policy Gradients (DDPG) algorithm, LSTDPG incorporates two key features. Firstly, it utilizes a Temporal Attention Network composed of Long Short-Term Memory (LSTM) networks, which facilitates high-quality state representation and function approximation. Secondly, an Episode-Based Prioritized Experience Replay (ePER) method is introduced to expedite and stabilize the convergence of model training. Experimental results demonstrate that the proposed LSTDPG outperforms several state-of-the-art DRL agents in terms of task completion time and energy consumption.

Keywords: Mobile Edge Computing · Computational offloading ·
Deep reinforcement learning · Deep deterministic policy gradient ·
Resource allocation

1 Introduction

In recent years, Mobile Edge Computing (MEC) has become a popular topic due to its ability to offload tasks from resource-constrained mobile devices (MDs) to edge servers [1]. This reduces latency, saves energy consumption, and improves security, making it superior to traditional mobile cloud computing [2].

This work is supported by Heilongjiang Provincial Science and Technology Program (No. 2022ZX01A16) and Sichuan Science and Technology Program (No. 2022 YFG0148).

© The Author(s), under exclusive license to Springer Nature Singapore Pte Ltd. 2024
Z. Tari et al. (Eds.): ICA3PP 2023, LNCS 14492, pp. 329–348, 2024.
https://doi.org/10.1007/978-981-97-0811-6_20

Determining the optimal task offloading strategy in the context of edge servers and resource-constrained mobile devices is an urgent problem. As the scale and complexity of tasks increase, the efficiency of task offloading also varies [3]. Therefore, an effective strategy is needed to offload tasks and optimize resource utilization in resource-constrained edge computing environments [4]. To address this problem, we need to consider multiple aspects such as alleviating computational burden, extending the battery life of mobile devices, minimizing task offloading latency and power consumption. A balanced solution that takes into account various factors needs to be found.

Furthermore, it is important to recognize that both computational and wireless resources are limited in mobile edge computing environments [5]. Efficient resource allocation is vital for mobile edge computing. Effective coordination and allocation prevent delays and excessive power use. Deep reinforcement learning offers effective solutions for such optimization challenges.

In the traditional DDPG model, fully connected networks (FCNs) are used as feature extraction networks in actor and critic networks [6]. However, FCNs have a large number of trainable weights, leading to high computational complexity. FCNs are inadequate for capturing the temporal dynamics and global features of tasks, particularly in computational tasks with complex temporal data. Concentrated task offloading from mobile devices to MEC servers can lead to server overload and significant delays, potentially causing dropped tasks. The dynamic workload of MEC servers requires decentralized offloading decisions, which poses challenges in task modeling and decision-making for each device.

To address the aforementioned limitations, the present article proposes the Long Short-Term Deterministic Policy Gradient (LSTDPG) as an improved DDPG proxy to resolve the decentralized JCORA problem within dynamic Mobile MEC environments. Executing the LSTDPG agents on each MD reduces control costs between the MD and its corresponding MEC server. Thus, compared to centralized JCORA methods, LSTDPG is better equipped for large-scale scenarios. The chief contributions of this paper are outlined below:

- This study investigates the JCORA problem in a MD and MEC system. The objective is to minimize a composite objective function by enabling complete task offloading to the MEC server. Each MD employs an LSTDPG agent, which autonomously makes decisions based on local JCORA information and environmental factors. The research focuses on deriving optimal JCORA policies for individual tasks under varying network conditions.

- LSTDPG combines the TAN and ePER. TAN, an attentional LSTM network, enhances the algorithm's representation and learning capabilities by capturing critical temporal features. ePER prioritizes significant episodic experiences during training, leading to accelerated learning and improved stability.

- The evaluation of the JCORA mechanism with LSTDPG is based on performance metrics such as long-term reward, completed tasks, task completion rate, and average computation time and energy consumption. Experimental results demonstrate the superiority of LSTDPG over other DDPG variants and state-of-the-art DRL agents, achieving better convergence and improved performance in various test scenarios.

The rest of thr paper is organized as follows. Section 2 investigates related works. Section 3 presents the system modeling and problem formulation. Section 4 introduces the proposed method in details. Section 5 provides the simulation results and Sect. 6 concludes this work.

2 Related Work

In order to tackle the challenges of computation offloading and resource allocation in MEC systems, researchers have proposed a range of methodologies, which can be broadly categorized into optimization-based and machine learning-based techniques. These approaches aim to optimize the allocation of computational tasks and resources in MEC systems to enhance overall system performance.

2.1 Optimization-Based Approach

The academic literature presents numerous approaches for task scheduling and offloading. One particular problem, known as the static JCORA problem, can be formulated as a mixed integer nonlinear programming problem. For example, Huang *et al.* [7] presents a comprehensive model for jointly constructing a system resource allocation scheme in MEC based on mixed integer nonlinear programming. The model aims to optimize the distribution of computational resources in a wholesale manner. Furthermore, the authors propose a methodology to convert the original non-linear formulation into a linear one to address challenges. Ahani *et al.* [8] proposed a task offloading method utilizing the Lagrangian dual algorithm. However, these precise methods often involve computationally intensive operations, such as matrix inversion and singular value decomposition, which can lead to slow decision-making for offloading tasks.

Certain researchers employ game theory as an approach to address the optimization problem of multi-user JCORA in static scenarios. For example, Wang *et al.* [9] propose a task offloading decision model using game theory-based task migration. The objective of this model is to maximize the system utility by considering the task offloading decisions of multiple end users. He *et al.* [10] introduced a game theory approach known as EUAGame, which is utilized to formalize the marginal user allocation problem as a potential game. And conducted an analysis of the game and provided proof of the existence of a Nash equilibrium within the game. However, these methods exhibit high computational complexity, particularly when dealing with large-scale scenarios.

In scenarios where the network environment is static or quasi-static, such as when the wireless channel state remains constant over time, meta-heuristic algorithms are often regarded as suitable options. For example, Vijayaram *et al.* [11] proposed a method for offloading task computation and resource allocation based on a hyperheuristic framework, which aims to minimize the delay and energy consumption of wireless devices by considering the two indicators of computation time and energy consumption. Real MEC networks are dynamic and require efficient responses to offload requests. Meta-heuristic algorithms, although effective in some scenarios, may not be suitable due to their computational limitations and slow convergence. Faster and more adaptable approaches are desired to meet the needs of real MEC networks.

2.2 Machine Learning Based Approach

Dynamic offloading decision-making is indeed a complex task due to the presence of multiple dimensions and time-varying factors. In recent years, there has been a growing interest in leveraging machine learning-based techniques, particularly deep learning (DL) and deep reinforcement learning (DRL), to address this challenge. These approaches have shown promise in capturing complex patterns and making intelligent offloading decisions that adapt to changing network conditions and user requirements.

Deep Learning Method. Deep learning models [12] have the ability to learn from labeled historical data, allowing them to predict future computation offloads. By leveraging these predictive capabilities, MEC systems can make informed decisions regarding task offloading and resource allocation. This enables the system to optimize its operations and allocate resources in a rational and efficient manner. Zhao et al. [13] developed a deep learning model that utilizes multiple LSTM networks to predict real-time traffic patterns of Small Base Stations (SBS) in a mobile network. Based on the predictions obtained from this model, they propose a mobile data offloading strategy using the cross-entropy (CE) method. Indeed, the training time required for deep learning models can be a significant challenge in real-time decision-making scenarios. The lengthy response time associated with model training is a limitation that needs to be addressed.

Deep Reinforcement Learning Methods. DRL is well-suited for dynamic MEC systems as it enables the learning of optimal policies and facilitates rapid decision-making through interactions with time-varying environments. There are primarily two types of DRL methods: value-based methods and policy-based methods.

Value-based methods can effectively estimate the value of different actions or states in a given environment. For example, Sadiki et al. [14] introduced an algorithm that utilizes DQN to address the challenge of state space explosion and computational offloading in massive multiple-input multiple-output (MIMO) MEC systems. The algorithm efficiently manages the large state space and optimizes the computational offloading process in MEC systems with massive MIMO configurations. Lin et al. [15] developed a computational offloading model utilizing the Markov decision process (MDP). They proposed a computational offloading strategy based on DRL, leveraging the deep Q network (SA-DQN) algorithm with simulated annealing to optimize joint objectives. The aforementioned DRL-based methods demonstrate satisfactory performance without the need for prior knowledge of environment statistics.

Value-based methods encounter challenges with continuous action spaces in dynamic JCORA problems. Methods like DQN require careful selection of discretization level. High discretization may lose valuable behavioral information, while low discretization increases complexity. Striking the right balance is crucial for effective and efficient application of value-based methods in dynamic JCORA.

Policy-based methods can efficiently approximate a policy function that maps states to actions in a given environment. For instance, Cao et al. [16] proposed a hybrid computing offloading framework based on the DDPG algorithm. Aimed to optimize the overall overhead of the MEC system by considering both network load and computing load. This approach effectively balances the computational resources and network conditions, leading to enhanced system performance and improved user experience in MEC environments. Ebrahim et al. [17] formulating the offloading problem as a Markov decision problem (MDP), their method aims at achieving the optimization goal of minimizing latency and energy consumption. By applying DRL, their method can make intelligent decisions during offloading, thereby improving the efficiency and resource utilization of MEC systems.

Policy-based approaches have demonstrated greater effectiveness in addressing dynamic JCORA problems compared to value-based approaches. Specifically, the decentralized JCORA mechanism based on DDPG has exhibited promising capabilities in handling large-scale MEC scenarios [18]. By leveraging policy-based methods, the DDPG-based JCORA mechanism enables efficient decision-making in complex and dynamic MEC environments, leading to improved resource allocation, task offloading, and system performance.

Traditional DDPG methods have limitations in extracting temporal information from task data and can experience convergence issues with uniform experience replay techniques. To address these, we propose LSTDPG, an improved algorithm for decentralized JCORA in multi-MD and multi-MEC networks. LSTDPG incorporates LSTM networks to capture rich temporal features and enhance performance and convergence speed in dynamic MEC environments.

3   System Model and Problem Formulation

In this section, we will provide an overview of the system. We will then delve into the details of the processing delay and energy consumption in the computing models. Following that, we will discuss the transmission delay and energy consumption in the communication model. Lastly, we will present the problem formulation.

3.1   System Model

As shown in Fig. 1, there is a group of n MDs $M = \{m_1, \cdots, m_n\}$ during a given time slot t. The MDs rely on a set of k MEC servers $S = \{s_1, \cdots, s_k\}$ to process tasks through computing offloading. The task of the ith user can be

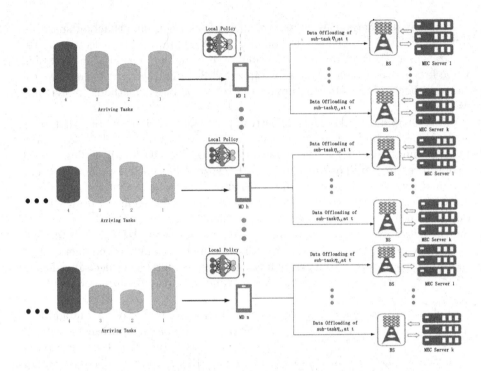

Fig. 1. An example MEC system.

defined as $J_i = \{\pi_i, d_i, c_i, \tau_{\max}\}$. Among them π_i, d_i, c_i, τ_i represent the time slot when the task arrives, the data volume of the task, the CPU cycle required to calculate the task and the maximum tolerance delay (expiration time) of the task. Further, tasks can be divided into smaller sub-tasks and offloaded to different MEC servers for parallel processing.

Especially in the system model of this system, we assume that the task can be divided into multiple independent subtasks by means of data segmentation and does not consider the relationship of data dependencies. So a task J_i can be divided into several smaller sub-tasks, $\eta_i = \{\eta_{i,1}, \cdots, \eta_{i,j}, \cdots, \eta_{i,k}\}$ $i \leq n, j \leq k$, where j is the index of the sub-task, and the number of sub-tasks is not greater than the number k of MEC servers.It takes $\tau_{i,j}$ to complete each sub-task $\eta_{i,j}$, that is, the time cost of all sub-tasks of J_i can be expressed as a vector $\tau_i = (\tau_{i,1}, \cdots, \tau_{i,j}, \cdots, \tau_{i,k}) i \leq n, j \leq k$. When calculating the time cost, it is necessary to consider the involvement of task data and result transmission between the mobile device and the MEC server. In particular, when a task offloads a portion of its input data to the server, it introduces delays in the transmission from the corresponding mobile device to the server. Additionally, this offloading process incurs energy consumption due to the transmission of data from the mobile device. It is worth noting that in this context, we neglect the transmission delay between the BS and the MEC server. This is because the connection between the

BS and the MEC server ensures faster data delivery compared to the wireless channel, thus minimizing the impact of transmission delays on the overall time cost.

The result of executing a task at the MEC server is typically much smaller in size compared to the partial input data that is offloaded from the MD. As a result, this paper disregards the transmission delay and energy consumption associated with sending the result back to the corresponding MD. Additionally, this paper utilizes MIMO as the air interface technology between the MD and the BS, leveraging its advantages in terms of increased capacity and improved spectral efficiency.So the time cost $\tau_{i,j}$ of sub-task $\eta_{i,j}$ can be calculated by:

$$\tau_{i,j} = \tau_{i,j}^T + \tau_{i,j}^C + \tau_{i,j}^R \tag{1}$$

where $\tau_{i,j}^T, \tau_{i,j}^C, \tau_{i,j}^R$ denote the transfer time, computation time and waiting time before being processed on the MEC server for sub-task $\eta_{i,j}$, respectively. The transmission time $\tau_{i,j}^T$ can be calculated by:

$$\tau_{i,j}^T = \frac{d_{i,j}}{\varsigma_{i,j}} \tag{2}$$

$d_{i,j}$ represents the size of the jth sub-task, $\varsigma_{i,j}$ represents the data transfer rate between the ith md and the jth MEC server. In calculating the data rate $\varsigma_{i,j}$, we assume that the channel state information (CSI) i.e., is known perfect CSI scenarios.

Let $G_i = [g_{i,1}, \cdots, g_{i,j}, \cdots, g_{i,k}]$ denote the $N \times K$ channel matrix between the ith md and all MEC servers.$g_{i,j}$ is the channel gain vector of $N \times 1$, which represents the channel gain between the ith mobile device and the jth MEC server, where N is the number of antennas. The transmission rate $\varsigma_{i,j}$ can be calculated by [19]:

$$\varsigma_{i,j} = Blog_2(1 + \frac{P_{i,j}^{tra}|g_{i,j}|^2}{N_0}) \tag{3}$$

where B is the bandwidth, $P_{i,j}^{tra}$ means that the ith md sends the uplink transmission power of the subtask $\eta_{i,j}$ to the jth MEC server, N_0 is the power of Gaussian white noise.

Once sub-task $\eta_{i,j}$ is transmitted to the MEC server, it is placed in a queue and awaits processing. The server processes tasks in a sequential manner, meaning that sub-task j will need to wait until all the tasks preceding it in the queue have been processed by the server. Assuming that there are i^{th} tasks or sub-tasks unloaded to the jth MEC server before unloading sub-task $\eta_{i,j}$, the waiting time $\tau_{i,j}^R$ of the jth sub-task in the queue on the jth MEC server can be calculated as:

$$\tau_{i,j}^R = \sum_{k=1}^{i} \tau_k^C \tag{4}$$

Among these factors, the $\tau_k^C = \frac{c_k}{f_k}$ and c_k represent the number of CPU cycles required to compute the offloaded sub-task within the MEC service. Additionally, f_k represents the computation frequency allocated by the MEC server for performing the computation of the sub-task. Sub-task $\eta_{i,j}$ will be processed after all sub-tasks preceding the current task have been completed, and the processing time is given by:

$$\tau_{i,j}^C = \frac{c_{i,j}}{f_{i,j}} \quad c_{i,j} \in c_i \tag{5}$$

where $c_{i,j}$ is the CPU cycle required to compute the jth sub-task of J_i, and $f_{i,j}$ is the frequency at which the MEC server processes this sub-task. So for task J_i, the task completion time can be expressed by the following formula:

$$T_i = \max(\tau_i) \tag{6}$$

where τ_i is the time cost of the corresponding sub-task. Task J_i is considered successfully completed if all its sub-tasks are processed and completed before the maximum tolerated delay. If the last sub-task is processed beyond the expiration time, the task is considered expired and cannot provide a response to the user. The energy consumption of processing task J_i by the mobile device during offloading and computing operations can be calculated using the following formula:

$$E_i = E_i^C + E_i^T \tag{7}$$

The energy consumption E_i^T caused by transmission, which is the sum of the energy consumption of subtask transmissions, can be calculated using the following formula [20]:

$$E_i^T = \sum_{j=0}^{k} \tau_{i,j}^T P_{i,j}^{tra} \tag{8}$$

Meanwhile, for calculating energy consumption E_i^C can be calculated as [21]:

$$E_i^C = \sum_{j=0}^{k} c(f_{i,j})^2 C_{i,j} \tag{9}$$

where c is a constant capacitive load per clock cycle, and $f_{i,j}$ is the frequency used to calculate the subtask j^{th}.

3.2   Problem Formulation

For each J_i, define its decision cost as the following formula:

$$C_i = \varepsilon T_i + (1 - \varepsilon)E_i \tag{10}$$

where C_i is the trade-off between the completion time and energy consumption of task J_i, and ε is the weight coefficient.

Therefore, the objective function can be defined as the following formula:

$$\min_{a_i} \sum_{i}^{n} \varepsilon T_i + (1 - \varepsilon) E_i$$

$$s.t. \ \tau_{i,j}^T + \tau_{i,j}^C + \tau_{i,j}^R \leq \tau_{\max},$$

$$f_{i,j}^k \leq f_j^{\max}, f_{i,j}^k \in \Gamma_i,$$

$$P_{i,j}^{tra} \leq P_{i,j}^{\max}, \tag{11}$$

$$a_i = \{\Phi_i, \Gamma_i\},$$

$$\Phi_i = \{\phi_0, \cdots, \phi_j, \cdots, \phi_k\},$$

$$\sum_{j=0}^{k} \phi_j = 1, 0 \leq \phi_j \leq 1$$

The dynamic JCORA problem aims to optimize resource allocation and task offloading decisions in the MEC system to minimize average long-term costs. Two key constraints are considered: task completion time and resource limitations. The first constraint ensures that tasks are completed within their maximum tolerable delay, satisfying quality-of-service requirements. The second constraint limits the CPU cycles per second of MEC servers, ensuring efficient resource utilization and preventing overloading. The third constraint controls the power consumption of mobile devices during subtask transmission to the Base Station, preventing excessive energy usage. Each action a_i contains two vectors, Φ_i for task division and Γ_i for frequency control. Specifically for Φ_i and Γ_i, $\Phi_i = \{\phi_0, \cdots, \phi_j, \cdots, \phi_k\}$ is to divide tasks according to the number of MEC servers, ϕ_j indicates the percentage of tasks j^{th} offloaded to the MEC server; $\Gamma_i = \{f_0, \cdots, f_j, \cdots, f_k\}$ indicates the recommended frequency of processing each sub-task.

4 The Proposed LSTDPG Agent

In our proposed system model, both state and action spaces are continuous, requiring a DRL model capable of handling them. To address this, we adopt a policy-based approach in this study. Our algorithm, based on DDPG, incorporates two key features: TAN and ePER. Following DDPG principles, each MD agent maintains state, action, and reward as fundamental components.

4.1 Three Basic Elements

The agent works by getting the current MD state from the corresponding MD. The agent then generates an action based on S_t through policy guidance (e.g. a deterministic policy). Then, actions a_t are performed on MD, and these performed actions get immediate reward $r(S_t, a_t)$. By finding the optimal policy of action, the algorithmic agent aims to maximize long-term reward during the decision-making process R_t.

The DRL agent undergoes several steps during the learning process. It selects actions based on environment observations and receives feedback in the form of rewards and the next state. This interaction data is stored in an experience replay buffer for training the model. The agent continuously generates training datasets by interacting with the environment.

The training data is then used to update the learning network within the DRL model. Each network has a corresponding target network for stabilization. By updating the network's parameters, the agent improves its decision-making abilities. The state space, action space, and reward function are described as follows.

State: Although the DRL agent cannot fully observe the MEC environment, which means that observation is not equivalent to state, we assume that observation is the same as other reinforcement learning methods. The state S_t of the MD in the time slot t contains two attributes, one is the task attribute, including task arrival time slot π_i, input data size d_i, the number of CPU cycles required to process each input data bit c_i and the maximum tolerable time τ_i. The other is the environment attribute, including the maximum computing power of all MEC servers $F^{\max} = \{f_0^{\max}, \cdots, f_j^{\max}, \cdots, f_k^{\max}\}$, the maximum uplink transmission power $P_i^{\max}$, the load of all MEC servers $Z(t) = \{z_0, \cdots z_k\}$, and the uplink channel gain matrix $G_i(t)$ between MD and BS. In other words, we have $S_i(t) = \{\pi_i, d_i, c_i, \tau_i, F^{\max}, P_i^{\max}, Z(t), G_i(t)\}$.

Action: Each action has two vectors, one vector $\Phi_i = \{\phi_0, \cdots, \phi_j, \cdots, \phi_k\}$ is used to divide the task into multiple subtasks according to the number of MEC servers, and the other vector $\Gamma_i = \{f_0, \cdots, f_j, \cdots, f_k\}$, where k is the number of MEC servers. Specifically, DRL specifies that tasks of p_j percent size are offloaded to j^{th} MEC servers. At the same time, it recommends that the jth MEC server use f_j percent of the maximum CPU frequency to process subtasks. All elements are concatenated and in the range of $[0, 1]$, so we can define constraints as $\varphi_j \in [0, 1]$ and $f_j \in [0, 1]$. When $\varphi_j = 0$, j^{th} server does not accept subtasks. Additionally, the sum of task slice scales must equal 1. Therefore, action a_i can be set as:

$$a_i = \{\Phi_i, \Gamma_i\}$$

$$s.t.\, 0 \leq \varphi_j \leq 1, \sum_{j=0}^{k} \varphi_j = 1 \tag{12}$$

Reward: The long-term reward function R_i of MD m_i is defined in the following equation:

$$R_i = \sum_{t=1}^{T} \gamma^t r\left(S_i(t), a_i(t)\right) \tag{13}$$

where the discount factor $0 \leq \gamma \leq 1$ represents the importance of each instant reward $r\left(S_i(t), a_i(t)\right)$ obtained over time, and the instant reward $r\left(S_i(t), a_i(t)\right)$ can be given by:

$$r\left(S_i(t), a_i(t)\right) = \wedge - \varepsilon T_i(t) + (1 - \varepsilon)E_i(t) \tag{14}$$

The $\wedge+1$ when task is completed. The goal of our LSTDPG agent is to minimize R_t over a long time span.

4.2 Framework Outline

The structure of LSTDPG is shown in the Fig. 2. This agent is a variant of DDPG with two performance enhancements, including TAN and ePER.

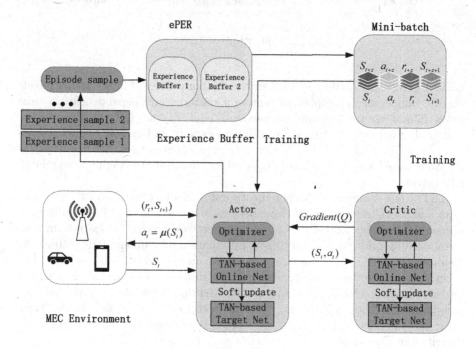

Fig. 2. Structure of the LSTDPG agent.

The LSTDPG utilizes actor and critic networks, each consisting of an online network and a target network. Both the actor and critic networks have identical architectures. However, instead of using a FCN for feature extraction and representation learning, a TAN is employed. The TAN functions as both the online and target network in the actor and critic networks. It extracts relevant features and learns effective representations for decision-making. By incorporating TAN, the algorithm captures temporal dependencies and patterns in the data, improving the agent's understanding of system dynamics. The actor network generates actions based on MD states, while the critic network evaluates actions using the action value function $Q(S_t, a_t)$.

The ePER buffer stores historical experiences for training the TAN-based network. It contains Z mini-batch samples selected from the buffer. TAN is an architecture that combines a LSTM network with a soft attention mechanism. This architecture enables the network to capture long-distance sequential patterns and focus on important reasoning information within these patterns. It effectively captures temporal dependencies and extracts meaningful features from sequences, making it suitable for dynamic and sequential data in the JCORA problem. The structure of TAN is shown in the Fig. 3.

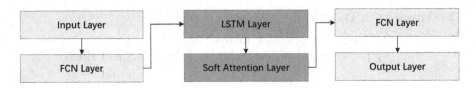

Fig. 3. Structure of TAN.

LSTM Layer. FCNs do not consider the temporal order of input data, neglecting the importance of capturing temporal patterns and dependencies in time series data. They treat inputs independently and may not effectively model the temporal dynamics, limiting their ability to explore and exploit temporal associations. In the context of time series data, alternative approaches like RNNs or attention-based architectures are preferred for effectively leveraging time-related information.

LSTM is a type of RNN that incorporates memory units to capture long-term and short-term temporal dependencies in sequential data [22]. Unlike traditional fully connected layers, LSTM can effectively model and learn from time series data by preserving and utilizing information over different time steps. As mentioned earlier, TAN combines the LSTM and the attention mechanism to effectively capture temporal dependencies in time series data. This is achieved by cascading an LSTM layer and a soft attention layer in the TAN network architecture. In the LSTM layer, the input vector O_t, the previous hidden state h_{t-1}, and the network's previous cell state c_{t-1} are used to update c_t and h_t, as described in Equation:

$$(c_t, h_t) = LSTM(O_t, c_{t-1}, h_{t-1}) \tag{15}$$

Soft Attention Layer. Traditional LSTM models update the hidden state based on previous states and current inputs, enabling them to capture long-term dependencies. However, they may struggle with capturing future information and identifying important events in the sequence.

To address this limitation, attention mechanisms have been proposed, enabling models to selectively focus on crucial parts of the sequence [23]. By incorporating attention mechanisms, models can effectively capture and leverage information from different sequence segments, including those with significant impact on future developments. Soft attention effectively assigns weights to critical information for accurate forecasts. It is commonly added after LSTM layers in the time direction to emphasize important temporal relations. Hence, we incorporate a soft attention layer after the LSTM layer to highlight significant temporal relationships in the input during critical moments.

In the soft attention layer, c_t and h_t are merged into a set of state summary vectors $U = [u_1, u_2, \cdots, u_T]$, where u_t is the feature vector in slot t. Let the temporal attention probability distribution in t be denoted by $A_{t,\hat{t}}$, where $\hat{t}$ is a slot earlier than t. Then, $A_{t,\hat{t}}$ is calculated based on u_t and $h_{\hat{t}}$, as defined in the equation:

$$A_{t,\hat{t}} = \frac{\exp(u_t^T \cdot h_{\hat{t}})}{\sum_{\hat{t}=1}^{T} \exp(u_t^T \cdot h_{\hat{t}})} \tag{16}$$

where u_t^T is the transpose of u_t, $A_{t,\hat{t}}$ indicates the extent to which the input in slot $\hat{t}$ contributes to the output of slot t, and $\hat{h}_t$ is the output hidden state, given by the equation:

$$\hat{h}_t = \sum_{t=1}^{T} A_{t,\hat{t}} \cdot h_{\hat{t}} \tag{17}$$

ePER. The ePER algorithm solves the issue of changing experience samples by employing a classification-based experience replay method. By storing experiences in two buffer pools based on their cumulative return value, higher-performing experiences are prioritized and replayed during training. This adaptive approach accommodates the dynamic nature of the buffer pool, allowing for continual optimization over time.

During initialization, the network model sets the average cumulative return of an episode to zero. After each episode, the average value is updated and compared with the cumulative return. Based on this comparison, the episode is stored in either experience buffer pool 1 or pool 2. During training, the algorithm prioritizes experience samples from the pool with higher cumulative returns. This weighting scheme enhances training quality, ensuring that experiences with better performance have a greater influence on the learning process and overall optimization of the network model.

The PER algorithm involves storing experience samples generated by the agent in the experience buffer pool, requiring frequent updates to their priority [24]. To alleviate the need for frequent priority assignment and updates, the ePER algorithm introduces a classification step based on the importance of the experiences. This classification enables more efficient training by selecting batches of experience samples from the different buffer pools during network training. This classification process helps reduce the computational overhead and improves the efficiency of managing experience samples.

Training Process. As aforementioned, there are N LSTDPG agents in the MEC system, each running on an MD. By interacting with a local MD, each agent makes independent JCORA decisions that are taken actions in that MD. Without loss of generality, we introduce the training process of the LSTDPG agent on MD n, $n = 1 \cdots N$, in Algorithm 1.

In the initialization phase, variables and the network are randomly initialized. Temporary and two experience buffers are created, and the network is copied to the target network. Two networks, actor and critic, are utilized with corresponding target networks to ensure training stability. The experience replay buffer stores the training data collected during interactions with the MEC network environment.

Algorithm 1. Training Process of the LSTDPG Agent on MD n

Input: $M, T, N_1, N_2, \gamma, \tau$

1: Initialize temporary replay memory D and replay memory D_1, D_2;
2: Randomly initialize θ^Q and θ^μ;
3: Initialize actor-network $\mu(s|\theta^\mu)$ with weights θ^μ;
4: Initialize critic-network $Q(S, a|\theta^Q)$
5: Initialize target networks Q' and μ' weights: $Q' \leftarrow \theta^Q$, $\mu' \leftarrow \theta^\mu$;
6: **for** episode=1 to M **do**
7:    Initialize cumulative return average:δ=0;
8:    Preprocess initial state: S;
9:    **for** t=1 to T **do**
10:       Select action according to the current policy and exploration noise:
         $a_t = \mu(s_t|\theta^\mu) + N_t$;
11:       Execute action a_t and observe reward r_t and next state S';
12:       Store experience (S, A, R, S') in D;
13:       **if** the D buffer is full **then**
14:          Calculate the average of cumulative returns: $\delta=$ Average(D);
15:          Store the episode experience samples in the temporary replay memory
            D in the replay memory D_1 or D_2 according to the comparison with
            the average cumulative return value;
16:          Empty temporary replay memory D;
17:       **end if**
18:       **if** the D_1 or D_2 buffer is full **then**
19:          Obtain random N_1 of experience samples from D_1 and N_2 of experi-
            ence samples from D_2 : mini-batch=$N_1 + N_2$;
20:          $y_i = r_i + \gamma Q'(s_{i+1}, \mu'(s_{i+1}|\theta^{\mu'}))$;
21:          Update critic by minimizing the loss:
            $L(\theta) = \frac{1}{N} \sum_i (y_i - Q(s, a|\theta_i^Q))^2$;
22:          Update critics:$\theta_i \leftarrow \min_{\theta_i} L(\theta)$;
23:          Update the actor policy using the sampled policy gradient:
            $\nabla_{\theta^\mu} J \approx \frac{1}{N} \nabla_a Q(s, a|\theta^Q) \nabla_{\theta^\mu} \mu(s|\theta^\mu)$;
24:          Update the target critic network:
            $\theta^{Q'} \leftarrow \tau\theta^Q + (1 - \tau)\theta^{Q'}$;
25:          Update the target actor network:
            $\theta^{\mu'} \leftarrow \tau\theta^\theta + (1 - \tau)\theta^{\mu'}$;
26:       **end if**
27:    **end for**
28: **end for**

The second block collects data by interacting with the MEC environment, generating training samples consisting of the current state, reward, and next state. These samples are stored in a temporary replay buffer. When the buffer is full, the average cumulative return is calculated. Samples with higher cumulative returns are stored in experience buffer 1, while samples below the average are stored in experience buffer 2. The empirical replay buffer has a fixed size and discards the oldest data when new data is added.

The third block trains the network model. When either experience buffer 1 or 2 is full, samples with a total size equal to the mini-batch size are retrieved from the respective buffer for training. The policy is optimized based on the critic's Q-value, following the actor-critic setup. The target network is updated using a combination of soft update and delayed update methods. Soft updates preserve a significant portion of the original weights, allowing for more frequent network updates without high variance. The update of the target network is controlled by the weight factor τ.

5   Simulation Results

In this section, we provide a comprehensive overview of the simulation setup and analysis of the obtained results. The implementation of the simulation involves the use of specific tools and frameworks for data preprocessing and DRL model construction. For data preprocessing, we utilize the Numpy [25] library. To build the DRL model, we employ PyTorch, a popular deep learning framework that provides a wide range of functionalities for building and training neural networks.

The simulation is divided into two main parts: the MEC network environment and the DRL model. The MEC network environment consists of various network entities, such as edge servers and mobile users. Given the heterogeneous nature of MEC networks, each edge server is configured with different computing resources, while mobile users generate diverse tasks for offloading. Additionally, the MEC network incorporates network properties such as channel gain and transmission speed matrices. To simulate these entities and their interactions, we employ processor-based models.

To assess the effectiveness of our proposed model, we conducted a comparative analysis with several existing methods in the field. The evaluated methods include DDPG, DDPG with ePER (DDPG-ePER), double-delay deep deterministic policy gradient (TD3) [26], and DDPG with LSTM (LSTM-DPG). The Table 1 summarizes the key parameter settings used in our experiments. We conducted a total of five experiments and averaged the results to account for the randomness present in the MEC network and DRL model. The experimental outcomes are presented in graphs. The graphs provide a visual representation of the trends and comparisons observed in the experimental results. Figure 4 specifically illustrates the convergence of LSTDPG on episode rewards.

The episode rewards represent the cumulative performance achieved by the LSTDPG algorithm over each episode. Each reward is composed of three components: task completion quantity, energy consumption, and time cost. These components can be weighted differently based on the specific requirements and objectives of the network providers.

Table 1. Parameter Settings.

Parameter	Value
Signal to Noise Ratio (dB)	100
Task Data Size (bits)	$[2 \times 10^5, 2 \times 10^7]$
Task Computing Size (cycles)	$[8 \times 10^6, 8 \times 10^7]$
Server Max Frequency (Hz)	$[2 \times 10^9, 8 \times 10^9]$
Number of Online Users	$[10, 100]$
Number of Edge Server s	$[5, 30]$
Batch Size	64
Learning Rate α	1×10^{-4}
Discount Factor γ	0.9

The convergence of LSTDPG to the optimal policy is observed to occur around 500 episodes, as depicted in Fig. 4. This indicates that the algorithm has learned an effective policy that balances the task completion quantity, energy consumption, and time cost according to the defined weights.

The ability to configure the weights of the reward components provides flexibility for network providers to align the algorithm with their specific application and business purposes. By adjusting these weights, providers can prioritize certain performance metrics over others, allowing for customization and adaptation to different scenarios. Figure 4 clearly demonstrates that DDPG-PER, LSTM-DPG, and TD3 exhibit superior convergence compared to the original DDPG in terms of reward. These findings suggest that the DDPG agents are more susceptible to getting trapped in local optima.

The limitations of DDPG can be attributed to the inadequate capability of the 2-layer FCN in effectively extracting local shapelets and capturing long-term dependencies from time-series data. This deficiency directly impacts the stability and effectiveness of the training process. We tackle this problem by introducing LSTM networks, which have a good performance in exploring long-term changes among time series data. Moreover, the uniform sampling approach employed in experience replay further compounds the issue by introducing substantial fluctuations during training. The lack of prioritization in the sampling process fails to appropriately assign importance to valuable experiences, potentially hindering the learning process. On the other hand, LSTM networks perform well in exploring long-range variations among time series data. This is also the reason why LSTM-DPG performs better than original DDPG. TD3 uses two independent critic networks to reduce the overestimation of the Q value and estimate the value function more accurately, so TD3 has better convergence performance than DDPG.

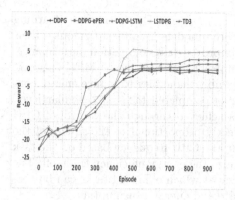

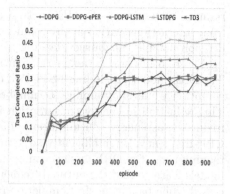

Fig. 4. Reward.   **Fig. 5.** Task Completed Ratio.

In Fig. 5, the completion rate of tasks before their deadlines is depicted. Initially, the edge servers exhibit a minimal task completion rate per episode since the model randomly selects actions and fails to allocate resources accurately. However, as the models iteratively interact with the MEC environment and learn from the experience, they gradually improve their policies. Consequently, the models become capable of efficiently allocating resources, resulting in a higher completion rate for offloaded tasks.

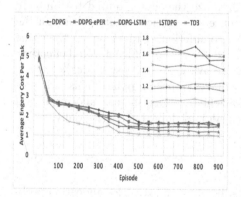

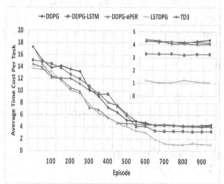

Fig. 6. Energy Consumption.   **Fig. 7.** Average Time Cost.

As the training progresses, the models optimize their policies and learn to make informed decisions regarding resource allocation. This enables them to effectively handle a greater number of offloaded tasks within the given time constraints. The increasing completion rate reflects the models' ability to adapt and improve their resource management strategies, ultimately maximizing task completion before their respective deadlines. The improved completion rate demonstrates the effectiveness of the DRL models, including LSTDPG, in optimizing resource allocation in the MEC network environment. By continuously learning from interactions with the environment, the models are able to enhance their decision-making capabilities, leading to improved task completion performance.

Figure 6 illustrates the energy consumption per task in the MEC network. The DRL models, including LSTDPG, exhibit the ability to reduce energy consumption while maintaining the number of completed tasks at a similar level. This reduction in energy consumption is achieved through the control of CPU frequency, as energy consumption is directly proportional to the square of the CPU frequency. By intelligently adjusting the CPU frequency based on the task requirements, the DRL models can optimize the energy-efficiency trade-off. The models learn to find the optimal frequency at which offloaded tasks can be processed while balancing the completion rate and energy consumption.

This allows the models to achieve a more efficient utilization of computing resources, resulting in reduced energy consumption without compromising the completion of tasks. Additionally, Fig. 6 also takes into account the energy consumed by overdue tasks. Considering the task completion rate depicted in Fig. 5, the wasted energy can be calculated by multiplying the energy consumption per task by the number of overdue tasks. This highlights the importance of efficient resource allocation and task scheduling in minimizing energy waste and improving overall energy efficiency in the MEC network.

Figure 7 depicts the average time cost in the MEC network. The DRL models, including LSTDPG, exhibit the capability to reduce time cost while maximizing the number of completed tasks. By optimizing the resource allocation and task scheduling, the models can effectively minimize the time required to complete tasks, thereby improving the user experience and service quality. Although the reward function already incorporates the completion of tasks before their due time, reducing the time cost further enhances the overall efficiency and effectiveness of the system. As shown in Fig. 7, the average time cost decreases as the DRL models converge towards an approximately optimal policy. This indicates that the models learn to make more efficient decisions in task allocation and processing, leading to reduced time overhead and improved service delivery.

6   Conclusions

This paper addresses the computational offloading and resource allocation problem in a scenario where multiple mobile devices (MDs) interact with multiple edge computing servers (MECs). To tackle this problem, the proposed approach is called Long Short-Term Deterministic Policy Gradient (LSTDPG). In LSTDPG, each MD is equipped with an agent that utilizes the LSTDPG algorithm to make dynamic decisions regarding task offloading and resource allocation.

The LSTDPG agent incorporates two key components: the Temporal Attention Network (TAN) and the Episodic Priority Experience Replay (ePER). TAN is responsible for capturing long-term dependencies and extracting relevant temporal patterns in the time series data associated with the MEC environment. This enables the agent to make informed decisions based on past experiences.

ePER, on the other hand, facilitates the storage and retrieval of experience samples, prioritizing those that are more relevant and important for training the agent. Experimental results demonstrate the effectiveness of the proposed LSTDPG approach. Compared to the original DDPG algorithm, LSTDPG achieves faster and more stable convergence. Furthermore, when compared to three other DDPG-based agents, LSTDPG consistently outperforms them in various performance metrics, including average long-term cost, average task completion time, and average energy consumption of the MDs.

As part of future work, the authors plan to explore the optimization of task partitioning and offloading based on subtasks that have dependencies and priorities. This will further enhance the efficiency and effectiveness of the proposed approach in handling complex scenarios where tasks have varying degrees of interdependencies and importance.

References

1. Hassan, N., Yau, K.L.A., Wu, C.: Edge computing in 5G: a review. IEEE Access **7**, 127276–127289 (2019)
2. Chen, Y., Zhang, N., Zhang, Y., Chen, X., Wu, W., Shen, X.: Energy efficient dynamic offloading in mobile edge computing for internet of things. IEEE Trans. Cloud Comput. **9**(3), 1050–1060 (2019)
3. Jošilo, S., Dán, G.: Wireless and computing resource allocation for selfish computation offloading in edge computing. In: IEEE INFOCOM 2019-IEEE Conference on Computer Communications, pp. 2467–2475. IEEE (2019)
4. Xiong, X., Zheng, K., Lei, L., Hou, L.: Resource allocation based on deep reinforcement learning in IoT edge computing. IEEE J. Sel. Areas Commun. **38**(6), 1133–1146 (2020)
5. Tran, T.X., Pompili, D.: Joint task offloading and resource allocation for multi-server mobile-edge computing networks. IEEE Trans. Veh. Technol. **68**(1), 856–868 (2018)
6. Lu, H., He, X., Du, M., Ruan, X., Sun, Y., Wang, K.: Edge QoE: computation offloading with deep reinforcement learning for internet of things. IEEE Internet Things J. **7**(10), 9255–9265 (2020)
7. Bi, S., Huang, L., Zhang, Y.J.A.: Joint optimization of service caching placement and computation offloading in mobile edge computing systems. IEEE Trans. Wireless Commun. **19**(7), 4947–4963 (2020)
8. Ahani, G., Yuan, D.: BS-assisted task offloading for D2D networks with presence of user mobility. In: 2019 IEEE 89th Vehicular Technology Conference (VTC2019-Spring), pp. 1–5. IEEE (2019)
9. Wang, S., Hu, Z., Deng, Y., Hu, L.: Game-theory-based task offloading and resource scheduling in cloud-edge collaborative systems. Appl. Sci. **12**(12), 6154 (2022)
10. He, Q., et al.: A game-theoretical approach for user allocation in edge computing environment. IEEE Trans. Parallel Distrib. Syst. **31**(3), 515–529 (2019)
11. Vijayaram, B., Vasudevan, V.: Wireless edge device intelligent task offloading in mobile edge computing using hyper-heuristics. EURASIP J. Adv. Sign. Proc. **2022**(1), 1–23 (2022)
12. LeCun, Y., Bengio, Y., Hinton, G.: Deep learning. Nature **521**(7553), 436–444 (2015)

13. Zhao, X., et al.: Deep learning based mobile data offloading in mobile edge computing systems. Futur. Gener. Comput. Syst. **99**, 346–355 (2019)
14. Sadiki, A., Bentahar, J., Dssouli, R., En-Nouaary, A., Otrok, H.: Deep reinforcement learning for the computation offloading in MIMO-based edge computing. Ad Hoc Netw. **141**, 103080 (2023)
15. Lin, B., Lin, K., Lin, C., Lu, Y., Huang, Z., Chen, X.: Computation offloading strategy based on deep reinforcement learning for connected and autonomous vehicle in vehicular edge computing. J. Cloud Comput. **10**(1), 33 (2021)
16. Cao, S., Chen, S., Chen, H., Zhang, H., Zhan, Z., Zhang, W.: HCOME: research on hybrid computation offloading strategy for MEC based on DDPG. Electronics **12**(3), 562 (2023)
17. Ebrahim, M.A., Ebrahim, G.A., Mohamed, H.K., Abdellatif, S.O.: A deep learning approach for task offloading in Multi-UAV aided mobile edge computing. IEEE Access **10**, 101716–101731 (2022)
18. Chen, Y., Han, S., Chen, G., Yin, J., Wang, K.N., Cao, J.: A deep reinforcement learning-based wireless body area network offloading optimization strategy for healthcare services. Health Inf. Sci. Syst. **11**(1), 8 (2023)
19. Hao, Y., Ni, Q., Li, H., Hou, S.: Energy-efficient multi-user mobile-edge computation offloading in massive MIMO enabled HetNets. In: ICC 2019–2019 IEEE International Conference on Communications (ICC), pp. 1–6. IEEE (2019)
20. Guo, S., Liu, J., Yang, Y., Xiao, B., Li, Z.: Energy-efficient dynamic computation offloading and cooperative task scheduling in mobile cloud computing. IEEE Trans. Mob. Comput. **18**(2), 319–333 (2018)
21. Wang, Y., Sheng, M., Wang, X., Wang, L., Li, J.: Mobile-edge computing: Partial computation offloading using dynamic voltage scaling. IEEE Trans. Commun. **64**(10), 4268–4282 (2016)
22. Graves, A., Graves, A.: Long short-term memory. Supervised sequence labelling with recurrent neural networks, pp. 37–45 (2012)
23. Vaswani, A., et al.: Attention is all you need. In: Advances in Neural Information Processing Systems, vol. 30 (2017)
24. Schaul, T., Quan, J., Antonoglou, I., Silver, D.: Prioritized experience replay. arXiv preprint: arXiv:1511.05952 (2015)
25. Harris, C.R., et al.: Array programming with NumPy. Nature **585**(7825), 357–362 (2020)
26. Fujimoto, S., Hoof, H., Meger, D.: Addressing function approximation error in actor-critic methods. In: International Conference on Machine Learning, pp. 1587–1596. PMLR (2018)

Query Optimization Mechanism for Blockchain-Based Efficient Data Traceability

Xu Yuan[1]([⊠]), Fangbo Li[1], Muhammad Zeeshan Haider[1], Feng Ding[1],
Ange Qi[1], and Shuo Yu[2]

[1] School of Software, Dalian University of Technology, Dalian 116620, China
{david,dingfeng}@dlut.edu.cn, 772888003@mail.dlut.edu.cn
[2] School of Computer Science and Technology, Dalian University of Technology,
Dalian 116024, China
yushuo@dlut.edu.cn

Abstract. Efficient traceability query processing is of utmost importance in blockchain systems. However, existing methods suffer from efficiency challenges, necessitating immediate optimization. Prior research primarily focuses on introducing complex data structures to alleviate storage system burdens in scenarios with frequent traceability requests. Unfortunately, these optimizations impose a substantial maintenance burden on full nodes, making it challenging to strike a balance between traceability efficiency and resource consumption. Through an analysis of traceability query processing methods in existing blockchain full nodes, this paper proposes a novel optimization scheme based on a multilevel cache. This scheme replaces repetitive and inefficient disk I/O operations with efficient in-memory operations, thereby significantly enhancing the efficiency of traceability queries. Additionally, the multilevel cache structure minimizes the memory overhead associated with maintaining the cache in full nodes, achieving an optimal trade-off between traceability efficiency and resource consumption while alleviating the burden on full nodes. Experimental validation confirms the effectiveness of the proposed efficiency optimization scheme for traceability queries in full nodes. The results demonstrate its ability to enhance the overall performance of traceability query processing, contributing to improved efficiency and scalability in blockchain systems.

Keywords: Blockchain · Query Optimization · Scalability · Data Traceability Queries

1 Introduction

Blockchain technology [1] has emerged as a foundational technology, revolutionizing the world of digital transactions by enabling decentralized peer-to-peer payments without the need for intermediaries. Its disruptive potential has extended far beyond the realms of cryptocurrencies, with applications spanning finance,

© The Author(s), under exclusive license to Springer Nature Singapore Pte Ltd. 2024
Z. Tari et al. (Eds.): ICA3PP 2023, LNCS 14492, pp. 349–367, 2024.
https://doi.org/10.1007/978-981-97-0811-6_21

supply chain management, healthcare, voting systems, and more [2]. By utilizing a robust and immutable chained data structure, combined with sophisticated cryptographic techniques, blockchain ensures the integrity and trustworthiness of stored data, thereby instilling confidence in the participants of the network [3]. However, as blockchain adoption continues to gain momentum, it has become evident that existing implementations face certain challenges, particularly in terms of data reading and writing efficiency. These limitations hinder the widespread utilization of blockchain technology across industries and necessitate the exploration of innovative solutions to optimize its performance and unlock its full potential. By addressing these efficiency issues, we can pave the way for broader adoption of blockchain and harness its transformative power in revolutionizing digital transactions and fostering trust in various sectors of the global economy [4].

To enhance the efficiency and practicality of blockchain database systems, researchers have dedicated extensive efforts to optimizing data read and write operations. Writing data within a blockchain network requires substantial computational and storage resources to achieve network consensus, commonly referred to as the consensus writing mechanism. Through various studies [5], significant advancements have been made in improving the efficiency of consensus mechanisms, reducing resource consumption, and enhancing consensus writing. Notably, the optimization of proof-of-work consensus [6,7] has mitigated resource wastage caused by mining competition and elevated consensus writing efficiency. Additionally, innovative strategies like chunking and grouping have narrowed the participation scope of nodes in the consensus process, streamlining network communication complexities and demonstrating superior performance in terms of efficiency and security.

Furthermore, the exploration of hybrid consensus [8] mechanisms has significantly expanded the scope and adaptability of consensus writing methods, catering to diverse application scenarios. Over the years, researchers have made notable progress in improving the efficiency and security of consensus writing through innovative approaches. However, amidst these advancements, there has been a noticeable lack of emphasis on data reading, particularly concerning blockchain data traceability queries. In the realm of blockchain technology, traceability queries hold a pivotal role in achieving comprehensive data traceability and ensuring the trustworthiness of information. The demand for efficient traceability queries is growing across various applications. Despite this, the prevailing approach in most blockchain systems revolves around the recursive retrieval of historical data versions by relying on hash values stored in the underlying storage system. Unfortunately, this approach introduces redundant disk read and write operations, which significantly compromise the efficiency of traceability queries. Additionally, the burden imposed on full nodes due to these operations has a direct impact on the overall throughput efficiency of the entire system.

This paper aims to address the limitations of traceability queries in the blockchain system by introducing a traceability query cache and designing a multi-level cache structure. The objective is to enhance data traceability efficiency while minimizing the memory footprint of the cache and reducing resource consumption for full nodes. Consequently, full nodes can optimize the allocation of computational and storage resources for network consensus, leading to improved traceability efficiency and overall system performance. By overcoming the challenges associated with traceability queries, this research contributes to the advancement of blockchain technology and its broader application in various domains.

2 Related Work

This section begins by presenting the fundamental blockchain concepts that are relevant to our research. We also discuss relevant technical literature to enhance the understanding of the research topic and provide a foundation for introducing our research proposal in later sections.

2.1 Blockchain Foundation

The existing Internet architecture is characterized by excessive centralization, resulting in numerous challenges and drawbacks [9]. These include the vulnerability of data security due to single points of failure [10] and the limited trustworthiness and circulation of information caused by closed information silos. Addressing these issues necessitates substantial resource investments. To tackle these challenges, blockchain technology has emerged as a decentralized paradigm, offering a promising technical solution. Researchers have embraced blockchain technology to mitigate the control exerted by central institutions over data, empowering data owners with control rights and thereby enabling decentralization and data tamper-proofing. Blockchain technology embodies key attributes [11,12] such as decentralization, multi-party sharing, non-tampering, verifiability, and traceability, thereby providing new avenues for addressing data security and trustworthiness [13–17] concerns. It serves as a distributed and trusted database with each blockchain full node maintaining a complete ledger. The inclusion of data within the blockchain requires multi-party consensus, and the security and trustworthiness of the data are ensured through redundant backup and consensus mechanisms.

The blockchain system can be categorized into two main types: public blockchains and consortium blockchains. Public blockchains are open to all nodes and are well-suited for scenarios that involve public participation. On the other hand, consortium blockchains restrict access to only members of an enterprise alliance and are more suitable for enterprise-level applications. When considering the logical layers of a blockchain system, it encompasses several layers: the data layer, network layer, consensus layer, incentive layer, contract layer, query layer, and application layer. These layers collectively contribute to the overall functioning of the blockchain system. The logical layer model of the blockchain [18–23] system is illustrated in Fig. 1.

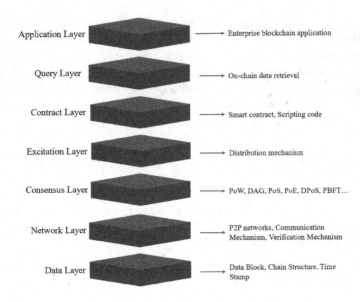

Fig. 1. The Blockchain Logic Layer Model

2.2 BCDC Model

The transaction information stored in the existing blockchain system is in a single format, and the application expansion is limited. To address this problem, the BCDB model [24, 25] modifies the transaction data structure in the Bitcoin system, enabling it to accommodate any record format and providing an improved specification for blockchain data formats.

Data Model Definition. To make the blockchain data more general, the existing transaction data structure is redefined. In the BCDB model, the data record consists of a Data Head and Data, as shown in Fig. 2. The Data part contains the data content, where the Key is used to uniquely identify the data record. The data attribute is represented by the Field,which can adopt any format to enhance the extensibility of the data record.

The data structure within the blockchain system consists of various components that enable data modification traceability and tamper-proofing. The PreHash attribute is utilized to reference the previous historical version of the same Key, ensuring the ability to trace data modifications. The Time attribute records the publication time of the data record. ScriptPubK specifies the public key of the next owner, granting write permissions, while ScriptSig represents the signature of the current owner, utilized to validate the authenticity of the record.During the data writing process, several checks are conducted. First, the existence of a data record with the corresponding Key is verified. Subsequently, the ScriptSig of the current data record is compared with the ScriptPubK of the previous data record. Only if these two components match, the data record

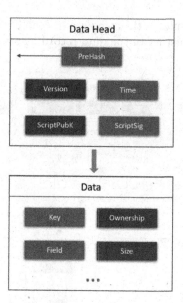

Fig. 2. Data Logging Structure Diagram

is deemed valid. In the case of the Key appearing for the first time, the Pre-Hash attribute is set to null. In situations where a Key data record undergoes modification, a new data record with the same Key is appended in a new block, thereby preserving all historical record versions in the historical blocks. These historical versions cannot be deleted. When querying a specific data record, a content retrieval process is performed based on the Key, resulting in the retrieval of the latest version of the data record. Additionally, by examining the PreHash attribute in the data header, it is possible to trace the modification history of a particular Key data record. The fundamental data structure operations within the blockchain system remain as adding, modifying, querying, and deleting data records.

Traceability Query Process. In the BCDB model, the data traceability query is a recursive query on Key. Each historical version of this data record is extracted recursively from the underlying storage system by using the PreHash field in the data to obtain the complete modification track of the data record identified by the Key.

As we can see in Fig. 3 , the process of querying the data traceability of Key=1 involves locating the latest version (Version=1.0) of the corresponding data record starting from the current latest block. By utilizing the PreHash field of that version, the previous version of the data record (Version=0.9) is retrieved from the underlying storage system (e.g., LevelDB). This procedure is repeated iteratively, extracting earlier versions of the data record (Version=0.8) based on the PreHash field of each version, until the original version is reached. This sequence of iterations results in the complete traceability chain, which illustrates the entire modification history of the data record associated with Key.

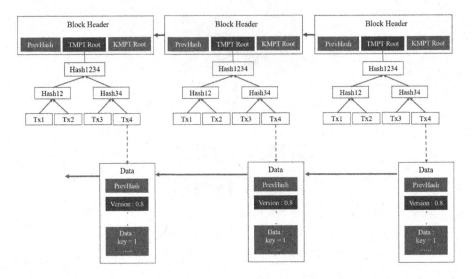

Fig. 3. Schematic Diagram of Traceability Retrieval

As the traceability queries involve retrieving data multiple times from the underlying storage system, which introduces inefficiencies in the current implementation. Moreover, the reliance on a full-node database system in these queries results in resource-intensive operations. There is a need for an optimized solution that not only improves the efficiency of traceability queries but also reduces resource consumption for full nodes. However, existing methods are burdened by complexity and high maintenance costs, which negatively impact the overall system efficiency.

3 Efficient Query Optimization Traceability Model

Ensuring the trustworthiness of data, blockchain data traceability stands as a crucial feature. In typical blockchain systems, hash pointers are employed to navigate through the entire data traceability chain, which originates from underlying storage systems such as LevelDB. Designed for write-intensive applications, LevelDB is a key-value database that prioritizes write performance over read performance. Given that traditional blockchain systems involve a higher frequency of data writing compared to data querying, the use of LevelDB is deemed appropriate. However, as blockchain technology progresses, the need for data traceability queries in various scenarios has grown. In cases where traceability queries occur frequently, LevelDB's insufficient read performance can lead to excessive and inefficient disk I/O retrieval processes in the underlying storage system. Consequently, the efficiency of traceability queries suffers.

According to the traceability specification of the BCDB model, a typical traceability query can be divided into two stages: the first stage determines the set of hash values of the Key of the traceability data; the second stage queries and returns the complete set of traceability data in levelDB. The study analyzed the retrieval time consumption of these two stages, which is shown in Fig. 4.

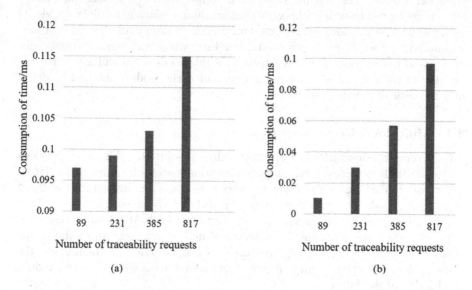

Fig. 4. The First and Second Stages of Traceability Query Take Time to Search

Figure 4-(a) shows the average time consumption for the first stage of traceability query retrieval after using the built-in indexing method, and Fig. 4-(b) shows the average time consumption for the second stage. From the figures, it can be seen that in the scenario of frequent traceability queries, as the number of query requests and the set of results increase, the time consumption of the two stages also increases, which leads to a decrease in the efficiency of traceability queries.

Further analysis shows that the built-in indexing method can only improve the first stage of the traceability query, while it has no effect on the second stage. The read efficiency of the underlying storage system LevelDB determines the time consumption of the second stage. Therefore, improving the reliance on the underlying storage system in the second stage becomes the key to improve the efficiency of the traceability query.

Numerous research endeavors have focused on enhancing the efficiency of traceability queries by proposing optimized structures that improve the retrieval capabilities of full nodes in data traceability. However, these approaches often introduce complex data structures, leading to increased maintenance costs for the full node. Moreover, they have the potential to consume computational and storage resources utilized by the full node for network consensus, thereby impacting the overall performance of the blockchain system. Additionally, these methods fall short in addressing the reliance on the underlying storage system during the second stage of traceability queries and are incapable of meeting the demands of scenarios involving frequent traceability requests.

Therefore, we propose a multi-level cache-based traceability query optimization scheme. The scheme reduces the reliance on the underlying storage system by introducing a caching mechanism and replacing inefficient disk IO retrieval with in-memory operations. Our goal is to design an efficient and low-maintenance traceability query cache structure, which can improve query efficiency and reduce resource consumption. In this section, we will introduce the multi-level cache structure, query process and cache update method in detail and theoretically analyze them.

3.1 Multi-level Cache Structure

To overcome the limitations of existing technical solutions, this paper presents a blockchain data traceability query optimization method.By introducing traceability query caching and efficient memory operations, it can reduce the times of inefficient disk IO, lower the reliance on the underlying storage system LevelDB, and improve the efficiency of traceability retrieval. At the same time, this paper also designs an optimized structure of multi-level cache to minimize the cost of computation and storage resources of the full node in maintaining the cache, balance the traceability efficiency and resource consumption, and reduce the burden of the full node.

Figure 5 illustrates the multi-level cache optimization structure proposed in this paper. The cache in the full node includes a first-level cache and a second-level cache. The first-level cache stores the hash values and data of each historical version identified by Key, i.e., (Hash_n, Data_n) → (Hash_n-1, Data_n-1)... (Hash_0, Data_0). The secondary cache stores only the hash values of each historical version of the data, i.e. (Hash_n) → (Hash_n-1)... (Hash_0), to reduce the amount of cached data storage. The cache elimination and downgrading mechanism can be based on a strategy of regular updates or setting a storage upper limit. Take setting the storage limit threshold as an example, when the primary cache reaches the storage limit, the elimination and downgrading mechanism is triggered to downgrade the Key data with lower query frequency to the secondary cache. Similarly, when the second-level cache reaches the upper limit, all data in the second-level cache will be emptied. The execution algorithm process is the same for both cases, only the time of execution is different. The crucial of the multi-level caching technology solution is to save the queried data traceability links. When the full node receives the traceability request for repeated Key value, it can obtain the data directly from the cache through the efficient memory operation and return the traceability result after verification. This can reduce the redundant disk operations for tracing repeated Key value in the tracing process and improve the tracing efficiency. It allows the full node to devote more computational resources to network consensus work, thus saving resources.

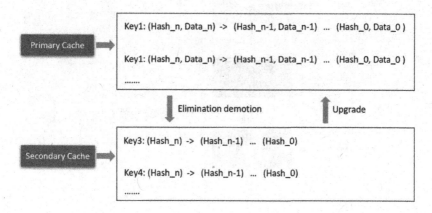

Fig. 5. Multi-level Cache Structure Diagram

The crucial of the multi-level caching technology solution is to save the queried data traceability links. When the full node receives the traceability request for repeated Key value, it can obtain the data directly from the cache through the efficient memory operation and return the traceability result after verification. This can reduce the redundant disk operations for tracing repeated Key value in the tracing process and improve the tracing efficiency. It allows the full node to devote more computational resources to network consensus work, thus saving resources.

3.2 Traceability Query Process

With the introduction of multi-level cache in the full node, the traceability query processing process is changed to improve query efficiency and maintain data consistency. Figure 6 illustrates the process. For the traceability goal marked by Key, the processing of the full node can be divided into three steps. First, the latest version data (LatestHash, LatestData) corresponding to this Key is retrieved by content retrieval. Then, the multi-level cache is queried to obtain the data traceability chain stored in the cache identified by this Key(Hash_n, Data_n)→(Hash_n-1, Data_n-1)...(Hash_0, Data_0).Thus, the complete traceability link of the data marked by this Key is obtained(LatestHash, LatestData)→...(Hash_n, Data_n)→(Hash_n-1, Data_n-1)...(Hash_0, Data_0).

The data hash value appears in pairs with the data during the traceability query, so tamper-proof and correctness verification can be easily performed by the data hash value.The traceability query process itself serves as a verification process for data integrity and correctness.

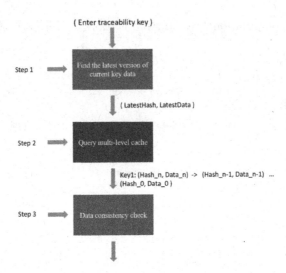

Fig. 6. Full Node Traceability Flow Chart

3.3 Cache Update Method

In the technical solution of this paper, the traceability query retrieval algorithm process is performed simultaneously with the cache update maintenance algorithm in order to ensure the consistency of data in the multi-level cache and blockchain ledger. The specific full-node traceability query and cache update algorithm process is illustrated in Fig. 7.

Step 1: Input the target data marked by the key, and the nodes in the network launch a traceability query request to the full node of the blockchain. The full node queries the LatestHash and LatestData of the data corresponding to the Key in the current blockchain ledger, noted as (LatestHash, LatestData).

Step 2: Query the Key in the cache of the blockchain full node.

Step 2.1: If the cache of the blockchain full node does not contain the data corresponding to the Key, the complete traceability chain (LatestHash, LatestData)$\rightarrow$...$\rightarrow$(Hash_0, Data_0) is traversed in the current blockchain ledger. Finally, the obtained complete traceability chain is added to the cache corresponding to the key, and the traceability query for the Key ends.

Step 2.2: If the data corresponding to the key exists in the cache, retrieve the complete traceability chain(Hash_n, Data_n)$\rightarrow$(Hash_n-1, Data_n-1)...(Hash_0, Data_0) corresponding to the Key from the cache and execute Step 3.

Step 3: The full node performs consistency verification by comparing the LatestHash obtained in Step 1 with Hash_n retrieved from the cache in Step 2, ensuring traceability integrity.

Step 3.1: If LatestHash matches Hash_n, the traceability chain obtained from the cache (Hash_n, Data_n)$\rightarrow$(Hash_n-1, Data_n-1)$\rightarrow$...$\rightarrow$(Hash_0, Data_0)

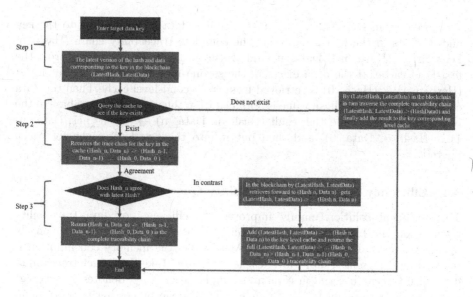

Fig. 7. Search Flow Chart

is the complete traceability chain. Return this result and end the traceability query for the Key.

Step 3.2: If the LatestHash doesn't match Hash_n, it will be retrieved from (LatestHash, LatestData) forward to (Hash_n, Data_n) in the blockchain to get (LatestHash, LatestData) → ... (Hash_n, Data_n). Add the obtained result to the cache corresponding to the Key to ensure the consistency of the cache with the data in the blockchain ledger. At the same time, the result is concatenated with the traceability chain retrieved from the cache in step 2 (Hash_n, Data_n)→(Hash_n-1, Data_n-1)...(Hash_0, Data_0) to form a complete traceability chain (LatestHash, LatestData) →... (Hash_n, Data_n) → (Hash_n-1, Data_n-1)... (Hash_0, Data_0). Return this result and end the traceability query for the Key.

The full-node cache has a multi-level structure, including a first-level cache and a second-level cache. The first-level cache stores the historical version hash values and data of the data marked by key, i.e., (Hash_n, Data_n) → (Hash_n-1, Data_n-1)... (Hash_0, Data_0). The second-level cache stores only the historical version hash values of the data, i.e. (Hash_n) → (Hash_n-1)... (Hash_0). When the first-level cache reaches its limit, it will trigger the elimination and downgrading mechanism to downgrade the lower query frequency Key data to the second-level cache. When the second-level cache space reaches the limit, all the data in it will be emptied to save memory space and improve the cache hit ratio.

Furthermore, in step 2.1, the complete traceability chain obtained during traversal will be added to the first-level cache corresponding to the key.

Moreover, in step 2.2, if the cache contains data corresponding to the key and it exists in the first-level cache, the complete traceability chain (Hash_n, Data_n) $\rightarrow$ (Hash_n-1, Data_n-1) ... (Hash_0, Data_0) is retrieved from the first-level cache. If the data exists in the second-level cache, the (Hash_n) $\rightarrow$ (Hash_n-1) ... (Hash_0) is retrieved from the second-level cache.Then the data corresponding to each hash value is obtained from the underlying database of the blockchain in bulk, and the result (Hash_n, Data_n) $\rightarrow$ (Hash_n-1, Data_n-1)... (Hash_0, Data_0), and then put it into the first-level cache for cache upgrade.

3.4   Efficiency Analysis of Traceability Query

This technical solution mainly improves the efficiency of data traceability retrieval on the blockchain by introducing caches and using efficient memory operations instead of disk operations. Meanwhile, the memory consumption is reduced by designing a multi-level cache structure. Take the target traceability data with traceability length N as an example, where t_{disk} denotes the average access time to disk and t_{memory} denotes the average processing time to memory. After locating the latest version of the data without cache optimization, the average required traceability time T is:

$$T = N * (t_{disk} + t_{mp})$$

In general, memory access speed is nano-second (-9th power of 10) and hard disk access speed is micro-second (-3th power of 10). In the case of sequential access, memory access speed is only 6 to 7 times faster than hard disk access speed while in the case of random access, memory access speed is more than 100,000 times faster than hard disk access speed, thus:

$$T \approx N * t_{disk}$$

After introducing a multi-level cache, let A be its cache hit rate (0<A<1). It is assumed that the data in the blockchain ledger and the cached data have been consistent, which means that if the cache hits, the complete traceability chain data or the complete traceability chain hash value is stored in the cache. The average traceability time needs discussion:

When the required traceability chain data are all in the first-level cache, which means that the average traceability time for the best-case multi-level cache is

$$T_{better} = (1 - A) * N * t_{disk} + A * N * t_{memory} \approx (1 - A) * N * t_{disk}$$

When all the required traceability chain data is in the second-level cache, that is, the worst-case scenario. In this case, since the complete traceability chain hash is available in the second-level cache, a batch and concurrent disk operation can be performed to improve efficiency. Therefore, the worst-case average traceability time for the multi-level cache is:

$$T_{worst} = (1 - A) * N * t_{disk} + A * N * t_{batch}$$

And:
$$T_{worst} < N * t_{disk} = T$$

And:
$$T_{better} < T_{mlevel} < T_{worst}$$

In summary:
$$T_{better} < T_{mlevel} < T_{worst} < T$$

That is, the multi-level cache traceability technology scheme proposed in this paper is theoretically better than the original traceability method in terms of traceability efficiency. In addition, the multi-level cache improves the storage space occupation by introducing a second-level cache structure and storing only its traceability hash chain for traceability data with low query frequency while maintaining the same cache hit rate. It alleviates the storage pressure on the full nodes of the blockchain, which is a significant improvement compared with the common caching method.

4   Experimental Results and Analysis

In this experimentation, a dedicated machine is employed to construct a simulated full node for the purpose of testing and verification. The hardware setup of the dedicated machine comprises an Intel(R) Core(TM) i5-7300HQ CPU operating at 2.50 GHz, 8 GB of RAM, and Windows 10 as the operating system. The experiment leverages the Java language to implement a traceability method utilizing a multi-level cache structure. To simulate traceability within the full node, the experiment makes use of the Redis cache component.

In addition, the experiments implement the original traceability method when there is no cache, and introduce the traceability method with a common single-level cache. Through comparative experimental tests, the improvement of traceability retrieval efficiency after the introduction of multi-level cache is verified, and the space saving and maintenance cost of multi-level cache, etc. are evaluated.

4.1   Comparison of Traceability Efficiency

To explore the optimization of blockchain data traceability efficiency through the introduction of multi-level caching, the experiment conducted 1000 traceability queries. These queries were performed with different percentages of repeated trace requests (requesting the same key), 0%, 30%, and 60%, respectively. The average time for single tracing query was calculated to compare the efficiency of the original tracing method (tracing PreHash fields sequentially) with the multi-level cache tracing method, as shown in Fig. 8.

From the experimental results, it can be seen that when the repeat percentage of traceability requests is 0%, the cache in the multi-level cache traceability method does not work, and the original traceability method is used for traceability, so the traceability efficiency is almost not different from the original

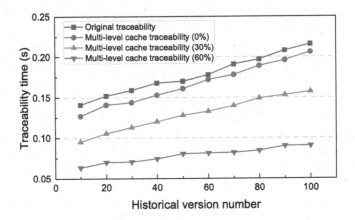

Fig. 8. Comparison Chart of Traceability Query Time

method. This also means that this is the worst case for traceability in a multi-level caching method. As the percentage of repeat traceability requests increases, the multi-level cache traceability method starts to work and its traceability efficiency is significantly better than that of the original traceability method, and this advantage becomes more obvious as the percentage of repeats increases. This is basically in line with the theoretical analysis in Sect. 3.4. The effectiveness of the multi-level caching method proposed in this paper in improving the efficiency of full-node traceability is verified.

4.2 Comparison of Traceability Time for Single/Multi-level Structures

The second-level cache in the multi-level cache structure stores the hash chain of the traceability, which not only saves space but also maximizes the read performance of the underlying Level DB. This, to some extent, improves the efficiency of retrieving the complete traceability chain.In this experiment, we made modifications based on the multi-level cache scheme by replacing the cache downgrade part with direct deletion elimination to implement the regular single-level cache method. We set the ratio R to represent the distribution of cached data in different application scenarios. In a multi-level cache, R = the amount of data cached at one level/the total amount of data; in a single-level cache, R = the amount of data cached/the total amount of data. We test the traceability of 1000 different Keys with 100 data history versions for each Key, and calculate the average traceability time of each Key data to compare the traceability efficiency of the multi-level caching approach with the common single-level caching method (Fig. 9).

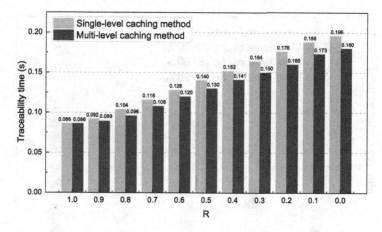

Fig. 9. Comparison of Single/Multi-level Traceability Cache Time Consumption

According to the experimental results in Fig. 4.6, we can see that as the R value decreases (i.e., the percentage of cached data decreases), the average traceability time increases for both methods due to the decrease of cache hit rate. However, the rising trend of the traceability time of the multi-level cache method is smoother than that of the normal single-level cache, showing better traceability efficiency. This is benefited from the traceability hash links stored in the second-level cache in the multi-level cache. When the first-level cache does not hit, the multi-level cache can use the traceability hash link in the second-level cache to query the traceability data in Level DB in bulk. Compared with a single-level cache that can only query the complete traceability links using recursive methods when the cache does not hit, the multi-level cache exploits the read performance of Level DB more fully and shortens the average traceability time. As a result, the R distribution of cached data varies in different application scenarios due to the different distribution of traceability query requests. With the same R distribution, the multi-level caching approach has higher traceability query efficiency compared with the common single-level caching method.

4.3   Comparison of Storage Space Consumption

In the Bitcoin system, the average transaction data size is about 250 bytes, and each transaction uses the SHA-256 algorithm to calculate the hash value with a fixed size of 32 bytes. In order to eliminate the influence of single data size on the traceability test and make the experimental results closer to the real scenario, this experiment sets the single traceability data size to 250 bytes and uses SHA-256 algorithm to calculate the hash value. The test is conducted by comparing the storage space occupation of multi-level cache and ordinary single-level cache under different traceability data. The experimental result shows in Fig. 10.

The memory space occupation in the multi-level cache method is determined by both the amount of stored data and the data distribution (the ratio of the

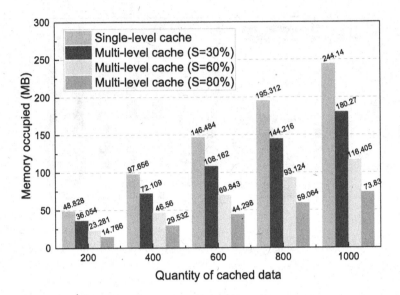

Fig. 10. Comparison of Storage Capacity

first-level and second-level cache occupancy). Therefore, S in Fig. 10 is set to be the proportion of the second-level cache in the multilevel caching scheme. Figure 10 shows the space consumption in the multi-level cache under different distribution S of data. From this figure, it is clear that after the introduction of the multilevel cache structure, the memory space occupied by the multi-level cache is better than that of the normal single-level cache. The reason for this is that the multi-level cache method can timely process the elimination and down-grading of the traceability data in the cache. As the proportion of traceability data eliminated and downgraded to the second-level cache increases, the advan-tage of the multi-level cache scheme in saving memory consumption becomes more and more obvious. The experimental results verify the advantages of the multi-level caching approach over the common single-level caching approach in terms of memory space consumption. It also proves that the multi-level cache traceability optimization scheme proposed in this paper achieves memory con-sumption reduction while improving traceability efficiency, thus reducing the memory storage burden of the full node.

4.4   Comparison of Cache Maintenance Cost

This technical solution uses a multi-level cache optimization strategy, which requires an additional process of cache elimination and downgrading for regu-lar updates compared to the common single-level cache method. To evaluate the additional consumption caused by this process, comparative experiments are designed to further research its computational resource consumption. The experiments store 100, 200, 300, and 1000 pieces of data into the normal single-level cache and multi-level cache structures, respectively. After the storage is completed, the multilevel cache starts the elimination and downgrading algorithm to downgrade the data in the first-level cache to the second-level cache to test the time consumption limit of the multi-level cache elimination algorithm. Then it is compared with the normal single-level cache method to evaluate the maintenance cost consumption of the multi-level cache.

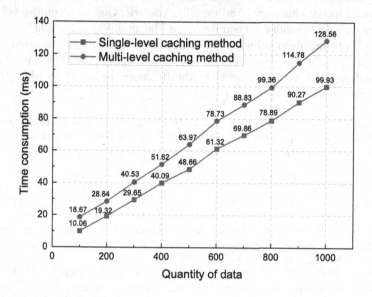

Fig. 11. Comparison of Time Consumption for Multi-level Cache Maintenance

As shown in Fig. 11, the time consumption of both methods grows essentially linearly as the amount of stored data increases. The multi-level caching scheme requires an additional process of elimination and downgrading, so it consumes more time than the normal single-level cache method. However, its additional cost remains within one-third of the time required to maintain a single-level cache. In addition, this experiment downgrades all first-level caches to second-level caches to test the limition of time consumption, but not all data in the first-level cache needs to be downgraded in the real environment, so the real effect will be better. Overall, the maintenance cost of the multi-level cache structure is in the acceptable range, so it has some application value for real traceability scenarios.

5   Conclusion

This study presents an optimization scheme to address the limitations of existing blockchain systems in traceability query processing. The objective is to enhance the traceability query capability of blockchain full-nodes without compromising the fundamental characteristics of the blockchain. The scheme introduces a multi-level cache-based optimization approach to tackle the challenge of balancing efficiency and resource usage in full-node traceability queries, especially in scenarios involving frequent traceability requests. Through experimental evaluations, this study demonstrates the replacement of redundant and inefficient disk IO traceability retrievals with efficient in-memory operations, resulting in improved query efficiency. Additionally, the adoption of a multi-level cache structure reduces memory consumption in the maintenance cache of the full node. This optimization achieves the dual effect of enhancing traceability efficiency while minimizing resource consumption, thereby alleviating pressure on the underlying disk storage system of the full node in scenarios with frequent traceability requests.

Acknowledgements. We sincerely thank Mr. Lihuang Huang for his help with experiments. This work is supported by the "High-level Talent Team" Project of Dalian Science and Technology Talent Innovation Support Policy Program under Project No. 2022RG11.

References

1. Gad, A.G., Mosa, D.T., Abualigah, L., Abohany, A.A.: Emerging trends in blockchain technology and applications: a review and outlook. J. King Saud Univ. Comput. Inf. Sci. **34**(9), 6719–6742 (2022)
2. Noreen, Qiufen Xia, M.Z.H.: Advanced dag-based ranking (ADR) protocol for blockchain scalability. Comput. Mater. Continua **75**(2), 2593–2613 (2023). https://doi.org/10.32604/cmc.2023.036139, http://www.techscience.com/cmc/v75n2/51970
3. Chen, Z., et al.: Reputation-based partition scheme for IoT security. Secur. Priv. **6**(3), e287 (2023). https://doi.org/10.1002/spy2.287, https://onlinelibrary.wiley.com/doi/abs/10.1002/spy2.287
4. Yuan, X., Luo, F., Haider, M., Chen, Z., Li, Y.: Efficient byzantine consensus mechanism based on reputation in IoT blockchain. Wirel. Commun. Mob. Comput. **2021**, 1–14 (2021). https://doi.org/10.1155/2021/9952218
5. Bamakan, S.M.H., Motavali, A., Bondarti, A.B.: A survey of blockchain consensus algorithms performance evaluation criteria. Expert Syst. Appl. **154**, 113385 (2020)
6. Karpinski, M., Kovalchuk, L., Kochan, R., Oliynykov, R., Rodinko, M., Wieclaw, L.: Blockchain technologies: probability of double-spend attack on a proof-of-stake consensus. Sensors **21**(19), 6408 (2021)
7. Saleh, F.: Blockchain without waste: proof-of-stake. Rev. Financ. Stud. **34**(3), 1156–1190 (2021)
8. Xie, M., Liu, J., Chen, S., Lin, M.: A survey on blockchain consensus mechanism: research overview, current advances and future directions. Int. J. Intell. Comput. Cybern. **16**(2), 314–340 (2023)
9. Da Xu, L., Lu, Y., Li, L.: Embedding blockchain technology into IoT for security: a survey. IEEE Internet Things J. **8**(13), 10452–10473 (2021)
10. Kreß, F., et al.: Cnnparted: an open source framework for efficient convolutional neural network inference partitioning in embedded systems. Comput. Netw. **229**, 109759 (2023)

11. Viriyasitavat, W., Hoonsopon, D.: Blockchain characteristics and consensus in modern business processes. J. Ind. Inf. Integr. **13**, 32–39 (2019)
12. Bhambhwani, S., Delikouras, S., Korniotis, G.M.: Blockchain characteristics and the cross-section of cryptocurrency returns. SSRN 3342842 (2021)
13. Kumi, S., Lomotey, R.K., Deters, R.: A blockchain-based platform for data management and sharing. Proc. Comput. Sci. **203**, 95–102 (2022)
14. Jaiman, V., Urovi, V.: A consent model for blockchain-based health data sharing platforms. IEEE Access **8**, 143734–143745 (2020)
15. Paik, H.Y., Xu, X., Bandara, H.D., Lee, S.U., Lo, S.K.: Analysis of data management in blockchain-based systems: From architecture to governance. IEEE Access **7**, 186091–186107 (2019)
16. Wei, J., Yi, X., Yang, X., Liu, Y.: Blockchain-based design of a government incentive mechanism for manufacturing supply chain data governance. Sustainability **15**(8), 6968 (2023)
17. Lemieux, V., Feng, C.: Building Decentralized Trust: Multidisciplinary Perspectives on the Design of Blockchains and Distributed Ledgers. Springer International Publishing, Cham (2021). https://books.google.com/books?id=Qh8SEAAAQBAJ
18. Stodt, F., Reich, C.: Introducing a fair tax method to harden industrial blockchain applications against network attacks: a game theory approach. Computers **12**(3), 64 (2023). https://doi.org/10.3390/computers12030064, https://www.mdpi.com/2073-431X/12/3/64
19. Chen, C., Huang, H., Zhao, B., Shu, D., Wang, Y.: The research of AHP-based credit rating system on a blockchain application. Electronics **12**(4), 887 (2023)
20. Sethaput, V., Innet, S.: Blockchain application for central bank digital currencies (CBDC). Cluster Computing, pp. 1–15 (2023)
21. Javaid, M., Haleem, A., Pratap Singh, R., Khan, S., Suman, R.: Blockchain technology applications for industry 4.0: a literature-based review. Blockchain Res. App. **2**(4), 100027 (2021). https://doi.org/10.1016/j.bcra.2021.100027, https://www.sciencedirect.com/science/article/pii/S2096720921000221
22. Kiu, M., Chia, F., Wong, P.: Exploring the potentials of blockchain application in construction industry: a systematic review. Int. J. Constr. Manag. **22**(15), 2931–2940 (2022)
23. Kalajdjieski, J., Raikwar, M., Arsov, N., Velinov, G., Gligoroski, D.: Databases fit for blockchain technology: a complete overview. Blockchain Res. App. **4**(1), 100116 (2023). https://doi.org/10.1016/j.bcra.2022.100116, https://www.sciencedirect.com/science/article/pii/S2096720922000574
24. Deepa, N., et al.: A survey on blockchain for big data: approaches, opportunities, and future directions. Future Gener. Comput. Syst. **131**, 209–226 (2022)
25. Tseng, L., Yao, X., Otoum, S., Aloqaily, M., Jararweh, Y.: Blockchain-based database in an IoT environment: challenges, opportunities, and analysis. Clust. Comput. **23**, 2151–2165 (2020)

Research on the Evolution Path
of Network Hotspot Events Based
on the Event Evolutionary Graph

Peiguo Fu[1], ZhiTao Huang[2], Ming Liu[1], Zhiyun Zhao[1], and Wen Jiang[3(✉)]

[1] National Computer Network Emergency Response Technical Team/Coordination
Center of China, Beijing 100190, China
[2] Shandong Branch of National Computer Network Emergency Technology
Processing Coordination Center, Jinan 250002, Shandong, China
[3] International Education School, Hunan University of Medicine, Huaihua 418000,
Hunan, China
jiangwencici221@163.com

Abstract. This paper researches the evolution path method of network
public opinion based on the event evolutionary graph. Taking the network
public opinion of network hotspot events as an example, collect relevant
typical events as research samples, identify the event relationship, build
the network public opinion event evolutionary graph and abstract network public opinion event evolutionary graph respectively, and analyze
the evolution path of network public opinion event risk from two levels.
This study strives to clearly present the evolution path of network public opinion of special network hotspot events, reveal the subject, node,
situation, trend and hidden information involved in the relevant events,
construct the evolution path of network public opinion based on the
rational graph, and reveal the characteristics and practical significance
of the network public opinion transmission of hotspot events.

Keywords: Event evolutionary graph · Network public opinion ·
Hotspot events · Evolution path

1 Introduction

With the continuous expansion of the scale of Internet users, the social media
platform dominated by Sina Weibo has become the main channel for people to
publish and obtain information. The openness, anonymity and interactivity of
the social media platform enable people to publicly express their ideas. Different
opinions promote the evolution of public opinion in different directions. At the
same time, network rumors and false information will also promote the evolution of public opinion in an uncontrollable direction. It is necessary to be vigilant

Supported by the National Natural Science Foundation of China under Grant 62106060,
the Social Science Foundation of Huaihua under Grant HSP2023YB68, the Philosophy
and Social Foundation of Hunan University of Medicine under Grant 2023SK24.

© The Author(s), under exclusive license to Springer Nature Singapore Pte Ltd. 2024
Z. Tari et al. (Eds.): ICA3PP 2023, LNCS 14492, pp. 368–381, 2024.
https://doi.org/10.1007/978-981-97-0811-6_22

about the new trends and new methods of hotspot events in social media. It is necessary to deeply study the characteristics, conditions and mechanisms of the formation of such network public opinion and to study the various communication strategies of hotspot events network public opinion. How to gain insight into the evolution characteristics and path of online public opinion and understand the development trend of public opinion has become a common topic of academic concern.

In addition, the current research methods of hotspot events network public opinion have not really realized the reason, and can not make regular judgments on why and how it occurs. It is difficult to conduct in-depth analysis of the evolution path of network public opinion from multiple dimensions such as persons, events and emotions, and it is also difficult to reveal the relationship between events and subjects. As the emergence and application of the next generation of knowledge graph, they can depict the logical evolution relationship between events, and its event relationship can fully explain the evolution path of network public opinion and clearly show the evolution direction of network public opinion. Therefore, this paper studies the evolution of hotspot events network public opinion based on event evolutionary graph.

2   Problem Statement

Scholars have made a lot of researches on online public opinion in the early stage. Ser-vi [1], Jin et al. [2] proposed as to improve the "smooth dynamic linear model" and "integrated crisis graph model" to analyze the emotional evolution of netizens in emergencies. With the increasingly serious interference and harm of network information noise, effective identification of the turning point of network public opinion has gradually become the research focus of network public opinion prediction in emergencies. In recent years, researchers have mined the key nodes of the evolution of network public opinion in emergencies based on static identification methods such as degree analysis [3], K-kernel analysis [4], core-edge analysis [5], burst words H-index [6], and realized the monitoring and prediction of the development direction of network public opinion. Chen Sijing et al. [7] proposed the dynamic identification method of user behavior characteristics and network global information to reveal the dynamic characteristics of key nodes in the network information dissemination life cycle of emergencies. Liu et al. [8] found that the cause-and-effect relationship and the sequential relationship of public opinion can be clearly described by using the causal and sequential relationship of the event evolutionary graph; Li et al. [9] proposed that the event evolution law and subsequent events can be predicted through the event graph.

The other latest research findings also include: 1. Research on the evolution characteristics of online public opinion. Xia Lixin et al. [10] proposed to use the visualization, multi-dimensional, fine-grained and other features of the event graph to generate a multi-dimensional summary of network public opinion events, and describe the development of network public opinion events. Chen Jianyao [11] proposed a model for constructing the topic graph of network public opinion. 2. Research on evolution algorithm of network public opinion. Tian

Yilin [12] used hierarchical clustering method to generalize the events with higher similarity into one category, and analyze the generated public opinion evolution path. Sun Zhuo [13] used content analysis and social network analysis to analyze the topic correlation and explore the evolution path of network public opinion. Qi kai [14]constructed government affairs short video network public opinion multi-agent simulation model, combined with the specific case, simulation subsystem in the interaction relationship between various influencing factors, to explore the dynamic mechanism of public opinion in its short video network transmission. 3. Research on the evolution path of network public opinion. Shan Xiaohong et al. [15] realized the generalization of public opinion events by improving the clustering algorithm, and predicted the network public opinion events according to the evolution direction and probability of events in the abstract event graph. Xu Hailing [16] used the succession relationship and causality relationship to construct a network public opinion affair graph, to reveal the propagation characteristics and practical significance of the evolution of network public opinion.

Through combing the relevant literature, it can be seen that many good research results have been achieved in the evolution characteristics of network public opinion, the evolution algorithm and the evolution path of network public opinion, but few scholars have paid attention to the model of the evolution path of network public opinion based on the events graph, and few scholars have been able to visually reveal the evolution path of network public opinion. This research is based on the research method of the evolution path of hotspot events network public opinion based on the event evolutionary graph, collecting relevant typical events as research samples, identifying the event relationship, constructing the network public opinion event evolutionary graph and abstract network public opinion event evolutionary graph respectively, and analyzing the evolution path of network public opinion event risk from two levels. This study strives to clearly present the evolution path of event risk in the hotspot events network public opinion, reveal the subject, node, situation, trend and hidden information involved in the relevant events, and provide reference for the study of the evolution path of network public opinion risk.

3   Construction of the Network Public Opinion of Event Evolutionary Graph

In recent years, the evolution path of hotspot events network public opinion shows uncertainty and complexity. The event evolutionary graph provides the ability to analyze problems from the perspective of "reason". The intuitive representation method can help us more effectively analyze specific potential risks in complex relationships. The purpose of this study is to achieve four aspects based on the event evolutionary graph: First, extract relevant events in the network to form a domain event database, and the source of extraction is mainly from network public media data; Second, the various attributes of a single hotspot event are structured and marked in a human-machine combination way; Third, mining and matching the dissemination data of a single hotspot event, including but not limited to emotional data, volume data, etc.; Fourth, sort out and determine the relationship between hotspot events. The event relationship is constructed as follows (Fig. 1):

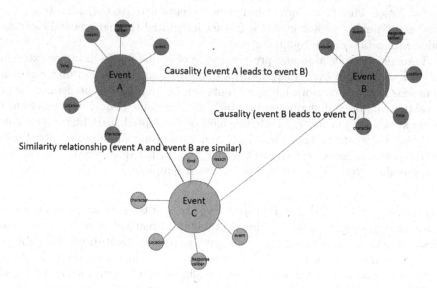

Fig. 1. Construction of the relationship of the event evolutionary graph

This study extracts and identifies the hotspot events in the network information by capturing the online public data, links the events in the event evolutionary graph, constructs the network public opinion of event evolutionary graph and the abstract network public opinion of event evolutionary graph respectively, and analyzes the evolution path of the network public opinion event risk from two levels. Combine emotion analysis technology, text tag technology, and text importance determination technology to filter influential information, and finally realize the screening of warning information based on user-defined tags and the impact tracing exploration based on this warning information.

3.1 Construction of Event Evolutionary Graph of Hotspot Events Network Public Opinion

Network public opinion reflects the occurrence of events in the objective world, taking events as nodes and the relationship between events as edges, and using the method of complex network to construct the network public opinion event evolutionary graph, which can depict the relationship and influence between events in the objective world. Therefore, constructing the network public opinion event evolutionary graph is the basis for analyzing the public opinion event chain and studying the evolution path of public opinion.

Extraction of Relevant Events of Network Hotspot Events. Collect media data based on the hotspot events domain key thesaurus, detect and cluster the text similarity through NLP (Natural Language Processing) method to form a similar article cluster, and extract the event key entities from the similar article cluster to form the hotspot events list. At the same time, the database of public opinion events is further integrated through manual cleaning and identification of the rough corpus.

Through the collection and pre-processing of online public data, we extracts the network hotspot events and forms a domain event database. Public opinion event extraction of network hotspot events can be divided into predefined event extraction and open domain event extraction. This topic mainly uses predefined event extraction. The pattern matching method is adopted, including three steps: preparing the event trigger vocabulary and extracting candidate events; Finding the sentence and event element recognition with trigger words; Extracting corresponding elements according to the event template.

Extraction of Attribute of Hotspot Events. For the network hotspot events in the database, the human-machine combination method is adopted. Among them, the entities involved in the event, such as people, institutions and regions, can be automatically extracted by machine methods. The attributes of the event such as category, emotion, and heat of transmission can be identified and marked by NLP technology. Other subjective judgment attributes are mainly based on manual judgment and marking. The attribute tag of network hotspot public opinion events is mainly realized by machine, partly by human. There are four main methods to add tags: intrinsic event basic attribute tags, tags obtained by basic information processing, tags inferred by user behavior, and tags output by data mining models.

Extraction of Relationship of Hotspot Events. Extract and mark the relationship between hotspot events. It is preliminarily divided into the nine relationships shown. The relationships of hotspot events are used to describe the association relationship between various data abstractly modeled as entities, so as to support the event association analysis, and some of them require manual judgment assistance (Table 1).

3.2 Construction of Abstract Event Evolutionary Graph of Hotspot Events Network Public Opinion

The events in the network public opinion event evolutionary graph are specific, which can clearly describe how an event evolves. In order to further discover the evolution path of network public opinion in a certain domain, it is necessary to generalize the events in the domain, build an abstract event evolutionary graph of network public opinion, and summarize the evolution path of network public opinion in the domain.

Table 1. Main types of rational relationships.

number	Events	meaning	formalization
1	Cause and effect	An event causes another event to occur	A causes B
2	Reasonable conditions	The occurrence of another event is in the condition of a event	If A then B
3	Reverse reason	One event is in opposition to another	Although A but B
4	Follow the principle	An event occurs immediately after another event	A followed by B
5	The upper and lower levels are reasonable	An event is the upper or lower event of another event	A is a class of B
6	Composition	An event is part of another event	A consists of B
7	Concurrent reasoning	An event occurs simultaneously with another event	A and B
8	Similarity relation	An event is similar to another event property	A is similar to B
9	Influence relationship	An event has a certain level of impact on another event	A impact B

Event Generalization. The key to the construction of abstract event evolutionary graph of network public opinion is event generalization. Select clustering method to select and generalize similar events. First, represent the event as an event that the machine can recognize. Second, get the vector representation of the whole event. Third, select the Euclidean distance calculation method. Fourth, event clustering. Fifth, generalize event representation.

The key to the construction of abstract event evolutionary graph of network public opinion is event generalization. $K - means$ clustering method is selected to generalize similar events. The specific algorithm is as follows:

Step 1: represent the event as an event that the machine can recognize.

The event in this paper is composed of words. Consider using the method of word vector to structure the event. The training corpus includes Wikipedia and sample data. $Word2vec$ is used to train the word vector. The training window is 5. The words with frequency less than 5 are deleted ($min\_count = 5$). The embedded dimension of the word vector is 60 dimensions.

Step 2: Get the vector representation of the whole event.

Given an event

$$e = \{\omega_i \| \omega_i \in (Verbs \bigcup Nouns)\} \tag{1}$$

via the graph function

$$f(e_i) = \{f(\omega_i)\|\omega_i \in (Verbs \bigcup Nouns)\} \tag{2}$$

to obtain the vector representation of the whole event. In this paper, the vector representation of the whole event is mapped using the average method, that is, the average value of the vectors of all words in the event is taken. Through this transformation method, all events are ultimately 60-dimensional vectors.

Step 3: Select the Euclidean distance calculation method.

For two $m-$ dimensional samples as $x_i = (x_{i1}, x_{i2}, ..., x_{im})$ and $x_j = (x_{j1}, x_{j2}, ..., x_{jm})$, the Euclidean distance is calculated as follows:

$$dist = \sqrt{\sum_{k=1}^{m} (x_{ik} - x_{jk})^2} \tag{3}$$

Step 4: Event clustering.

Use the $K - means$ algorithm to select the initial centroid, calculate the distance between events according to formula (3), and divide the events close to the centroid into the same cluster.

Step 5: Generalize event representation.

$K - means$ clustering nodes can only indicate which events are in the same class, but can not directly give the class name of each class. Use $Jieba$ word segmentation to segment events in the same category, and use the word with higher frequency as the class name of the event.

Build a Event Evolutionary Graph. The event set is the node, the causal relationship is the directed edge, and the frequency of causal events is the weight of the edge. Build a network graph based on $Gephi$ or $PythonNetwork$ library to form an abstract event evolutionary graph, and manually adjust the drawing structure.

The rational graph can be expressed as $Graph = \{Nodes, Edges\}$, for example, $Nodes = \{e_1, e_2, ..., e_p\}$ is a node set, node $e_i = \{\omega_i\|\omega_i \in (Verbs \bigcup Nouns)\}$, $Edges = \{l_1, l_2, ..., l_q\}$ is the set of edges, p and q are the number of nodes and edges respectively. Each l_i is a directed edge $e_i \rightarrow e_j$ and weight ω_{ij}. The weight can be calculated by the following formula:

$$\omega_{ij}(e_i\|e_j) = \frac{count(e_i, e_j)}{\sum_k count(e_i, e_k)} \tag{4}$$

Where $count(e_i, e_j)$ represents the number of occurrences of event pair (e_i, e_j) in the data.

Construction of Abstract Event Evolutionary Graph. Events can be generalized to represent the relationship between events at a higher level. The generalized event is taken as the node, and the relationship between the generalized nodes is taken as the edge to construct the abstract event graph. The weight calculation method in the abstract network public opinion event evolutionary graph is the same as that in the event evolutionary graph. So at last we can easily construct the abstract event evolutionary graph.

The construction of the abstract event evolutionary graph as Fig. 2 shows. Part (a) represents the event evolutionary graph composed of specific events, (b) is the event evolutionary graph formed after event generalization, and (c) is the final generated abstract event evolutionary graph. The events e_{b1} and e_{b2} are divided into e_b cluster after clustered by $K-means$ method. Similarly, events e_{e1} and e_{e2} can be represented by e_e, events e_{d1} and e_{d2} can be represented by e_d. The frequency on the edge in (b) is formed by adding the frequency on the edge in (a), and the weight on the edge in (c) is calculated based on the method mentioned in the previous section.

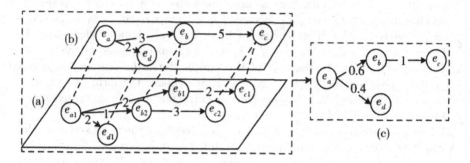

Fig. 2. The construction of the abstract event evolutionary graph

4   Experiment

In this experiment, taking the network hotspot public opinion in the field of historical nihilism as an example, analyzing the evolution path of network public opinion event risk from two levels. However, due to the large amount of data in this experiment and being unique to our department, it is not possible to compare the experimental results with relevant research algorithms. In this article, frontline business personnel were used to sample and analyze the data of relevant experiments, and the results were compared with the experimental results of the algorithm in this article. The results shows that the experimental results of the manual and algorithm were basically consistent, but the experimental time was much less.

4.1   Data Acquisition

We use *Python* to crawl the data of related events. Data collected from WeChat, Weibo, APP clients and other network platforms from 2021 to 2022. Use *jieba* library of *Python* to process sentence segmentation, word segmentation and part of speech tagging of data.

4.2 Event Extraction

The event extraction of historical nihilism related network public opinion events from the processed data is carried out in three steps.

1. Causal event extraction: extract events based on causal sentence integration rules: Causal sentences have special identifiers. This paper uses rule templates to judge causal sentences. The rule template is in the form of$<$ $Pattern, Constraint, Priority$ $>$, where $Pattern$ represents the matching rule of the sentence, $Constraint$ represents the matching constraint, and $Priority$ represents the matching priority. The $priority$ is determined by the number of times the designed template appears in the "Peking University Modern Chinese Corpus" (CCL). The higher the frequency, the higher the priority of rule matching. For example, the netizens were angry because the teacher made wrong remarks. Because, as a trigger word, $regularization$ is extracted to $cause$: "The teacher of Shanghai Vocational College told the history of the Nanjing Massacre," $effect$: "netizens expressed anger."

2. Gerund extraction: In the text of the sample data, the sentence structure of subject-verb-object (vob) is not clear. We choose the form of gerund to represent the event. The above can be extracted as: $cause$: "teacher tells historical speech", $effect$: "netizen anger".

3. Event generalization: event generalization is based on the word vector model, which converts words into machine-learning vector patterns. The $Python$ $word2vec$ library can help to build a word vector model. Through learning the corpus, we can obtain the high-dimensional vector of each word. After obtaining the vector of each word, based on $k-means$ clustering, many events are fused and generalized into a controllable number of event sets by averaging the word vector of each event.

4.3 Ten Events of Network Public Opinion in Historical Nihilistic

Through the analysis of the data collected, 10 events of historical nihilistic network public opinion are selected, and summarized as Table 2 shows (Fig. 3):

The project generalizes events into 13 categories, and each node is shown in Fig. 2. The two ends of the directed edge represent the cause event and the result event respectively, and the weight on the edge represents the probability that the cause event occurs and the result event also occurs. It can be seen that the behavior or remarks of the characters will lead to the occurrence of the behavior of apology when the netizens pay a lot of attention, reprimand and anger. The further result is detention, fine and administrative penalty. The insults to Mao Anying, the nuclear waste water from washing the ground in Japan and Zhang Zhehan's visit to the Yasukuni Shrine in Japan have all led to the occurrence of netizens' anger. Among them, Zhang Zhehan's visit to the Yasukuni Shrine in Japan has the highest probability of causing netizens' anger.

Table 2. The information of ten events of network public opinion in historical nihilistic.

number	Events	Data volume after de-duplication	Number of causal pairs of events
1	The teacher of Shanghai Sindan Vocational College made a wrong statement about the Nanjing Massacre	45912	45404
2	Actor Zhang Zhehan was accused of visiting the Yasukuni Shrine in 2019	255910	26038
3	Jinshan company bully pushes pop-up window of the insulting Qiu Shaoyun	1080	3696
4	Influencer Luo Changping slandered the volunteer army ice sculpture company	57680	22034
5	Netizen said that China Everbright Bank invited Yuan Tengfei to give a lecture	142	278
6	Liu Xiaoyan, the "famous teacher for postgraduate entrance examination", taunts Dong Cunrui with online jokes	1324	6582
7	The film "Ace Agent: Origin" portrays Lenin as a spy	4372	2751
8	Influencer dances in front of the Lvshun Museum	543	5076
9	The official microblog of Jiangsu Unicom released remarks the insulting of Mao Anying	9699	10306
10	The Third Historical Resolution stands out against historical nihilism	178	898

4.4 Experimental Result

Through the processing and analysis of data and the construction of an abstract graph of network public opinion of historical nihilism events, we can see the two major characteristics of historical nihilism network public opinion events:

1. The hot public opinion of historical nihilism in network is frequent, with "defamation of models", "disputes involving Japan" and "academic coat" as the main characteristics.

   According to the analysis of online information related to historical nihilism in China, public opinion events in this field are basically evenly distributed throughout the year, and hot public opinion events occur all the time, of which five events form public opinion peaks, respectively: in June 2021, netizens reported that Everbright Bank invited Yuan Tengfei, the "pioneer of historical nihilism", to give lectures; In August 2021, actor Zhang Zhehan was exposed to have visited the militarist shrine and other improper behaviors; In October 2021, influencer Luo Changping slandered the Volunteer Army Ice Sculpture Company as "Sand Sculpture Company"; In December 2021, Song Gengyi, a teacher of Shanghai Sindan Vocational College, made a wrong statement about the Nanjing Massacre in class; In February 2022, netizens questioned that the American film "Secret Agent Ace: The Origin" portrayed Lenin as a spy to control the situation in Russia.

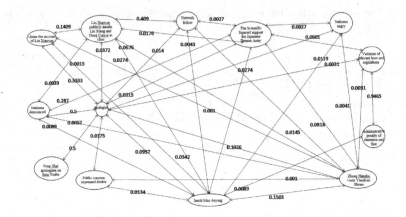

Fig. 3. The abstract event evolutionary graph of historical nihilism events

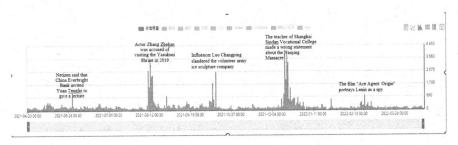

Fig. 4. Total distribution of historical nihilistic information related to the internet since 2021

2. Social platforms represented by microblog and WeChat official account have become the main forum for discussion on historical nihilism.

The total amount of information related to historical nihilism collected online of this time is 180900, and the relevant topics are generally stable and hot. From the distribution of information channels, the channel with the largest amount of historical nihilistic information released online is Weibo, with a total of more than 100000 messages (excluding comments in the comment area), accounting for 55.56% of the total information. The second is the articles on WeChat official account, with a total of more than 69000 articles, accounting for 38.34% of the total information. In addition, the number of information from news websites, app clients and forums was 8745, 1153 and 59, accounting for 4.83%, 0.64% and 0.03% respectively (Figs. 4, 5 and 6).

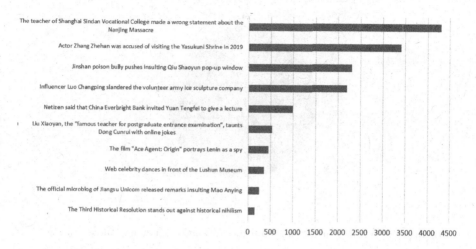

Fig. 5. The top ten topics of historical nihilism in China since 2021

4.5 Result Analysis

This study focuses on 10 online public opinion events in the field of historical nihilism, extracts relevant public opinion events from the online public opinion text of WeChat, Weibo and other media platforms, and constructs 10 rational graphs of online public opinion according to the causal relationship between events. According to the evolution path of online public opinion, it can be clearly found that some events will have a relatively large degree of variation, such as "netizen anger", "administrative fines" and "detention penalties". The research results show that most public opinion events related to historical nihilism can cause "netizen anger" and other events. As a relatively frequent event in this field, paying attention to these events is of great help to the control the online public opinion; At the same time, due to the multi-directional nature of the evolution of network public opinion, the size of the weight on the evolution path represents the possibility of the path. For example, insulting Mao Anying, supporting the nuclear waste water of Japan and Zhang Zhehan's visit to the Yasukuni Shrine in Japan have all led to netizens' anger. Among them, Zhang Zhehan's visit to the Yasukuni Shrine in Japan has the highest probability of causing netizens' anger.

In addition, through the analysis of specific events, we can find that the subject of the event can explain and apologize in a timely and public manner, which can effectively inhibit the derivation of public opinion. In the face of major events involving revolutionary history, netizens have a clear attitude towards boycott, which often results in spontaneous boycott without official intervention.

According to the analysis of 10 events of network public opinion in the selected field of historical nihilism in this experiment, the current network communication of historical nihilism reflects three characteristics: First, the mode of communication tends to be covert. The historical nihilism in the Internet envi-

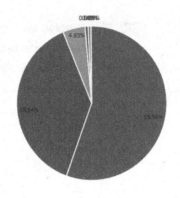

Fig. 6. Distribution of historical nihilism information channels online since 2021

ronment uses covert and imperceptible means to let people receive the wrong historical information transmitted by the historical nihilism invisibly. The second is that the communication content presents the characteristics of literature and art. Third, the target audience is becoming younger. The Internet is a comprehensive platform for gathering social information.

5   Conclusion

This paper studies the evolution path method of network public opinion based on the event evolutionary graph. By focusing on 10 network hotspot public opinion events, relevant public opinion events are extracted from the network public opinion text of WeChat, Weibo and other media platforms. According to the relationship between the events, 10 network public opinion event evolutionary graphs and abstract network public opinion event evolutionary graphs are constructed respectively. And analyzes the evolution path of network public opinion event risk from two levels. The research strives to clearly present the evolution path of network hotspot public opinion events, reveal the subject, node, situation, trend and hidden information involved in the relevant events, construct the evolution path of network public opinion based on the rational graph, and reveal the characteristics and practical significance of the transmission of network hotspot public opinion events.

This study has a wide range of data sources and a large amount of text. Text processing and data cleaning take a long time and may have omissions. The results of story extraction and generalization may not be ideal. In the future, the text can be refined and then studied again to mine deep conclusions.

Acknowledgements. This work is financially supported by the National Natural Science Foundation of China under Grant 62106060, the Social Science Foundation of Huaihua under Grant HSP2023YB68, the Philosophy and Social Foundation of Hunan University of Medicine under Grant 2023SK24.

References

1. Servi, L., D.: Analyzing social media data having discontinuous underlying dynamics. Oper. Res. Lett. **41**(6), 581–585 (2013)
2. Jin, Y., Pang, A., Cameron, G.T.: Developing a publics-driven, emotion-based conceptualization in crisis communication: final stage testing of the integrated crisis mapping (ICM) model (2009)
3. Tyshchuk, Y., Wallace, W.A.: Actionable information during extreme events-case study: Warnings and 2011 Tohoku earthquake. In: ASE/IEEE International Conference on Social Computing ASE/IEEE International Conference on Privacy (2012)
4. Yao, C., He, G.: Key nodes analysis of microblogging public opinion spread about public emergencies-such as the missed Malaysia flight MH370. In: 2015 International Conference on Modeling, Simulation and Applied Mathematics (2015)
5. Chuanlei, W., ZhangYan, Yidi, W., Fengyun, Y.: Comparison of keyword topology networks for Weibo and WeChat emergency information based on SMISC. E-Government **10**(6), 1–7 (2017)
6. Xiaoxia, Z., Mingyang, W., Chongchong, J.: DongXu: micro-blog emergencies detection approach based on the h-index of burst words. J. Intell. **000**(002), 37–41 (2015)
7. Chen Sijing, M.J. Li Gang, Zhichao, B.: Dynamic identification of key nodes in information propagation networks during emergencies. J. China Soc. Sci. Techn. Inf. **38**(2), 13 (2019)
8. Liu, T., Cui, Y., Yin, Q., Zhang, W., Wang, S., Hu, G.: Generating and exploiting large-scale pseudo training data for zero pronoun resolution. arXiv e-prints (2016)
9. Li, Z., Ding, X., Liu, T.: Constructing narrative event evolutionary graph for script event prediction (2018)
10. Lixin, X., Jjanyao, C., Huajuan, Y.: Research on the visual summary generation of network public opinion events based on multi-dimensional characteristics of event evolution graph. Inf. Stud. Theory App. **43**(10), 8 (2020)
11. Jianyao, C., Lixin, X., Xingyue, L.: Visual analysis of network public opinion feature evolution based on topic map. Inf. Sci. (2021)
12. Yilin, T., Xing, L.: Analysis on the evolution path of COVID-19 network public opinion based on the evolutionary graph. Theory Application, Information studies (2021)
13. SunZhuo, Hong, Z., Zongshui, W.: Analysis on the association and evolution path of internet public opinion. Libr. Inf. Serv. **65**(7), 12 (2021)
14. Rui, Q., Xiaoyu, W., Rui, Z.: Dynamic evolution analysis of government short video network public opinion based on SD mode. Inf. Stud. Theory App. **044**(003), 115–121130 (2021)
15. Xiaohong, S., Shihong, P., Xiaoyan, L.: YangJuan: analysis on the evolution path of internet public opinions based on the event evolution graph: taking medical public opinion as an example. Inf. Stud. Theory Appl. **43**(10), 7 (2020)
16. Hailing, X.: The evolution path of multi-dimensional feature network public opinion based on the event evolutionary graph. Inf. Sci. **40**(7), 7 (2022)

Task Offloading in UAV-Assisted Vehicular Edge Computing Networks

Wanjun Zhang, Aimin Wang, Long He, Zemin Sun$^{(\boxtimes)}$ (ID), Jiahui Li (ID), and Geng Sun (ID)

College of Computer Science and Technology, Jilin University, Changchun 130012, China

{wangam,sunzemin,sungeng}@jlu.edu.cn, wanjun21@mails.jlu.edu.cn

Abstract. As a promising architecture for supporting various intelligent vehicle applications, Vehicle Edge Computing (VEC) has received extensive research attention. However, during peak hours, the limited computing resources of VEC servers can make it difficult to meet the needs of delay-sensitive and computation-intensive tasks generated by a large number of vehicles. To overcome this challenge, we propose a UAV-assisted vehicle edge network that deploys a UAV equipped with mobile edge computing (MEC) capabilities as an aerial edge to alleviate the overload of VEC servers. To evaluate the performance of the network, the processing latency and energy consumption of tasks are incorporated into a system overhead construction. Moreover, we formulate a joint resource allocation and task offloading problem aimed at minimizing the system overhead. Since the formulated problem is proven to be NP-hard, we propose a hybrid algorithm based on genetic and simulated annealing algorithms (HGSAA), which can obtain a sub-optimal solution in polynomial time complexity. Simulation results demonstrate that HGSAA outperforms other benchmark schemes, achieving superior system performance. Simulation results show that HGSAA can achieve superior system performance compared to the other benchmark schemes.

Keywords: Mobile edge computing · Genetic algorithm · Simulated annealing algorithm · Resource allocation · Vehicular networks

1 Introduction

With the rapid development of wireless communication technology and artificial intelligence, various intelligent applications, such as autonomous driving, augmented reality and language processing, are explosively emerging in vehicular networks (VNs) [1]. These intelligent applications usually generate computation-intensive and delay-sensitive computational tasks that need to be executed by

This study is supported in part by the National Natural Science Foundation of China (62172186, 62002133, 61872158, 62272194), and in part by the Science and Technology Development Plan Project of Jilin Province (20230201087GX).

© The Author(s), under exclusive license to Springer Nature Singapore Pte Ltd. 2024
Z. Tari et al. (Eds.): ICA3PP 2023, LNCS 14492, pp. 382–397, 2024.
https://doi.org/10.1007/978-981-97-0811-6_23

vehicles and require extensive computation resources. However, it is challenging to fulfill these computation tasks due to the limited computation capability of vehicles. In order to overcome this challenge, mobile edge computing (MEC), which provides cloud-computing capabilities and the radio access network (RAN) in close proximity to mobile devices, is integrated into the VNs forming a new paradigm of vehicle edge computing (VEC) [2]. Specifically, by deploying lightweight VEC servers to the road side units (RSUs), vehicles can offload tasks to these VEC servers through wireless access networks to expand their computation capabilities.

However, the computing resources of the VEC server are also limited. Especially during peak hours, the task offloading requests generated by a large number of vehicles may lead to severe overloads of the VEC servers, which would increase the task completion delay. A lot of related work has been carried out to explore the performance of VEC network. For example, to minimize energy consumption, Yang et al. [3] propose a joint optimization strategy for computing offloading, subcarrier allocation, and resource allocation. Guo et al. [4] design a new optimization algorithm based on genetic algorithm and particle swarm algorithm to reduce energy consumption. However, these studies only focus on energy consumption without considering the delay. In [5], the cost of compute offloading and content caching id minimized through convex optimization. However, this study only consider the case of a single edge server.

Several studies focus on partial task offloading in MEC networks. In [6], a computational offloading method is proposed considering both fixed and flexible CPU frequencies of UTDs. An SDR-based algorithm is used to find the optimal solution. The authors in [7,8] employ a partial offloading approach to divide the computational tasks into two parts that are executed locally on the user and remotely on the MEC server, respectively. Studies of [9,10] achieve the goal of minimizing the weighted sum of offloading latency and energy consumption by using deep reinforcement learning methods. However, the network model needs to be trained before executing the algorithm, which can take a long time. Many studies innovatively integrate genetic algorithms into the design of task offloading in MEC. Du et al. [11] minimize the total system overhead of the MEC system by jointly and optimally computing offloading decisions and communication channel assignments with a GA. Li et al. [12] solve the cost minimization problem of the MEC problem using an improved genetic algorithm , but did not consider and test offloading delays. Wang et al. [13] propose a distributed a priori offloading mechanism, which is based on a genetic algorithm and a joint proportional offloading transfer (PROMOT) energy algorithm. PROMOT uses a priori information provided by past offloading feedback to determine whether the task should be processed locally or remotely. This mechanism has been adopted in many MEC-related studies. However, a drawback of genetic algorithms is their uncertain convergence. If the parameters are not set properly, it can easily lead to local optima or slow convergence. However, the currently available studies seem not to take into account this drawback of genetic algorithms.

Several studies attempt to address this issue by incorporating cloud computing into VEC networks. For example, Zhang et al. [14] combine MEC and cloud
computing to provide task offloading services for vehicles, which tasks generated
by vehicles can be offloaded to the remote cloud server to alleviate the overloads
of the MEC server. Sun et al. [15] present a vehicle-edge-cloud hierarchical architecture to improve the resource utilization for servers by jointly optimizing the
intra-VEC server resource allocation and inter-VEC server load-balanced offloading. Although the remote cloud server is powerful in terms of computational and
storage capabilities, it is difficult to satisfy the requirements of delay-sensitive
tasks due to the large communication delay. Different from the above works,
we propose a UAV-assisted vehicle edge network, where a UAV equipped with
MEC capabilities are deployed as aerial edges to alleviate the overloads of VEC
servers during peak hours. According to the network, we propose a fusion algorithm based on genetic and simulated annealing algorithm (HGSAA) for joint
optimization of task offloading and resource allocation to maximize the system
performance. The main contributions are summarized as follows:

- First, we employ a UAV-assisted vehicle edge network to coordinate the UAV
  and VEC servers to cooperatively perform a large number of tasks generated
  by vehicles during peak hours.
- Second, we formulate a joint task offloading and resource allocation problem
  (JTORAP), with the aim of minimizing the system overhead. Specifically,
  the system overhead is theoretically constructed by synthesizing the delay
  and energy consumption in processing the tasks.
- Third, due to the NP-hardness of JTORAP, we propose a fusion algorithm
  HGSAA that combines the advantages of the genetic algorithm and simulated annealing algorithm to achieve faster convergence and better solution.
  Moreover, the proposed HGSAA is proved to be polynomial computation
  complexity through theoretical analysis.
- Finally, simulation results demonstrate that the proposed HGSAA is able to
  achieve superior performance compared to several benchmark schemes.

The rest of this paper is organized as follows. Section 2 presents the proposed system model and problem formulation. In Sect. 3, we detail the proposed
HGSAA algorithm. Section 4 shows the simulation results. Finally, Sect. 5 concludes the overall paper.

2   System Model

As shown in Fig. 1, we consider a UAV-assisted VEC network consisting of
N vehicles with a set $\mathcal{N} = \{1, 2, \ldots, N\}$, M VEC servers with a set $\mathcal{M} = \{1, 2, \ldots, M\}$ and a UAV u on a research road segment of length L. The VEC
servers are deployed alongside the road with coverage radius R, which can pro-
vide both radio access and computing offload services to vehicles within their
communication range. For each VEC server $m \in \mathcal{M}$, its characteristics can be
represented by $\mathbf{St}_m = \{\mathbf{P_m}, f_m^{\max}\}$, where in $\mathbf{P_m} = [x_m, y_m, 0]$ and $f_m^{\max}$ denote

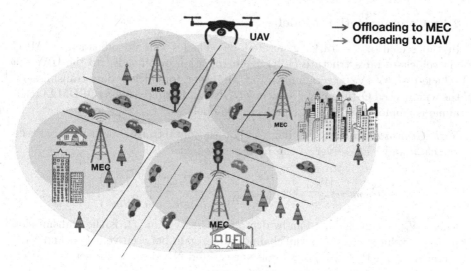

Fig. 1. UAV-assisted VEC network.

the position and the maximum computing resources of VEC server m, respectively. The UAV equipped with MEC capability is deployed above the road as an aerial edge to relieve the overload of the VEC network during peak hours. The UAV u is characterized by $\mathbf{St}_u = \{\mathbf{P}_u, f_u^{\max}\}$, wherein $\mathbf{P}_u = [x_u, y_u, H]$ and $f_u^{\max}$ represent the position and the maximum computing resources of the UAV, respectively. The vehicles driving on the road with two traffic streams follow homogeneous Poisson spatial distribution with densities ρ_1 and ρ_2 [16]. Each vehicle $n \in \mathcal{N}$ is characterized by $\mathbf{St}_n = \{\mathbf{P}_n, d_n, f_n\}$, wherein $\mathbf{P}_n = [x_n, y_n, 0]$ and f_n is the position and local computing capability of vehicle n, respectively. $d_n \in \{0, 1\}$ is a binary variable representing the moving direction of vehicle n, where $d_n = 0$ means the vehicle is moving toward the right. We assume each vehicle n has a delay-sensitive computing task $\mathbf{\Phi}_n = \{D_n, \eta_n, T_n^{\max}\}$, where D_n denotes the input data size of the task (i.e., bit), η_n represents the computational density of the task (i.e., cycles/bit) and $T_n^{\max}$ is the maximum tolerable delay of the task. Due to the limited computing resources of vehicles, each task can be divided into two independent sub-tasks with one executed at the vehicle and the other offloaded to the VEC server or the UAV for execution. To this end, we first define $\lambda_n \in [0, 1]$ as the offloading ratio of task $\mathbf{\Phi}_n$, i.e., sub-task $\lambda_n D_n$ is offloaded and sub-task $(1 - \lambda_n)D_n$ is executed locally on vehicle n. Second, a binary variable $a_n \in \{0, 1\}$ is used to represent the offloading decision of vehicle n, where $a_n = 0$ means that the sub-task is offloaded to the VEC server connected to the vehicle and $a_n = 1$ denotes that the sub-task is offloaded to the UAV.

2.1 Communication Model

In the scenario, the tasks generated by vehicles can be offloaded to MEC via vehicle-to-infrastructure (V2I) communication or offloaded to the UAV via vehicle-to-UAV (V2U) communication. Moreover, for efficient data transmission, the widely used Orthogonal Frequency Division Multiple Access (OFDMA) technique is employed in the communication models.

V2I Communication. Similar to [17], the data transmission rate between vehicle n and VEC server m can be calculated as:

$$R_{n,m} = \frac{W_{n,m}}{\mathbf{N}_m} \log_2 \left(1 + \frac{P_n g_{n,m}^{\mathrm{v2i}}}{N_0 + \sum_{i \in \mathbf{N}_m, i \neq n} P_i g_{i,m}^{\mathrm{v2i}}} \right), \tag{1}$$

where W_m is the channel bandwidth of VEC server m, P_n is the transmission power of vehicle n, $g_{n,m}^{\mathrm{v2i}}$ is the channel power gain between vehicle n and VEC server m, N_0 is the background noise power and $\mathbf{N}_m$ denotes the set of interference vehicles within the range of VEC server m.

V2U Communication. The V2U communication is modeled as a free-space path loss model since it is dominated by LoS links. Therefore, the data transmission rate between vehicle n and the UAV can be calculated as [18]:

$$R_{n,u} = \frac{W_u}{\mathbf{N}_u} \log_2 \left(1 + \frac{P_n \beta_0 d_{n,u}^{-2}}{N_0} \right), \tag{2}$$

where W_u is the channel bandwidth of the UAV, $\mathbf{N}_u$ represents the set of vehicles that offload tasks to the UAV, β_0 denotes the channel gain at the reference distance $d_0 = 1$ m and $d_{n,u} = \|\mathbf{P_m} - \mathbf{P}_u\|$ denotes the distance between vehicle n and the UAV.

2.2 Computation Model

Local Computing. Vehicle n locally computes sub-task $(1 - \lambda_n)D_n$. The delay caused by local computing can be calculated as:

$$T_n^{\mathrm{loc}} = \frac{(1 - \lambda_n)\eta_n D_n}{f_n}. \tag{3}$$

Correspondingly, the energy consumption generated by local computing can be calculated as:

$$E_n^{\mathrm{loc}} = \kappa(f_n)^3 T_n^{\mathrm{loc}}, \tag{4}$$

where κ denotes the effective switched capacitance depending on the CPU architecture of vehicle n [19].

Therefore, the system overhead caused by local computing can be expressed as:

$$L_n^{\mathrm{loc}} = \gamma_n^{\mathrm{T}} T_n^{\mathrm{loc}} + \gamma_n^{\mathrm{E}} E_n^{\mathrm{loc}}, \tag{5}$$

where γ_n^{T} and γ_n^{E} represent the weight coefficients of delay and energy consumption respectively.

Terrestrial Edge Computing. Sub-task $\lambda_n D_n$ is offloaded to VEC server m computing. The total service delay of terrestrial edge computing is composed of the transmission delay and the computation delay of the VEC server, which can be calculated as:

$$T_n^{\mathrm{tec}} = \frac{\lambda_n D_n}{R_{n,m}} + \frac{\lambda_n \eta_n D_n}{f_{m,n}^{\mathrm{tec}}}, \tag{6}$$

where $f_{m,n}^{\mathrm{tec}}$ denotes the computation resources allocated to vehicle n by VEC server m. Note that we omit the delay of the result feedback because the data size of the computation result is much smaller than that of the computation input for most mobile applications [20]. Correspondingly, the energy consumption including the transmission energy consumption and the computation energy consumption of VEC server m can be calculated as:

$$E_n^{\mathrm{tec}} = P_n \frac{\lambda_n D_n}{R_{n,m}} + \kappa (f_n^{\mathrm{tec}})^3 \frac{\lambda_n \eta_n D_n}{f_{m,n}^{\mathrm{tec}}}. \tag{7}$$

Therefore, the system overhead caused by terrestrial edge computing can be given as:

$$L_n^{\mathrm{tec}} = \gamma_n^{\mathrm{T}} T_n^{\mathrm{tec}} + \gamma_n^{\mathrm{E}} E_n^{\mathrm{tec}}, \tag{8}$$

Aerial Edge Computing. Sub-task $\lambda_n D_n$ is offloaded to the UAV u computing. The total service delay of aerial edge computing is composed of the transmission delay and the computation delay of the UAV, which can be calculated as:

$$T_n^{\mathrm{aec}} = \frac{\lambda_n D_n}{R_{n,u}} + \frac{\lambda_n \eta_n D_n}{f_{u,n}^{\mathrm{aec}}}, \tag{9}$$

where $f_{u,n}^{\mathrm{aec}}$ denotes the computation resources allocated to vehicle n by the UAV. The energy consumption of aerial edge computing can be given as:

$$E_n^{\mathrm{aec}} = P_n \frac{\lambda_n D_n}{R_{n,u}} + \kappa (f_n^{\mathrm{aec}})^3 \frac{\lambda_n \eta_n D_n}{f_{u,n}^{\mathrm{aec}}}. \tag{10}$$

Therefore, the system overhead caused by aerial edge computing can be given as:

$$L_n^{\mathrm{aec}} = \gamma_n^{\mathrm{T}} T_n^{\mathrm{aec}} + \gamma_n^{\mathrm{E}} E_n^{\mathrm{aec}}, \tag{11}$$

2.3 Problem Formulation

This work aims to minimize the system overhead by jointly optimizing the task offloading decision $\mathcal{A} = \{a_n\}_{n \in \mathcal{N}}$, the offloading percentage $\Lambda = \{\lambda_n\}_{n \in \mathcal{N}}$ and the computation resource allocation $\mathcal{F} = \{f_{m,n}^{\text{tec}}, f_{u,n}^{\text{aec}}\}_{m \in \mathcal{M}, n \in \mathcal{N}}$. Therefore, the optimization problem can be formulated as follows:

$$\mathbf{P} : \underset{\mathcal{A}, \Lambda, \mathcal{F}}{\&\min} \sum_{m=1}^{M} \sum_{n=1}^{N} \left(L_n^{\text{loc}} + (1 - a_n) L_n^{\text{tec}} + a_n L_n^{\text{aec}} \right) \tag{12a}$$

$$\text{s.t.} \& C1 : a_n \in \{0, 1\}, \forall n \in \mathcal{N}, \tag{12b}$$

$$C2 : \lambda_n \in [0, 1], \forall n \in \mathcal{N}, \tag{12c}$$

$$C3 : 0 \le f_{m,n}^{\text{tec}} \le f_m^{\max}, \forall m \in \mathcal{M}, \forall n \in \mathbf{N}_m, \tag{12d}$$

$$C4 : \sum_{n \in \mathbf{N}_m} f_{m,n}^{\text{tec}} \le f_m^{\max}, \forall m \in \mathcal{M}, \tag{12e}$$

$$C5 : 0 \le f_{u,n}^{\text{aec}} \le f_u^{\max}, \forall n \in \mathbf{N}_u, \tag{12f}$$

$$C6 : \sum_{n \in \mathbf{N}_u} f_{u,n}^{\text{aec}} \le f_u^{\max}, \tag{12g}$$

Constraint $C1$ represents the values of task offloading decisions of vehicles. Constraint $C2$ denotes the offloading percentage of vehicle n. Constraints $C3$ and $C4$ are the constraints on the resource allocation of the VEC servers. Constraints $C5$ and $C6$ are the constraints on the resource allocation of the UAV. The above problem $\mathbf{P}$ contains both binary variables (i.e., task offloading decision $\mathcal{A}$) and continuous variables (i.e., offloading percentage Λ and computation resource allocation $\mathcal{F}$) is a mixed-integer nonlinear programming (MINLP) problem, which is non-convex and NP-hard [21, 22]. To this end, we design an HGSAA algorithm that can achieve a sub-optimal solution in polynomial time complexity.

3   The Proposed Algorithm

In this section, the joint resource allocation and task offloading problem based on the UAV-assisted vehicle edge network with the proposed minimizes the system overhead. The problem is divided into two phases, and the first phase groups tasks to VEC or UAV based on a preference metric that considers both VEC and UAV proximity and workload. The original problem is divided into several subproblems that prove to be NP-hard, thus HGSAA is designed to solve these subproblems in parallel. The capability of the genetic algorithm is the strong global search capability. The capability of the annealing algorithm is to avoid falling into local optimum. The HGSAA combines the advantages of both algorithms by introducing the temperature parameters of the simulated annealing algorithm into the genetic algorithm. GASS can inherit the powerful global search capability of the genetic algorithm and reduce the convergence speed to avoid falling into partial optimum upfront [23].

3.1   Task Offload Decision

To determine vehicles' offloading decisions, we propose a preference-based task offloading scheme. Specifically, vehicles tend to offload tasks to the nearest server (the VEC servers or the UAV) because closer distance means better communication conditions. However, due to the uneven spatial distribution of vehicles, this scheme could lead to overloading of servers with high vehicle densities. Therefore, by jointly considering the distance and the load of servers, we design the vehicle's preference function for the servers as follows:

$$pre_j^i = \frac{1}{\frac{d_j^i}{C_j} + \frac{load_j}{n}},$$ (13)

where d_j^i denotes the distance between user i and the VEC servers or the UAV j, C_j denotes the signal coverage of the UAV or VEC, and $load_j$ calculates the load of the UAV or VEC, which is the number of users assigned to it. n is the total number of users. We iterate through all users and use a grouping policy to select the offload destination, the following is a detailed description of the offload destinations selected by users based on their preferences.

- First, each vehicle calculates the distance to all reachable servers, and queries the workload of these servers.
- Second, each vehicle calculates its preference for all reachable servers according to Eq. (13).
- Finally, each vehicle selects the server with the maximum preference value to offload the task.

3.2   Task Offload Ratio and Resource Allocation

Given the task offloading decisions of all vehicles, the optimization problem **P** can be transformed into a jointly optimizing task offload ratio and resource allocation problem (JOTRP). Due to the advantages of being simple and effective, metaheuristic algorithms have been successfully applied to solve various optimization problems in recent years [24]. To this end, we propose the HGSAA to solve the JOTRP, which combines the better global search ability of the genetic algorithm and the better local search ability of the simulated annealing algorithm to obtain a much better solution and faster convergence. The main steps of HGSAA are detailed as follows.

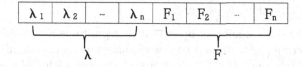

Fig. 2. Individual.

Initialize Population. A random group of individuals is generated as the initial population, each individual contains a separate set of λ and F denoting the task offload ratio and the allocated resource amount of the server, respectively, and the total amount of resources allocated for tasks offloaded to the same VEC or UAV cannot exceed its total resources. As shown in Fig. 2, assuming that there are N tasks to be processed, an overall matrix of x rows and $2N$ columns is generated. The number of rows is set to be 4 to 6 times the number of columns to make the algorithm converge more easily and reduce the iteration complexity, so $x = 8N$. In addition to the population size, the initialization algorithm parameters are the initial temperature, termination temperature, cooling parameters, mutation rate and crossover rate. The population K equation is as follows:

$$K = \begin{bmatrix} k_{1,1} & k_{1,2}... & k_{1,N} & k_{1,N+1} & k_{1,N+2}... & k_{1,2N} \\ k_{2,1} & k_{2,2}... & k_{2,N} & k_{2,N+1} & k_{2,N+2}... & k_{2,2N} \\ ... & & & & ... \\ k_{x,1} & k_{x,2}... & k_{x,N} & k_{x,N+1} & k_{x,N+2}... & k_{x,2N}, \end{bmatrix}, \tag{14}$$

$k_{i,j}$ $(i \in x, j \in N)$ denotes the percentage of vehicle task j unloaded to the selected device in each policy combination i. $k_{i,N+j}$ $(i \in x, j \in N)$ denotes the computational resources allocated to the vehicle i by the offloading device (VEC or UAV) selected by the vehicle task j in each policy combination i.

Selection. The process of selecting high quality parents based on the size of the fitness function and randomly eliminating a fraction of inferior chromosomes. The selected chromosomes are considered as parental chromosomes.

Crossover. The parental chromosomes are recombined to produce new individuals, where F is constrained to select the same MEC or UAV parental crossover and the resulting individuals satisfy the constraints. Thereby computing the parental and child chromosome overheads according to the fitness function(system overhead), we define Δf as

$$\Delta f = f(child) - \min(f(parent_1, f(parent_2))), \tag{15}$$

$f(child)$ represents the child chromosome overhead, if $\Delta f < 0$, Then the newly generated child chromosome is better, and the child chromosome replaces the parents. Otherwise, compute the probability $\exp(-\Delta f/(\xi \cdot T))$, where ξ is a constant and T is the current temperature. Randomly generate a random number rand between 0 and 1 , if $rand < \exp(-\Delta f/(\xi \cdot T))$ add the child chromosome to the population, otherwise discard it.

Mutation. Mutation is another operation that produces a new chromosome by randomly selecting a gene from a parent chromosome and randomly changing it to another feasible gene. If $\Delta f < 0$ replace the parent with child, else calculate the probability value $\exp(-\Delta f/(\xi \cdot T))$. Generate the random number rand. If $rand < \exp(-\Delta f/(\xi \cdot T))$ replace the parent with child, otherwise keep the parent and discard the child.

The method ensures that chromosomes with lower fitness values have a higher probability of retention. In the traditional genetic algorithm, it is desired to increase the ability of the system to explore the solution and be able to avoid falling into local optimum solutions as much as possible. In the improved algorithm, the global optimal solution is found faster by adjusting the retention probability so that the better chromosomes are more easily inherited to the next generation. When the temperature becomes low, that is, it enters the later convergence stage, the convergence speed and efficiency of the algorithm are further improved because the probability of worse chromosomes being filtered out is very low. The detailed procedure is summarized in Algorithm 1.

Algorithm 1: HGSAA

Input: $I_i = \{D_n, \eta_n, T_n^{max}\}$: Vehicle calculation tasks, Initial temperature T, Termination temperature T_{min}, Cooling parameters α, Crossover probability p_c, Mutation probability p_m.

Output: $BEST_\lambda$, $BEST_F$, $BEST_{fitness}$

1 Randomly generate a population K as population;
2 **while** $T > T_{min}$ **do**
3   Selection parents from population ;
4   **if** $rand < p_c$ **then**
5   generate the *child* chromosome by crossover operation;
6   **if** $f_{child} < f_{parents}$ **then**
7            $parents \leftrightarrow child$;
8   **else if** $rand < \exp\left(-\Delta f / (\xi \cdot T)\right)$ **then**
9            $parents \longleftarrow child$;
10   **end**
11   **end**
12   **if** $rand < p_m$ **then**
13   generate the *child* chromosome by mutation operation;
14   **if** $f_{child} < f_{parents}$ **then**
15           $parents \leftrightarrow child$;
16   **else if** $rand < \exp\left(-\Delta f / (\xi \cdot T)\right)$ **then**
17           $parents \longleftarrow child$;
18   **end**
19   **end**
20     $population \longleftarrow parents$ Replace the bad according to the fitness function;
21     $T = T * \alpha$;
22 **end**

Result: $BEST_{pop}$, $BEST_{fitness}$

3.3   Time Complexity Analysis

In the optimal task offloading decision stage, each vehicle calculates its preference to all reachable servers with a time complexity of $O(N \times M)$, where N represents the number of vehicles, and $M + 1$ represents the number of servers. In the optimal task offload ratio and resource allocation stage, the time complexity is mainly related to the parameters of the HGSAA algorithm. G and P denote the

Table 1. System Parameters in Simulation

Parameter	value	symbol
The background noise power	-100 dB	N_0
The maximum computing resources of VEC	20 GHz	f^m_{max}
The channel bandwidth	10 MHz	W
The maximum computing resources of the UAV	50 GHz	f^u_{max}
Local computing capability of vehicle n	0.8 GHz	f^n
The transmission power of vehicle n	0.5W	P_n
The computational density of the task	100–1500 cycles/bit	η_n
Channel gain	140.7+36.7log10d	g
The effective switched capacitance	10^{-27}	κ
The input data size of the task	600Kb	D_n
The weight coefficients of delay	0.5	γ^T
The weight coefficients of energy consumption	0.5	γ^E
Termination temperature	2	T_{min}
Initial temperature	100	T
Cooling temperature	0.98	η

number of iterations and the size of the population, respectively. The size of the individuals is equal to the number of users. The length of each individual is $2N$. Therefore, the time complexity can be summarized as $O(2 \times G \times P \times N)$.

4   Experiments

In this section, the performance of the algorithm proposed is evaluated by Matlab simulation to test the impact of various metrics on task utility and delay, and the convergence of the algorithm is also evaluated. The MEC coverage radius is 100m and the UAV is fixed at an altitude of 200m. The detailed parameters are shown in Table 1 .

The simulation results are compared with the following three benchmark schemes:

- Entire local offloading (ELO) strategy: all tasks of vehicles are completed locally.
- Local and MEC offloading (LMO): each vehicle processes part of the task locally and offloads the other part to the VEC server.
- Random offloading (RO): vehicles randomly select
  means that all of the users who generate tasks randomly choose to offload to the UAV and MEC servers where communication can be established. At this moment, due to the lack of reasonable resource allocation, when the server resources are overloaded, the remaining users have to choose local computing. Therefore, after reaching a certain amount of users, the utility increases significantly, at a similar rate as the ELO increases.

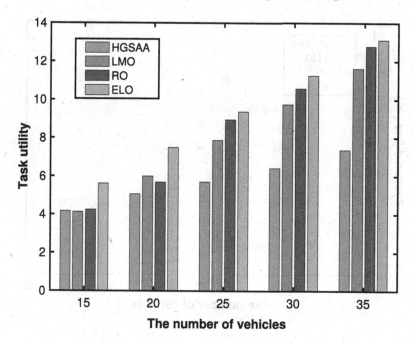

Fig. 3. Task utility with respect to the number of vehicles.

4.1   Optimization Results

Figure 3 illustrates the task utility with respect to the number of vehicles. Among the four strategies, HGSAA shows superior performance compared to ELO, LMO, and RO when all other parameters are held constant. The results indicate that ELO incurs the highest system cost among all strategies, which increases linearly with the number of tasks due to limited local processing capacity. On the other hand, although RO shows similar growth to ELO, it performs better; however, it lacks reasonable resource allocation, leading to easy server overloading and the last tasks being processed locally. The cost of the HGSAA strategy is lower than that of the LMO strategy, proving UAV's effectiveness in reducing the system cost. Furthermore, the cost of the LMO strategy is lower than that of both ELO and RO, confirming the necessity of distributed offloading optimization for computational tasks.

Figure 4 compares the system delay with respect to the number of vehicles. The results show that due to the limited local computing resources, the system delay of ELO grows linearly with the increase of vehicles and has the highest latency. The RO strategy, on the other hand, has the second highest latency after ELO because part of the tasks are offloaded to MEC or UAV without reasonable resource allocation, resulting in another part of the tasks cannot be offloaded and can only be processed locally, while the system delay of the strategy using the partial offloading model with HGSAA to optimize the task offloading ratio and resource allocation is much lower than the above two. Observing HGSAA

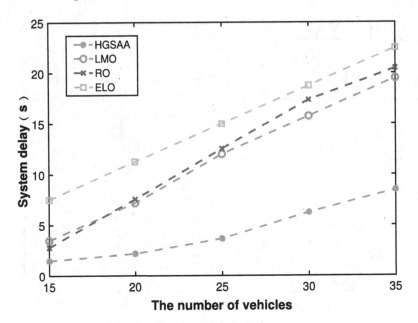

Fig. 4. System delay with respect to the number of vehicles.

and LMO shows that the latency of LMO is higher than HGSAA, which is the reason for not including UAV. The results show that HGSAA is particularly effective when there are more vehicles. This once again proves the significance of UAV assistance, with the excellence of HGSAA algorithm.

Figure 5 shows the task utility with respect to the task size. It can be observed that all the curves are almost linearly increasing except for HGSAA. When resources of MEC and UAV are not allocated reasonably and some tasks occupy all of them, most of the remaining vehicle tasks can only be executed locally. Therefore, for HGSAA, the overall system cost is optimal when task size increases gradually.

At the end, the maximum fitness in each generation is analyzed when P_c and P_m are set as 0.8 and 0.01, respectively. The x-axis of Fig. 6 represents the algorithm iterations and the y-axis represents the maximum fitness in each generation of the population. The maximum fitness individual in the population corresponds to a specific offload ratio and MEC or UAV resource allocation size. The smaller its value is, the lower the user overhead is. Moreover, it can be concluded that the algorithm converges in the 85th generation. Furthermore, based on the analysis of P_c and P_M, we can ensure that the algorithm has converged to the global optimum.

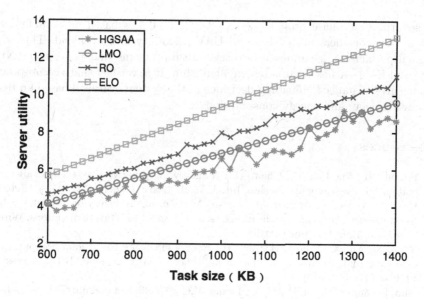

Fig. 5. Impact of task data size on task utility.

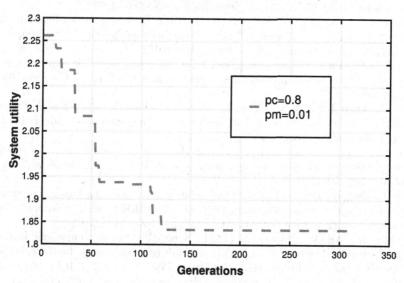

Fig. 6. The convergence performance of proposed HGSAA.

5   Conclusion

In this paper, a UAV-assisted VEC network is proposed to alleviate the overload of VEC servers by deploying UAV with MEC computing capability as an aerial edge. In addition, a joint resource allocation and task offloading problem is proposed to minimize the system overhead. The problem is divided into two phases, the first phase groups users of VECs or UAVs based on a preference metric that considers both VEC and UAV proximity and workload. Then the original problem is decomposed into several subproblems that prove to be NP-hard and HGSAA based on a genetic algorithm. It is verified that the proposed scheme gives balanced offloading decisions in the system overhead based on task processing delay and energy consumption.

References

1. Fourati, H., Maaloul, R., Chaari, L.: A survey of 5g network systems: challenges and machine learning approaches. Int. J. Mach. Learn. Cybern. **12**, 385–431 (2021)
2. Raza, S., Wang, S., Ahmed, M., Anwar, M.R., et al.: A survey on vehicular edge computing: architecture, applications, technical issues, and future directions. Wirel. Commun. Mob. Comput. (2019)
3. Yang, X., Yu, X., Huang, H., Zhu, H.: Energy efficiency based joint computation offloading and resource allocation in multi-access MEC systems. IEEE Access **7**, 117054–117062 (2019)
4. Guo, F., Zhang, H., Ji, H., Li, X., Leung, V.C.: An efficient computation offloading management scheme in the densely deployed small cell networks with mobile edge computing. IEEE/ACM Trans. Netw. **26**(6), 2651–2664 (2018)
5. Wang, C., Liang, C., Yu, F.R., Chen, Q., Tang, L.: Computation offloading and resource allocation in wireless cellular networks with mobile edge computing. IEEE Trans. Wireless Commun. **16**(8), 4924–4938 (2017)
6. Dinh, T.Q., Tang, J., La, Q.D., Quek, T.Q.: Offloading in mobile edge computing: task allocation and computational frequency scaling. IEEE Trans. Commun. **65**(8), 3571–3584 (2017)
7. Wu, Y., Qian, L.P., Ni, K., Zhang, C., Shen, X.: Delay-minimization nonorthogonal multiple access enabled multi-user mobile edge computation offloading. IEEE J. Sel. Topics Signal Process. **13**(3), 392–407 (2019)
8. Wang, F., Xu, J., Wang, X., Cui, S.: Joint offloading and computing optimization in wireless powered mobile-edge computing systems. IEEE Trans. Wireless Commun. **17**(3), 1784–1797 (2017)
9. Li, J., Gao, H., Lv, T., Lu, Y.: Deep reinforcement learning based computation offloading and resource allocation for MEC. In: IEEE Wireless Communications and Networking Conference (WCNC) 2018, pp. 1–6. IEEE (2018)
10. Zhang, H., Wu, W., Wang, C., Li, M., Yang, R.: Deep reinforcement learning-based offloading decision optimization in mobile edge computing. In: IEEE Wireless Communications and Networking Conference (WCNC), pp. 1–7. IEEE 2019 (2019)
11. Du, C., Chen, Y., Li, Z., Rudolph, G.: Joint optimization of offloading and communication resources in mobile edge computing. In: 2019 IEEE Symposium Series on Computational Intelligence (SSCI), pp. 2729–2734. IEEE (2019)
12. Kuang, L., Gong, T., OuYang, S., Gao, H., Deng, S.: Offloading decision methods for multiple users with structured tasks in edge computing for smart cities. Futur. Gener. Comput. Syst. **105**, 717–729 (2020)
13. Wang, J., et al.: A probability preferred priori offloading mechanism in mobile edge computing. IEEE Access **8**, 39758–39767 (2020)
14. Zhao, J., Li, Q., Gong, Y., Zhang, K.: Computation offloading and resource allocation for cloud assisted mobile edge computing in vehicular networks. IEEE Trans. Veh. Technol. **68**(8), 7944–7956 (2019)

15. Sun, Z., Sun, G., Liu, Y., Wang, J., Cao, D.: BARGAIN-MATCH: a game theoretical approach for resource allocation and task offloading in vehicular edge computing networks. IEEE Trans. Mob. Comput. 1–18 (2023)

16. Zhang, J., Guo, H., Liu, J., Zhang, Y.: Task offloading in vehicular edge computing networks: a load-balancing solution. IEEE Trans. Veh. Technol. **69**(2), 2092–2104 (2019)

17. Zhang, J., Liu, J., Guo, H., Zhang, Y.: Task offloading in vehicular edge computing networks: a load-balancing solution. IEEE Trans. Veh. Technol. **69**(2), 2092–2104 (2019)

18. Hu, Q., Cai, Y., Yu, G., Qin, Z., Zhao, M., Li, G.Y.: Joint offloading and trajectory design for UAV-enabled mobile edge computing systems. IEEE Internet Things J. **6**(2), 1879–1892 (2019)

19. Liao, Z., Peng, J., Xiong, B., Huang, J.: Adaptive offloading in mobile-edge computing for ultra-dense cellular networks based on genetic algorithm. J. Cloud Comput. **10**(1), 1–16 (2021)

20. Guo, H., Liu, J.: Collaborative computation offloading for multiaccess edge computing over fiber-wireless networks. IEEE Trans. Veh. Technol. **67**(5), 4514–4526 (2018)

21. Guo, S., Liu, J., Yang, Y., Xiao, B., Li, Z.: Energy-efficient dynamic computation offloading and cooperative task scheduling in mobile cloud computing. IEEE Trans. Mob. Comput. **18**(2), 319–333 (2019)

22. Belotti, P., Kirches, C., Leyffer, S., Linderoth, J., Luedtke, J., Mahajan, A.: Mixed-integer nonlinear optimization. Acta Numer. **22**, 1–131 (2013)

23. Wu, H., Deng, S., Li, W., Fu, M., Yin, J., Zomaya, A.Y.: Service selection for composition in mobile edge computing systems. In: 2018 IEEE International Conference on Web Services (ICWS), pp. 355–358. IEEE (2018)

24. Dokeroglu, T., Sevinc, E., Kucukyilmaz, T., Cosar, A.: A survey on new generation metaheuristic algorithms. Comput. Indus. Eng. **137**, 106040 (2019)

Path Planning of Coastal Ships Based on Improved Hybrid A-Star

Zhiying Cao[1(✉)], Hongkai Wang[1], Xiuguo Zhang[1], Yiquan Du[2],
and Dezhen Zhang[1]

[1] School of Information Science and Technology, Dalian Maritime University,
Dalian 116026, China
{czysophy,wanghk,zhangxg,dezhen}@dlmu.edu.cn
[2] Netease Youdao Information Technology (Beijing) Co., Ltd., Beijing 100089, China

Abstract. The path planning of ships along the coast is a key topic in
the field of shipping and is the core basis for the intelligent development
of ships. As a well developed algorithm for path planning, the Hybrid A-
star algorithm is time-efficient and economical. However, in the marine
sector, the proximity of the planned paths to obstacles makes it too dan-
gerous to be used in a practical environment. To tackle the problem, this
paper proposes an improved Hybrid A-star algorithm for path planning
of coastal ships. The method, on the one hand solves the problem of planning path close to obstacles by introducing repulsive gain to ensure the
safety of the planned path, on the other hand introduces gravitational
gain to speed up the planning of the path and ensure the timeliness of the
path planning. In addition, as the actual navigation of the ship requires
the path to have no curves and as few turning points as possible, this
paper designs a method of path optimization. By compressing waypoints
and eliminating redundant turning points, the method further improves
the economy and safety of the planned paths, making the final optimized
paths fit the actual navigation requirements of the ship. The results of
the simulation experiments show that the path planned by this method
outperforms other algorithms in terms of economy, safety and time.

Keywords: Coastal waters · Ship path planning · Hybrid A-star
algorithm · Path optimization

1 Introduction

With the rapid development of economy, the traffic on the water surface is
becoming more and more complex, and the requirements for ship navigation are
becoming higher and higher [1,2]. Ship navigation safety is a long-term research
topic of the field of ships in the world. Statistical data show that 80% of ship
collision accidents are caused by human factors [3], while the ship path planning
provides a generic solution to reduce the occurrence of ship collision [4]. By considering environmental factors in the process of path planning, the planned path
can meet the safety and economy of ship navigation [5–7].

© The Author(s), under exclusive license to Springer Nature Singapore Pte Ltd. 2024
Z. Tari et al. (Eds.): ICA3PP 2023, LNCS 14492, pp. 398–417, 2024.
https://doi.org/10.1007/978-981-97-0811-6_24

Compared with narrow and open waters, coastal waters have no coastline restrictions, but temporary obstacle areas such as shipwreck area, prohibited area and military exercise area may appear. If these temporary obstacles are on the navigation path of the ship, local path optimization is required and has higher requirements for the safety, economy and timeliness of the planned path.

Scholars around the world have proposed many methods for coastal ship path planning. At present, they are mainly divided into bionic intelligent algorithm, machine learning related algorithm and traditional method.

Bionic intelligent algorithm mainly transforms path planning problem into path optimization problem, which takes path distance and ship navigation angle as constraints and combines collision risk to carry out path planning. However, this kind of method tends to fall into local optimum and can't find the optimal path before Many iterations. The amount of calculation and planning time rapidly increase with the size of the environment.

Machine learning algorithm needs to use a large number of actual data for model training, and then use the model for path planning [8–10]. This kind of algorithm has good self-learning ability and strong migration, but it heavily depends on the data samples during training, the safety of the planned path can not be guaranteed, and it is prone to latitude disasters.

Compared with bionic intelligent algorithm and machine learning algorithm, the traditional algorithm does not plan path by iteration, does not rely on model training, and the planning time is very short, so it is still widely used in coastal ship path planning [11]. As one of the most classical methods, A-star algorithm is widely used in ship path planning [12–14]. However, the path planned by A-star algorithm is not combined with the motion model, and does not meet the kinematic requirements. Hybrid A-star algorithm [15] proposed by Dolgov combines A-star with kinematics model to solve the defects of A-star kinematics. It is one of the most mature algorithms used in industry. However, the Hybrid A-star algorithm uses the Manhattan distance as the estimated cost, which causes a large number of unnecessary calculation when the next optimal node it selects is in an obstacle notch, and there is a situation where the planned path passes between two obstacles that are close together, which makes the planned path less safe and does not meet the actual navigation requirements of the ship. Besides, if the granularity of environment grid division is not appropriate, the planning time overhead of the algorithm will be increased.

Due to the large volume of the ship, the planned path should try to avoid the problems of large number of turning points and radian. However, most of the paths planned by the current path planning algorithms have too many turning points, which is not suitable for the actual voyage of the ship.

To solve the above problems, an improved hybrid A-star algorithm is proposed and applied to ship path planning. The main contributions of this paper can be summarized as follows:

Introducing repulsion gain and attraction gain improves the heuristic function of the Hybrid A-star algorithm, addressing the issue of excessive unnecessary iterative calculations when the selected next optimal node falls within an obstacle crevice. Additionally, it speeds up the Hybrid A-star algorithm.

This paper designs a path optimization method to compress and optimize the path which removes the redundant turning points, so as to make the planned path safer and more economical and better suited to the actual navigation requirements of the ship.

The rest of the paper is arranged as follows. The Sect. 2 introduces the related research of ship path planning. The Sect. 3 introduces the ship path planning method and experimental comparison of the improved hybrid A-star algorithm. The conclusions and future work will be introduced in the Sect. 5.

2   Related Research

In ship path planning, many scholars have carried out a variety of research and achieved excellent results. At present, the path planning algorithms of USVs mainly include traditional algorithms, bionic intelligent algorithms and machine learning algorithms.

Traditional algorithms mainly include velocity obstacles algorithm, RRT algorithm and A-star algorithm etc. For example, Kuwata Y et al. [16] combine velocity obstacles algorithm with Convention on the International Regulations for Preventing Collisions at Sea (COLREGS), proposed an USV motion planning algorithm for safe navigation in dynamic and chaotic environment. But this method is easy to fall into local optimization, and the planned path can not ensure the minimum economic cost. Lazarowska A et al. [17] divides the environment into several units. When planning the path, the next unit will be selected from adjacent nodes each time. Although this algorithm can plan the path, the potential energy value of each unit is difficult to calculate, and when the potential energy value is not allocated properly, it will fall into local optimization. Xiang et al. [18] proposed an improved bidirectional RRT algorithm. Although this method can quickly plan the path, it can not guarantee the optimal path. Gao Feng et al. [19] proposed a global path planning method for USVs based on improved A-star algorithm. On the basis of the traditional A-star algorithm, the original 24 search field is expanded to 48 search field to find the global optimal solution in a larger range. However, the planned path is too close to obstacles and does not meet the safety requirements. Reference [12] combines the hybrid A-star algorithm with the visible graph algorithm. The visible graph is used to explore the shortest waypoint for the hybrid A-star algorithm and hybrid A-star algorithm plane the path by these waypoints. The path is close to the obstacle and the safety is insufficient. A hybrid path planning algorithm based on A-star algorithm and improved APF method is proposed in reference [13]. The surface water environment model with unknown area is established by using the grid method. By using the optimized repulsive potential function with direction random strategy, the problems of unreachable goal point and easy to fall into local minimum in the original APF algorithm are solved. However, the path planned by this method has redundant turning points and radians, which does not meet the actual navigation requirements of ships.

Bionic intelligent algorithms mainly include genetic algorithm, particle swarm optimization algorithm and ant colony algorithm. Xiong et al. [20] proposed a path planning method based on Improved Particle Swarm Optimization (PSO) algorithm, which uses sinusoidal function dynamics to reduce particle weight and chaotic algorithm to avoid falling into local minimum. But this method still has the problem of insufficient path safety and does not meet the actual navigation requirements of ships. Inspired by APF algorithm, Lazarowska et al. [21] proposed a new ship path planning method, which considers both static (land and shallow water) and dynamic obstacles, and finds the safe path by calculating the collision risk of ships at sea. The method can plan a safe path, but the planned path has more turning points, which reduces the economic benefits in the process of ship navigation.

Machine learning algorithms are represented by deep reinforcement learning algorithms [22,23]. For example, Zhang [24] proposed a ship path planning model based on deep reinforcement learning. In this method, the ship motion space and reward function are designed to learn and train the planning strategy in the quantitative scene. The improved DRL algorithm can plan a feasible path, but there are still some unnecessary waypoints in the path planned by this method, which increases the cost of ship navigation. Bhopale et al. [25] adopted the improved Q-learning algorithm for path planning of underwater vehicles. This method detects obstacles through sensors. When obstacles are detected by sensors, the algorithm will forcibly select actions that can leave the unsafe. This method can make the underwater vehicle avoid the detected obstacles. However, this method relies too much on Q-table and has poor fitting ability.

By comparing with the existing methods and referring to the actual navigation specifications of ships, this paper proposes a ship path planning method based on improved hybrid A-star, introduces repulsive gain and gravitational gain to improve the cost calculation method of the exploration area when the algorithm plans the path, so as to solves the problem of multiple iterations of the algorithm at the obstacle groove area, and improves the planning speed of the algorithm. In addition, a path optimization method is designed to compress and optimize the redundant turning points, so that the final planned path can meet the actual navigation requirements of the ship on the premise of ensuring safety.

3   Ship Path Planning Based on Improved Hybrid A-Star Algorithm

In order to improve the safety and practicability of coastal ship path planning, this paper uses the grid method to extract the marine environmental information and complete the environmental modeling. Firstly, the cost calculation method of hybrid A-star exploring nodes is improved. Next, an optimization method is designed to delete the redundant turning points to ensure that the final planned path meets the actual navigation requirements of the ship.

3.1 Framework Description

The coastal ship path planning framework is shown in Fig. 1. The framework includes two parts: coastal marine environment modeling and coastal ship path planning. Firstly, process the marine environment information and construct the grid marine environment. Secondly, aiming at the problem that the path planned by hybrid A-star algorithm is close to obstacles, combined with the requirements for path economy and safety, repulsive gain and gravitational gain are introduced to improve hybrid A-star algorithm, so as to improve the safety and timeliness of the planned path. Further, according to the actual navigation requirements of ships, a ship path optimization method is designed to optimize the planned path, so as to make the final path safer and more economical.

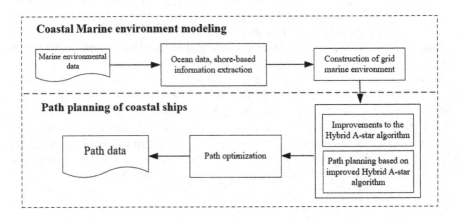

Fig. 1. Coastal ship path planning framework.

3.2 Rasterization of Marine Environment

During path planning, it is necessary to plan an effective obstacle avoidance path according to the specific location of obstacles, the goal point and starting point of the planned path. Because the grid method is simple, easy to implement and less computation, it can intuitively represent the path planning by the algorithm. Besides, the longitude and latitude coordinate system of the sea area environment is generally two-dimensional coordinates, which can be effectively converted into grid coordinate system, which can locate accurately and is conducive to the realization of path planning. Therefore, this paper uses grid method for environmental modeling.

The surface environment area of ship navigation is divided into several rectangular grid areas, which contains environmental information. The obstacle area is obstacle grid(unreachable area) and the obstacle free area is free grid (reachable area). After establishing the environment model, the information of each grid needs to be identified from bottom to top and from left to right. The sequence number of the first grid in the lower left corner is 1, 0 represents free grid and 1 represents obstacle grid. The connectivity between grids is consistent with the actual environment, and the mapping relationship of grid sequence number is shown in formulas (1) and (2).

$$N = x + \lambda(y - 1) \tag{1}$$

$$\begin{cases} x = \mod((N - 1), \lambda) + 1 \\ y = fix((N - 1), \lambda) + 1 \end{cases} \tag{2}$$

where N is the grid number, λ is the number of grids in a row, x, y represents the coordinates of the grid respectively, $\mod(x, y)$ represents the remainder after x is divided by y, and $fix(x, y)$ represents the rounding after x is divided by y.

3.3 Path Planning Based on Improved Hybrid A-Star Algorithm

Hybrid A-Star Algorithm and Its Shortcomings. Hybrid A-star algorithm adds a motion model to the conventional A-star algorithm, which makes the path planned by hybrid A-star algorithm smoother and more in line with kinematics. When planning ship path by hybrid A-star algorithm in the grid ship navigation environment, first check its adjacent grids from the starting point and expand around to find all feasible grids until the goal point is found, and then find the path with the lowest mobile price in the feasible grids.

In the path planning, for each explored node n, the actual cost of node n is calculated according to Equation (3), the estimated cost $h(n)$ from the current node n to the end point is calculated according to Equation (4), and the total cost $f(n)$ of node n is calculated according to Eq. (5). In which ct is a constant, ϕ_n is the ϕ value of node n, ϕ_{n-1} is the ϕ value of the parent node of node n, dis is the Manhattan distance between node n and node $n - 1$, (x_n, y_n) is the coordinates of node n, and (x_{goal}, y_{goal}) is the coordinates of the goal point. Save all explored nodes to a list and select the node with the lowest total cost from the list as the current node after each exploration and start exploring again until reach the goal point.

$$g(n) = ct(\phi_n - \phi_{n-1}) + g_{n-1} + dis \tag{3}$$

$$h(n) = |x_n - x_{goal}| + |y_n - y_{goal}| \tag{4}$$

$$f(n) = g(n) + h(n) \tag{5}$$

The Hybrid A-star algorithm's search time and how well it plans the path are directly related to the calculation of the estimated cost. When the estimated cost $h(n)$ is greater than the actual distance from node n to the goal node, the number of search nodes decreases and the search efficiency increases, but it often does not result in the optimal path. When the estimated cost $h(n)$ is less than the actual distance from node n to the goal node, the number of search nodes increases, the search range increases, and the optimal solution can be obtained, but the search time is longer. Only if the estimated cost $h(n)$ is close to the actual distance from node n to the goal node can the optimal path be obtained quickly and accurately. In most studies, $h(n)$ has been calculated using the Manhattan distance formula, Equation (4), which leads to a bias towards choosing the node with the smallest Manhattan distance when pick the optimal node. This leads to a number of unnecessary iterations when the next optimal node selected by the algorithm is in an obstacle notch, as the nodes near the obstacle notch have similar estimated costs. In addition, the use of the Manhattan distance formula also results in the final planned path being too close to the obstacle and possibly passing between two obstacles that are close together, significantly increasing the risk during navigation.

In summary, the Hybrid A-star algorithm suffers from several problems:

The planned path is close to the obstacle and selecting the next best node in the obstacle notch leads to multiple unnecessary iterations, which increasing the time to plan the path. In addition, the final planned path may pass between two obstacles that are close together, which increasing the risk during navigation.

When the environmental data is relatively large, the time required for route planning will increase dramatically and the timeliness will be insufficient.

The redundant turning points in the planned path will reduce the safety and economic efficiency in the navigation of the ship which does not fit the actual navigation of the ship.

Improvement of Hybrid A-Star. The Hybrid A-star algorithm incorporates a kinematic model as shown in Equation (6), where D is the distance of the ship at each movement, L is the length of the ship, θ is the yaw angle, ϕ is the steer angle of the ship and (x, y) is the position information of the ship. However, due to the large size of the ship itself, the selection range of sailing angle is not as wide as the angle selection range of vehicles and robots. Considering the actual sailing process of the ship for the range of sailing angle, is $[-35°, 35°]$ this paper also adopts this range as the angle constraint in the planning of the sailing process.

$$\begin{cases} x = D * \cos\theta \\ y = D * \sin\theta \\ \theta = \dfrac{D * \tan\phi}{L} \end{cases} \tag{6}$$

The node information contains the ship's yaw angle θ, the ship's steer angle ϕ, the set of X coordinates XL and the set of Y coordinates YL of each motion track point generated by reaching the node, as shown in Fig. 2. When planning the path, in order to change the ship's steer angle ϕ as small as possible each time, in this paper the actual cost $g(n)$ of node n is calculated instead by Equation (7), which is used to select the ϕ with the smallest cost. Where ϕ_n is the ϕ value of node n, ϕ_{n-1} is the ϕ value of node n's parent node. $\lambda \in [1, 3]$ is the penalty factor for changing steer angle, with the larger the change in steer angle, the larger the value of λ taken. And the smaller the change in steer angle, the smaller the value of λ taken.

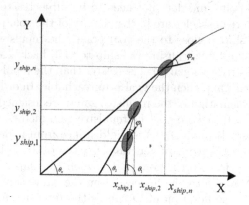

Fig. 2. The relationship between nodes.

$$g(n) = \lambda(\phi_n - \phi_{n-1}) + g_{n-1} + dis \qquad (7)$$

Moreover, this paper takes reference from the APF algorithm and redesigned the way of calculating the estimated cost when Hybrid A-star algorithm explores the nodes. Added gravitational gain and repulsive gain, solving the problems of poor timeliness and proximity to obstacles, multiple iterations and insecurity in planning path due to the original estimated costing method, which adds gravitational gain and repulsive gain, solving the problems of poor timeliness and proximity to obstacles, multiple iterations and insecurity in planning path due to the original estimated costing method. The new estimated cost is calculated as shown in Equation (8), where $D(n)$ is the distance from node n to the goal node, $U_{att}(n)$ is the gravitational gain from node n to the goal node, and $U_{rep}(n)$ is the repulsive gain from node n to the nearest obstacle. The gravitational gain can guide the algorithm to explore quickly towards the goal point, speeding up the algorithm's path planning. The repulsive gain ensures that when an obstacle has a notch or two obstacles are close to each other, the repulsive gain will be large, causing the algorithm not to explore the surrounding nodes.

$$h(n) = D(n) + U_{att}(n) + U_{rep}(n) \qquad (8)$$

$$U_{att}(n) = \frac{1}{2}\xi d_{tar}^2(n) \qquad (9)$$

Equation (9) is the gravitational gain calculation formula, where ξ is the gravitational gain factor, $d_{tar} = ||P_n - P_g||$ is the distance between node n and the goal point, and P_g is the coordinate of the goal point. The gravity generated by the goal point covers the whole area detected by the ship. When the distance d_{tar} between node n and the goal point is large, the ship will be subjected to a large gravitational force, conversely, when d_{tar} is small, it means that the ship is close to the goal point and does not need to be subjected to a large gravity.

When one or more obstacles are in the vicinity of the goal point, at a certain moment when the ship is close to the goal point, the repulsive force is greater than the gravitational force, causing the ship to "fall back", and after the ship has "fall back", the gravitational force is greater than the repulsive force, leading the ship to "advance", and then the goal is unreachable. In order to prevent this phenomenon, the calculation of the repulsive gain is redesigned in this paper, as shown in Equation (10), where η is the repulsive gain factor, ρ_0 is the radius of influence of the obstacle and $d_{obs}(n)$ is the distance from node n to the nearest obstacle. Since the repulsive gain generated by the obstacle is proportional to the distance from node n to the goal point, when the ship is close to the goal point, the gravitational force is decreasing at the same time as the repulsive force is decreasing. The addition of D causes the repulsive force generated by the obstacle near the goal to decrease significantly and at the goal point the repulsive force drops to zero and no longer rejects the ship towards the goal point.

$$U_{rep}(n) = \begin{cases} \dfrac{1}{2}\eta(\dfrac{1}{d_{obs}(n)} - \dfrac{1}{\rho_0}) * d_{tar}(n), & d_{obs}(n) \leq \rho_0 \\ 0, & d_{obs}(n) > \rho_0 \end{cases} \tag{10}$$

Path Planning Algorithm. The improved hybrid A-star retains the two sets O and C of the original hybrid A-star algorithm. O is used to save the nodes extended but not selected yet, and C is used to save the nodes selected or out of the boundary or colliding with obstacles. During path planning, if O is empty, the path planning fails and the algorithm ends. Otherwise, select the node N with the lowest $f(n)$ from O and move it from O to C. If node N is the goal point G, generate the path and the algorithm ends, otherwise make extension. Set node N as the starting point, and for each value within the steer angle ϕ range, calculate expansion node N' through Equation (5). If node N' out of the boundary or collides with an obstacle, it will be added to C, and this extension will end. Otherwise, judge whether there is node P in O and the position of node P is the same as that of node N'. If there is, update the one with the lower actual cost to O, and set the node as the child node of N, and this extension ends. Otherwise, calculate the $h(N')$ and $g(N')$, set node N as the parent node of node N', and add node N' to O, and this extension ends. Circle the above steps until the path planning is completed.

In navigation practice, too many turning points in the path will cause additional resource consumption. Therefore, it is necessary to ensure that the fewer turning points of the optimized path, the better [26,27]. The improved hybrid A-star algorithm can efficiently and accurately plan a path connecting the starting node and the goal point, but there are too many redundant turning points in the path, and the cost of changing direction is high for the ship, so it is necessary to remove the redundant turning points from the path as much as possible. In order to increase the availability and economy of the path, this section designs a path optimization method to remove redundant turning points, so that the optimized path can more meet the actual navigation requirements of the ship.

In order to ensure that the final planned path meets the actual navigation requirements of the ship, it is necessary to optimize the planned path. The basic idea of optimization is to maximize the removal of redundant turning points in the path and ensure that the planned path can bypass obstacles. Since there are many waypoints in the initial planned path, it is necessary to compress the path first, and then optimize the turning point. The specific optimization methods are as follows:

Path Compression. Due to the large number of waypoints in the planned path, it would be more computationally intensive to perform turning point elimination directly, so waypoint compression would be performed first. Path compression is an operation to compress path data by extracting some waypoints from the original path without changing the direction of the original path.

Figure 3 shows the schematic First set the compression threshold d_{max} (to ensure that the direction of the path is not changed), the compression step L and the array A that holds the compressed waypoints. The path is then divided into several segments by selecting a segment point every L length from the starting point, setting these points as $p_1, p_2, p_3, ..., p_n$, and then compressing each segment. Set the waypoints between p_n and p_{n-1} as $q_1, q_2, q_3, ..., q_k$, define the connecting line segment between point p_n and point p_{n-1} as l. If l does not crosses through the obstacle, calculate the distance from $q_1, q_2, q_3, ..., q_k$ to l respectively, keep the waypoint whose distance is greater than d_{max} , otherwise discard it, and finally save the retained waypoints into the array A. If l crosses the obstacle, pick the point q_i between p_{n-1} and p_n which is furthest from l. Divide the waypoint between p_{n-1} and p_n into two segments according to q_i, and then perform the above operation separately for each segment until the end of this segment compression. Figure 3 shows the case where l crosses the obstacle. Since neither of the two parts of the split crosses the obstacle and the waypoints in between can all be discarded, the point p_{n-1}, p_n and q_i are finally save in the array A. Then perform the next segment of the compression operation, which is exactly the same as the above steps, represented in Fig. 3 as the section between point p_{n-1} and point p_{n-2}. As the connecting line segment m in this segment does not cross the obstacle, it is only necessary to calculate which waypoints need to be retained. Eventually the point p_{n-1}, the point q_j and the point p_{n-2} are saved in array A. Because the point p_{n-1} already exists in A, then only the point q_j and the point p_{n-2} need to be saved.

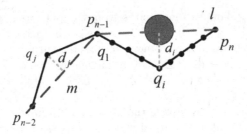

Fig. 3. Schematic diagram of path compression.

Turning Point Elimination. Set the last waypoint of the path as the current point, connect with the current point successively from the first waypoint of the path, and judge whether there are obstacles on the connection. If there are no obstacles, update the planned path and delete all waypoints between the two points on the path, otherwise no operation will be carried out. Then, set the penultimate point on the path as the current point, and repeat the above operation until the second point of the path is set as the current point. Finally, the optimized path can be obtained by connecting the remaining trajectory points on the path in turn to form a broken line.

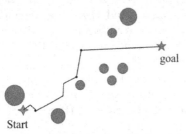

(a) Before path optimization.

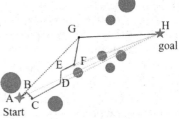

(b) Turning point elimination.

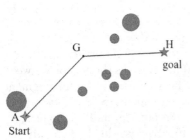

(c) Final path after optimization.

Fig. 4. Turning point elimination idea.

As shown in Fig. 4, (a) is the path after path compression, (b) is the optimization process, and (c) is the final path after elimination. In (b), point H is connected with point A, point B, point C, point D, point E and point F respectively. All the lines pass through obstacles, so point G is changed to the current point. Point G is connected with point A. The connection does not pass through the obstacle. The intermediate points B, C, D, E and F are directly removed. At this time, point G is the second point and the algorithm ends. Finally, the optimized path is shown in (c). Compared with (a), there is only 1 turning point in (c) and 6 in (a), which greatly improves the economic benefits of ship navigation.

4   Experimental Comparison and Analysis

This section verifies the reliability and effectiveness of the improved hybrid A-star algorithm in coastal waters through simulation experiments. For ship path planning, the length of path and the number of turning points of path determine the economy of path, the distance between the path and obstacles determines the safety of the path, the time required to plan the path determines the timeliness of the path. Therefore, this section compares it from three aspects: economy, safety and timeliness.

4.1   Environmental Construction

The data of a sea area is selected from the electronic chart as the experimental environment data. Because the safe distance of the ship is related to the size of the ship itself, large ships often need a longer safe distance and take action earlier. Therefore, this study expands the obstacles in the marine environment and increases the boundary of the obstacles by 0.5 nautical mile on the basis of the original proportion.

4.2   Baseline and Evaluation Indicators

In order to verify the effectiveness of this method, this paper will prove the applicability of this method by comparing the path planned by this method, the original hybrid A-star algorithm, A-star algorithm, the algorithm in reference [12], the algorithm in reference [13] and the algorithm in reference [14]. Besides, through the analysis and summary of the characteristics of coastal ship path planning, this section takes the following three factors as the evaluation indicators of the experimental results:

(1) Security: The primary condition of path planning is to ensure the safety of the navigation. Safety is mainly reflected in whether the planned path passes through obstacles and whether the path is too close to obstacles, resulting in collision risk. Therefore, in the process of path planning, the safety of path planning is measured by the distance between path and obstacles.
(2) Economy: In the actual environment, the path is usually a broken line segment composed of several turning points. The economy of path planning is measured by calculating the overall path length. The shorter the path length,

the less fuel is consumed and the more economical the path is. On the other hand, redundant turning points will also cause ships to take unnecessary actions, thus increasing navigation consumption. Therefore, the economic indicators in this paper are compared from two aspects: path length and number of turning points.

(3) Timeliness: The less time required for path planning means the better timeliness of the algorithm, which also reflects that ships can take measures to ensure the safety and economy of navigation in a short time. Therefore, the time consumption of path planning is selected as an indicator of timeliness. Besides, the larger the area explored by the algorithm, the longer the path planning time. Therefore, the size of the algorithm exploration area is selected as another index of path timeliness.

4.3   Experimental Comparison

In this section, we verifies the ship path optimization method proposed in this paper, and illustrates the effectiveness of the optimization method by comparing the non optimized path with the optimized path first. Then, illustrates the superiority of this method by comprehensively comparing the path planned by this method with that planned by other algorithms.

Comparison and Analysis. Figure 5 shows the paths planned by different algorithms in the same marine environment. The position of the red point is the goal point of the ship, the position of the green point is the starting point of the ship, the white line is the final planned path, the yellow is the obstacle part, the color change around the obstacle part represents the change of the obstacle height, the dark blue part is the navigable area of the ship, and the light blue convex part is the area explored by the algorithm. In Fig. 5, (a) is the path planned using the method in this paper, (b) is the path planned using the original hybrid A-star algorithm, (c) is the path planned using the A-star algorithm, (d) is the path planned using the algorithm in reference [12], (e) is the path planned using the algorithm in reference [13], and (f) is the path planned using the algorithm in reference [14].

In terms of path safety, it can be seen from Fig. 5 (a) that the cost of areas near obstacles increases due to the addition of repulsive gain which means the algorithm will not explore these areas. Thus, the path has a high level of safety. Because the path is optimized by optimization method and redundant turning points in the path are removed, therefore, the planned path is relatively smooth as a whole, which meets the actual navigation requirements of the ship.

It can be seen from Fig. 5(b) that the original hybrid A-star algorithm adds a kinematic model, so the path is relatively smooth as a whole, but the path is not optimized by the optimization method, so there are curves and redundant turning points in the path, which increases the economic cost of navigation. Besides, it only takes the shortest distance as the constraint, so the planned path is close to obstacles, with low safety, which does not meet the actual navigation requirements of the ship.

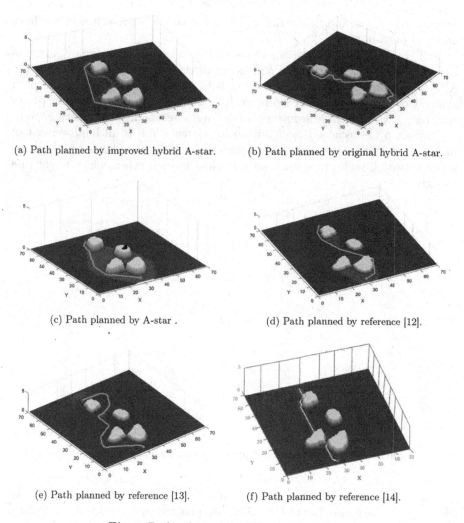

(a) Path planned by improved hybrid A-star.

(b) Path planned by original hybrid A-star.

(c) Path planned by A-star .

(d) Path planned by reference [12].

(e) Path planned by reference [13].

(f) Path planned by reference [14].

Fig. 5. Paths planned by different algorithms.

The A-star algorithm in Fig. 5(c) does not include a kinematic model, so the planned path does not include curve segments. However, because the algorithm only takes the shortest distance as the constraint condition, the algorithm will explore the area near the obstacle during planning, resulting in the path being close to the obstacle and increasing the risk of ship navigation.

In Fig. 5(d), because the visual graph algorithm is added to the algorithm in reference [12], the "viewable point" will be found as the middle turning point before planning the path, so there are few redundant turning points in the path. However, because the two points are connected to see whether to pass through the obstacle as the selection basis of "viewable point", the path passes between two obstacles close to each other, it increases the collision risk during the actual navigation of ships, and the overall safety of the path is poor, so it does not have practical operability.

In Fig. 5(e), because the algorithm in reference [16] is an improvement of the APF algorithm, the planned path is smoother than the path planned by the APF algorithm. However, due to the defects of the APF algorithm, the path planned by the algorithm in reference [13] also has redundant turning points and some paths are curve segments, so the overall economy of the path is poor.

In Fig. 5(f), because the algorithm in reference [14] is an improvement of the D* Lite algorithm, so the planned path is short, but the planned path passes between two obstacles that are relatively close to each other, which is not safe enough.

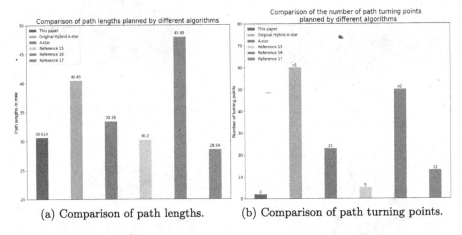

(a) Comparison of path lengths.   (b) Comparison of path turning points.

Fig. 6. Comparison of path lengths and turning points planned by different algorithms.

Figures 6(a) and 6(b) show the comparison in terms of path economy. It can be seen from the figure that the shortest path planned by reference [14] is 28.54 n miles, and the path planned by the algorithm of reference [12] and the algorithm of this paper are 30.2 n miles and 30.614 n miles. Although the path planned by the algorithm of reference [12] and the algorithm of reference [14] is not different from that of the algorithm of this paper, some path segments of their paths are close to obstacles and have poor safety, not applicable to actual ship navigation. The path planned by A-star algorithm is 33.38 n miles, the path planned by the original hybrid A-star is 40.45 n miles, and the path planned by the algorithm in reference [13] is 47.95 n miles. In terms of the number of turning points, the minimum number of turning points of the path planned by the algorithm of this paper is 2. There are 5 turning points in the path planned by the algorithm of reference [12], there are 23 turning points in the path planned by A-star algorithm, and there are 13 turning points in the path planned by the algorithm of reference [14]. Although there are no redundant turning points in the algorithm of reference [12], the number of turning points is more than that of the algorithm of this paper, The increase of turning points increases the economic cost of the ship in the actual navigation process, while there are a large number of redundant turning points in the path planned by A-star algorithm, which greatly reduces the economic benefits obtained in the actual navigation process. There are many curve segments in the paths planned by other methods, and the number of turning points is much larger than those of the above methods, so $N1$ and $N2$ are used to mark in Fig. 7. However, it can be seen from the figure that the overall smoothness of the path planned by the algorithm of reference [13] is better than that planned by the original hybrid A-star. To sum up, compared with the above algorithms, the planned path conforms to the actual navigation specifications and is safer and more economical.

From the comparison of exploration area indicators in timeliness, the exploration area of this method in Fig. 5 (a) is less as a whole. When the obstacle groove is explored, due to the effect of repulsive gain, the algorithm can quickly explore the appropriate position and reduce the number of explorations. At the same time, the gravitational gain will increase the tendency of the algorithm to the end point and shorten the time for the algorithm to plan the path.

Figure 5(b) the exploration area of the original hybrid A-star algorithm is concentrated near the obstacle. Because there is no repulsive gain and gravitational gain, the exploration area is much larger than that of the method in this paper. When the algorithm is exploring near the obstacle, it will iteratively calculate the nodes near the obstacle for many times, which increases the time for the algorithm to plan the path.

It can be seen from Fig. 5(c) that the exploration area of A-star algorithm is much larger than that of the method in this paper and the original hybrid A-star algorithm. The exploration area of A-star algorithm is mainly concentrated near obstacles. This is because A-star algorithm takes the shortest distance as the constraint condition and can only find the shortest distance through multiple explorations, so it will explore the area near obstacles for many times.

In Figs. 5(d) and 5(e), the method in reference [12] and the method in reference [13] have greatly improved the overall exploration area compared with the original hybrid A-star algorithm, but they will still explore many times near the goal point, increasing the time of path planning.

In Fig. 5(f), the exploration area of the improved D* Lite algorithm is smaller than that of other algorithms, because it uses the minimum binary heap to optimize the priority queue of D* Lite, which significantly reduces the exploration area of the route.

The time comparison of paths planned by different algorithms is shown in Fig. 7. As can be seen from the figure, the shortest time for path planning in reference [14] is 1.031 s, and the time for path planning by this method is 1.247 s. Then the time for path planning in reference [13] is 1.432 s, the time for path planning in reference [12] is 1.505 s, the time for path planning in the original hybrid A-star algorithm is 1.513 s, and the longest time for path planning in A-star algorithm is 1.709 s. The comparison shows that the timeliness of this method is also better than other algorithms except the algorithm in reference [14].

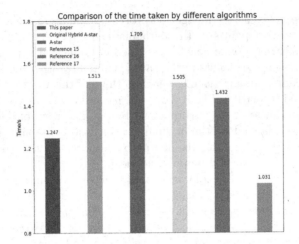

Fig. 7. Time of path planning with different algorithms.

Universal Verification. In order to illustrate the generality of the algorithm in this paper, the algorithm verification is carried out in different environments. Figure 8 shows two environmental data with different complexity selected from the chart. Both environments use the method in this paper and other algorithms for path planning.

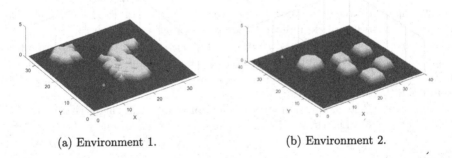

(a) Environment 1.   (b) Environment 2.

Fig. 8. Two environments with different complexity.

Compare the paths planned by different algorithms in environment 1 and environment 2 from three aspects: length, time and number of turning points. Table 1 shows the comparison of different algorithms for environment 1, and Table 2 shows the comparison of different algorithms for environment 2.

As can be seen from Table 1, in terms of length, the A-star algorithm, the algorithm of reference [12] and the algorithm of reference [14] are shorter than the path planned by the algorithm of this paper, and other algorithms are longer

Table 1. Comparison of different algorithms in environment 1.

Algorithms	Length/ n mile	Time/s	Turning points
This paper	34.243	0.45	4
Original hybrid A-star	39.900	0.61	N1
A-star	28.3137	0.54	7
Reference [15]	32.046	0.83	3
Reference [16]	47.900	0.71	N2
Reference [17]	28.321	0.53	N3

Table 2. Comparison of different algorithms in environment 2.

Algorithms	Length/ n mile	Time/s	Turning points
This paper	27.031	1.41	1
Original hybrid A-star	43.900	1.63	N4
A-star	39.799	1.61	8
Reference [15]	44.235	1.52	2
Reference [16]	55.900	1.75	N5
Reference [17]	38.433	1.43	N6

than the path planned by the method in this paper. In terms of the number of turning points, the number of turning points of the path planned by the algorithm of reference [12] is less than that planned by the algorithm of this paper, the number of turning points of the path planned by A-star algorithm is more than that planned by the algorithm of this paper, and some paths planned by other methods are curve segments, which does not meet the requirements of the path in the actual navigation of the ship. Although the path planned by the algorithm of reference [12] is better than the method in terms of the number of turning points and path length, one of the paths is too close to obstacles and the risk is too high, so it is not suitable for the actual navigation of ships. In terms of time, the algorithm of this paper takes at least 0.45 s. Similarly, it can be seen from Table 2 that the shortest path planned in this paper is 27.031 n miles, the minimum number of turning points is 1, and the shortest time is only 1.41 s.

Through comparison, it is found that the path planned by this method is better than other algorithms in path length, number of turning points and planning time, and the path planned by this method has higher safety and meets the requirements of the path in the actual navigation process.

5   Conclusions

This paper studies the path planning of coastal ships, and puts forward an improved path planning method of hybrid A-star. Firstly, this paper introduces repulsive gain and gravitational gain to improve the calculation method of node cost when planning path, solves the problems of multiple iterations when the path planned by the algorithm is close to the obstacle and when the obstacle has a groove, and improves the efficiency of the algorithm. Then, in view of the fact that the planned path may have more turning points, which reduces the economic benefits of ship navigation and increases the collision risk in the process of ship navigation, this paper proposes a path optimization method to compress and optimize the path and remove the redundant turning points, so as to make the planned path safer and more economical and more in line with the actual navigation requirements of ships.

Compared with other path planning algorithms and verified in different environments, the results show that using this method to plan path takes less time, short path length and less turning points, which meets the actual navigation requirements of ships. At the same time, this paper still has the following problems to be solved, which are also the key research contents of this paper.

Taking dynamic obstacles into account, the algorithm can avoid not only static temporary obstacles, but also temporary dynamic obstacles.

The collision avoidance rules are added to the algorithm, so that the algorithm can avoid not only obstacles, but also ships, and increase the scope of application of the algorithm.

Acknowledgements. This work is supported by the National Natural Science Foundation of China (Grant No. 52231014) and Liaoning Province Applied Basic Research Program Project (Grant No.2023JH2/101300195).

References

1. Zhou, Z., Zhang, Y., Wang, S.: A coordination system between decision making and controlling for autonomous collision avoidance of large intelligent ships. J. Mar. Sci. Eng. **9**(11), 1202 (2021)
2. Wang, S., Zhang, Y., Zheng, Y.: Multi-ship encounter situation adaptive understanding by individual navigation intention inference. Ocean Eng. **237**, 109612 (2021)
3. Cheng, X., Liu, Z.: Trajectory optimization for ship navigation safety using genetic annealing algorithm. In: Third International Conference on Natural Computation (ICNC 2007), vol. 4, pp. 385–392. IEEE (2007)
4. Zhang, D., Zhang, Y., Zhang, C.: Data mining approach for automatic ship-route design for coastal seas using AIS trajectory clustering analysis. Ocean Eng. **236**, 109535 (2021)
5. Shaobo, W., Yingjun, Z., Lianbo, L.: A collision avoidance decision-making system for autonomous ship based on modified velocity obstacle method. Ocean Eng. **215**, 107910 (2020)
6. Shah, B.C., Gupta, S.K.: Long-distance path planning for unmanned surface vehicles in complex marine environment. IEEE J. Ocean. Eng. **45**(3), 813–830 (2019)
7. Du, L., Goerlandt, F., Banda, O.A.V., Huang, Y., Wen, Y., Kujala, P.: Improving stand-on ship's situational awareness by estimating the intention of the give-way ship. Ocean Eng. **201**, 107110 (2020)
8. Guo, S., Zhang, X., Zheng, Y., Du, Y.: An autonomous path planning model for unmanned ships based on deep reinforcement learning. Sensors **20**(2), 426 (2020)

9. Shen, H., Hashimoto, H., Matsuda, A., Taniguchi, Y., Terada, D., Guo, C.: Automatic collision avoidance of multiple ships based on deep q-learning. Appl. Ocean Res. **86**, 268–288 (2019)

10. Wang, Y., Tong, J., Song, T.Y., Wan, Z.H.: Unmanned surface vehicle course tracking control based on neural network and deep deterministic policy gradient algorithm. In: 2018 OCEANS-MTS/IEEE Kobe Techno-Oceans (OTO), pp. 1–5. IEEE (2018)

11. Sheng, W., Li, B., Zhong, X.: Autonomous parking trajectory planning with tiny passages: a combination of multistage hybrid a-star algorithm and numerical optimal control. IEEE Access **9**, 102801–102810 (2021)

12. Sedighi, S., Nguyen, D.V., Kuhnert, K.D.: Guided hybrid a-star path planning algorithm for valet parking applications. In: 2019 5th International Conference on Control, Automation and Robotics (ICCAR), pp. 570–575. IEEE (2019)

13. Yu, J., Deng, W., Zhao, Z., Wang, X., Xu, J., Wang, L., Sun, Q., Shen, Z.: A hybrid path planning method for an unmanned cruise ship in water quality sampling. IEEE Access **7**, 87127–87140 (2019)

14. Zhu, X., Yan, B., Yue, Y.: Path planning and collision avoidance in unknown environments for USVs based on an improved d* lite. Appl. Sci. **11**(17), 7863 (2021)

15. Dolgov, D., Thrun, S., Montemerlo, M., Diebel, J.: Practical search techniques in path planning for autonomous driving. Ann Arbor **1001**(48105), 18–80 (2008)

16. Kuwata, Y., Wolf, M.T., Zarzhitsky, D., Huntsberger, T.L.: Safe maritime autonomous navigation with COLREGS, using velocity obstacles. IEEE J. Oceanic Eng. **39**(1), 110–119 (2013)

17. Lazarowska, A.: A discrete artificial potential field for ship trajectory planning. J. Navig. **73**(1), 233–251 (2020)

18. Xiang, J., Wang, H., Ouyang, Z., Yi, H.: Local path planning algorithm of unmanned vehicle based on improved two-way RRT. Chinese Shipbuild. **61**(1), 157–166 (2020)

19. Gao, F., Zhou, H., Yang, Z.: Global path planning of surface unmanned ship based on improved a-star algorithm. App. Res. Comput. **37**(S1), 120–121 (2020)

20. Xiong, Q., Zhang, H., Rong, Q.: Path planning based on improved particle swarm optimization for AUVs. J. Coast. Res. **111**(SI), 279–282 (2020)

21. Lazarowska, A.: A new potential field inspired path planning algorithm for ships. In: 2018 23rd International Conference on Methods & Models in Automation & Robotics (MMAR), pp. 166–170. IEEE (2018)

22. Cao, X., Sun, C., Yan, M.: Target search control of AUV in underwater environment with deep reinforcement learning. IEEE Access **7**, 96549–96559 (2019)

23. Wang, T., Wu, Q., Zhang, J., Wu, B., Wang, Y.: Autonomous decision-making scheme for multi-ship collision avoidance with iterative observation and inference. Ocean Eng. **197**, 106873 (2020)

24. Zhang, X., Wang, C., Liu, Y., Chen, X.: Decision-making for the autonomous navigation of maritime autonomous surface ships based on scene division and deep reinforcement learning. Sensors **19**(18), 4055 (2019)

25. Bhopale, P., Kazi, F., Singh, N.: Reinforcement learning based obstacle avoidance for autonomous underwater vehicle. J. Mar. Sci. Appl. **18**, 228–238 (2019)

26. Guo, S., Zhang, X., Du, Y., Zheng, Y., Cao, Z.: Path planning of coastal ships based on optimized DQN reward function. J. Mar. Sci. Eng. **9**(2), 210 (2021)

27. Du, Y., et al.: An optimized path planning method for coastal ships based on improved DDPG and DP. J. Adv. Transp. **2021**, 1–23 (2021)

Data Augmentation Method Based on Partial Noise Diffusion Strategy for One-Class Defect Detection Task

Weiwen Chen[1], Yong Zhang[1,2(✉)], and Wenlong Ke[1]

[1] School of Information Engineering, Huzhou University, Huzhou, China
`zhyong@zjhu.edu.cn`
[2] School of Computer and Information Technology, Liaoning Normal University, Dalian, China

Abstract. One-class defect detection has proven to be an effective technique. However, the performance of complex models is often limited by existing data augmentation methods. To address this issue, this paper proposes a novel data augmentation method based on a denoising diffusion probability model. This approach generates high-quality image samples using partial noise diffusion, eliminating the need for extensive training on large-scale datasets. Experimental results demonstrate that the proposed method outperforms current methods in one-class defect detection tasks. The proposed method offers a new perspective on data augmentation and demonstrates its potential to tackle challenging computer vision problems.

Keywords: Defect detection · Denoising diffusion probability model · Data augmentation · Deep learning · Image generation

1 Introduction

Anomaly detection (AD) is a crucial task in computer vision that involves automatically detecting and identifying anomalous situations in images. This technique has numerous valuable applications, such as industrial defect detection, medical imaging, quality control, and security monitoring. In the context of industrial defect detection, the goal is to identify regions or pixels in the input image that do not conform to expected patterns. These anomalies may be caused by defects, scratches, stains, abnormal areas, or other visual changes. By detecting these anomalies, industrial defect detection systems can provide corresponding feedback or alerts, enabling users to take necessary actions and mitigate potential issues effectively.

As the capabilities of computers continue to advance, deep learning has found significant applications across various domains, including network flow scheduling, image recognition, and particularly emerging as the predominant technique for image anomaly detection [13–15]. However, in the field of industrial defect detection, acquiring anomalous data can be challenging due to the presence of

© The Author(s), under exclusive license to Springer Nature Singapore Pte Ltd. 2024
Z. Tari et al. (Eds.): ICA3PP 2023, LNCS 14492, pp. 418–433, 2024.
https://doi.org/10.1007/978-981-97-0811-6_25

unknown types of anomalies, and the distribution between normal and abnormal classes may be highly imbalanced. These factors make supervised deep learning methods expensive and difficult to apply in practical industrial environments. As a result, semi-supervised defect detection methods [15] have become increasingly popular in industrial defect detection tasks. This method only uses normal samples during the training process, and is referred to as a one-class defect detection task. Previous studies have shown that this method can successfully extract rich semantic representations of images. However, its ability to detect fine-grained anomalies is limited. To address this issue, several practical datasets have been proposed, with the MVTec-AD dataset [4] being the most widely used. Researchers have proposed numerous techniques to enhance the performance of anomaly detection based on the MVTec dataset. According to Table 1, the most advanced methods currently achieve an exceptionally high area under the receiver operating characteristic curve (AUROC) of 99.5% on the MVTec-AD dataset, with the best-performing methods all being one-class defect detection methods.

Table 1. AUROC of mainstream models for defect detection.

Model	AUROC	Model	AUROC
CFA [11]	99.5	CutPaste [12]	96.1
FastFlow [26]	99.4	Padim [6]	95.0
PatchCore [19]	99.1	DFM [1]	94.3
CFLOW [8]	98.3	STFPM [23]	90.3
DRAEM [27]	98.0		

Semi-supervised deep learning methods have proven to be effective in defect detection tasks, with generative models being particularly valuable. AnoGAN [20] was one of the pioneering methods to utilize generative adversarial networks (GANs) for defect detection, learning the distribution of image scores by mapping from image space to latent space. Another advanced model, f-AnoGAN [21], further accelerated the process using a more sophisticated GAN. However, GANs require large training datasets. In contrast, CutPaste [12] demonstrated that data augmentation on the training set can improve model performance. Nevertheless, the data augmentation approach used by CutPaste is manually designed and cumbersome. To address these issues, the denoising diffusion probabilistic model (DDPM) [10] has emerged as one of the most advanced generative models. DDPM offers superior image quality and faster speed compared to GANs and variational auto-encoders (VAEs). By carefully adjusting the iteration number and noise level of DDPM, it is possible to generate simulated images that effectively improve positive results. This paper proposes the utilization of DDPM for data augmentation on defect detection datasets, accompanied by an image generation algorithm and appropriate noise levels for augmentation.

Experimental results demonstrate that the DDPM-based image augmentation strategy performs well in enhancement tasks on defect detection datasets. The main contributions of this paper are listed as follows:

(1) Designing an image generation algorithm based on partially denoising diffusion strategy to construct defect detection datasets comprising both normal and abnormal samples. These datasets will exclusively consist of images generated by the diffusion probability model.

(2) Conducting extensive experiments to investigate the effectiveness of various image augmentation methods based on different iteration steps and noise levels for semi-supervised learning tasks. This research aims to provide concise guidance to other researchers, enabling them to achieve optimal results using DDPM-based image augmentation strategies promptly.

2 Related Work

In defect detection tasks, data augmentation methods play a crucial role as part of the data preprocessing stage, especially since datasets are typically small in scale. These methods can be broadly categorized into two types: traditional data augmentation methods and deep learning-based data augmentation methods.

2.1 Traditional Data Augmentation Methods

Traditional data augmentation methods mainly include six types: geometric transformations, color space transformations, sharpness transformations, noise injection, local erasure, and multi-data mixing [16]. These methods manipulate the original data to generate new data by changing its representation. Traditional data augmentation methods are fast and simple, making them the primary data augmentation method in the field of image processing. Among them, Mixup [28] integrates different traditional data augmentation methods by using a linear interpolation-based image blending technique to generate blended images in the image space, which improves model generalization ability. Mosaic [5] mixes four training images, applies random scaling, random cropping, and random arrangement to splice them for data augmentation, and has been successfully applied to the YOLO series models in object detection. CutPaste [12] is a simple data augmentation strategy that can crop image patches and randomly paste them into any position in a large image, resulting in significant performance improvement on the MVTec-AD dataset. MemSeg [25], from the perspective of image differences, uses Perlin noise and foreground targets to generate masked images, extracts noise foregrounds from ROIs defined in early images, and finally superimposes them on the original images to manually generate simulated abnormal images. Both CutPaste and MemSeg have proved to be effective in defect detection tasks, demonstrating the effectiveness of data augmentation strategies. However, these methods all have a common problem: manually generating simulated abnormal images through various traditional data augmentation methods is laborious and complex, and they do not provide additional information. Therefore, they cannot significantly improve model accuracy.

2.2 Data Augmentation Methods Based on Deep Learning

Deep learning-based data augmentation methods utilize prior knowledge to train and learn the feature space distribution of the dataset, which can result in better generated data. One common approach is to use GAN models to fit the data distribution and sample from the fitted distribution to generate simulated images [2]. In addition to GANs, there are also strategies based on meta-learning, reinforcement learning, and other techniques. For example, CycleGAN [3] can be used to generate different facial expression images of the same identity, thereby enhancing the diversity of facial expression data. In the field of medical image defect detection, AnoDDPM [24] utilizes the DDPM to generate images. The DDPM is a generative model that learns the data distribution through both forward and backward processes. Both processes can be viewed as parameterized Markov chains, and the backward process can be used to generate new images.

However, it's worth noting that deep learning-based data augmentation methods often require large training sets and longer training times compared to traditional methods. They rely on complex models and may need substantial computational resources.

3 Data Set Generation Method Based on DDPM

In this section, we will provide a detailed description of the DDPM. This includes an explanation of the forward and backward propagation processes of DDPM, the variational lower bound of the marginal likelihood, and the utilization of neural networks to learn the parameters involved in the backward propagation process.

3.1 DDPM and Its Advantages

DDPM is a generative model that has certain advantages over GAN in terms of sample quality [7]. Based on the DDPM method, this paper uses non-homogeneous Markov chains and single-step transition densities to describe the forward process of noise-corrupted images. The inverse process of recovering the original image is learned through backpropagation. Compared with GAN-based methods, the proposed method has the following advantages:

(1) It has a closed-form expression that can be optimized, making it easier to control and optimize by adjusting specific parameters when compared to GANs.
(2) It can generate better samples with faster generation speed. DDPM has been shown to produce higher quality images than GANs and VAEs in terms of sharpness and visual quality. Additionally, DDPM achieves faster generation speed as it does not rely on a discriminator network like GANs do.

Forward

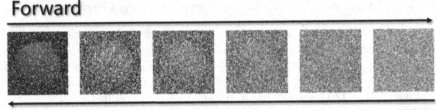

Back

Fig. 1. DDPM forward propagation and back propagation diagram.

3.2 Forward and Backward Propagation Processes of DDPM

As shown in Fig. 1, the primary objective of DDPM's forward diffusion process is to transform the initial distribution $q(x_0)$ into a normal distribution $q(x_T)$, where T represents the number of steps. In the gradual pursuit of this goal, the data undergoes a series of transformations, including sampling noise from the model and employing the single-step transition density $q(x_t \mid x_{t-1})$ to transform the data from its previous time step x_{t-1} to the current time step x_t. As time steps proceed, the data becomes increasingly infused with noise, ultimately converging towards a normal distribution. The calculation formula for the forward diffusion process can be expressed as:

$$q(x_t \mid x_{t-1}) = \mathcal{N}(x_t \mid x_{t-1}\sqrt{1-\beta_t}, \beta_t I) \tag{1}$$

The conditional distribution is a multivariate normal distribution with a mean vector of $x_{t-1}\sqrt{1-\beta_t}$ and a covariance matrix of $\beta_t I$, where I is the identity matrix and β_t is a learnable parameter used to control the amount of noise added at each time step. Usually, β_t is set to a scalar value and remains constant across all time steps. This process is repeated T times until the data converges to the desired normal distribution.

In the inverse process of learning noise-corrupted images in DDPM, the conditional distribution $p_\theta(x_{t-1} \mid x_t)$ models the relationship, with the noise being sampled from the prior distribution $p(x_0)$. The generative model, parameterized by θ, begins with $x_T \sim \mathcal{N}(0, I)$. Subsequently, samples are generated following formula (2).

$$p_\theta(x_{t-1} \mid x_t) = \mathcal{N}(x_{t-1} \mid \mu_\theta(x_t, t), \widetilde{\beta}_t I) \tag{2}$$

For $t = 1, ..., T$ and $\widetilde{\beta}_t = \frac{1-\alpha_{t-1}}{1-\alpha_T}\beta_t$, μ_θ can be implemented using a U-Net-like architecture [18]. The variational lower bound, L_{vlb} on the marginal likelihood $p_\theta(x_0)$ is used as the loss function for training $p_\theta(x_{t-1} \mid x_t)$, which is computed according to the following formulas (3–6) shown below:

$$L_{vlb} = L_0 + L_1 + \cdots + L_{T-1} + L_T \tag{3}$$

$$L_0 = -\log p_\theta(x_0 \mid x_1) \tag{4}$$

$$L_{t-1} = D_{KL} \left(q(x_{t-1} \mid x_t, x_0) \mid p_\theta(x_{t-1} \mid x_t) \right) \qquad (5)$$

$$L_T = D_{KL} \left(q(x_T \mid x_0) \mid p(x_T) \right) \qquad (6)$$

where D_{KL} is the Kullback-Leibler (KL) divergence. Since the distribution $q(x_{t-1} \mid x_t, x_0)$ has a closed-form expression, the above expression can be optimized to generate improved samples and accelerate the generation process. This allows for sampling at any time step without the necessity of finding intermediate samples x_0.

The DDPM [10] gives detailed derivations of the formulas and mathematical proofs.

3.3   Using DDPM to Generate Synthetic Defect Detection Dataset

In this section, we will outline how DDPM can be employed to generate a synthetic defect detection dataset. Algorithm 1 shows the specific steps involved in this approach. By fine-tuning and optimizing the parameters of DDPM, we aim to generate high-quality images specifically tailored for one-class defect detection tasks. The effectiveness of this approach has been verified through several experiments.

Algorithm 1: Image Generation Algorithm

   Data: Training images x_{train}, noise scalar β_t, iteration steps s, number of images N

   Result: Generated image dataset X_{gen}

1 $X_{\text{gen}} \leftarrow \{\}$;   // Initialize the generated image dataset as an empty set
2 **for** i *in range (N)* **do**
3      $x_0 \leftarrow x_{\text{train}}[i \bmod N]$;   // Select the i-th training image as the initial image
4   **for** t *in range (s)* **do**
5          $q(x_t \mid x_{t-1}) \leftarrow \mathcal{N}(x_t \mid x_{t-1}\sqrt{1-\beta_t}, \beta_t I)$;   // forward propagation sampling of the image based on noise scalar
6          $Loss \leftarrow \nabla_\theta \left[\parallel \epsilon - \epsilon_\theta \left(x_0\sqrt{\overline{\alpha}_t} + \sqrt{1-\overline{\alpha}_t}\epsilon, t \right) \parallel^2 \right]$;   // calculate the loss function through gradient descent, ϵ and ϵ_θ can be obtained through the neural network.
7          $p_\theta(x_{t-1} \mid x_t) \leftarrow \mathcal{N}(x_{t-1} \mid \mu_\theta(x_t, t), \widetilde{\beta}_t I)$; // reverse propagate based on learning the forward sample process
8          $x_{t-1} \leftarrow x_t$;   // Update previous image with current image
9   **end**
10    $X_{\text{gen}} \leftarrow X_{\text{gen}} \cup \{x_s\}$;   // Add x_s to the generated image dataset
11 **end**

---

4   Experimental Results and Analysis

4.1   Dataset

In this paper, the MVTec-AD dataset's training set is utilized as the normal sample dataset. The MVTec-AD dataset [4] was introduced by MVTec in 2019 and is specifically designed for unsupervised segmentation tasks in the context of industrial product inspection. It contains 15 categories of products, each exhibiting different defect types. The images in the dataset range in resolution from 700×700 to 1024×1024 pixels. The training and validation subsets of the MVTec-AD dataset consist of 3629 normal images, while the testing subset comprises 1725 images. For the defective images in the dataset, pixel-level annotations are provided, covering approximately 1900 annotated regions in total. These annotations precisely identify the regions containing defects within the images.

4.2   Mainstream Model Selection

In this paper, we reproduced and evaluated several popular semi-supervised models, including CFA, PatchCore, CFlow, PaDim, and STFPM, using the MVTec-AD dataset. The evaluation metric used in this study is AUROC. To ensure fair and efficient comparison, the ResNet-18 architecture [9] was adopted as the backbone network for all the models. Table 2 presents the AUROC values obtained by these seven models on the 15 sub-datasets within the MVTec-AD dataset.

The highest average AUROC value obtained was 0.973 for the PatchCore model, followed by 0.958 for the CFlow model. The lowest average AUROC value was 0.424 for the GANnomaly model. The performance of each model varied across different product categories and defect types. For example, the highest AUROC value obtained for the Carpet sub-dataset was 0.975 by the CFlow model, while the highest AUROC value obtained for the Leather sub-dataset was 1.000 by the Patchcore model. Overall, the results suggest that the PatchCore model performs the best on average, followed by the CFlow model. The results also indicate that the effectiveness of each model varies depending on the product category and defect type.

4.3   Implementation and Usability

In all DDPM experiments of this paper, the same U-Net architecture [18] as described in [7] was used for the approximation. Additionally, Transformer sine positional embeddings [22] were used to encode the time steps. The hyperparameters of the model used in the experiments are listed in Table 3. The experiments were conducted using PyTorch framework and a single NVIDIA GEFORCE RTX3090 GPU with 24 GB GDDR6 memory. The Python version utilized was 3.8, and the hardware setup comprised an AMD R9 5900X CPU with 64 GB memory.

Table 2. Reproduction results of mainstream models using ResNet-18 as the backbone

Model	PatchCore	CFlow	CFA	PaDiM	DFM	STFPM	GANnomaly
Carpet	0.971	0.975	0.954	0.948	0.820	0.960	0.203
Grid	0.944	0.952	0.939	0.854	0.769	0.981	0.404
Leather	1.000	0.998	0.998	0.980	0.990	0.977	0.413
Tile	0.995	0.999	1.000	0.949	0.967	0.952	0.412
Wood	0.987	0.991	1.000	0.994	0.982	0.961	0.744
Bottle	1.000	0.998	0.990	0.998	1.000	0.979	0.252
Cable	0.983	0.873	0.946	0.881	0.965	0.801	0.477
Capsule	0.967	0.942	0.854	0.925	0.943	0.784	0.683
Hazelnut	0.999	0.999	0.994	0.967	0.989	0.999	0.537
MetalNut	0.992	0.992	0.933	0.988	0.918	0.958	0.271
Pill	0.926	0.904	0.887	0.937	0.957	0.711	0.472
Screw	0.944	0.901	0.692	0.846	0.894	0.855	0.234
Toothbrush	0.932	0.888	0.996	0.944	0.960	0.994	0.385
Transistor	0.997	0.936	0.900	0.971	0.940	0.857	0.449
Zipper	0.960	0.977	0.918	0.887	0.969	0.651	0.437
Average	0.973	0.958	0.934	0.938	0.937	0.894	0.424

4.4   Results of the Generated Images

In this section, the experimental results of the generated images at different iteration steps and noise strengths on mainstream semi-supervised defect detection models are presented. These results are visualized in Fig. 2 and Fig. 3, which show partially generated samples.

Figure 2 shows the sample generation results using the DDPM-generated images for subsets such as Hazelnut, Capsule, Pill, Cable, and Toothbrush. The images were generated under different iteration steps (1, 50, 75, 100, 150 from left to right) and with a noise intensity of 10. Through this manipulation of iteration steps and noise intensity, an analysis was conducted to assess their impact on the quality and performance of the generated images. The experimental results indicate that the finer-grained performance of the generated images improves as the number of iteration steps increases with a certain level of noise intensity. This suggests that training the model for more iterations leads to enhanced image quality and accuracy in defect detection. However, it is worth noting that an increase in the number of iteration steps also incurs a higher time cost. Therefore, considering the trade-off between performance and efficiency, this paper determines the optimal number of iteration steps to be 50. This value strikes a balance between achieving high-quality generated images and maintaining reasonable computational costs.

Table 3. Hyperparameter Settings.

Hyperparameter	Values
Max iterations for diffusion model	150
Noise schedule	Linear
Selected iteration steps	1, 50, 75, 100, 150
Selected noise values	0, 5, 10, 15, 20, 25, 30
Learning rate	1e−4
Generated image sizes	512×512, 1024×1024

Furthermore, as shown in Fig. 3, the experimental results on Hazelnut, Capsule, Pill subsets show that the finer-grained differences in the images vary according to different levels of noise strength, provided that the number of iteration steps is set to 50. Specifically, the results indicate that the finer-grained changes in the generated images become more noticeable and discernible to the naked eye when the noise intensity exceeds 10. This observation suggests that the DDPM-generated images can effectively simulate various levels of defects, thereby enabling a robust evaluation of defect detection models' generalization ability and resilience.

4.5   Sample Separability Display

To evaluate the effectiveness of the anomaly simulation strategy, we employed ResNet-18 as the classifier and utilized t-SNE [17] for visualizing the model's output. This evaluation was conducted on subsets including Capsule, Leather, MetalNut, and Pill. The evaluation process involved comparing three types of samples: simulated anomaly samples, real anomaly samples, and normal samples.

The results, as depicted in Fig. 4, revealed that in the majority of categories, there was an overlapping spatial distribution between the simulated anomaly samples and the real anomaly samples. Conversely, the anomaly samples and the normal samples exhibited distinctive feature regions. Although certain categories, like the Leather category, displayed poor two-dimensional separability of features, we obtained excellent classification results for categories that are particularly sensitive to fine-grained information, such as Capsule and Pill. These results demonstrate the effectiveness of our anomaly simulation strategy in generating challenging samples for deep learning models. By creating realistic anomaly samples that closely resemble real anomalies, our strategy enhances the robustness and generalization capability of these models.

4.6   Different Performance Comparisons

To verify the effectiveness of the image augmentation strategy based on DDPM, multiple experiments were conducted on the MVTec-AD dataset. The performance of mainstream models under different iteration numbers is shown in

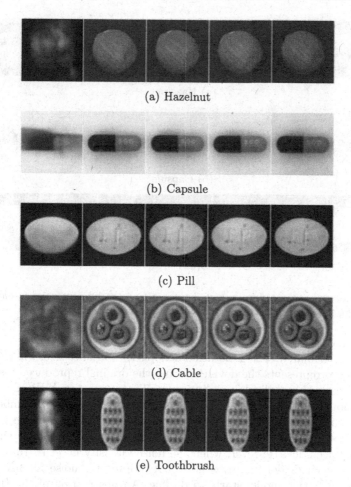

(a) Hazelnut

(b) Capsule

(c) Pill

(d) Cable

(e) Toothbrush

Fig. 2. Simulated generated images of the dataset under different iteration steps (1, 50, 75, 100, 150 from left to right) and with a noise intensity of 10.

Table 4. For each model, the experiment involved generating a dataset using our image augmentation strategy and testing it ten times, following which an average performance metric was calculated. The ± notation indicates the variance size, while the data in parentheses represents the deviation from the original results.

The experimental results indicate that when using ResNet-18 as the backbone network and setting the noise intensity to 10 (as depicted in Fig. 5), the model accuracy improves with an increase in the iteration number. The highest model accuracy is achieved when the iteration number reaches 150.

Table 5 presents the experimental results for different noise intensities, utilizing ResNet-18 as the backbone network with an iteration number of 50. The dataset was generated and tested on each model ten times, and the average

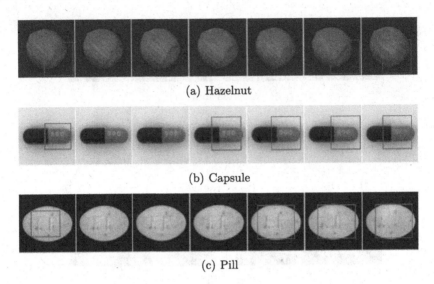

(a) Hazelnut

(b) Capsule

(c) Pill

Fig. 3. Simulated generated images of the dataset under different noise intensities (0, 5, 10, 15, 20, 25, 30 from left to right) and 50 iteration steps.

results were calculated. The ± notation denotes the variance size, while the data in parentheses represents the deviation from the original reproduced results.

Additionally, Fig. 6 displays a line chart illustrating the AUROC values of different models under various noise intensities, with an iteration number of 50. According to the observations from Fig. 6, it can be seen that when the noise intensity is set to 0, 5, or 20, the accuracy of each model remains similar to the original accuracy. However, when the noise intensity is set to 10 or 15, the accuracy of each model improves noticeably. Beyond a noise intensity of 25, the accuracy of each model starts to decline. An observation of the Hazelnut, Capsule, and Pill sub-datasets shown in Fig. 3 also reveals that the images lose details as the noise intensity exceeds 25. Under low noise intensity, the differences

Table 4. AUROC of different models under different iteration numbers.

Model	PatchCore	CFlow	CFA	PaDiM	DFM	STFPM
50 iters	0.980 ± 0.001 (↑0.007)	0.961 ± 0.001 (↑0.002)	0.938 ± 0.003 (↑0.004)	0.944 ± 0.001 (↑0.006)	0.940 ± 0.002 (↑0.003)	0.899 ± 0.002 (↑0.005)
75 iters	0.979 ± 0.002 (↑0.006)	0.959 ± 0.001 (↑0.001)	0.938 ± 0.002 (↑0.004)	0.943 ± 0.001 (↑0.005)	0.941 ± 0.001 (↑0.004)	0.900 ± 0.002 (↑0.006)
100 iters	0.981 ± 0.002 (↑0.008)	0.962 ± 0.002 (↑0.004)	0.940 ± 0.003 (↑0.006)	0.943 ± 0.002 (↑0.005)	0.939 ± 0.002 (↑0.002)	0.901 ± 0.003 (↑0.007)
150 iters	0.981 ± 0.002 (↑0.008)	0.963 ± 0.002 (↑0.005)	0.939 ± 0.002 (↑0.005)	0.945 ± 0.001 (↑0.007)	0.939 ± 0.003 (↑0.002)	0.901 ± 0.003 (↑0.007)
Original	0.973	0.958	0.934	0.938	0.937	0.894

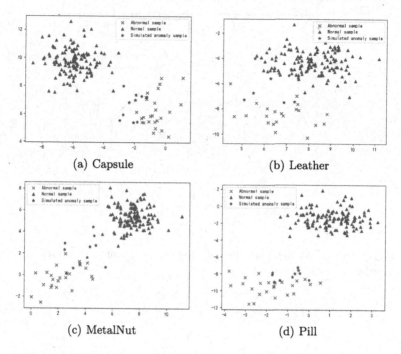

(a) Capsule

(b) Leather

(c) MetalNut

(d) Pill

Fig. 4. The separability of normal samples, simulated abnormal samples and real abnormal samples in the MVTec-AD dataset.

compared to the original images may not be obvious to the naked eye, but they are still subtly present. This allows for achieving diversity at a fine-grained level within the dataset.

4.7   The Time to Generate Images

Figure 7 shows the time taken to generate images of different sizes with noise intensity ranging from 0 to 30 under various iteration numbers. The notation '5_512' in the figure represents a noise intensity of 5 and an image size of 512×512. As shown in Fig. 7, the time required to generate images of resolution $1,024 \times 1,024$ and 512×512 is almost identical, even when the iteration numbers vary. As the iteration numbers increase, there is a linear increase in the time cost. Although generating 512×512 images is faster, there is a slight decrease in overall accuracy compared to the $1,024 \times 1,024$ images. Considering that the original image sizes range from 700×700 to $1,024 \times 1,024$, the experiments utilized images of size $1,024 \times 1,024$.

The experimental results demonstrate that the DDPM-generated synthetic dataset can effectively improve the performance of semi-supervised defect detection models. The synthesized images demonstrate various defect types, shapes, and sizes, which makes the generated dataset suitable for training deep learning models in defect detection tasks. Moreover, using this synthetic dataset can

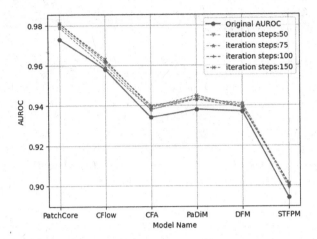

Fig. 5. Line chart of AUROC of different models under different iteration numbers with a noise intensity of 10.

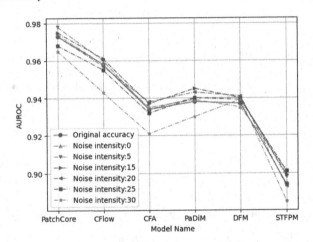

Fig. 6. Line chart of AUROC of different models under different noise intensities with an iteration number of 50.

significantly reduce the cost and time required for acquiring and annotating real-world image datasets. Overall, this method provides a practical solution for developing and evaluating robust and effective defect detection models.

Table 5. AUROC of different models under different noise intensities.

Model	Patchcore	CFlow	CFA	PaDiM	DFM	STFPM
Noise:0	0.974 ± 0.002 (↑0.001)⁻	0.958 ± 0.002 (–)	0.935 ± 0.002 (↑0.001)	0.939 ± 0.002 (↑0.001)	0.935 ± 0.002 (↓0.002)	0.901 ± 0.002 (↑0.007)
Noise:5	0.978 ± 0.001 (↑0.005)	0.960 ± 0.002 (↑0.002)	0.934 ± 0.002 (–)	0.940 ± 0.002 (↑0.002)	0.940 ± 0.002 (↑0.003)	0.898 ± 0.002 (↑0.004)
Noise:10	0.980 ± 0.002 (↑0.007)	0.961 ± 0.002 (↑0.003)	0.938 ± 0.002 (↑0.004)	0.944 ± 0.002 (↑0.006)	0.940 ± 0.002 (↑0.003)	0.899 ± 0.002 (↑0.005)
Noise:15	0.975 ± 0.004 (↑0.002)	0.961 ± 0.002 (↑0.003)	0.937 ± 0.002 (↑0.003)	0.945 ± 0.002 (↑0.007)	0.940 ± 0.002 (↑0.003)	0.899 ± 0.002 (↑0.005)
Noise:20	0.973 ± 0.005 (–)	0.957 ± 0.002 (↑0.001)	0.938 ± 0.002 (↑0.004)	0.943 ± 0.002 (↑0.005)	0.941 ± 0.002 (↑0.004)	0.893 ± 0.002 (↓0.001)
Noise:25	0.968 ± 0.005 (↓0.005)	0.955 ± 0.002 (↓0.003)	0.932 ± 0.002 (↓0.002)	0.940 ± 0.002 (↑0.002)	0.939 ± 0.002 (↑0.002)	0.901 ± 0.002 (↑0.007)
Noise:30	0.965 ± 0.006 (↑0.008)	0.943 ± 0.002 (↓0.015)	0.921 ± 0.002 (↓0.013)	0.930 ± 0.002 (↓0.008)	0.939 ± 0.002 (↑0.002)	0.885 ± 0.002 (↓0.009)
Original	0.973	0.958	0.934	0.938	0.937	0.894

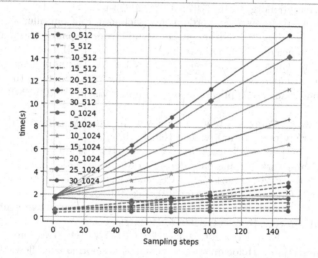

Fig. 7. Line chart of time taken to generate images of different sizes with noise intensity ranging from 0 to 30 under different iteration numbers.

5   Conclusion

The aim of this paper is to evaluate the one-class defect detection task, explore mainstream models, and construct different simulated datasets based on the DDPM image augmentation method. In comparison to the traditional approach of manually introducing noise, the proposed image augmentation method can generate high-quality samples close to the original image at a fine-grained level, enhance sample diversity, and improve model accuracy when reaching a bottleneck. By leveraging these new datasets and employing various augmentation strategies, the performance of the models can be effectively enhanced. The data augmentation method performs well in generating high-quality and diverse image samples and also has efficiency and generalization performance.

Acknowledgements. This work was supported by the National Natural Science Foundation of China under Grant 61772252, the Scientific Research Foundation of the Education Department of Liaoning Province under Grant LJKZ0965, and the Huzhou Science and Technology Plan Project under Grants 2022GZ08 and 2023ZD2004.

References

1. Ahuja, N.A., Ndiour, I., Kalyanpur, T., Tickoo, O.: Probabilistic modeling of deep features for out-of-distribution and adversarial detection. arXiv preprint arXiv:1909.11786 (2019)
2. Akcay, S., Atapour-Abarghouei, A., Breckon, T.P.: GANomaly: semi-supervised anomaly detection via adversarial training. In: Jawahar, C.V., Li, H., Mori, G., Schindler, K. (eds.) ACCV 2018. LNCS, vol. 11363, pp. 622–637. Springer, Cham (2019). https://doi.org/10.1007/978-3-030-20893-6_39
3. Almahairi, A., Rajeshwar, S., Sordoni, A., Bachman, P., Courville, A.: Augmented CycleGAN: learning many-to-many mappings from unpaired data. In: International Conference on Machine Learning, pp. 195–204. PMLR (2018)
4. Bergmann, P., Fauser, M., Sattlegger, D.: MVTec AD - a comprehensive real-world dataset for unsupervised anomaly detection. In: Proceedings of the IEEE/CVF Winter Conference on Applications of Computer Vision (WACV 2019), pp. 9592–9600 (2019)
5. Bochkovskiy, A., Wang, C., Liao, H.: Yolov4: optimal speed and accuracy of object detection. arXiv preprint arXiv:2004.10934 (2020)
6. Defard, T., Setkov, A., Loesch, A., Audigier, R.: PaDiM: a patch distribution modeling framework for anomaly detection and localization. In: Del Bimbo, A., et al. (eds.) ICPR 2021, Part IV. LNCS, vol. 12664, pp. 475–489. Springer, Cham (2021). https://doi.org/10.1007/978-3-030-68799-1_35
7. Dhariwal, P., Nichol, A.: Diffusion models beat GANs on image synthesis. Adv. Neural. Inf. Process. Syst. **34**, 8780–8794 (2021)
8. Gudovskiy, D., Ishizaka, S., Kozuka, K.: CFLOW-AD: real-time unsupervised anomaly detection with localization via conditional normalizing flows. In: Proceedings of the IEEE/CVF Winter Conference on Applications of Computer Vision, vol. 12999, pp. 98–107. Springer (2022)
9. He, K., Zhang, X., Ren, S., Sun, J.: Deep residual learning for image recognition. In: Proceedings of the IEEE Conference on Computer Vision and Pattern Recognition, pp. 770–778 (2016)
10. Ho, J., Jain, A., Abbeel, P.: Denoising diffusion probabilistic models. In: Advances in Neural Information Processing Systems, vol. 33, pp. 6840–6851 (2020)
11. Lee, S., Lee, S., Song, B.: CFA: coupled-hypersphere-based feature adaptation for target-oriented anomaly localization. In: Proceedings of the 2022 IEEE International Conference on Robotics and Automation (ICRA), pp. 78446–78454. IEEE (2022)
12. Li, C., Sohn, K., Yoon, J., Pfister, T.: CutPaste: self-supervised learning for anomaly detection and localization. In: Proceedings of the IEEE/CVF Conference on Computer Vision and Pattern Recognition, pp. 9664–9674. Springer (2021)

13. Li, W., Liu, D., Chen, K., Li, K., Qi, H.: Hone: mitigating stragglers in distributed stream processing with tuple scheduling. IEEE Trans. Parallel Distrib. Syst. **32**(8), 2021–2034 (2021)

14. Li, W., et al.: Efficient coflow transmission for distributed stream processing. In: IEEE INFOCOM 2020 - IEEE Conference on Computer Communications, pp. 1319–1328 (2020)

15. Luo, J., Dong, T., Song, D.: Review of surface defect detection. J. Comput. Sci. Explor. **8**(9), 1041–1048 (2014)

16. Ma, D., Tang, P., Zhao, L., Zhang, Z.: A review of research on depth learning image data augmentation methods. Chin. J. Image Graph. **26**(03), 487–502 (2021)

17. Van der Maaten, L., Hinton, G.: Visualizing data using t-SNE. J. Mach. Learn. Res. **9**(11), 2579–2605 (2008)

18. Ronneberger, O., Fischer, P., Brox, T.: U-Net: convolutional networks for biomedical image segmentation. In: Navab, N., Hornegger, J., Wells, W.M., Frangi, A.F. (eds.) MICCAI 2015. LNCS, vol. 9351, pp. 234–241. Springer, Cham (2015). https://doi.org/10.1007/978-3-319-24574-4_28

19. Roth, K., Pemula, L., Zepeda, J.: Towards total recall in industrial anomaly detection. In: Proceedings of the IEEE/CVF Conference on Computer Vision and Pattern Recognition (CVPR), pp. 14318–14328. IEEE (2022)

20. Schlegl, T., Seeböck, P., Waldstein, S.: Unsupervised anomaly detection with generative adversarial networks to guide marker discovery. In: Information Processing in Medical Imaging: 25th International Conference, pp. 146–157 (2017)

21. Schlegl, T., Seeböck, P., Waldstein, S.: f-AnoGAN: fast unsupervised anomaly detection with generative adversarial networks. Med. Image Anal. **54**, 30–44 (2019)

22. Vaswani, A., Shazeer, N., Parmar, N.: Attention is all you need. Adv. Neural. Inf. Process. Syst. **30**, 5998–6008 (2017)

23. Wang, G., Han, S., Ding, E., Huang, D.: Student-teacher feature pyramid matching for anomaly detection. arXiv preprint arXiv:2103.04257 (2021)

24. Wyatt, J., Leach, A., Schmon, S., Willcocks, C.: AnoDDPM: anomaly detection with denoising diffusion probabilistic models using simplex noise. In: Proceedings of the IEEE/CVF Conference on Computer Vision and Pattern Recognition, pp. 650–656 (2022)

25. Yang, M., Wu, P., Feng, H.: MemSeg: a semi-supervised method for image surface defect detection using differences and commonalities. Eng. Appl. Artif. Intell. **119**, 105835 (2023)

26. Yu, J., Zheng, Y., Wang, X.: Fastflow: unsupervised anomaly detection and localization via 2D normalizing flows. arXiv preprint arXiv:2111.07677 (2021)

27. Zavrtanik, V., Kristan, M., Skočaj, D.: DRAEM - a discriminatively trained reconstruction embedding for surface anomaly detection. In: Proceedings of the IEEE/CVF International Conference on Computer Vision, vol. 140, pp. 8330–8339. Springer (2021)

28. Zhang, H., Cisse, M., Dauphin, Y., Lopez-Paz, D.: Mixup: beyond empirical risk minimization. arXiv preprint arXiv:1710.09412 (2017)

K Asynchronous Federated Learning with Cosine Similarity Based Aggregation on Non-IID Data

Shan Wu[1], Yizhi Zhou[1], Xuesong Gao[2,3], and Heng Qi[1(✉)] [ID]

[1] School of Computer Science and Technology, Dalian University of Technology, Dalian, China
hengqi@dlut.edu.cn
[2] College of Intelligence and Computing, Tianjin University, Tianjin, China
[3] State Key Laboratory of Digital Multimedia Technology, Hisense Co., Ltd., Qingdao, China

Abstract. In asynchronous federated learning, each device updates the model independently as soon as it becomes available, without waiting for other devices. However, this approach is confronted with two critical challenges, namely the non-IID data and the staleness issue, which can adversely impact the performance of the model. To address these challenges, we propose a novel framework called Class-balanced K-Asynchronous Federated Learning (CKAFL). In this framework, we adopt a two-pronged approach, aiming to resolve the problems of non-IID and staleness separately on the client and server side. We give a novel evaluation method that employs cosine similarity to measure the staleness of a delayed gradient to optimize the aggregation algorithm on the server side. We introduce a class-balanced loss function to mitigate the non-IID data in the client side. To evaluate the effectiveness of CKAFL, we conduct extensive experiments on three commonly used datasets. The experimental results show that even when a large proportion of devices have stale updates, the proposed CKAFL framework presents its effectiveness by outperforming baselines on both non-IID and IID cases.

Keywords: Federated Learning · Asynchronous Learning · Non-IID Data

1 Introduction

Mobile and edge devices have become widely adopted and generate a tremendous amount of valuable data for various applications. These devices also have increased the need for Machine Learning to enable personalized and low-latency AI applications [2,8]. However, centralized data collection and training are not feasible due to privacy and bandwidth constraints. Therefore, Federated Learning (FL) [11] has been introduced as a paradigm that enables collaborative machine learning across a large number of edge devices without sharing their data.

FL is a type of Machine Learning that allows multiple, decentralized participants to train and share models on their datasets, while keeping the local data

© The Author(s), under exclusive license to Springer Nature Singapore Pte Ltd. 2024
Z. Tari et al. (Eds.): ICA3PP 2023, LNCS 14492, pp. 434–452, 2024.
https://doi.org/10.1007/978-981-97-0811-6_26

private. This form of distributed Machine Learning can be used for collaborative model training, where each node could contribute something unique to the overall model [1]. FL can also be used in settings where user data must remain confidential or cannot leave its original environment [7], such as in healthcare and finance settings.

FL operates by enabling edge devices to send updates to a central server, which in turn distributes updated models to each edge device for further training. This distributed approach allows the edge devices to perform the majority of the computation, while the central server updates the model parameters based on the descending directions provided by the edge devices. Moreover, FL has three unique characteristics that distinguish it from the standard parallel optimization: decentralized training data, fault tolerance and privacy preservation [18]. To address the unique challenges of FL, many algorithms have been proposed for FL [4,8,11,16]. Among these algorithms, FedAvg is one of the most classical FL algorithms and runs in a synchronous manner which suffers from delays in waiting for the slowest learners (stragglers).

In addition to the synchronous methods, asynchronous FL (AFL) has also attracted considerable attention [2,21,23,25], where the server updates the global model after receiving the gradients from first K clients. This allows the clients who miss the current iteration to continue their training and reduce the runtime in the next iteration. AFL can cope with the straggler effects in synchronous FL and is suitable for scenarios with a large number of participants. When AFL is trained on a non-IID dataset, there are two theoretical downside named the non-IID data and staleness issue. First, the global model will not be able to capture the full distribution of the data due to some local data distributions being over represented or underrepresented. This can lead to significant bias in the final model. Secondly, since AFL embraces asynchronous updates that relax the constraint of model consistency, a discrepancy can occur between the model used for training and the model being updated. This misalignment means that a client's update may be outdated, resulting in a suboptimal gradient.

To address the non-IID data challenge, existing solutions have been proposed, such as data resampling [33] and class-balanced loss functions [14]. These approaches aim to alleviate training biases across clients by rebalancing the data distribution or adjusting the loss function accordingly. Regarding the staleness issue, several studies have highlighted the importance of adapting the learning rate [23,31] in response to the staleness level. By negatively correlating the learning rate with the staleness, the impact of outdated gradients can be mitigated. Another approach involves employing a two-stage training process [3], where the initial stage focuses on reducing staleness through accelerated updates, followed by a refinement stage for fine-tuning the model.

A potential solution that addresses the challenges of staleness and non-IID data in AFL is Weighted K-Asynchronous Federated Learning (WKAFL) proposed by Zhou et al. [30]. WKAFL adopts a two-stage approach with adaptive learning rate to mitigate these issues. However, WKAFL primarily tackles these challenges at the server side, which places a significant computational burden on the central server. Additionally, WKAFL incorporates momentum to alleviate the impact of non-IID data, which can potentially conflict with the staleness issue. To address these concerns, this paper introduces a novel framework for AFL called Class-balanced K-Asynchronous Federated Learning (CKAFL). The key objective of CKAFL is to provide solutions that do not inherently contradict each other when addressing the challenges of staleness issue and non-IID data, while also reducing the computational burden on the server. In light of these goals, our contributions are outlined as follows:

1) Considering the aforementioned challenges, we approach the non-IID data and the staleness issue as distinct problems, addressing them individually from both the client and server sides. To tackle these challenges, we propose a novel framework called Class-balanced K-Asynchronous Federated Learning (CKAFL). CKAFL incorporates a evaluation method for measuring the staleness of delayed gradients, leveraging cosine similarity. Additionally, we optimize the aggregation algorithm on the server side. To mitigate the impact non-IID data, we introduce a class-balanced loss function that is resilient to variations in class distributions [14], facilitating the training of a generalized classifier with a consistent objective across clients.

2) To evaluate the effectiveness of CKAFL in terms of training speed, prediction accuracy, and training stability, we conduct extensive experiments using three commonly used datasets: MNIST, CIFAR-10, and Shakespeare text data. Our experimental results indicate that even when a large proportion of devices have stale updates, the proposed CKAFL framework presents its effectiveness by outperforming baselines on both non-IID and IID cases.

The subsequent sections of this paper are structured as follows. Section 2 provides an overview of related works in this field. Section 3 presents the motivation behind our proposed framework, CKAFL, and provides a detailed description of its components. To validate the effectiveness of our proposed framework, we evaluate its performance in various scenarios in Sect. 4. Finally, Sect. 5 summarizes the key contributions and findings of this paper.

2 Related Work

In this section, we provide an overview of the related work in the field of FL concerning non-IID data and staleness issue. We also discuss several approaches proposed to address both these challenges.

Several studies have proposed new methods to mitigate the impact of non-IID data on the performance and fairness of FL models, especially for classification tasks [6]. These methods can be broadly categorized into three types: data-level, algorithm-level, and hybrid methods. Data-level methods involve techniques such as data resampling [33] and data augmentation [34], which aim to balance the data distribution across clients. Algorithm-level methods modify the training algorithm or network structure [17], for example, through knowledge transfer [15] or generative adversarial networks (GANs) [10]. Hybrid methods combine data-level and algorithm-level techniques to achieve improved performance.

For instance, Zhao et al. [28] demonstrated the negative impact of highly skewed non-IID data on the accuracy of FedAvg and proposed a data-level method that creates a small subset of data shared globally to improve training on non-IID data. Similarly, Li et al. [8] introduced FedProx, an algorithm-level method that incorporates a proximal term in the local objective function to handle non-IID data. Additionally, Xiao et al. [22] effectively alleviate biases in model training by dynamically adjusting the weight of local training samples for each round across all participants. Shang et al. [32] explored the problem of federated learning on heterogeneous and long-tailed data, proposing a hybrid method that retrains the classifier using federated features extracted from local data.

The staleness issue in AFL arises when the global server receives outdated local parameters from certain clients due to communication delays or asynchronous updates. This can negatively impact the convergence and accuracy of federated learning, particularly in the presence of non-IID data. To address these challenges and enhance the efficiency of AFL, some researchers have proposed various methods to address these challenges and improve the efficiency in AFL. In prior work of A-PSGD(Asynchronous-Parallel SGD) [9,27], Lian et al. allow server to update global model using stale model parameters and modulate the learning rate according to the gradient staleness and provide theoretical guarantees for convergence of this algorithm. Xie et al. [24] estimate the descent of the loss value after the candidate gradient is applied, where large descent values indicate that the update results in optimization progress.

One effective approach to mitigate the impact of staleness is through learning rate adjustments [29]. For example, Xie et al. [23] proposed a new asynchronous federated optimization algorithm and employed three different strategies to measure the staleness of delayed models. Wang et al. [19] introduced a novel evaluation method for delayed gradients, measuring staleness based on the Euclidean distance between the stale model and the current global model. Nguyen et al. [12] utilized a receive buffer, where the server aggregates averaged local updates in the buffer once the number of received local updates exceeds the buffer limit.

While AFL schemes offer potential benefits in terms of system efficiency, the non-IID data distribution can often lead to lower model accuracy. To address straggler issues and alleviate the impact of non-IID data, Wu et al. [21] proposed fast-K or M-step aggregation schemes, which aim to mitigate the staleness issue and improve system efficiency. Wu et al. [20] classified participating clients into different classes to enhance efficiency and improve the quality of the global model. Chen et al. [3] divided the federated learning training process into two stages to maintain efficiency and robustness.

These previous works highlight the efforts made to overcome the challenges posed by non-IID data and staleness issue in AFL. Data-level, algorithm-level, and hybrid approaches have been proposed to balance data distribution, modify training algorithms, and leverage hybrid techniques for improved model performance. Additionally, strategies such as adjusting learning rates, incorporating receive buffers, and employing advanced aggregation schemes have been explored to address the staleness issue and enhance system efficiency. Despite these advances, there is still room for further research and development in the field of AFL to optimize model accuracy and convergence in the presence of non-IID data and staleness.

3   Proposed Framework

In this section, we provide a brief introduction to K-Asynchronous Federated Learning (KAFL). We then motivate our work by measuring and discussing the impact of non-IID data and correlation between the staleness and the directional deviation. Finally, we present our framework, which aims to address training efficiency concerns by introducing a class-balanced loss function to mitigate the non-IID issue and incorporating a novel evaluation method for delayed gradients.

3.1   Preliminary

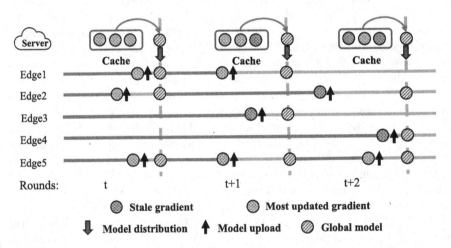

Fig. 1. Illustration of the training procedure of K-FedAsync algorithm with K = 3. The server executes the global aggregation when receiving three parameters from the fastest three edge devices.

In KAFL, as shown in Fig. 1, the central server collect and aggregate parameters uploaded by either the fastest K edge clients (K-FedAsync) or the first M updated parameters sent from any clients (M-step-FedAsync) [21]. This approach helps mitigate the stragglers effect that occurs in synchronous FL.

The training process in KAFL begins with the central server broadcasting the initial parameter w_0 to all edge clients. Upon receiving the parameters from the central server, each edge device independently trains its local model using its own set of data samples. The objective of the central server is to obtain a global shared model that minimizes the global objective function defined as follows:

$$\min_{\mathbf{w} \in \mathbb{R}^d} f(\mathbf{w}) := \sum_{m=1}^{M} \frac{|\mathcal{D}_m|}{|\mathcal{D}|} F_m(\mathbf{w}), \quad where \quad F_m(\mathbf{w}) = \frac{1}{|\mathcal{D}_m|} \sum_{\xi_i \in \mathcal{D}_m} \ell(\mathbf{w}; \xi_i) \quad (1)$$

Here, ξ_i denotes the i-th data point (x_i, y_i) sampled from the local dataset $\mathcal{D}_m$, and $|\mathcal{D}_m|$ represents the total number of data samples on client m. The parameter M refers to the number of edge participants.

The server waits for the first K out of M clients to complete their updates, while the remaining clients continue computing gradients. As a result, in each iteration, the gradients received by the server might be computed based on stale parameters. For K-FedAsync, the updating formula is expressed as follows:

$$w_{j+1} = w_j - \frac{\eta_0}{K} \sum_{i=1}^{K} \alpha * g(w_{j,i}, \xi_{j,i}) \quad (2)$$

Here, $g(w_{j,i}, \xi_{j,i})$ is the stale gradient received by the server in the j-th round, $\xi_{j,i}$ is the data sample of client i in the j-th round and $w_{j,i}$ is the stale model parameters used by client i to compute the gradient vector and α is the adaptive aggregation weight according staleness of delayed gradient in the server. η_0 is the initial learning rate.

3.2 Motivation

Impact of Non-IID Data. As outlined in [30], in order to mitigate the impact of non-IID data, the server necessitates an increased number of gradients. However, in order to prevent the detrimental effects of stale gradients on model utility, it is preferable to restrict the aggregation process to low-staleness gradients. As a result, a conundrum arises. It is our contention that the underlying reason for the server's requirement of a greater number of gradients is due to the weight divergence [28] which can be explained by the absence of consistent objects shared among clients. One way to mitigate the weight divergence is to make local training object of clients align with each other. The empirical loss (e.g., cross entropy), which is the local model's attempt to optimize to fit its own dataset, though used in most classification tasks, gives a biased gradient estimation under non-IID data distribution scenario.

$$weight \ divergence = \|w^{Fed} - w^{SGD}\| / \|w^{SGD}\| \quad (3)$$

By replacing the conventional empirical loss with the class-balanced loss, clients are able to share a more consistent objective during the federated learning process. This leads to a reduction in the degree of gradient divergence, eliminating the need for an increased number of gradients to be processed on the server.

To demonstrate the weight divergence of non-IID data in FL, we conducted experiments using a CNN model on the CIFAR-10 dataset with a 2-class non-IID data distribution. Specifically, the training sets were evenly partitioned into 10 clients, and the data was sorted into 20 partitions. Each client was randomly assigned 2 partitions from the 2 classes.

As depicted in Fig. 2, it displays the weight divergence of the local models trained using the conventional cross entropy loss or the class-balanced loss 4. The weight divergence refers to the discrepancy in model weights between different clients' local models. Under the non-IID data distribution, we observe that the weight divergence of local models trained with the cross entropy loss is higher compared to those trained with the class balance loss at around 40%–60%.

The reduced weight divergence observed when using the class-balanced loss indicates that the shared objective becomes more consistent across the participating clients. This alignment in objectives helps to mitigate gradient divergence, ultimately enhancing the efficiency and effectiveness of the FL process. The experimental results depicted in Fig. 2 provide empirical evidence supporting the advantages of utilizing the class-balanced loss in the context of non-IID data distributions.

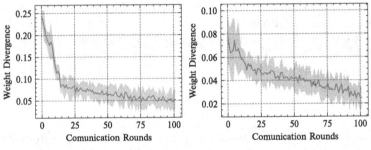

(a) Statistics for empirical loss   (b) Statistics for class-balanced loss

Fig. 2. Weight divergence of CNN with empirical loss or class balance loss, the shaded region is the standard deviation.

Correlation Between the Staleness and the Directional Deviation. To understand the detrimental effects of delayed gradients on model aggregation and highlight the limitations of measuring staleness solely based on the number of iteration lag τ, we delve into this issue.

Previous research [5] emphasizes that delayed gradients can impede model convergence. This is because the delayed gradient is computed based on the model $w_{t-\tau}$, which may not align with the current optimal gradient. In the AFL setting, multiple clients independently update the global model. Thus, when a client uploads its local results, the global model has already progressed to w_t. Consequently, the stale gradient deviates from the current optimal gradient in view of stochastic gradient descent (SGD), and naively aggregating such gradients can adversely impact the model's utility.

Existing staleness measurement strategies, such as using lag τ or local training time, have demonstrated effectiveness in addressing the stale gradient issue in experimental settings. However, these approaches possess notable limitations when applied in practical scenarios. For example, gradients with low latency may align the descending directions well with the current optimal gradient, while gradients with high latency may still exhibit minimal deviation from the optimal gradient direction. Consequently, employing a fixed threshold for staleness based on iteration lag τ or local training time may erroneously discard gradients that could still contribute to the convergence of the global model. This can hinder the training process and impede model convergence. Thus, in practice, these methods do not accurately assess whether a stale gradient is beneficial to the global model's convergence.

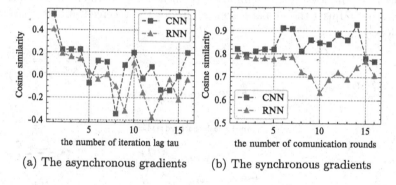

(a) The asynchronous gradients   (b) The synchronous gradients

Fig. 3. (a) The cosine similarity between asynchronous gradients. (b) The cosine similarity between synchronous gradients.

To investigate the directional deviation between asynchronous gradient vectors and synchronous gradient vectors in FL, we follow the previous work [23] and train two models on common datasets: CNN for MNIST and RNN for Shakespeare text data, each with 20 clients. Our aim is to measure the directional deviation by computing the cosine similarity between the delayed gradient and the current updated gradient as the iteration lag τ increases. We also compute the cosine similarity between synchronous gradients across different communication rounds in Fig. 3.

In the AFL scenario, we observed that the deviation was linearly proportional to lag τ when τ was smaller than 5. However, when τ exceeded a certain threshold, the deviation became independent of τ, indicating a weak correlation between the staleness and the deviation. Furthermore, we noticed that some delayed gradients with large staleness still had small deviation from the current updated gradient. In the synchronous FL scenario, we observed that the cosine similarity between synchronous gradients was stable around 0.8 and did not change significantly with the number of training rounds.

Based on these observations, we conclude that the iteration lag τ is not a reliable indicator of the directional deviation between the delayed gradient and the updated gradient. Also, the synchronous gradients had strong and consistent alignment. Therefore, we proposed a new evaluation method for staleness based on the directional consistency between the delayed gradient and the current global gradient.

3.3   Cosine Similarity Based Federated Aggregation

In this section, we present the system overview of CKAFL framework in Fig. 4 and then introduce the detailed component of CKAFL. First, we introduce a class-balanced loss function to mitigate the non-IID issue, which can handle non-identical class distributions. Furthermore, we integrate a novel evaluation method that employs cosine similarity to measure the staleness of a delayed gradient and adjusts the global learning rate for the received updates based on its staleness.

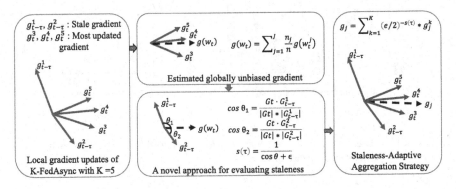

Fig. 4. Illustration of the aggregation rule of CKAFL with K = 5.

Improving AFL with Balanced Loss Minimization. In FL settings, it is common for a client to possess a non-IID dataset distribution, leading to local datasets that are not representative of the overall dataset population. Therefore, researchers have delved into the study of class-imbalanced learning within this context. The primary objective of class-imbalanced learning is to train a classifier that exhibits high performance across all classes, regardless of the training class distribution. To accomplish this objective, we propose treating each client's local training as a class-imbalanced learning problem and applying techniques developed within this sub-field. While re-weighting and re-sampling techniques are commonly employed, certain strategies are not applicable in the FL scenario due to the unavailability of local client data.

Recently, many class-imbalanced works have proposed to replace the cross entropy loss with a class-balanced loss, which shows more promising results than re-weighting or re-sampling. As discussed in [14], the softmax function which is used in most classification tasks gives a biased gradient estimation due to non-IID data distribution of clients. We adopt this idea by replacing the empirical loss L_m with the balanced softmax loss L_{BSM} which effectively handles the distribution shift between the local side and global side.

$$\ell^{\text{BSM}}(x, y; \boldsymbol{w}) = -\log \left(\frac{N_y^\gamma \exp\left(g_y(\boldsymbol{x}; \boldsymbol{w})\right)}{\sum_{i=1}^c N_i^\gamma \exp\left(g_i(\boldsymbol{x}; \boldsymbol{w})\right)} \right) \tag{4}$$

here, $g_i(x; w)$ is the logit for class i, N_i is the number of instances of class i, γ is the hyper-parameter for this loss function. The BSM loss is an unbiased extension of softmax intended to compensate for the shift in class distribution between training and testing. It promotes a minor-class instance to claim a larger logit $g_y(x; w)$ in training in order to overcome feature deviation in testing.

Estimated Globally Unbiased Gradient. As discussed in the previous section (see Sect. 3.2), it is evident that the number of iteration lag τ is an inadequate measure of a gradient's staleness. Notably, gradients with low latency may align closely with the current optimal gradient, while gradients with high latency may not significantly deviate from the current optimal gradient direction. Ideally, in each communication round, before comparing the directional deviation between stale gradients and the globally optimal gradient of the current update, the server should estimate the globally unbiased gradient.

Gradients derived from outdated models often exhibit inconsistent directions, prompting us to posit that gradients with lower staleness possess a greater likelihood of consistency. Grounded in this hypothesis, in K-asynchronous Federated Learning, where the server node updates the global model upon receiving gradients from the first K clients, we advocate for selecting the gradient that is synchronized with the server from this set of K gradients as the foundation for estimating the global unbiased gradient. To accomplish this, we propose assigning weights to gradients based on their respective sample counts, ensuring that the estimation of the global unbiased gradient takes into account the varying contributions of each gradient. The estimating rule is

$$g\left(w_t\right) = \sum_{j=1}^J \frac{n_j}{n} g\left(w_{t,j}\right) \tag{5}$$

where, J is the number of most updated gradient among these K gradients, n is the total amount of data for the synchronous clients, and n_j is the amount of data for the client j. $g\left(w_{t,j}\right)$ is the gradient of local client j. $g\left(w_t\right)$ is the estimated globally unbiased gradient.

Algorithm 1. Training procedure of CAKFL

Input: Initial model weight x_0; The maximum number of communications K; Hyper-parameters γ, ε;

Output: Global model parameters x_T;

  Global Server Process

  Initialize $Q = \varnothing$;

  Send x_j, j to the newly connected client;

  for all $j = 1, 2, \cdots, globalIter$ **do**

    Receive a gradient update from an arbitrary work node;

    if $|Q| < M$ **then**

      Push $w_{j,i}$ to Q;

    else

      Estimate the globally unbiased gradient according to Equation 5;

      Adaptively aggregate the stale gradient according to Equation 8;

      $w_{j+1} = w_j - \frac{\eta_0}{K} \sum_{i=1}^{K} \alpha * g\left(w_{j,i}, \xi_{j,i}\right)$;

    end if

    Send w_{j+1} to connected clients;

    $j = j + 1$;

  end for

  Client Process:

  Receive global model w from the central server;

  for all $i = 1, 2, \cdots, localIter$ **do**

    $w_{i+1} = w_i - \eta_i g\left(w_i\right)$; //update local model according to Equation 4;

    Upload local results to server;

  end for

A Novel Approach for Evaluating Staleness. In the context of the AFL algorithm, a notable discrepancy arises between the update directions of synchronous clients and delayed clients during the parameter aggregation process. This disparity can be attributed to the directional error that arises from the difference between their gradient vectors. This directional error quantifies the degree to which the original update direction of stochastic gradient descent (SGD) is deviated from due to staleness. The accumulation of direction errors can significantly impact the global convergence direction of the objective function, impeding its ability to reach the optimal point.

To quantitatively estimate this deviation, we propose employing the estimated globally unbiased gradient at the current iteration as a surrogate for the gradient vector of the synchronous client. Subsequently, we calculate the cosine similarity between this vector and the local stale gradient of the delayed client. Cosine similarity is defined as the ratio of the dot product and the product of the lengths of the two vectors. Notably, it solely relies on their angle and disregards their magnitudes.

By utilizing this cosine similarity measure, we can effectively assess the degree of directional error between the synchronous client's gradient vector and the local stale gradient of the delayed client. This evaluation method provides insights into the correlation between the staleness and the directional deviation, enabling us to better understand and address the challenges associated with parameter aggregation in asynchronous settings.

The cosine similarity is computed as follows:

$$\cos(G_t, G_{t-\tau}) = \frac{G_t \cdot G_{t-\tau}}{\|G_t\| * \|G_{t-\tau}\|} \tag{6}$$

To adaptively adjust the aggregation weight based on staleness, we introduce the function $s(\tau)$ defined as:

$$s(\tau) = \frac{1}{\cos(G_t, G_{t-\tau}) + \epsilon} \tag{7}$$

In this equation, ϵ is a small constant to avoid division by zero. The function $s(\tau)$ assigns higher weights to gradients with lower staleness, ensuring that delayed gradients with a smaller deviation from the globally unbiased gradient receive more emphasis during the aggregation process.

The novel approach proposed for evaluating staleness in federated learning offers several advantages over existing strategies that measure staleness based on the number of iteration lag τ or local training time. Relying solely on the number of iteration lag or local training time as metrics for staleness overlooks the fact that the evaluation of staleness is not solely determined by temporal factors. In asynchronous federated learning, gradients from delayed clients may exhibit inconsistency in direction even if the delay duration is relatively short. On the other hand, gradients from stale clients may align well with the current optimal gradient direction despite longer delays. The proposed approach takes into account the actual deviation in update directions, capturing the true effect of staleness on the convergence behavior.

The new approach utilizes cosine similarity to measure the deviation between the gradient vectors of the synchronous and delayed clients. This measure focuses on the angle between the vectors, rather than their magnitudes. By doing so, it provides a more robust and meaningful assessment of the staleness impact. The proposed approach leverages the estimated global gradient at the current iteration as a proxy for the gradient vector of the synchronous client. This estimation serves as a reliable reference point for evaluating the staleness of delayed gradients. It allows for a more accurate assessment of the extent to which the SGD deviates from its original update direction due to staleness, thus providing insights into the convergence behavior of the federated learning algorithm.

Staleness-Adaptive Aggregation Strategy. In order to further optimize the optimization process, we introduce an aggregation strategy that takes into account the staleness of each delayed gradient. This strategy aims to adaptively adjust the aggregation weight based on the level of staleness, thereby promoting optimal convergence behavior. The aggregation formula is defined as follows:

$$g(w_j) = \sum_{i=1}^{K} (e/2)^{-s(\tau)} * g(w_{j,i}, \xi_{j,i}) \tag{8}$$

In this formula, $g(w_j)$ represents the aggregated gradient at iteration j, $g(w_{j,i}, \xi_{j,i})$ represents the gradient from client i, and τ represents the staleness of the gradient of client i. The function $s(\tau)$ determines the aggregation weight based on the staleness, with the exponent $(e/2)^{-s(\tau)}$ reflecting the importance assigned to each delayed gradient. For the most updated gradient, the function $s(\tau)$ is assigned a value of 0, indicating that it should receive the highest weight in the aggregation process. By incorporating this staleness-adaptive aggregation strategy, we aim to improve the convergence behavior of the optimization process in the context of federated learning.

The staleness-adaptive aggregation strategy outlined in Eq. (8) aims to address the challenge of varying staleness levels in federated learning. By assigning different weights to delayed gradients based on their staleness, the aggregation process can effectively adapt to the changing conditions and promote convergence towards an optimal solution.

4   Experiments

In this section, we present the experimental results that validate the performance of our proposed method CKAFL in terms of convergence rate, prediction accuracy, and efficiency with varying degrees of non-IIDness. We compare CKAFL with four existing algorithms, as discussed in Sect. 4.1. Furthermore, we utilize three benchmark datasets with diverse levels of non-IIDness. Detailed results are provided in Sect. 4.2 to demonstrate the efficiency of CKAFL.

4.1   Experimental Settings

Datasets and Models. We select three commonly used datasets for federated learning: MNIST, CIFAR-10, and Shakespeare text data. To simulate both independent and identically distributed (IID) and non-IID data distributions among the clients, we adopt the method described by Hsu et al. [6] to generate heterogeneous partitions for M clients. We employ different network architectures for each dataset, catering to their specific characteristics. Specifically, we use CNN for MNIST, ResNet18 for CIFAR-10, and RNN for Shakespeare text data. The mini-batch sizes is set to 50, 50, and 20, respectively [26].

To simulate non-IID data distributions for MNIST, CIFAR-10, and Shakespeare text data, we create an M-dimensional vector called q_c using the $Dir(\alpha)$ function for class c. We then distribute data of class c to client m proportionally to $q_c[m]$. Consequently, each client possesses a different number of total images and a distinct class distribution. As α approaches infinity, all clients have identical distributions according to the prior. Conversely, as α approaches zero, each client contains examples from only one randomly chosen class. A larger value of α leads to a more balanced data distribution across clients.

Baselines. We compare the proposed method CKAFL with several baseline algorithms. First, we include FedAvg [11], which is a synchronous federated learning (SFL) algorithm based on iterative model averaging. FedAvg is the first SFL method proposed for non-IID data and remains an efficient and standard baseline for FL research. Additionally, we consider two asynchronous federated learning (AFL) algorithms: KAFL [21] and WKAFL [30]. KAFL allows the global server to iteratively collect and aggregate the parameters uploaded by the fastest K edge clients designed to enhance the flexibility and scalability of AFL. WKAFL, on the other hand, is a two-stage weighted K-asynchronous FL algorithm aimed at improving the model utility of AFL. By comparing CKAFL with these baselines, we can evaluate its performance in terms of training speed, prediction accuracy, and stability.

Training Environment. In our experiments, we utilize PyTorch [13] to construct the model architectures. Following the approach of previous works [21], we employ a 64-client setting to simulate a large number of participating edge devices, with K = 8 as typically observed in practical federated learning (FL) scenarios. The training set is partitioned across n = 64 clients. For the CNN model trained on the MNIST dataset, we set the initial learning rate (η) to 0.05. The local training epoch (E) was set to 3. For the ResNet18 model trained on the CIFAR-10 dataset, we set η to 0.01 and E to 3. In the case of the RNN model trained on the Shakespeare dataset, we set η to 0.01 and E to 3. To simulate the suspension probability, we introduce a probability parameter (P) of 50%, indicating that each client had a 50% chance of being suspended. We simulate the hang time by randomly sampling the staleness ($t - \tau$) from a uniform distribution.

4.2 Results and Discussions

Efficiency with Non-IIDness. We test the efficiency of CKAFL on convergence rate. Figure 5 shows the test accuracy versus running time curves of our method and three baselines on non-IID and IID data with 50% stale devices. Table 1 shows the final prediction accuracy. As described in Sect. 4.1, we use the $Dir(\alpha)$ to measure the non-IID level. We set the non-IID hyperparameter α of the client data distribution to 0.1, 0.5 and ∞ which means all clients have identical distributions to the prior.

Our method outperforms other baselines in both convergence rate and final accuracy, especially in non-IID cases. Notably, our method exhibits enhanced training stability in non-IID cases, while existing algorithms such as KAFL and FedAvg experience significant degradation in performance compared to IID cases. This highlights the robustness of our algorithm in addressing the challenges posed by non-IID data. Although our method is specifically designed for non-IID cases by transforming the empirical risk L_m into a class-balanced risk L_{BR}, it also yields promising results in IID scenarios. Despite asynchronous training introduces bias and model staleness, CKAFL do not suffer from much

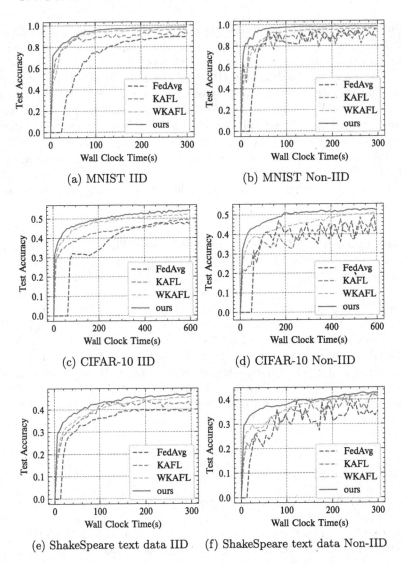

Fig. 5. Test accuracy v.s. training time curves of our method with different FL algorithms on both IID and Non-IID data settings.

accuracy loss compared to FedAvg algorithms. As such, it has the potential to be a valuable addition to the arsenal of optimization algorithms for federated learning.

Robustness Against Gradient Staleness. We evaluate the robustness of our method under varying levels of staleness. In the experiments, we compare the maximum test accuracy achieved by different algorithms within a given time

Table 1. Performance of our methods with three baseline methods on different degrees of Non-IIDness.

Methods	MNIST(%)			CIFAR-10(%)			Shakespeare text data (%)		
	Dir(0.1)	Dir(0.5)	IID	Dir(0.1)	Dir(0.5)	IID	Dir(0.1)	Dir(0.5)	IID
FedAvg	87.21	92.89	95.35	40.26	43.36	47.08	35.25	36.63	39.46
KAFL	91.20	92.37	93.04	45.55	47.02	48.93	39.26	41.38	42.52
WAKFL	95.31	95.43	95.90	50.08	50.32	52.21	42.18	42.53	45.83
ours	97.16	97.83	98.25	52.13	52.29	54.03	43.21	44.73	46.22

(500 s in these experiments) with respect to different maximum staleness degrees which is simulated by randomly sampling the staleness $(t - \tau)$ from a uniform distribution in Fig. 6. Our proposed method CKAFL outperforms the other baseline algorithms in terms of achieving higher test accuracy within the given time constraint. Furthermore, CKAFL demonstrates stability even when the maximum staleness value reaches 20 compared with the baselines. The test accuracy only exhibits an average drop of 8% when the staleness value increases from 12 to 20. These results illustrate the resilience of CKAFL to the effects of staleness, making it a reliable and effective solution for handling delayed updates in federated learning.

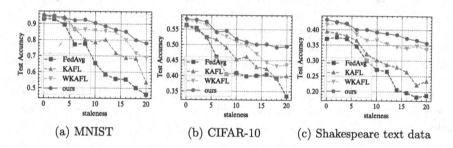

      (a) MNIST   (b) CIFAR-10   (c) Shakespeare text data

Fig. 6. Test accuracy of different methods with various degrees of staleness on MNIST, CIFAR-10 and Shakespeare text data.

Ablation Studies. As shown in Table 2, we investigate the impact of each component in the method through an ablation study. Our-w/o BSM indicates the performance of our method without balance softmax loss and our-w/o SA aggregation strategy indicates the performance of our method without staleness-adaptive aggregation strategy. $Dir(\alpha)$ indicates the degree of non-IIDness in the dataset. Avg is the average accuracy. We can see that the model accuracy decreases by 6% in the absence of balance softmax loss. In the absence of staleness-Adaptive aggregation strategy, the model accuracy decreases by 5.21%. The ablation experiment verifies that all modules are essential for CKAFL.

Table 2. Ablation experiments.

Methods	CIFAR-10(%)				
	Dir(0.1)	Dir(0.5)	Dir(1)	IID	Avg
FedAvg	42.26	45.36	46.21	50.21	46.01
our-w/o BSM	44.52	46.02	47.64	49.34	46.88
our-w/o SA aggregation strategy	45.73	45.87	48.20	50.85	47.67
ours	51.83	52.30	53.25	54.13	52.88

5 Conclusion

In this paper, we present a dual approach that effectively tackles the issues of non-IID data and staleness challenges separately on the client and server sides. Specifically, we investigate the impact of non-IID data and staleness in the context of AFL. To this end, we introduce a class-balanced loss function to mitigate the non-IID data problem. The balance loss function introduced in the client side not only addresses the non-IID issue but also helps to alleviate the computational burden on the server. Furthermore, we re-examine existing methods for tackling the staleness issue, which are commonly based on the number of iteration lag τ or local training time. We analyze these methods from the perspective of the gradient descent direction, specifically investigating the directional error that arises in gradient descent due to delayed gradients, as compared to synchronous gradients. Finally, we propose to use this directional error as a metric for the staleness of delayed gradients. Our research thus presents a comprehensive approach that effectively addresses the non-IID and staleness challenges in AFL, providing insights into the relationship between non-IID data and staleness. Our empirical evaluation validated both fast convergence and staleness tolerance.

Acknowledgments. This work was supported in part by the NSFC under Grant 62072069, in part by Hisense Group Holdings Company.

References

1. Chai, Z., et al.: TiFL: a tier-based federated learning system. In: Proceedings of the 29th International Symposium on High-Performance Parallel and Distributed Computing (2020). https://doi.org/10.1145/3369583.3392686
2. Chai, Z., Chen, Y., Zhao, L., Cheng, Y., Rangwala, H.: FedAT: a communication-efficient federated learning method with asynchronous tiers under non-IID data (2020)
3. Chen, M., Mao, B., Ma, T.: FedSA: a staleness-aware asynchronous federated learning algorithm with non-IID data. Futur. Gener. Comput. Syst. **120**, 1–12 (2021)
4. Chen, Y., Sun, X., Jin, Y.: Communication-efficient federated deep learning with layerwise asynchronous model update and temporally weighted aggregation. IEEE

Trans. Neural Netw. Learn. Syst. 4229–4238 (2019). https://doi.org/10.1109/tnnls.2019.2953131

5. Dai, W., Zhou, Y., Dong, N., Zhang, H., Xing, E.: Toward understanding the impact of staleness in distributed machine learning (2018)
6. Hsu, H., Qi, H., Brown, M.: Measuring the effects of non-identical data distribution for federated visual classification. arXiv, Learning (2019)
7. Kairouz, P., et al.: Advances and open problems in federated learning. arXiv, Learning (2021). https://doi.org/10.1561/9781680837896
8. Li, T., Sahu, A., Zaheer, M., Sanjabi, M., Talwalkar, A., Smith, V.: Federated optimization in heterogeneous networks. arXiv, Learning (2018)
9. Lian, X., Zhang, W., Zhang, C., Liu, J.: Asynchronous decentralized parallel stochastic gradient descent. arXiv, Optimization and Control (2017)
10. Liu, Y., Wu, G., Zhang, W., Li, J.: Federated learning-based intrusion detection on non-IID data. In: Meng, W., Lu, R., Min, G., Vaidya, J. (eds.) ICA3PP 2022. LNCS, vol. 13777, pp. 313–329. Springer, Cham (2023). https://doi.org/10.1007/978-3-031-22677-9_17
11. McMahan, H., Moore, E., Ramage, D., Hampson, S., Arcas, B.: Communication-efficient learning of deep networks from decentralized data (2016)
12. Nguyen, J., et al.: Federated learning with buffered asynchronous aggregation (2021)
13. Paszke, A., et al.: Pytorch: an imperative style, high-performance deep learning library (2019)
14. Ren, J., et al.: Balanced meta-softmax for long-tailed visual recognition. In: Neural Information Processing Systems (2020)
15. Tong, G., Li, G., Wu, J., Li, J.: GradMFL: Gradient Memory-Based Federated Learning for Hierarchical Knowledge Transferring Over Non-IID Data, pp. 612–626 (2022). https://doi.org/10.1007/978-3-030-95384-3_38
16. Wang, H., Yurochkin, M., Sun, Y., Papailiopoulos, D., Khazaeni, Y.: Federated learning with matched averaging (2020)
17. Wang, L., Xu, S., Wang, X., Zhu, Q.: Addressing class imbalance in federated learning. In: Proceedings of the AAAI Conference on Artificial Intelligence, pp. 10165–10173 (2021). https://doi.org/10.1609/aaai.v35i11.17219
18. Wang, L., Wang, W., Li, B.: CMFL: mitigating communication overhead for federated learning. In: International Conference on Distributed Computing Systems (2019)
19. Wang, Q., Yang, Q., He, S., Shui, Z., Chen, J.: Asyncfeded: asynchronous federated learning with euclidean distance based adaptive weight aggregation (2022)
20. Wu, W., He, L., Lin, W., Mao, R., Maple, C., Jarvis, S.: Safa: a semi-asynchronous protocol for fast federated learning with low overhead. IEEE Trans. Comput. 655–668 (2020). https://doi.org/10.1109/tc.2020.2994391
21. Wu, X., Wang, C.L.: KAFL: achieving high training efficiency for fast-k asynchronous federated learning (2022)
22. Xiao, W., et al.: Fed-Tra: Improving Accuracy of Deep Learning Model on Non-IID in Federated Learning, pp. 790–803 (2022). https://doi.org/10.1007/978-3-030-95384-3_49
23. Xie, C., Koyejo, O., Gupta, I.: Asynchronous federated optimization. arXiv, Distributed, Parallel, and Cluster Computing (2019)
24. Xie, C., Koyejo, S., Gupta, I.: Zeno++: robust fully asynchronous SGD (2020)
25. Xu, C., Qu, Y., Xiang, Y., Gao, L.: Asynchronous federated learning on heterogeneous devices: a survey. arXiv preprint arXiv:2109.04269 (2021)

26. Yao, L., et al.: A benchmark for federated hetero-task learning (2022)
27. Zhang, W., Gupta, S., Lian, X., Liu, J.: Staleness-aware async-SGD for distributed deep learning (2016)
28. Zhao, Y., Li, M., Lai, L., Suda, N., Civin, D., Chandra, V.: Federated learning with non-IID data (2018)
29. Zhou, Z., Mertikopoulos, P., Bambos, N., Glynn, P., Ye, Y.: Distributed stochastic optimization with large delays. Math. Oper. Res. **47**(3), 2082–2111 (2021)
30. Zhou, Z., Li, Y., Ren, X., Yang, S.: Towards efficient and stable k-asynchronous federated learning with unbounded stale gradients on non-IID data. IEEE Trans. Parallel Distrib. Syst. **33**(12), 3291–3305 (2022)
31. Zhu, F., Hao, J., Chen, Z., Zhao, Y., Chen, B., Tan, X.: STAFL: staleness-tolerant asynchronous federated learning on non-IID dataset. Electronics **11**(3), 314 (2022)
32. Shang, X., Lu, Y., Huang, G., Wang, H.: Federated learning on heterogeneous and long-tailed data via classifier re-training with federated features (2022)
33. Ziang, J.: KNN approach to unbalanced data distributions: a case study involving information extraction (2003)
34. Lee, H., Park, M., Kim, J.: Plankton classification on imbalanced large scale database via convolutional neural networks with transfer learning. In: 2016 IEEE International Conference on Image Processing (ICIP) (2016). https://doi.org/10.1109/icip.2016.7533053

MPQUIC Transmission Control Strategy for SDN-Based Satellite Network

Jinyao Liu[1,2], Xiaoqiang Di[1,2,3], Weiwu Ren[1,2], Ligang Cong[1,2], and Hui Qi[1,2(✉)]

[1] Department of Computer Science and Technology, Changchun University of Science and Technology, Changchun, China
qihui@cust.edu.cn
[2] Key Laboratory of Network and Information Security, Changchun University of Science and Technology, Changchun, China
[3] Department of Information Center, Changchun University of Science and Technology, Changchun, China

Abstract. The combination of Software-defined networking and multi-path transmission technology can enhance the transmission efficiency of satellite networks. However, the current SDN controller does not support the emerging multipath QUIC protocol, and the routing algorithm based on minimum hop count is inadequate for meeting the demands of high real-time business. To address these issues, this study designs and implements an SDN controller that supports MPQUIC protocol, and proposes a routing algorithm based on multi-objective optimization. This algorithm not only ensures the transmission throughput, but also selects links with lower propagation delays to improve overall transmission efficiency. The performance of the proposed solution is validated through satellite network simulation. The experimental results demonstrate that the proposed solution can enhance network throughput and reduce link delay in MPQUIC data flow transmission.

Keywords: Satellite network · Software-defined networking · Multipath QUIC

1 Introduction

Traditional satellite networks plays a critical role in enabling 5G applications, offering extensive coverage and ubiquity [5]. Software-defined networking(SDN), as an innovative network architecture, provides centralized control through the separation of data and control planes, flexible interfaces, and automaterd configuration management [8]. Previous research [1] has introduced the concept of SDN into satellite networks, leading to the design of a software-defined satellite

This paper is supported by the National Natural Science Foundation of China (No. U21A20451) and the Science and Technology Planning Project of Jilin Province (No. 20220101143JC).

© The Author(s), under exclusive license to Springer Nature Singapore Pte Ltd. 2024
Z. Tari et al. (Eds.): ICA3PP 2023, LNCS 14492, pp. 453–464, 2024.
https://doi.org/10.1007/978-981-97-0811-6_27

network architecture, achieving efficient and flexible control over satellite network. However, existing studies have mainly focused on architectural aspects, leaving room for further exploration and optimization in the aspect of transmission control. The emergence of new multipath transport protocols has sparked research into optimization solutions at the transmission control level. The current focus of research lies in the integration of SDN with multipath transport protocols to achieve end-to-end collaboration. Existing researchs [4,6] leavearge SDN to provide the necessary network capabilities for Multipath TCP (MPTCP), guiding data transmission by periodically collecting network state, selecting transmission paths, and deploying flow tables. De Coninck [2] tested Multipath QUIC(MPQUIC) and MPTCP, finding that MPQUIC outperforms MPTCP in packet loss scenarios, making it more suitable for high packet loss environments like satellite networks. However, existing SDN controllers have not yet support MPQUIC protocol, rendering them ineffective in guiding data transmission for MPQUIC subflows.

Existing research on MPQUIC has primarily focused on performance testing and resource scheduling. Early studies compared the performance of MPQUIC and MPTCP, including performance comparisons in different network environments and variations in application performance on smartphones [2,3]. Additionally, some studys proposed solutions based on resource scheduling models to address performance degradation caused by bursty transmissions [9,10]. However, current research predominantly revolves around optimizing the MPQUIC protocol itself, without fully leveraging the global network state to assist MPQUIC in data communication. Experimental findings revealed a crucial issue when applying MPQUIC in an unoptimized SDN environment: the chosen transmission paths by the SDN controller led to congestion of all subflows on a single path, resulting in bandwidth contention and thus negating the advantage of aggregating bandwidth across multiple paths.

We have devised an SDN controller capable of recognizing the MPQUIC protocol and harnessing it to gather bandwidth and propagation delay information from the network topology. We have developed a path selection algorithm based on multi-objective optimization, allowing MPQUIC subflows to choose paths with lower propagation delays that are non-intersecting, thus maximizing the utilization of path resources. Simulation results reveal that our approach enhances throughput and reduces transmission latency. The key contributions of this paper are as follows:

(1) We developed an SDN controller capable of recognizing the MPQUIC protocol, incorporating protocol parsing, path selection, connection management, and flow table generation modules.
(2) We proposed a multi-objective optimization path selection algorithm that leverages the SDN controller to gather global network topology data and choose paths with low propagation delay and high available bandwidth, effectively utilizing path transmission capacity.

(3) We created an SDN-based satellite network architecture using STK to gather actual satellite inter-visibility data. Utilized visibility matrix sequences for building a satellite network simulation platform.

The organization of this paper is as follows: Sect. 2 discusses the related work on the transmission control aspects of SDN and MPQUIC. Section 3 introduces the SDN-based satellite network architecture. Section 4 elaborates on the design of the SDN controller supporting MPQUIC and the path selection algorithm. Section 5 presents the experimental evaluation of our approach. Section 6 concludes the paper.

2   Related Works

2.1   The Combination of SDN and Transport Layer Protocols

Existing research has predominantly focused on exploring the collaboration between SDN and transport layer protocols in terrestrial networks to optimize communication performance. Some approaches employ SDN-triggered intelligent false timeout detection and response to enhance TCP transmission in wireless networks [11]. Additionally, quality of service-based MPTCP subflow routing strategies and SDN-based multipath routing protocols have been proposed to address network congestion and performance issues [6]. However, many of these solutions require modifications to the endpoints. Given the characteristics of satellite networks, existing terrestrial network transmission control strategies are challenging to directly apply. Research has proposed SDN-assisted MPTCP architectures that extend control to end-hosts to enhance throughput of MPTCP connections in satellite networks [7]. Another study introduced MPTCP into LEO satellite networks and demonstrated the improvement in network performance through the collaboration of SDN and MPTCP [4].

2.2   MPQUIC

Research on MPQUIC primarily focuses on performance testing and resource scheduling. De Coninck et al. initially designed MPQUIC and conducted performance tests against MPTCP, encompassing both high and low BDP network environments [2]. In [3], MPQUIC was ported to smartphones, and application-based performance comparisons were conducted under various network conditions. Addressing performance degradation caused by bursty transfers, [9,10] assigned data priorities based on data resource characteristics and established resource scheduling models to optimize transmission quality. In the field of satellite communication, there has been limited MPQUIC research, with only [12] proposing a prediction scheme based on backward delay differences in satellite networks, albeit without considering link switching. Existing literature mainly optimizes the MPQUIC protocol to enhance transmission quality, lacking research into leveraging global network state to assist MPQUIC. This study directly conducts MPQUIC communication experiments in an SDN network, revealing that the SDN controller tends to select paths with higher propagation delay or intersections for transmission, resulting in performance degradation.

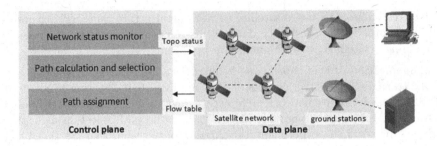

Fig. 1. SDN-based satellite network architecture.

3   Satellite Network Architecture Based on SDN

The SDN satellite network architecture is shown in Fig. 1 and consists of two main parts: the control plane (SDN controller) and the data plane (communication satellites, ground stations, and communication terminals).

3.1   Data Plane

The communication terminals, such as MPQUIC-enabled mobile devices, ground stations, and communication satellites, constitute crucial components of the data plane. In this architecture, both ground stations and communication satellites are equipped with switches that support the OpenFlow protocol. The Open-Flow protocol plays a pivotal role in SDN, facilitating signaling exchange and data delivery between controllers and switches. Through the OpenFlow proto-col, key header information of each packet, such as IP addresses and ports, can be accurately matched, enabling various functionalities including forwarding, access management, multicast, and virtual networking. What sets this architec-ture apart is the decoupling of communication satellites and ground stations from the control plane, introducing granularity-based flow table flexibility and controllability to the data plane. It's important to note that this architectural introduction brings new possibilities to the field of satellite communication. Through data plane optimization, it becomes possible to better address the unique requirements of satellite communication, while simultaneously enhanc-ing overall performance and flexibility.

3.2   Control Plane

In the SDN-based satellite network architecture, the core of the control plane is the controller, which primarily deals with two types of OpenFlow messages: Packet_In messages and FlowRemoved messages. Packet_In messages are gener-ated when a switch receives the first packet of a subflow, and they forward the packet to the controller. FlowRemoved messages are sent when a flow table entry in the switch expires, notifying the controller of the removal of the entry.

The controller continuously monitors the topology of the satellite network, including links and traffic conditions, and performs operations such as routing, switching, and resource allocation. The key idea of the architecture is to generate flow table entries in the control plane and then transmit these entries to the switches on each satellite through the OpenFlow channels of the satellite network. This allows satellites to quickly and easily forward packets to the next node.

In the actual communication process, when the sender and receiver initiate communication, the sender's ground station switch receives packets with source IP, destination IP addresses, as well as source and destination ports. The ground station switch extracts this information and matches it with internal flow table entries. If a match is found, the switch forwards the packet to the next satellite node according to the rules in the flow table entry. If there is no matching flow table entry, the switch encapsulates the packet in a Packet_In message and transmits this message to the controller through the OpenFlow channel of the satellite network. At the same time, the packet is cached in the waiting queue of the ground station switch.

4   SDN Controller Supporting MPQUIC

This paper addresses the MPQUIC protocol by developing an SDN controller encompassing packet parsing, path selection, flow table generation, and connection management modules. As illustrated in Fig. 2a, the parsing module within the controller is responsible for decoding the header information of MPQUIC data packets. The path selection module selects non-overlapping paths for each MPQUIC subflow, while the connection management module stores the parsed packet header information as subflow metadata and maintains a mapping table. The flow table generation module generates flow table entries based on the subflow transmission paths and subsequently deploys these entries to the switches along the paths. The switches forward packets to the next nodes according to the rules defined by the flow table entries.

4.1   MPQUIC Protocol Parsing Module

In order to aggregate path resources and enhance connection recovery, De Coninck [2] designed the MPQUIC protocol, which builds upon QUIC by introducing the "path" parameter to identify the data transmission path of MPQUIC subflows. The MPQUIC packet (Fig. 2b) comprises a 64-bit Connection ID, an 8-bit Path ID, and a variable-sized Packet Number. After establishing a connection, MPQUIC enables multipath communication, distinguished by the M flag where $M = 1$ indicates that the data packet includes a Path ID, and $M = 0$ signifies that the Path ID is 0. These field details are visible to intermediate network devices.

In this module, the fields to be parsed include Path ID, source IP, source port number, destination IP, destination port number, and Connection ID. The

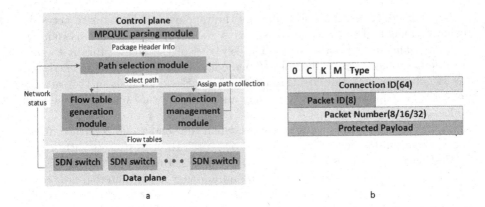

Fig. 2. a SDN Controller Module Diagram, b MPQUIC packet header.

controller initially extracts the UDP packet from the data packet to acquire the source IP, source port number, destination IP, and destination port number. Subsequently, the UDP header is removed to parse the UDP payload (MPQUIC data packet), and based on the structure of the MPQUIC data packet, the required fields such as Path ID and Connection ID are extracted.

4.2 Routing Module

Previous approaches that solely considered hop count-based routing schemes are not suitable for application in satellite networks. This paper takes into account the propagation delay and available bandwidth of satellite links and proposes a multi-objective optimization-based routing algorithm for satellite networks.

Suppose we use the graph $G = (V, E)$ to model a satellite network topology. Here, V represents the non-empty vertex set of $n = |V|$ network nodes, and $E = \{< v_i, v_j > |i \neq j, v_i \in V, v_j \in V\}$ is a set of edges representing a series of direct communication links connecting two nodes v_i and v_j. The set of available bandwidth for the links in the satellite network topology is denoted by W, and the set of propagation delay for the links is denoted by D. Here, $w_{i,j} \in W(i \neq j, 1 \leqslant i, j \leqslant n)$ represents the available bandwidth of $< v_i, v_j >$, where $\forall v_i \in V, w_{i,i} = \infty$; $d_{i,j} \in D(i \neq j, 1 \leqslant i, j \leqslant n)$ represents the propagation delay of $< v_i, v_j >$, where $\forall v_i \in V, d_{i,i} = 0$.

Define the path $p_{s,t}$ as a series of edges from source node v_s to target node v_t,

$$p = \{< v_s, v_1 >, < v_1, v_2 >, \cdots, < v_{m-1}, v_m >, < v_m, v_t >\} \qquad (1)$$

where $v_s \neq v_1, v_m \neq v_t$, and $\forall v_i, v_j$, if $i \neq j$, then $v_i \neq v_j$ where $i, j \in (1, \cdots, m-1)$. Based on the definition of path p, further define the set of k disjoint paths between v_s and v_t, denoted as $P_{s,t}^k$,

$$P_{s,t}^k = \{P_{s,t}^i\}_{i=1}^k \qquad (2)$$

where $\forall i \neq j,\ p^i_{s,t} \cap p^j_{s,t} = \varnothing\ (1 \leqslant i, j \leqslant k)$.

The bottleneck bandwidth of path $p_{s,t}$ from v_s to v_t can be defined as follows:

$$BW(p_{s,t}) = min(w_{s,1}, w_{1,2}, \cdots, w_{m-1,m}, w_{m,t}) \tag{3}$$

Assuming that there are multiple paths from the same source node v_s to the same target node v_t in $P^k_{s,t}$, let SBW denote the sum of bottleneck bandwidths of all paths in $P^k_{s,t}$,

$$SBW(P^k_{s,t}) = \sum_{i=1}^{k} BW(p^i_{s,t}) \tag{4}$$

Based on the definition of path p, the propagation delay of path $p_{s,t}$ from v_s to v_t can be defined as follows:

$$PD(p_{s,t}) = d_{s,1} + \sum_{i=1}^{m-1} d_{i,i+1} + d_{m,t} \tag{5}$$

The total propagation delay of all paths in $P^k_{s,t}$ can be defined as follows:

$$SPD(P^k_{s,t}) = \sum_{i=1}^{k} PD(p^i_{s,t}) \tag{6}$$

To increase the throughput and decrease the transmission delay in a satellite network, multiple-objective optimization model on k disjoint paths between source node v_s and target node v_t is defined based on the available bandwidth and propagation delay of the links.

$$Maxmise : SBW(P^k_{s,t}) \tag{7}$$

$$Minimise : SPD(P^k_{s,t}) \tag{8}$$

These are used as the multiple-objective optimization model for k disjoint paths between source node v_s and target node v_t, subject to the constraint that

$$\forall p^i_{s,t}, p^j_{s,t} \in P^k_{s,t}, p^i_{s,t} \cap p^i_{s,t} = \varnothing \tag{9}$$

To efficiently solve this multi-objective optimization problem, an approximate solution method is used. First, the Dijkstra algorithm is used repeatedly to select the shortest path $p_{s,t}$ from node v_s to v_t in the graph G. Each time a loop is executed, the edges included in the shortest path are deleted from graph G. When the loop ends and there is no path between v_s and v_t, a set of disjoint paths M can be obtained.

Next, each path $p^i_{s,t}$ in the set is evaluated based on both its propagation delay and available bandwidth:

$$PD(p^i_{s,t}) * \frac{f_B(p^i_{s,t})}{1 + BW(p^i_{s,t})} \tag{10}$$

where $1 \leqslant i \leqslant n$, and n is the number of paths in the path set. The function f_B is the maximum design bandwidth, designed using rules similar to BW. Formula 10 is the path cost model designed in this paper, where propagation delay is the basis and bandwidth usage is used as the coefficient. If the bottleneck bandwidth of the path is larger, the coefficient is smaller, resulting in less influence on propagation delay. Conversely, a smaller bottleneck bandwidth significantly lengthens the propagation delay. This transforms the multi-objective optimization problem into a single-objective optimization problem, simplifying the solution process.

Finally, k paths with the lowest cost are selected from the path set and are allocated to k sub-flows. The detailed steps of the above process are shown in Algorithm 1. The core of Algorithm 1 is to repeatedly invoke the Dijkstra algorithm. Assuming that the time complexity of the Dijkstra algorithm is O_D, and there are m disjoint paths between v_s and v_t, the time complexity of Algorithm 1 is $mO_D + km$, which is a linear combination of O_D. Therefore, the execution efficiency of Algorithm 1 is of the same order as that of the Dijkstra algorithm.

Algorithm 1: Path Selection Based on Propagation Delay and Bandwidth

---

Input: network topo G, source node v_s and target node v_t

Output: best paths set $P_{s,t}^k$

$M = \varnothing$;

while *exit paths between v_s and v_t* **do**

 $p_{s,t} = Dijkstra(G, v_s, v_t)$

 $M = M \cup p_{s,t}$

 $G = G - p_{s,t}$

end

for $p_{s,t} \in M$ **do**

 $c_{s,t} = PD(p_{s,t}) * \dfrac{f_B(p_{s,t}^i)}{1 + BW(p_{s,t}^i)}$

end

Select the smallest k paths from M to a new set $P_{s,t}^k$ according to the value of $c_{s,t}$.

return $P_{s,t}^k$

4.3   Flow Table Generation Module

The flow table generation module ensures the normal forwarding of packets from sub-flows along specified paths by creating forward and reverse flow tables for all switches on the path. Upon packet arrival at a switch, flow table rules instruct the switch on packet forwarding; if a sub-flow's packet arrives at a switch without a corresponding flow table rule, the packet is redirected to the controller. The controller calculates the sub-flow's path and installs flow table rules across all switches on the path, which ensures that subsequent packets of the same sub-flow are transmitted along the same path, avoiding redirection to the controller.

Assuming that the controller of MPQUIC computes a path, denoted as $p_{s,t}^i$, from v_s to v_t for a certain subflow, where $p_{s,t}^i = \{<v_s, v_1>, <v_1, v_2>, \cdots, <v_{m-1}, v_m>, <v_m, v_t>\}$ and v_j represents the j th switch on the i th available path, where $i \in [1, k]$ and $j \in [1, m]$. The SDN controller uses the source IP, source port number, destination IP, and destination port number as the flow table entries for all switches on the path $p_{s,t}^i$ to install the forward flow tables. For switch v_j, the flow table entry is to forward the UDP packet from the source IP and source port number to the destination IP and destination port number to v_{j+1}.

5 Experiment

5.1 Network Topo

The hardware environment used in this study consisted of an Intel(R) Core(TM) i5-10400 CPU @2.90GHz with 10GB of memory. The operating system used was Ubuntu 16.04. We built a satellite network simulation environment based on software-defined networking (SDN) using Mininet and STK. We employed Floodlight as the SDN controller and Open vSwitch to simulate the space-to-ground switching nodes. The connection protocol between the controller and switches was OpenFlow 1.3. The version of the multipath transport protocol MPQUIC used was MPQUIC-go. The experiments were debugged and executed on the Ubuntu 16.04 platform.

This paper constructs a hybrid constellation comprising 38 LEO (Low Earth Orbit) satellites and 2 GEO (Geostationary Orbit) satellites. The parameters of the LEO satellites include an orbit altitude of 780 km, 3 near-earth orbital planes, a 100Mbit/s bandwidth for LEO inter-satellite links, and a packet loss rate of 0.1%. The GEO satellites have parameters including an orbit altitude of 35786 km, 1 orbital plane, 2 satellites per orbital plane, an orbital inclination of 0.0°, a 100Mbit/s bandwidth for GEO inter-satellite links, and a packet loss rate of 0.1%. Additionally, two ground stations are set up, located in Beijing, China, and Melbourne, Australia. The STK software is employed to analyze satellite orbital parameters and export the time-varying distance matrix L between nodes. Elements of matrix L represent distances between any two satellite nodes, where 0 indicates no connection, and non-zero values represent actual distances. The network topology is dynamically updated based on the distance matrix.

5.2 Result and Analysis

In the simulation experiments, this study performs performance testing and comparison among Disjoint [13], QSMPS [6], and the proposed SDN_MPQUIC solution. The evaluation primarily employs the following metrics: (1) Average file transmission completion time, assessing the time required for transmitting fixed-size files; (2) Throughput variations, obtained by analyzing captured packets through the Wireshark traffic monitoring tool to gather throughput data.

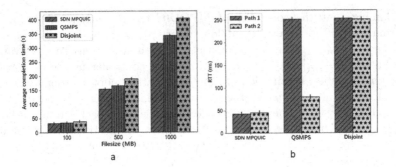

Fig. 3. a Average completion time for transferring files of different sizes, b The RTT of the path.

Firstly, the average completion time of each scheme is evaluated based on transferring files of different sizes. The experiments use different routing algorithms to route sub-flows, and each algorithm is tested ten times, and the experimental results are statistically analyzed. As shown in Fig. 3 a, the blue diagonal bar chart, green vertical bar chart, and yellow asterisk bar chart represent the time required for SDN MPQUIC, QSMPS, and Disjoint to transfer files of different sizes, respectively. According to the experimental data, when transferring a file of size 100 MB, the average completion time of SDN MPQUIC is reduced by 6.8% compared to QSMPS and 15.0% compared to Disjoint; when transferring a file of size 500MB, the average completion time of SDN MPQUIC is 7.1% less than that of QSMPS and 19.0% less than that of Disjoint; when transferring a file of size 1000 MB, the average completion time of SDN MPQUIC is 8.4% less than that of QSMPS and 22.0% less than that of Disjoint.

Statistical analysis of the results reveals that in the context of a mixed high-low orbit satellite network, SDN_MPQUIC optimizes transmission time and enhances efficiency compared to the QSMPS and Disjoint approaches. The statistical analysis of selected path latencies is depicted in Fig. 3b. Disjoint solely considers hop count as the sole criterion for path selection. In the satellite scenario, it chooses the path with the least number of hops for MPQUIC subflows, but experimental results indicate that Disjoint tends to favor paths containing GEO satellites. As a result, the propagation latency of paths chosen by Disjoint is generally larger, leading to longer transmission times for terminals. QSMPS prioritizes paths with higher bandwidth, and experiments show a significant time difference of 171 milliseconds between the two paths selected by QSMPS for MPQUIC subflows. This notable time difference between paths results in a substantial number of out-of-order data packets in the receiving buffer, affecting network transmission efficiency. SDN_MPQUIC employs a path selection algorithm based on latency and bottleneck bandwidth. The assigned latencies for the two paths are 42.4 milliseconds and 45.6 milliseconds, with both paths having ample bandwidth. The strategy presented in this paper allocates MPQUIC subflows to paths with lower propagation latency and higher bandwidth, effectively leveraging the characteristics of the satellite network and enhancing transmission efficiency.

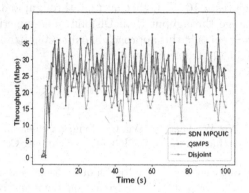

Fig. 4. Server throughput of route selection schemes.

Figure 4 illustrates the variation in throughput for SDN_MPQUIC, QSMPS, and Disjoint when client and server employ the MPQUIC protocol to transmit data in a mixed high-low orbit scenario. The blue line, green line, and yellow line respectively represent the throughput variations at the server end for SDN_MPQUIC, QSMPS, and Disjoint during the first 100 s. The average throughput of SDN_MPQUIC is 27.10 Mbps, QSMPS records an average throughput of 23.21 Mbps, and Disjoint achieves an average throughput of 21.3 Mbps. Within the initial 100 s, SDN_MPQUIC outperforms QSMPS by 16.7% and Disjoint by 27.0% in terms of average throughput. Since QSMPS prioritizes available bandwidth in path selection, and Disjoint uses hop count as its criterion, both schemes tend to select higher satellites in a multi-layer network. Consequently, QSMPS and Disjoint forward data via GEO satellites. The approach presented in this paper takes both latency and bandwidth into account and leans towards paths with shorter latency during route selection, resulting in enhanced transmission performance. This indicates that the method proposed in this paper is more suitable for satellite networks characterized by significant temporal and spatial variations.

6   Conclusion

SDN's separation of control and data planes enables flexible and efficient control over satellite networks. However, SDN currently lacks the ability to interpret the MPQUIC protocol and overlooks factors such as propagation delay during routing in satellite networks. This paper introduces the first SDN controller that supports MPQUIC and proposes a multi-objective optimization-based routing algorithm to select disjoint paths for subflows with smaller propagation delays. Experimental results demonstrate that, in tasks involving the transmission of files of varying sizes, the SDN_MPQUIC approach reduces the average completion time by 6.8% to 8.4% compared to QSMPS and by 15.0% to 22.0% compared to Disjoint. In a scenario involving a mix of high and low orbit satellites, SDN_MPQUIC achieves a 16.7% higher average throughput than QSMPS and a 27.0% higher average throughput than Disjoint. The experimental outcomes underscore the effectiveness of this approach in enhancing bandwidth utilization, throughput, and reducing transmission delays.

References

1. Bao, J., Zhao, B., Yu, W., Feng, Z., Wu, C., Gong, Z.: Opensan: a software-defined satellite network architecture. ACM SIGCOMM Comput. Commun. Rev. **44**(4), 347–348 (2014)
2. De Coninck, Q., Bonaventure, O.: Multipath quic: design and evaluation. In: Proceedings of the 13th International Conference on Emerging Networking Experiments and Technologies, pp. 160–166 (2017)
3. De Coninck, Q., Bonaventure, O.: Multipathtester: Comparing MPTCP and MPQUIC in mobile environments. In: 2019 Network Traffic Measurement and Analysis Conference (TMA), pp. 221–226. IEEE (2019)
4. Du, P., Nazari, S., Mena, J., Fan, R., Gerla, M., Gupta, R.: Multipath TCP in SDN-enabled LEO satellite networks. In: MILCOM 2016–2016 IEEE Military Communications Conference, pp. 354–359. IEEE (2016)
5. Gaber, A., ElBahaay, M.A., Mohamed, A.M., Zaki, M.M., Abdo, A.S., AbdelBaki, N.: 5g and satellite network convergence: survey for opportunities, challenges and enabler technologies. In: 2020 2nd Novel Intelligent and Leading Emerging Sciences Conference (NILES), pp. 366–373. IEEE (2020)
6. Gao, K., Xu, C., Qin, J., Yang, S., Zhong, L., Muntean, G.M.: QoS-driven path selection for MPTCP: a scalable SDN-assisted approach. In: 2019 IEEE Wireless Communications and Networking Conference (WCNC), pp. 1–6. IEEE (2019)
7. Jiang, Z., Wu, Q., Li, H., Wu, J.: scMPTCP: SDN cooperated multipath transfer for satellite network with load awareness. IEEE Access **6**, 19823–19832 (2018)
8. Li, R., Lin, B., Liu, Y., Dong, M., Zhao, S.: A survey on laser space network: terminals, links, and architectures. IEEE Access **10**, 34815–34834 (2022)
9. Shi, X., Wang, L., Zhang, F., Liu, Z.: FStream: flexible stream scheduling and prioritizing in multipath-QUIC. In: 2019 IEEE 25th International Conference on Parallel and Distributed Systems (ICPADS), pp. 921–924. IEEE (2019)
10. Shi, X., Wang, L., Zhang, F., Zhou, B., Liu, Z.: Pstream: priority-based stream scheduling for heterogeneous paths in multipath-QUIC. In: 2020 29th International Conference on Computer Communications and Networks (ICCCN), pp. 1–8. IEEE (2020)
11. Singh, K.V.K., Pandey, M.: An SDN-based true end-to-end TCP for wireless LAN. Wireless Netw. **27**, 1413–1430 (2021)
12. Xie, C., Hu, H., Liu, Y.: Shared bottleneck detection for multipath transmission in high latency satellite network. In: 2019 IEEE 7th International Conference on Computer Science and Network Technology (ICCSNT), pp. 38–42. IEEE (2019)
13. Zannettou, S., Sirivianos, M., Papadopoulos, F.: Exploiting path diversity in datacenters using MPTCP-aware SDN. In: 2016 IEEE symposium on computers and communication (ISCC), pp. 539–546. IEEE (2016)

An Energy Prediction Method for Energy Harvesting Wireless Sensor with Dynamically Adjusting Weight Factor

Zhenbo Yuan[ORCID], Yongqi Ge[✉], Jiayuan Wei, Shuhua Yuan, Rui Liu, and Xian Mo

School of Information Engineering, Ningxia University, Yinchuan 750021, China
{geyongqi,mxian168}@nxu.edu.cn

Abstract. Energy Harvesting wireless sensors (EHWS) experience non-linear variations in energy collection over time, leading to potential energy waste or inadequate energy supply. These factors can result in diminished Energy Neutral Operation (ENO) performance and even system failure. To address this issue, this paper proposes a novel EHWS energy prediction method that dynamically adjusts weight factors. Building upon the EWMA method, the proposed approach introduces dynamic weight factors and devises a Dynamic Error-Exponentially Weighted Moving Average (DE-EWMA) algorithm to dynamically balance the significance of predicted values and actual values. Two experimental scenarios are designed, and a comparative analysis is conducted with the classical EWMA, WCMA, and Pro-Energy algorithms. The experimental findings demonstrate that DE-EWMA achieves prediction accuracy improvements of 25.08%, 6.38%, and 31.98% compared to the classical EWMA, WCMA, and Pro-Energy approaches, respectively.

Keywords: Energy Harvesting wireless sensors · Energy Prediction · Weighting factor

1 Introduction

Wireless sensor networks find wide application in various domains, including biomedical [11], environmental monitoring [8], and military defense [10], where they play a pivotal role. As the fundamental building blocks of wireless sensor networks, wireless sensors are primarily responsible for tasks such as data collection, processing, and transmission. Typically powered by batteries, wireless sensors face significant limitations in terms of their lifespan dictated by battery capacity. Once the battery is depleted, the sensor ceases to function, necessitating battery replacement. However, a substantial number of wireless sensor nodes are deployed in harsh or inaccessible environments, making battery replacement costly or even infeasible. To address this challenge, researchers have proposed the utilization of energy harvesting techniques for wireless sensors [12].

© The Author(s), under exclusive license to Springer Nature Singapore Pte Ltd. 2024
Z. Tari et al. (Eds.): ICA3PP 2023, LNCS 14492, pp. 465–477, 2024.
https://doi.org/10.1007/978-981-97-0811-6_28

Energy harvesting technology enables sensor nodes to acquire energy from the surrounding environment and convert it into electrical energy. This enables the sensor nodes to operate continuously without the need for battery replacement, thereby avoiding node failure. In this paper, we refer to wireless sensor nodes powered by energy harvesting technology as Energy Harvesting Wireless Sensors (EHWS).

Due to being a strictly resource-constrained system [13], EHWS imposes more demanding requirements on prediction methods in terms of computing power and storage space. Machine learning prediction methods achieve higher accuracy by processing and training on large amounts of data, but they also bring greater computational burden, resulting in increased energy consumption. This study focuses on a data-driven energy prediction method that offers advantages in reducing computational complexity and data processing requirements, thus improving energy prediction efficiency.

The main contribution of this paper lies in the proposal of a dynamic weight factor-based energy prediction algorithm, named Dynamic Error—Exponentially Weighted Moving Average (DE-EWMA). DE-EWMA dynamically calculates weight factors based on the previous time period's error and the maximum error of the past day. The dynamic weight factor will dynamically change according to the change of weather conditions, giving different proportions to the predicted value and the actual value, so as to improve the prediction accuracy. Two distinct experimental scenarios are designed to compare this method with three existing classical algorithms: Exponentially Weighted Moving Average (EWMA), Weather Conditioned Moving Average (WCMA), and Pro-Energy. Experimental results demonstrate that the DE-EWMA method achieves prediction accuracy improvements of 25.08%, 6.38%, and 31.98% compared to the classical algorithms on the selected dataset.

The structure of this paper is as follows: Section 2 presents a review of the related work on energy prediction methods. In Sect. 3, the DE-EWMA energy prediction method is described in detail. Section 4 provides a comparative analysis of the DE-EWMA method with three classical prediction methods. Finally, Sect. 5 summarizes the findings of this paper.

2   Related Work

Currently, energy prediction methods can be classified into three categories: data-driven prediction methods [3–6,9], mechanism-based prediction methods [7], and machine learning-based prediction methods [2]. Mechanism-based models are built on physical principles or laws, providing strong interpretability but demanding high modeling requirements. Machine learning methods have advantages in prediction performance but exhibit high computational complexity, significant resource requirements, and poor interpretability. Consequently, both these approaches are unsuitable for practical scenarios involving energy-constrained EHWS. This paper primarily focuses on data-driven prediction methods.

Data-driven prediction methods leverage historical data to forecast future energy availability, reducing computational complexity and data processing demands, thereby enhancing prediction efficiency. Among these approaches, three stand out: EWMA, WCMA, and Pro-Energy. EWMA, introduced by Kansal et al. [6], assumes that energy availability in a specific time slot resembles past data. It calculates the predicted energy by averaging historical values from previous days, considering current-day data with weighted factors. While it's computationally efficient, EWMA may have limitations in prediction accuracy. WCMA, proposed by Piorno et al. [9], improves accuracy by factoring in weather conditions. It introduces a "GAP" factor atop EWMA, quantifying differences in current and past weather conditions. Significant weather changes result in larger GAP values, prioritizing current data, while stability gives precedence to historical averages. This adaptation to weather changes enhances energy prediction accuracy. Cammarano et al. [3] introduced Pro-Energy, a multi-source EHWS method. It enhances accuracy by matching current weather conditions to past configuration files and real energy data. Pro-Energy selects the most fitting configuration file from a pool of past files, combining it with real energy data to predict future energy availability. While this method excels in accuracy, it heavily relies on stored configuration files, which may lead to deviations if files for specific conditions are missing.

This study conducted research and analysis on the aforementioned classic data-driven energy prediction methods. It was found that they all employ fixed weighting factors to adjust the importance between real energy values and historical averages, overlooking the dynamic nature of environmental energy. Therefore, this paper adopts dynamic weighting factors to adjust the relevance between different information. These weighting factors consider the relationship between collected energy information and errors, contributing to a more accurate energy prediction.

3   Energy Prediction Algorithm

3.1   EWMA

The EWMA method as shown in Eq. (1). Here, E_{pre} represents the predicted energy, $E(d, n-1)$ denotes the actual collected energy in the $(n-1)$th time slot of the dth day, and $M_D(d, n)$ represents the average of the energy collected in the nth time slot over the past D days, which can be computed using Eq. (2).

$$E_{pre} = aE(d, n-1) + (1-a)M_D(d, n) \tag{1}$$

$$M_D(d, n) = \frac{1}{D} \sum_{i=1}^{D} E(d-i, n) \tag{2}$$

EWMA is a widely utilized data-driven energy prediction method that introduces a weighting factor α, which ranges between 0 and 1. This factor is employed to balance the relative importance of actual energy and historical averages. By taking the weighted average of these two energy sources, EWMA method offers reliable short-term energy predictions.

3.2 DE-EWMA

In this section, the DE-EWMA method is proposed as an energy prediction technique for EHWS. The approach enhances the existing EWMA method by introducing a weighting factor, W, to integrate the adjusted real energy information with the adjusted predicted energy information. The prediction formula for DE-EWMA is represented as Eq. (3).

$$E_{pre} = W * E_{re} + (1 - W) * E_{cor} \tag{3}$$

The weighting factor, W, is introduced with a range of values between 0 and 1. The adjusted energy information, E_{re}, and the adjusted predicted energy value, E_{cor}, are combined using W.

Calculate the Weight Factor W. The calculation formula for W is presented as Eq. (4).

$$W = max(0, min(1, \frac{Err_{last}}{max_{err}})) \tag{4}$$

The calculation of W involves the utilization of Err_{last} and max_{err}. Err_{last} can be computed using Eq. (5), which represents the absolute difference between the predicted energy information obtained using EWMA method for the (n − 1)th time slot and the actual collected energy information. max_{err} denotes the maximum error between the energy information predicted using EWMA method and the actual collected energy information within the past day.

$$Err_{last} = |EWMA(d, n - 1) - E(d, n - 1)| \tag{5}$$

The Corrected Energy Information. In DE-EWMA, the index is defined as the optimal matching day, which corresponds to the day in the energy information matrix E that exhibits the most similar energy trend to day d. The value of index can be calculated using Eq. (6). $\bar{d}$ is the number of days stored in matrix E.

$$index = \max_{(\forall \bar{d} \in E)} \frac{1}{k} \sum_{i=1}^{k} |E(d, n - i) - E(\bar{d}, n - i)| \tag{6}$$

The adjusted actual energy value E_{re} and the difference between the nth time slot and the (n − 1)th time slot of the optimal matching day, denoted as Err_{index}, are respectively represented in Eqs. (7) and (8). E_{re} can be calculated by adding the previous time slot's actual energy information to Err_{index}.

$$E_{re} = E(d, n-1) + Err_{index} \tag{7}$$

$$Err_{index} = E(index, n) - E(index, n-1) \tag{8}$$

The Corrected Predicted Energy Information. EWMA is subject to certain prediction errors during practical applications. To correct the predicted values obtained using EWMA, an error value $Error$ is introduced and computed using the formula presented as Eq. (9).

$$Error = EWMA(d, n-1) - E(d, n-1) \tag{9}$$

By subtracting the calculated error value from the current time slot's EWMA predicted value, and the adjusted prediction value E_{cor} is represented as Eq. (10).

$$E_{cor} = EWMA(d, n) - Error \tag{10}$$

In cases where the prediction error is significant, a larger weight is assigned to E_{re} to emphasize its importance in energy prediction. Conversely, when the prediction error is small, indicating better prediction performance, a larger weight is assigned to the adjusted EWMA predicted value.

4   Experimental Setup and Results

4.1   Dataset and Experimental Environment

This study utilized solar radiation data from the National Renewable Energy Laboratory (NREL) [1] as the dataset, specifically selecting data from November 1, 2022, to December 31, 2022, for processing.

To address situations where solar radiation is close to zero during nighttime, a threshold value called *"no_light_threshold"* is set. When the value falls below this threshold, the corresponding time slot is categorized as a period with no sunlight. For subsequent metric calculations, data from these no-light periods are excluded to ensure data accuracy and reliability. The value of *no_light_threshold* can be calculated using Eq. (11).

$$no\_light\_threshold = max(E(d, n)) * 0.05 \tag{11}$$

After data processing, a high-quality solar radiation dataset was obtained through filtering and processing. This dataset covers a total of 61 days of solar radiation data, considering each day as one cycle for energy prediction. Each day is divided into N time slots, with each slot representing a half-hour interval, i.e., N = 48. Therefore, a total of 87,840 solar radiation data points were obtained. Subsequently, the collected energy information from the past is stored in a matrix of size $D \times N$, where D represents the number of days of energy observations, and N represents the number of time slots in a day. This transformation of time series data into a two-dimensional matrix facilitates further processing and analysis.

4.2   Regulation of the Weighting Factor

In addition to the aforementioned EWMA method, this paper also compares DE-EWMA with two classic energy prediction methods: WCMA and Pro-Energy.

Among the four methods mentioned in this paper, the number of past energy information days D is set to 4, and the number of past time slots K is set to 3. These parameter values will be used in all the experiments in the subsequent comparative analysis section.

In the EWMA, WCMA, Pro-Energy, and the proposed method in this paper, weight factor α is introduced to adjust the proportion between the modified real energy information and the modified predicted energy information. The range of α should be between 0 and 1 to ensure an appropriate balance. However, the optimal value of weight factor α may vary for different datasets. Therefore, in practical applications, it is necessary to choose the optimal value of α based on specific circumstances. One commonly used approach is to utilize the Mean Absolute Percentage Error (MAPE) to evaluate the deviation between the energy prediction values and the actual values and determine the value of weight factor α based on MAPE. MAPE is a widely applied evaluation metric that directly explains the error results in percentage form. The formula for calculating MAPE is as follows (see Eq. (12)), where E represents the energy collection matrix, P represents the predicted energy matrix, and n represents the number of time slots considered.

$$MAPE = \frac{1}{n} \sum_{i=1}^{n} |\frac{E-P}{E}| * 100\% \tag{12}$$

Based on the adopted dataset, the optimal values for the three algorithms were tested, and the results are shown in Fig. 1. In the EWMA, WCMA, and Pro-Energy methods, the adjustment process of weight factor α exhibits significant fluctuations. However, in DE-EWMA, the fluctuation of α during the adjustment process is smaller. This is because in this method, α only adjusts a portion of the prediction, while a dynamic weight factor W is set, which can be automatically updated based on new prediction results to self-regulate the proportion between the two parts. Therefore, DE-EWMA can better adapt to changes in different datasets and has better stability and adaptability.

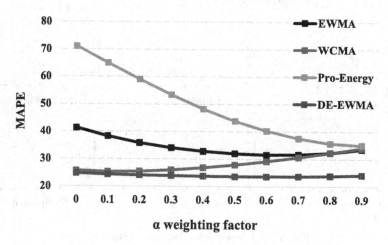

Fig. 1. Adjusting weight factor.

According to Fig. 1, when the value of α is 0.6, the MAPE of the EWMA method reaches its minimum value at 31.57%. The WCMA method achieves the minimum MAPE value of 25.29% when α is set to 0.1, while the Pro-Energy method reaches the minimum MAPE value of 35.07% at $\alpha = 0.9$. Considering that the DE-EWMA method utilizes EWMA predictions, when the EWMA error is minimized at $\alpha = 0.6$, this method also achieves the optimal result of 23.65%. When α approaches 1, the MAPE values of EWMA, WCMA, and Pro-Energy tend to converge. This is because when α is equal to 1, the predicted values of these three methods are equal to the real energy values collected in the previous time slot. When the α values of all four algorithms are optimized, the proposed method in this paper shows an improvement of 25.08% compared to the EWMA method, 6.38% compared to the WCMA method, and 31.98% compared to the Pro-Energy method.

4.3   Experimental Scenario Design

Stable Solar Radiation. Six days with relatively stable solar radiation data were selected from the dataset for analysis, as shown in Fig. 2. Among them, Profile1 and Profile3 exhibit minor fluctuations in the morning phase while remaining stable in other time periods. Profile2 and Profile6 show minor fluctuations during midday while remaining stable in other time periods. Profile5 exhibits minor fluctuations in the afternoon while remaining stable in other time periods. Profile4 represents stable weather conditions throughout the day.

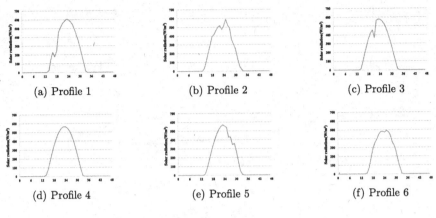

Fig. 2. Harvested solar radiation (Stable).

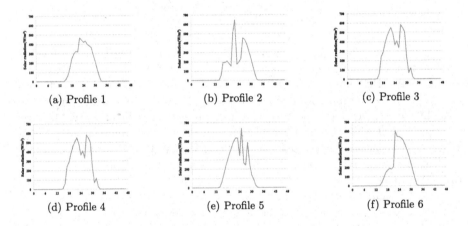

Fig. 3. Harvested solar radiation (Fluctuation).

Fluctuating Solar Radiation. Figure 3 displays six days with significant fluctuations in solar radiation selected from the dataset for analysis. These six plots show pronounced fluctuations in solar radiation collection during sunlight periods, posing a more challenging task for prediction methods.

4.4   Analysis of Experimental Results

Energy prediction was performed using EWMA, WCMA, Pro-Energy, and the proposed DE-EWMA methods, considering different weather conditions. The prediction frequency was set to every 30 min, resulting in 288 predicted time slots. Subsequently, an analysis was conducted based on different evaluation metrics. To accurately assess the performance of the prediction methods, the MAPE and prediction error were chosen as evaluation metrics. MAPE can be calculated using formula (12), while the prediction error is defined by formula (13), which is the absolute difference between the predicted value and the actual value.

$$Error = |E(d, n) - E_{pre}(d, n)| \tag{13}$$

Stable Solar Radiation. Figure 4 presents the solar radiation prediction results obtained during the six days with relatively stable solar radiation. It is evident from the figure that the proposed method provides a better fit to the actual values and is applicable to all six plots.

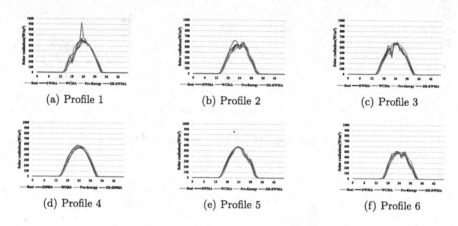

(a) Profile 1   (b) Profile 2   (c) Profile 3

(d) Profile 4   (e) Profile 5   (f) Profile 6

Fig. 4. Prediction effect (Stable).

Figure 5 illustrates the prediction errors of different methods under relatively stable solar radiation conditions. By observing Fig. 5, it is clear that DE-EWMA exhibits significantly lower prediction errors than the other three algorithms for the majority of time slots, with most errors below 100 W/m². Compared to the other three algorithms, the proposed method demonstrates better adaptation to stable weather conditions, yielding more accurate energy predictions. The prediction errors primarily cluster narrowly, highlighting the proposed method's effectiveness in capturing subtle energy variations during predictions, resulting in greater precision. Figure 6 offers a comprehensive view of daily MAPE in stable solar radiation conditions, comparing four prediction algorithms. Results reveal that the Pro-Energy algorithm exhibits the poorest daily MAPE performance, exceeding 20%. This can be attributed to its core approach of finding historical weather data most akin to the current day; however, significant disparities between identified historical data and current weather can lead to diminished prediction accuracy.

In contrast, DE-EWMA consistently achieves the best performance each day, with MAPE values below 15% for all dates. Particularly, the fourth day demonstrates the best performance with a MAPE of only 3.66%. This indicates that the proposed method provides highly accurate energy predictions under relatively stable solar radiation data conditions.

The average MAPE was calculated for the six days with relatively stable solar radiation, and the results shows a total of 108 valid time slots across these six days. Under these conditions, DE-EWMA achieves an average MAPE of 10.21%.

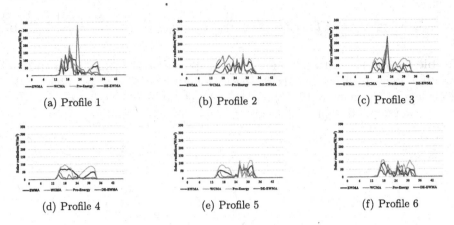

| (a) Profile 1 | (b) Profile 2 | (c) Profile 3 |

| (d) Profile 4 | (e) Profile 5 | (f) Profile 6 |

Fig. 5. Prediction error (Stable).

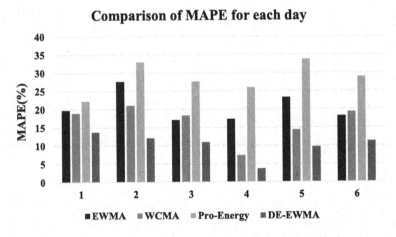

Comparison of MAPE for each day

Fig. 6. MAPE per day (Stable).

Compared to the EWMA algorithm, the proposed method improves accuracy by 50% and outperforms the WCMA algorithm by 38%. These results further validate the ability of the proposed method to make more accurate predictions under relatively stable solar radiation data conditions.

Fluctuating Solar Radiation. Figure 7 presents the curve graphs of the results obtained by four algorithms for energy prediction and the actual energy information within six days of high solar radiation fluctuations. It is evident from the figure that significant differences exist between the prediction curves of the four algorithms and the actual energy information in this scenario. This discrepancy arises because data-driven energy prediction methods employ past energy information to forecast the future, resulting in larger errors during periods of high weather fluctuations.

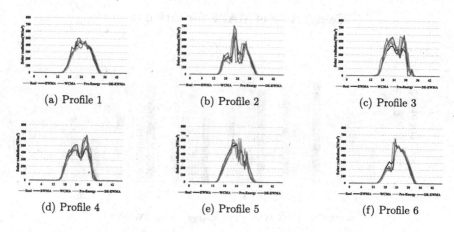

(a) Profile 1   (b) Profile 2   (c) Profile 3

(d) Profile 4   (e) Profile 5   (f) Profile 6

Fig. 7. Prediction effect (Fluctuation).

Figure 8 evaluates the performance of the four algorithms in terms of prediction errors during high solar radiation fluctuations. Notably, for the majority of time slots, the proposed method exhibits lower prediction errors than the other three algorithms, with comparable errors in a few individual time slots. Overall, the majority of prediction errors remain below 200 W/m².

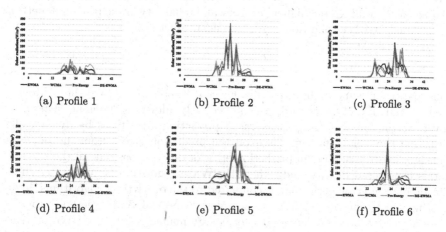

(a) Profile 1   (b) Profile 2   (c) Profile 3

(d) Profile 4   (e) Profile 5   (f) Profile 6

Fig. 8. Prediction error (Fluctuation).

Figure 9 illustrates the daily MAPE under conditions of high solar radiation fluctuations. The DE-EWMA method proposed in this study yields significantly lower MAPE values on the first and sixth days compared to the other three methods. Even in the worst-case scenario (the second day), DE-EWMA achieves a MAPE of 40%, which is still superior to the other three methods.

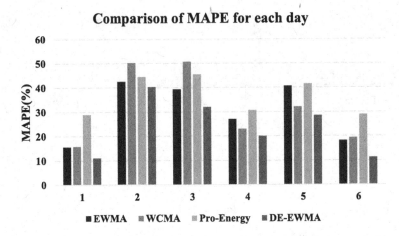

Fig. 9. MAPE per day (Fluctuation).

For the case of high solar radiation fluctuations, the average MAPE and the number of valid time slots were computed for the six days, and the results are summarized. It is observed that there are 104 valid time slots for solar radiation values across these six days. Within this context, the DE-EWMA method achieves an average MAPE of 23.95%, representing a 23% improvement over the EWMA algorithm and a 25% improvement over the WCMA algorithm. These findings further affirm the higher accuracy and reliability of the proposed method in handling weather conditions with significant fluctuations.

5   Conclusion

This paper introduces the Dynamic Error - Exponentially Weighted Moving Average (DE-EWMA) method, which effectively addresses the issues caused by static weight factor settings in classical prediction algorithms by dynamically adjusting the weight factors based on the prediction errors from the previous time slot and the maximum prediction error of the previous day. In the two scenarios considered in this study, the DE-EWMA algorithm demonstrates superior prediction error performance compared to the other three algorithms, highlighting its significant advantages. Therefore, the DE-EWMA method provides more accurate and reliable support for energy management in EHWS, presenting crucial practical applications. However, there are still potential areas for improvement in this research. For instance, the calculation method for weight factors could be further optimized to enhance prediction accuracy and stability. Moreover, it's crucial to validate the DE-EWMA algorithm's applicability and performance across different regions and seasons. Future research will include experiments on diverse datasets to accomplish this.

Acknowledgements. This work is supported in part by the National Natural Science Foundation of China (Grant number 62162052, 62262052), in part by the Natural Science Foundation of Ningxia Province (Grant number 2021AAC03041, 2022AAC03004, 2020AAC03280), and in part by the Key R & D projects of Ningxia Province (Grant number 2021BEB04016, 2022BDE03007).

References

1. National Renewable Energy Laboratory. https://midcdmz.nrel.gov/
2. Al-Omary, M., Hassini, K., Fakhfakh, A., Kanoun, O.: Prediction of energy in solar powered wireless sensors using artificial neural network. In: 2019 16th International Multi-Conference on Systems, Signals & Devices (SSD), pp. 288–293. IEEE (2019)
3. Cammarano, A., Petrioli, C., Spenza, D.: Pro-energy: a novel energy prediction model for solar and wind energy-harvesting wireless sensor networks. In: 2012 IEEE 9th International Conference on Mobile Ad-Hoc and Sensor Systems (MASS 2012), pp. 75–83. IEEE (2012)
4. Cammarano, A., Petrioli, C., Spenza, D.: Online energy harvesting prediction in environmentally powered wireless sensor networks. IEEE Sens. J. **16**(17), 6793–6804 (2016)
5. Hassan, M., Bermak, A.: Solar harvested energy prediction algorithm for wireless sensors. In: 2012 4th Asia Symposium on Quality Electronic Design (ASQED), pp. 178–181. IEEE (2012)
6. Kansal, A., Hsu, J., Zahedi, S., Srivastava, M.B.: Power management in energy harvesting sensor networks. ACM Trans. Embedded Comput. Syst. (TECS) **6**(4), 32-es (2007)
7. Koirala, B., Dahal, K., Keir, P., Chen, W.: A multi-node energy prediction approach combined with optimum prediction interval for RF powered WSNs. Sensors **19**(24), 5551 (2019)
8. Lanzolla, A., Spadavecchia, M.: Wireless sensor networks for environmental monitoring (2021)
9. Piorno, J.R., Bergonzini, C., Atienza, D., Rosing, T.S.: Prediction and management in energy harvested wireless sensor nodes. In: 2009 1st International Conference on Wireless Communication, Vehicular Technology, Information Theory and Aerospace & Electronic Systems Technology, pp. 6–10. IEEE (2009)
10. Pragadeswaran, S., Madhumitha, S., Gopinath, S.: Certain investigation on military applications of wireless sensor network. Int. J. Adv. Res. Sci. Commun. Technol. **3**(1), 14–19 (2021)
11. Roy, S., Azad, A.W., Baidya, S., Alam, M.K., Khan, F.: Powering solutions for biomedical sensors and implants inside the human body: a comprehensive review on energy harvesting units, energy storage, and wireless power transfer techniques. IEEE Trans. Power Electron. **37**(10), 12237–12263 (2022)
12. Sanislav, T., Mois, G.D., Zeadally, S., Folea, S.C.: Energy harvesting techniques for internet of things (IoT). IEEE Access **9**, 39530–39549 (2021)
13. Sharma, H., Haque, A., Jaffery, Z.A.: Solar energy harvesting wireless sensor network nodes: a survey. J. Renew. Sustain. Energy **10**(2), 023704 (2018)

Bayesian Optimization for Auto-tuning Convolution Neural Network on GPU

Huming Zhu[✉], Chendi Liu, Lingyun Zhang, and Ximiao Dong

Key Laboratory of Intelligent Perception and Image Understanding, Ministry of Education, Xidian University, Xi'an 710071, China
zhuhum@mail.xidian.edu.cn

Abstract. GPU as a hardware processor plays an important role in the training of deep neural networks. However, when using GPUs for computation on convolutional neural network models, different combinations of GPU kernel configuration parameters have different performance. Therefore, this paper proposes BAGF, a bayesian auto-tuning framework for GPU kernels, which parameterizes the factors affecting the performance of GPU programs and uses bayesian optimization methods to search for the best parameters in the search space consisting of the parameters. Compared with other optimization algorithms, BAGF obtains excellent configuration parameters with fewer iterations. This paper analyzes the performance of BAGF on four benchmarks and compares with other common optimization algorithms. In addition, the performance improvement of each parameter configuration is analyzed. Finally, the BAGF was tested with the convolution layer of Alexnet, and the results of the Roofline model were analyzed. Compared with the original parameter configuration, the speed of BAGF was increased by 50.09%.

Keywords: CNN · Parallel computing · GPU · Auto-tuning

1 Introduction

Deep neural networks (DNNs) have excellent algorithm performance and have been widely used in self-driving, speech recognition, human face recognition, object detection and semantic segmentation domains [1–3]. Convolutional neural networks (CNNs) represent deep neural networks whose algorithmic accuracy typically increases with network parameters and computational effort. Therefore, GPUs are used for training and inference of CNN models with their powerful parallel computing capabilities [4–6]. Furthermore, in order to obtain an efficient combination of GPU kernel configuration parameters to utilise the optimal performance of the GPU, it is usually necessary to search a huge and discontinuous search space, and due to the data-intensive and computation-intensive characteristics of the neural network, the sampling cost of searching using brute force methods is unacceptable.

Supported by organization x.

© The Author(s), under exclusive license to Springer Nature Singapore Pte Ltd. 2024
Z. Tari et al. (Eds.): ICA3PP 2023, LNCS 14492, pp. 478–489, 2024.
https://doi.org/10.1007/978-981-97-0811-6_29

In recent years, many optimization methods, such as Genetic Algorithm (GA), Differential Evolution (DE), Basinhopping Algorithm (BS), have been applied to the optimization of GPU configuration parameters [7,8]. Unlike the population iteration mode of evolutionary algorithms, the bayesian optimization algorithm only produces one new sample point at a time, which greatly reduces the sampling cost, and has been widely used in many domains. Feurer, M et al. [9] proposed an automatic learning system based on bayesian optimization: Auto-skLearning. Snoek, J et al. applied bayesian optimization to automatically adjust hyperparameters in convolutional neural networks [10]. Mahendran, N et al. proposed an AdaGrad markov chain monte carlo algorithm based on bayesian optimization [11]. Wu, J et al. [12] used bayesian optimization to automatically adjust the hyperparameters in convolutional neural networks and recursive neural networks. They showed that the bayesian optimization algorithm based on Gaussian Process can achieve great accuracy in a few samples. Therefore, this paper will investigate the application of bayesian optimization algorithm to the optimization of GPU configuration parameter search space. thereby realizing a framework for automatic tuning of GPU configuration parameters based on bayesian optimization.

The main contributions of this paper are as follows:

(1) BAGF, a bayesian auto-tuning framework for GPU is proposed. Taking four typical applications as baselines, the performance of BAGF is analyzed. The convolutional algorithm is used as an example to analyze the effect of configuration parameters on BAGF.
(2) Compare the performance differences between BAGF-based CUDA and OpenCL.
(3) Based on the Roofline performance model, the tuning performance of BAGF in deep neural networks is analyzed.

2   Related Work

Due to the complexity of the GPU kernel configuration, manually finding the optimal kernel configuration parameters is time-consuming. So auto-tuning techniques are used to automatically adjust user-defined code parameterization.

Dao, T. T et al. [13] analyzed the factors affecting the performance for different work-group sizes in the OpenCL kernel code. After fully considering the occupancy, coalesced global memory accesses, cache contention, and the variation in the amount of workload of the kernel, etc., the auto-tuning techniques build a performance model and derive the best number of work-groups per GPU platform. Li, J et al. [14] proposed a fine-grained prefetching scheme to optimize the performance of generalized matrix multiplication (GEMM) on GPUs in order to maximize the performance of DGEMM under different problem sizes.

In addition to the conventional optimization algorithms, the machine learning method is also applied to auto-tuning. Petrovič, F et al. [15] implemented a benchmark set containing ten auto-tunable kernels for important computational problems implemented by OpenCL and CUDA. Using random search, simulated annealing, and Markov chain Monte Carlo searcher as an optimization strategy, they demonstrated that most kernels achieve near-peak performance on a variety of GPUs through auto-tuning. Cheema, S et al. [16]. proposed an automated GPU device tuning framework for tuning OpenCL application kernels. The GPU tuning framework uses a multi-objective optimization approach to improve performance and power consumption of an application by applying various algorithmic variations such as circular deconvolution, caching, work-group size and memory utilization.

However, these methods suffer from slow search speed and are based on specific kernels without experimenting, analyzing and optimizing the whole CNN network, while bayesian optimization is able to reduce the number of samples by incorporating prior knowledge to guide sampling. Therefore, this paper presents BAGF, a bayesian auto-tuning framework for GPU kernels.

3   BAGF: The Bayesian Auto-tuning Framework for GPU

There are usually two factors that affect GPU program performance. The first is related to parall granularity, such as work-group size settings and the amount of tasks per work-item. The second is whether to use performance optimization techniques and local memory, such as loop unrolling, and using local memory, etc.

When using brute force methods, the search space is too large, which will incur huge sampling costs. If parameter selection is based on experience, it is impossible to accurately determine the optimal configuration parameters for each platform. Therefore, this paper designs a auto-tuning framework based on GPU architecture.

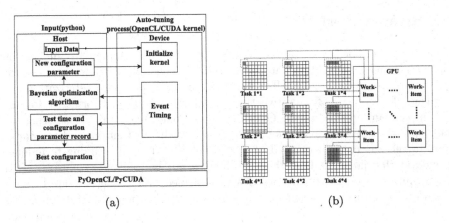

(a)   (b)

Fig. 1. Overall implementation of the BAGF and increase the amount of task in each work-item.

Figure 1(a) shows the specific implementation process of BAGF. BAGF builds a search space by parameterizing the key factors affecting GPU performance, setting the value range, and arranging and combining them. Then random initialization parameters configure the kernel function, and run the test to get the time. Pass the parameters and time to the host, add the historical observation set, and update the model. Select the next set of parameters according to the sampling function, and repeat the configuration test process. Iteration optimization until completion, and finally sort to get the optimal time and parameter configuration.

There are two types of parameters when parameterizing the platform configuration:

(1) The first is from the perspective of parall granularity, including the setting of work-group size in the OpenCL kernel and changes in the amount of tasks per work-item. The size of the work-group is determined according to the specific application.

(2) The second type is Boolean variables. For local memory, loop unrolling and other commonly used GPU code optimization methods, you can try to use them during tuning and judge the effectiveness of the tuning program.

The following of parameter variables are analyzed in detail:

(1) WG_size_x: work-group x-dimensional size, usually a multiple of 32, is optimized to access memory. Program performance improves due to aligning access to global memory on the GPU. When 32 work-items are aligned to access 128-byte addresses, they can be combined into one access request, making the most efficient use of memory bandwidth. This means that starting from a 128-byte-aligned address, 32 work-items will sequentially access 4-byte elements.

(2) WG_size_y: The y-dimensional size of the work-group, usually a multiple of 2, does not affect memory access.

(3) ELE_PER_X, ELE_PER_Y: increase the number of tasks for each work-item, reduce the number of work-groups, and improve data reusability. The specific implementation process is shown in Fig. 1(b). Assuming that there are currently many work-groups, the number of work-groups is halved under the premise that all tasks remain unchanged, so the workload of each work-item in the work-group will become twice the original workload.

(4) USE_LOCAL: Whether to use local memory. Local memory is visible to all work-items in the same work-group. The period of global memory access is 400–600 clock cycles, while the access speed of local memory is only 1–32 clock cycles.

(5) USE_PADDING: When mapping to GPU memory, the local memory is divided into multiple memory blocks of the same size, namely bank. In order to ensure the parallelism of memory reading and writing, each bank is only accessed by up to 1 thread at the same time, but when multiple threads access different addresses of the same bank at the same time, the access to memory is serial. This causes bank conflicts and seriously reduces performance.

Therefore, bank conflicts should be avoided as much as possible. When copying the data required for computation from global memory to local memory, the storage method of memory should be changed, which requires additional memory space.

(6) USE_UNROLL: Whether the loop is unrolled. Loop unrolling can reduce instruction consumption and add more independent scheduling instructions. When unrolling a loop, the index of the array will be converted from variable index to constant index. Arrays known to use constant index at compile time are usually located in registers, which will greatly increase the number of registers.

4   The Performance of BAGF

In this paper, three different GPUs are used as experimental platforms for OpenCL-based performance portability tests: the NVIDIA GeForce GTX 1070, NVIDIA GeForce RTX 2080Ti and AMD Radeon Vega Frontier Edition. Each of these GPUs has a single-precision computing power of 6.463 TFLOPS, 13.45 TFLOPS and 13.11 TFLOPS respectively.

4.1   Performance Analysis of BAGF on Four Benchmarks

In order to prove the effectiveness of BAGF, this paper adopts four baseline algorithms - template operation, matrix multiplication, vector addition and convolution operation. The computational grid size for template operations is set to 8192 × 4096. For matrix multiplication, two square matrices of size 4096 × 4096 are used for matrix multiplication. Vector addition uses two vectors of size 10 million to add. For the convolution operation, a 1024 × 1024 input image is selected and the convolution kernel is set to 11 × 11.

By comparing the auto-tuning with the original kernel (WG_size_x = 16, WG_size_y = 16), this paper uses the average of twenty execution results as the final execution result of the internal kernel function. The execution time of the kernel function in OpenCL is used as the test time. The configuration parameters used to auto-tune each baseline algorithm are shown in Table 1. "✔" means that the baseline algorithm will use this parameter when auto-tuning, and "✗" means that this parameter is not used.

Table 1. The used parameters on benchmark.

Benchmark	Stencil	Vector addition	Matrix	Multiplication	Convolution
WG_size_x	✔	✔	✔	✔	
WG_size_y	✔	✗	✔	✔	
ELE_PER_X	✗	✗	✔	✔	
ELE_PER_Y	✗	✗	✔	✔	
USE_LOCAL	✗	✗	✔	✔	
USE_PADDING	✗	✗	✗	✔	
USE_UNROLL	✗	✗	✔	✔	

As can be seen from Fig. 2(a), after auto-tuning the internal kernel functions running on each GPU platform, the performance of the four baseline tests has been improved, especially the two GPU-accelerated operations of matrix multiplication and convolution. Analyzing the computing power of the computing platform, the single-precision floating-point computing power of the RTX 2080Ti is 2.08 times that of the GTX 1070 and 1.02 times that of the AMD Radeon Vega Frontier Edition.

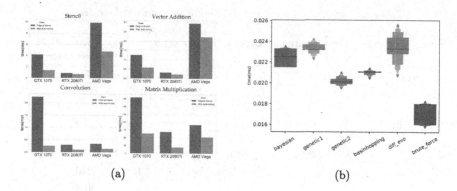

(a)   (b)

Fig. 2. Comparison between auto-tuning results and original kernel and performance distributions of 2D convolution kernel configurations benchmark using bayesian optimization and other strategies.

Before optimizing the baseline algorithm, the four different algorithms on the GTX 1070 took 2.60 to 7.41 times longer than the RTX 2080Ti, while the AMD Radeon Vega Frontier Edition was 1.14 to 10.84 times longer than the RTX 2080Ti. After optimization, the four different algorithms on the GTX 1070 took 1.94 to 3.25 times longer than the RTX 2080Ti, while the AMD Radeon Vega Frontier Edition was 1.36 to 9.77 times longer than the RTX 2080Ti. The actual performance of the GTX 1070 and RTX 2080Ti is positively correlated with single-precision floating-point computing power. However, the performance of AMD Vega does not match the computing power, mainly because its local memory per SM is only 16KB, which limits the degree of parallelism.

4.2   Comparison with GA, BS and DE Algorithms

In this experiment, taking a convolutional operation with an input of 512 × 512 and a convolutional kernel size of 3 × 3 as an example, BAGF is compared with Genetic Algorithm (GA), Differential Evolution (DE) and Basinhopping Algorithm (BS) and Brute Force(BF). The two genetic algorithms with different parameters are called GA1 and GA2, respectively. Table 2 lists the parameters and number of iterations for the different optimization strategy. Figure 2(b) shows the performance distribution of BAGF and other optimization strategies in matrix multiplication. It can be seen that BAGF has the least number of samples. BAGF uses an algorithm similar to GA1 for the number of samples. But the results are better than GA1. Although the best value of sampling is not as good as DE, the average is better and the distribution is more stable. Although GA2 and BS can get good results, the number of samples is high, which is unacceptable when the kernel function takes a long time. Now, the number of brute force samples is 10368, which is unacceptable in most cases.

Table 2. Default parameters for different optimization strategy.

Strategy	Population size	Iterations	Algorithmic parameters	Theoretical sampling times
bayesian	-	15	Initial point=15	30
GA1	12	10	$P_{crossover} = 1$, $P_{mutation} = 0.1$	120
GA2	20	100	$P_{crossover} = 1$, $P_{mutation} = 0.1$	2000
BS	-	1000	$T = 1.0$	1000
DE	15	1	tol $= 0.01$,F $= (0.5, 1)$,CR $= 0.7$	240
BF	-	-	-	10368

4.3 The Impact of Each Configuration Parameter on Performance in BAGF

In this experiment, the convolution operation is performed with a fixed-size image, the operational intensity is improved by increasing the size of the convolutional kernel, and the influence of different parameter configurations on the convolution performance is gradually tested. The experimental platform is NVIDIA GeForce GTX 1070, The kernel time is the average of the results of 5 tests with the serial time on the CPU. As shown in Table 3, the performance improvement is up to 91.7% compared to the original kernel time. According to the analysis results in Table 3, only adjusting the size of the x and y dimensional work-groups has no obvious effect. However, after increasing the number of tasks per thread in the x dimension, the performance is greatly improved, as much as 74.9%. This may be due to increased data reuse, reduced redundant instructions and thread overhead.

The above conclusion can be further deduced using the operational intensity. Assuming that the height of the input image is I_h, the width is I_w, and the height F_h and width of the convolution kernel is F_w, respectively. $F_w \times F_h$ multiplication operation and $F_w \times F_h - 1$ subtraction operation is required, the total operation amount is $2 \times F_w \times F_h - 1$, the required data amount is $2 \times F_w \times F_h \times 4$ byte, the operational intensity is as shown in Eq. (1). It indicates from the formula that the larger the convolution kernel size is, the greater the operational intensity of

Table 3. Improvement of performance for each parameter Configuration.

Convolution Kernel	WG_size_x	WG_size_y	ELE_PER_X	ELE_PER_Y	LOCAL
3 × 3	1.54%	9.01%	61.4%	71.6%	75.1%
5 × 5	19.4%	39.15%	74.9%	87.2%	88.4%
7 × 7	10.5%	27.7%	71.8%	88.1%	91.4%
9 × 9	19.4%	20.1%	71.1%	88.6%	91.7%
11 × 11	9.3%	10.3%	60.5%	86.1%	89.2%

each work-item is, and the maximum value is 0.25. If the amount of tasks for each work-item is n times compared with the previous one, then the total alculation volum is $(2 \times F_w \times F_h - 1) \times n$. In this case, the reuse of input image data is increased. Therefore, the required amount of data is $F_w \times F_h \times 4 \times 2 + F_w \times (n - 1)$ byte, then the operational intensity AI' is as shown in Eq. (2), The relationship between the AI' and AI without increasing the task amount is shown in Eq. (3). The values of n are set 1, 2, 4, 8, 16, 32, From the Eq. (3), increasing the workload of each work-item, the operational intensity increases and is proportional to n, F_h.

$$AI = \frac{2 \times F_w \times F_h - 1}{2 \times F_w \times F_h \times 4} = \frac{2 - \frac{1}{F_w \times F_h}}{8} \tag{1}$$

$$AI' = \frac{(2 \times F_w \times F_h - 1) \times n}{F_w \times F_h \times 4 \times 2 + F_w \times (n - 1)} = \frac{2 - \frac{1}{F_w \times F_h}}{\frac{8}{n} + \frac{1}{F_h}\left(1 - \frac{1}{n}\right)} \tag{2}$$

$$\frac{AI'}{AI} = \frac{1}{\frac{1}{n}\left(1 - \frac{1}{8 \times F_h}\right) + \frac{1}{8 \times F_h}} \tag{3}$$

Finally, the usage of local memory has been increased. On the basis of increasing the workload of each work-item in the Y dimension, the performance is improved by about 4 times. It can be seen that on the GPU platform, the use of local memory is a very effective means of improving performance and optimizing algorithms.

4.4 Parallel Convolution Operator Analysis Based on CUDA and OpenCL

In this experiment, we implemented the computational procedure for the convolutional layer. The input image size is 1024×1024, the experimental platform on the GTX 1070, and the results are averaged over 20 runs, and then the performance of the convolution operator is compared in CUDA and OpenCL. The original kernel test time is based on the original kernel(WG_size_x = 16, WG_size_y = 16, ELE_PER_X = 1, ELE_PER_Y = 1, LOCAL = 0), and the original kernel test time is before auto-tuning, Table 4 lists the performance improvements relative to the original kernel.

Table 4. CUDA/OpenCL performance improvement ratio.

Convolution Kernel	3 × 3	5 × 5	7 × 7	9 × 9	11 × 11	13 × 13	17 × 17
CUDA	62.74%	80.56%	88.19%	89.12%	88.70%	91.48%	89.04%
OpenCL	70.47%	83.57%	88.78%	88.06%	87.13%	89.36%	85.79%

As can be seen from Table 4, the performance of CUDA and OpenCL-based convolution operations has been greatly improved after kernel parameter tuning. Performance continues to increase as the size of the convolutional kernel increases. When the convolution kernel exceeds 7 × 7, the CUDA-based kernel code is stable at around 89%, and the OpenCL-based kernel code is stable at around 88%. The operation intensity of the convolution operation can be calculated by the equation (3). The 25.21 FLOP/Byte of the computing device is approximately equal to the operation intensity when the convolutional kernel is 7 × 7. The NVIDIA 1070 platform is limited by memory, so CUDA and OpenCL have similar performance. When the size of the convolution kernel exceeds 7 × 7, convolution is computationally limited as the size increases. However, since NVIDIA GPUs are better optimized for CUDA, the performance improvement based on CUDA auto-tuning is higher than that of OpenCL.

4.5   The Portability Analysis of BAGF for CNN Based on Roofline Model

As a representative of CNN, Alexnet is a milestone in the development of deep learning and is widely used in image classification and other domains. In this experiment, this experiment optimises AlexNet and tests the performance of the auto-tuning model in deep neural networks based on the Roofline model [17].

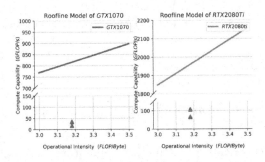

Fig. 3. Roofline models for GTX 1070 and RTX 2080Ti

Draw the Roofline model using the computational power and bandwidth of the GPU. The computational power of the GTX 1070 is 6463 GFLOPS, the bandwidth is 256.3 GB/s, and the maximum operation intensity is about 25 FLOPS/Byte. The computational power of the RTX 2080Ti is 13450 GFLOPS, the bandwidth is 616 GB/s, and the maximum operation intensity is about 22 FLOPS/Byte. Analyzing the Alexnet model parameters and calculation volum, the memory is about 229 MB, the computation volum is about 727 MFLOPS, and the operation intensity is about 3 FLOPS/Byte. As shown in Fig. 3, the actual operation intensity of Alexnet is less than the mode operation intensity of the GPU, which belongs to the bandwidth limitation area. For clarity, 3–3.5 FLOPS/Byte is used as the operation intensity range in the figure. The green triangle represents the performance under the original configuration of Alexnet, and the red triangle represents the performance after BAGF optimization.

Table 5. Portability analysis of AlexNet convolutional layers on different GPUs.

	Layer	Conv 1	Conv 2	Conv 3	Conv 4	Conv 5	FC1	FC2	FC3	Total
	FLOPs (M)	105	223	149	112	74	37	16	4	727
	Params (M)	0.035	0.37	0.884	1.3	0.442	37	16	4	60
NVIDIA GeForce GTX 1070	I (GFLOP/s)	11.50	21.11	33.48	25.92	40.21	4.41	4.28	2.32	16.36
	I-opt (GFLOP/s)	62.13	126.70	100	73.20	54.81	4.41	4.28	2.32	34.28
	Original kernel(ms)	9.13	10.56	4.45	4.32	1.84	8.38	3.73	1.72	44.13
	Optimal Time(ms)	1.69	1.76	1.49	1.53	1.35	8.38	3.73	1.72	21.65
	Speedup ratio	81.5%	83.4%	66.5%	64.8%	26.6%	–	–	–	50.9%
NVIDIA GeForce GTX 2080Ti	I (GFLOP/s)	30.97	154.86	117.32	98.24	66.07	16.22	15.68	14.81	65.45
	I-opt (GFLOP/s)	250	377.96	173.25	151.35	121.31	16.22	15.68	14.81	106.19
	Original kernel(ms)	3.39	1.44	1.27	1.14	1.12	2.28	1.02	0.27	11.92
	Optimal Time(ms)	0.42	0.59	0.86	0.74	0.61	2.28	1.02	0.27	6.78
	Speedup ratio	87.61%	59.02%	32.28%	35.09%	45.54%	–	–	–	43.12%

Auto-tuning Alexnet written by CUDA and the results are shown in Table 5. It can be seen from Table 5 that by tuning the parameters of the Alexnet convolution layer, the running time of the convolution layer can be significantly reduced and the operation intensity can be improved. Taking the first convolution layer as an example, the running time on the GTX 1070 is reduced from the original 9.13 ms to 1.69 ms, and the operation intensity is increased from 11.50 GFLOPS to 62.13 GFLOPS. After tuning, the total running time of Alexnet on the GTX 1070 is reduced from 44.13 ms to 21.65 ms, and the actual performance is increased from 16.36 GFLOPS to 34.28 GFLOPS.

However, there is still a significant gap with the GTX 1070's theoretical peak performance of 815 GFLOPS. Similarly, on the RTX 2080Ti, the runtime is reduced from 11.92 ms to 6.78 ms, and the actual performance is improved from 65.45 GFLOPS to 106.19 GFLOPS, which is still a gap from the theoretical peak performance of 1958.88 GFLOPS. The reason is that the operation intensity of Alexnet is only 3.18 FLOPS/byte, which is in the memory bottleneck area and is much lower than the mode operation intensity of the GPU. The Roofline model gives only the theoretical upper bound, and there are other influencing factors, such as cache size. In fact, the model cannot make full use of the GPU computing power. Therefore, the operation intensity of the model should be improved to keep it within the calculation range, and other influencing factors should be considered to improve the GPU occupancy rate to further approach the upper bound of Roofline performance.

5   Conclusion

This paper analyzes parallel program optimization and performance portability through four image processing algorithms. Taking convolution as an example, this paper analyzes the impact of parameters on BAGF performance, and verifies that the bayesian optimization sampling times are low but the effect is good, and compares the time performance differences between CUDA and OpenCL. Applying BAGF on Alexnet improves the performance by 50.9%, and analyzes its performance in deep network based on Roofline model. BAGF performs well on limited benchmarks and convolutional networks, but the kernel function in the test is still limited. The next step is to expand the test scope and understand the auto-tuning difficulty and performance portability more comprehensively. In addition, BAGF is only optimized for a single kernel and cannot tune networks with multiple kernels as a whole, which needs further research.

Acknowledgements. This work is funded in part by the Key Research and Development Program of Shaanxi (Program No. 2022ZDLGY01-09), GHfund A No. 202107014474, GHfund 202202036165, Wuhu and Xidian University special fund for industry- university- research cooperation (Project No. XWYCXY-012021013), and Cloud Computing Key Laboratory of Gansu Province.

References

1. Cao, Z.: Continuous improvement of self-driving cars using dynamic confidence-aware reinforcement learning. Nat. Mach. Intell. **5**(2), 145–158 (2023)
2. Mao, J.: 3D object detection for autonomous driving: a comprehensive survey. Int. J. Comput. Vision **131**(8), 1909–1963 (2023)
3. Aldarmaki, H.: Unsupervised automatic speech recognition: a review. Speech Commun. **139**, 76–91 (2022)
4. Kim, H.: Performance analysis of CNN frameworks for GPUs. In: ISPASS 2017 - IEEE International Symposium on Performance Analysis of Systems and Software, pp. 55–64. IEEE, Piscataway, NJ (2017)
5. Hu, Y.: A survey on convolutional neural network accelerators: GPU, FPGA and ASIC. In: 2022 IEEE 14th International Conference on Computer Research and Development. ICCRD 2022, pp. 100–107. IEEE, Piscataway, NJ (2022)
6. Wu, Y., Zhu, H., Zhang, L., Hou, B., Jiao, L.: Accelerating deep convolutional neural network inference based on OpenCL. In: Shi, Z., Jin, Y., Zhang, X. (eds.) Intelligence Science IV. ICIS 2022. IFIP Advances in Information and Communication Technology, vol. 659. Springer, Cham (2022). https://doi.org/10.1007/978-3-031-14903-0_11
7. Schoonhoven, R.A.: Benchmarking optimization algorithms for auto-tuning GPU kernels. IEEE Trans. Evol. Comput. **27**(3), 550–564 (2023)
8. van Werkhoven, B.: Kernel tuner: a search-optimizing GPU code auto-tuner. Futur. Gener. Comput. Syst. **90**, 347–358 (2019)
9. Feurer, M.: Efficient and robust automated machine learning. In: Advances in Neural Information Processing Systems, pp. 2962–2970. Neural Information Processing Systems Foundation, La Jolla, California (2015)
10. Snoek, J.: Practical Bayesian optimization of machine learning algorithms. In: Advances in Neural Information Processing Systems, pp. 2951–2959. Neural Information Processing Systems Foundation, La Jolla, California (2012)
11. Mahendran, N.: Adaptive MCMC with Bayesian optimization. In: 15th International Conference on Artificial Intelligence and Statistics, pp. 751–760. PMLR, New York, NY, USA (2012)
12. Wu, J.: Hyperparameter optimization for machine learning models based on Bayesian optimization. J. Electron. Sci. Technol. **17**(1), 26–40 (2019)
13. Dao, T.T.: An auto-tuner for OpenCL work-group size on GPUs. IEEE Trans. Parallel Distrib. Syst. **29**(2), 283–296 (2017)
14. Li, J.: A fine-grained prefetching scheme for DGEMM kernels on GPU with auto-tuning compatibility. In: 2022 IEEE International Parallel and Distributed Processing Symposium (IPDPS), pp. 863–874. IEEE, Piscataway, NJ (2022)
15. Petrovič, F.: A benchmark set of highly-efficient CUDA and OpenCL kernels and its dynamic autotuning with kernel tuning toolkit. Futur. Gener. Comput. Syst. **108**, 161–177 (2020)
16. Cheema, S.: GPU Auto-tuning framework for optimal performance and power consumption. In: Proceedings of the 15th Workshop on General Purpose Processing Using GPU, pp. 1–6. Association for Computing Machinery, New York, NY, USA (2023)
17. Lo, Y.J., et al.: Roofline model toolkit: a practical tool for architectural and program analysis. In: Jarvis, S.A., Wright, S.A., Hammond, S.D. (eds.) PMBS 2014. LNCS, vol. 8966, pp. 129–148. Springer, Cham (2015). https://doi.org/10.1007/978-3-319-17248-4_7

Author Index

© The Editor(s) (if applicable) and The Author(s), under exclusive license
to Springer Nature Singapore Pte Ltd. 2024
Z. Tari et al. (Eds.): ICA3PP 2023, LNCS 14492, pp. 491–492, 2024.
https://doi.org/10.1007/978-981-97-0811-6

Printed in the United States
by Baker & Taylor Publisher Services

Printed in the United States
by Baker & Taylor Publisher Services